California Department of
Division of Mines and

Simplified Fault Activity Map of California

Compiled by C.W. Jennings , 1995

0 30 62 Miles
0 50 100 Kilometers

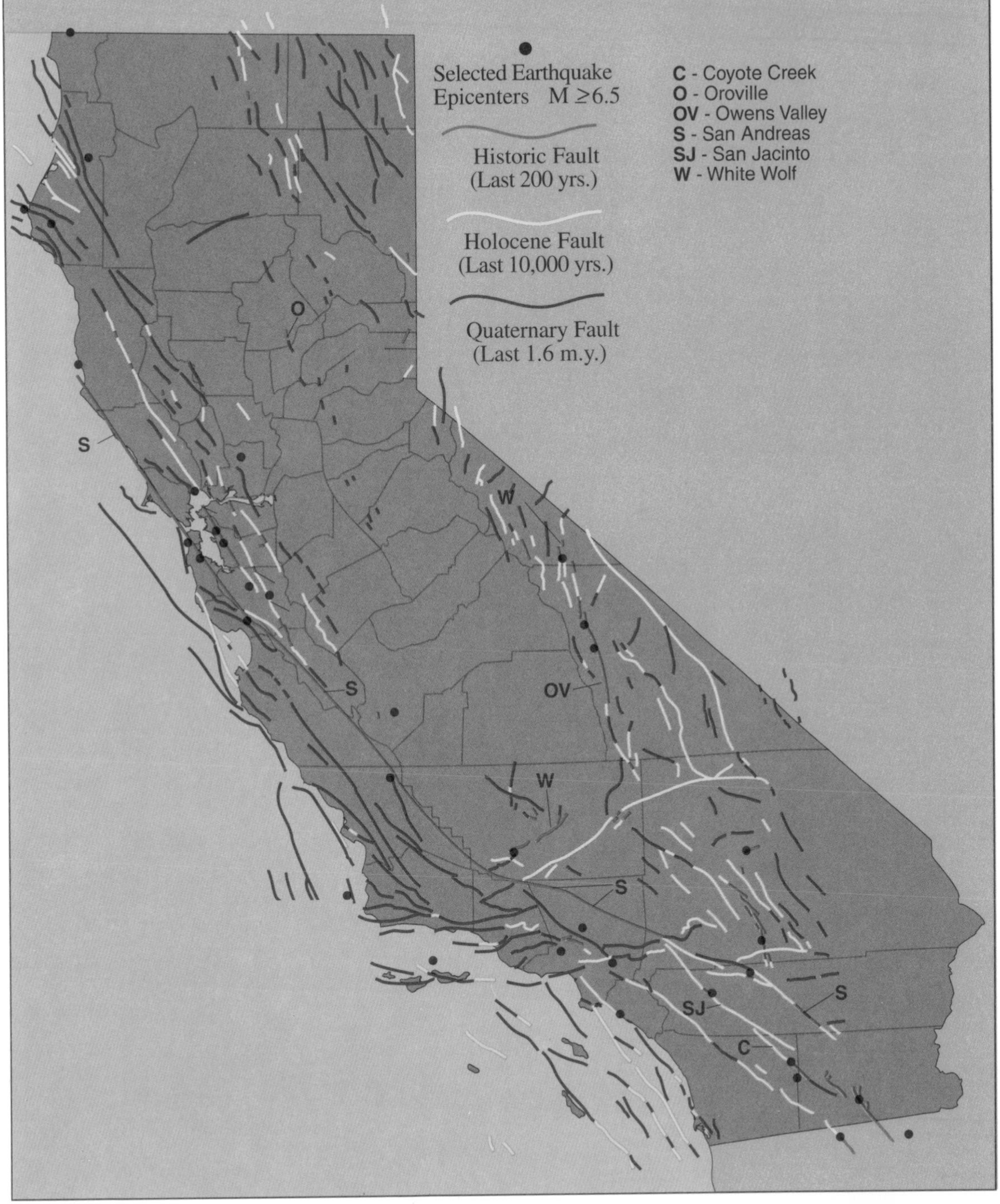
Selected Earthquake Epicenters M ≥6.5
Historic Fault (Last 200 yrs.)
Holocene Fault (Last 10,000 yrs.)
Quaternary Fault (Last 1.6 m.y.)
C - Coyote Creek
O - Oroville
OV - Owens Valley
S - San Andreas
SJ - San Jacinto
W - White Wolf
O
S
W
S
OV
W
S
SJ
S
C

California Geology

California Geology

Deborah R. Harden

San José State University

Prentice Hall, Inc.
Upper Saddle River, New Jersey 07458

Library of Congress Cataloging-in-Publication Data

Harden, Deborah Reid
California geology / Deborah R. Harden.
p. cm.
Includes bibliographical references and index.
ISBN 0–02–350042–5
1. Geology—California. I. Title.
QE89.H37 1997
557.94—dc21 97–38180
CIP

Executive Editor: Robert A. McConnin
Editor-in-Chief: Paul F. Corey
Editorial Director: Tim Bozik
Assistant Vice President of Production and Manufacturing: David W. Riccardi
Executive Managing Editor: Kathleen Schiaparelli
Manufacturing Manager: Trudy Pisciotti
Full Service Liaison/Manufacturing Buyer: Benjamin D. Smith
Creative Director: Paula Maylahn
Art Director: Jayne Conte
Art Manager: Gus Vibal
Full-Service Design/Production, Editing, Composition, Layout: Accu-color, Inc., St. Louis, Missouri
Marketing Manager: Leslie Cavaliere
Cover Design: Patricia Wosczyk
Cover Image: *Face of the Earth,*™ ARC Science Simulations; copyright © 1997

Printed in the United States of America

10 9 8 7 6 5 4 3 2 1

ISBN 0-02-350042-5

Prentice-Hall International (UK) Limited, *London*
Prentice-Hall of Australia Pty. Limited, *Sydney*
Prentice-Hall Canada, Inc., *Toronto*
Prentice-Hall Hispanoamericana, S.A., *Mexico*
Prentice-Hall of India Private Limited, *New Delhi*
Prentice-Hall of Japan, Inc., *Tokyo*
Simon & Schuster Asia Pte. Ltd., *Singapore*
Editora Prentice-Hall do Brasil, Ltda., *Rio de Janeiro*

To the memory of
Hope and Bill Harden

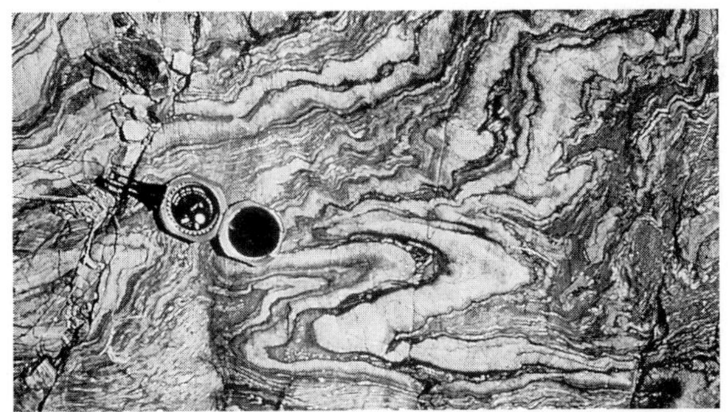

Contents

PART II GEOLOGIC HIGHLIGHTS OF CALIFORNIA

10 Water in California 205

11 The Great Valley: Sediments and Soils 233

12 The Coast Ranges: Mountains of Complexity 252

Preface

Perhaps nowhere else in the United States does geology play as large a role in everyday life as in California. Californians are frequently reminded of our position along the active boundary between the Pacific and North American plates, and for the past 230 million years, plate tectonics has been the dominant force in California's geologic history. Today, plate tectonics is responsible for some obviously unwelcome aspects of our lives—earthquakes, landslides, volcanic eruptions—as well as for some of the attributes that attract people to the state—rugged mountain ranges and a rocky and dramatic coastline. Many people are unaware that plate tectonics is a major reason that California is so different from the East Coast or the Great Plains. Without geologic resources such as gold, petroleum, and fertile soils, the face of California's history and economy would be entirely different. The sufficiency, distribution, and quality of water all of which are in part controlled by geology, continue to be driving issues in California's political life.

For some of these reasons, students in California are attracted to beginning-level geology courses. Many come to these courses without prior knowledge of the basics of geology or familiarity with its complex vocabulary. Students may not be particularly interested in detailed information about formations, rock types, and ages of materials. It is the intent of this book to acquaint nongeologists with California's geology and to cover the topics that most interest these students. Examples of catastrophic natural disasters, excerpts from California history that relate to geology, illustrations of mining methods, and discussions of the societal impacts of geologic processes bring the lessons of geology closer to the everyday context of California life.

PURPOSE

This book is intended as a text for a one-semester course in California Geology, but readers and teachers alike should be aware that to cover all of the material in detail during one semester will not be feasible. It is my intention that a college student, precollege teacher, or any interested person could comprehend the material in this book without having to take any prerequisite course in general or physical geology. However, the content of the book will also be informative to geology majors, graduate students, and geologists not intimately acquainted with California geology.

After the reader has completed this book, a few major concepts should be thoroughly understood:

- The effects of plate tectonics on California's geology and landscape are overwhelming.
- Most of California is geologically recent because of plate tectonics.
- Maps and cross sections are the language of geology.
- Geologists study present geologic environments to better understand past geologic events as they are recorded by rocks, sediments, and structures. In turn, understanding the types, magnitude, and frequency of past events allows geologists to predict future hazardous events.
- Many of California's natural resources and natural hazards arise from its geologic history and present geologic setting.

ORGANIZATION

California Geology is divided in three broad sections. At the outset, readers with no background in geology are introduced to the basic principles of geology. In Chapters 1 through 4, these principles are reviewed specifically as they apply in California. The tools for expressing geologic information—geologic maps and cross sections—are also introduced. Throughout the book, the portrayal of information on maps, cross sections, and diagrams will develop the reader's skills in using these tools.

In Chapters 5 through 17, readers are introduced to the major features of specific provinces or important regional processes (because some features and processes cross over province boundaries). The intent is not to give a comprehensive listing or description of all rock types and ages in each province, but rather to discuss the most important and interesting geologic features of each. Examples of geologic events that have shaped California's history, such as the 1906 earthquake, illustrate geologic processes.

Part II presents the provinces of California so that less complicated areas and processes come first, and subsequent chapters logically build from them. However, Chapters 5 through 17 could logically be covered in a variety of orders, depending on the preference of the instructor or reader. Because Chapters 16 and particularly 17 come at the end of the regional chapters, their content does rely heavily on earlier chapters. Those who wish to begin their geologic tour in southern California are advised to use the subject index and the place index to locate items of particular interest that pertain specifically to southern California.

The final chapters assemble a unifying picture of the state's geology. Chapter 18 synthesizes the geologic history of California as a whole, emphasizing the role of plate tectonics in producing the current geologic history of California, emphasizing the role of plate tectonics in producing the current geologic picture. This discussion provides a review of the important aspects of individual provinces, allows a comprehensive geographic picture to emerge from those parts, and ties California's geologic history into that of the rest of western North America. The final chapter in the book examines the relevance of geology in the everyday lives of Californians. At the end of my own course each semester, I schedule a day entitled "So what?" Chapter 19 reviews the geologic processes that create hazards to people living in California, reviews the causes, and discusses prevention, mitigation, or preparation measures. I believe that as Earth scientists this is perhaps our most important task, because societal decisions about natural hazards are not made by the Earth scientists. It is my hope that this book will leave the reader better equipped to deal with the complex issues of public safety, land use, and environmental protection that are facing California.

ACKNOWLEDGMENTS

Many people have been critical to the making of this book. The driving force behind the entire project has been the hundreds of students taking Geology of California at San José State University. Their encouragement, helpful suggestions, justified confusion, and continued requests for simplifications and explanations have provided invaluable feedback. I also acknowledge the helpful comments from instructors Barbara Callison, Paula Jeffries, and Matt McMackin, who have been willing to use drafts of various chapters in their courses. Precollege teachers attending the Bay Area Earth Science Institute have also provided helpful suggestions, and I am grateful to Ellen Metzger for distributing materials from this book for teacher's critique.

Many of my colleagues have provided helpful suggestions, interesting and relevant data and references, and good-humored corrections concerning matters in which my own expertise is woefully inadequate. I am especially grateful to Sue DeBari, Richard Sedlock, John Williams, and particularly to Cal Stevens at San José State University and to David Harwood of the U.S. Geological Survey and Hardenwood Farms. Michael Moore, Mike Rymer, and King Huber of the U.S. Geological Survey were very generous with their time and their slide and photo collections.

I also acknowledge the substantial improvements resulting from the review of this book by Prentice Hall's peer reviewers; all had helpful comments and suggestions, and many supplied valuable information to enhance its contents. My thanks go to Bruce A. Carter, Pasadena City College; Joanne Danielson, Shasta Community College; Jeffrey A. Grover, Los Osos; Glenn R. Roquemore, Irvine Valley College; and David L. Schwartz, Cabrillo College. I am particularly grateful to Mary E. Templeton of California State University, Fullerton, who went far beyond the limits of reviewer duty to supply thorough, thoughtful, and substantive comment on every page of this manuscript, including all of the illustrations.

This book would never have attained reality without the help and encouragement of Robert McConnin, Executive Editor, at Prentice Hall. I thank Bob for knowing when to prod and when to wait patiently and for his wisdom regarding the workings of the publishing world.

Ben Smith has coordinated the production of the book, helped with the sometimes baffling permission process, and given me an insight into the modern world of publication. Karin Sandberg, Prentice Hall's excellent sales representative in this area, has been an active advocate for the book and has provided many instructors with materials in advance.

Suzanne Wakefield at Accu-color has been the unifying force behind the illustrations in this book. With patience, good humor, persistence, and attention to detail, she has performed the nearly impossible task of bringing sense and simplicity to geologic diagrams and maps. I am most grateful to her, to Wendy Cope, to Bill Drone for his eye-pleasing, reader-friendly page layouts, and to many others at Accu-color.

Special thanks are also owed to Lena Dida, Ross Martin, and Gary Taylor at the California Division of Mines and Geology for their extensive help with illustrations.

Many others throughout California have assisted by providing slides, maps, and diagrams for use in this book.

Throughout my career as a geologist, I have been very fortunate to have some of California's finest geologists and teachers as role models. Although they are now departed, the late Edgar Bailey, Richard Janda, Clyde Wahrhaftig, and Bill Bradley were all influential mentors who introduced me to the complexities of California geology and to the importance of communication between geologists and society. The results of their teachings appear throughout this book, both in content and in matters of style, and I hope that my efforts go a small way toward repaying their support. Other prominent California geologists, some of whom I have not known personally, also deserve credit for their contributions to California geology. These include the late Gordon Oakshotte, authors Robert Norris and the late Robert Webb, Robert E. Wallace, and many others.

Finally, one person deserves acknowledgment verging on co-authorship for his contributions to this book. Andrei Sarna-Wojcicki of the U.S. Geological Survey assisted in many ways, including editing the first draft, supplying many of the photographs, helping with original illustrations, sharing unpublished work, hunting down resources at the U.S.G.S. library, and solving last-minute crises ranging from printers to house details to child-shuttling. As many can verify, a two-geologist family is not always ideal, but without Andrei's help and support, this book would never have seen the light of day.

Deborah R. Harden

Plate 1

Plate 2

Plate 3

Plate 4

Rocks

Plate 1 Layers of lake sediments and volcanic ashes at Wilson Creek, just north of Mono Lake. The youngest layers in the sequence, at the top of the cliffs, are composed of rare basaltic ash, erupted from nearby Black Point about 13, 000 years ago. The snow-capped eastern Sierra Nevada can be seen in the background. (Source: Sarna-Wojcicki, A. U.S. Geological Survey.)

Plate 2 Tilted and faulted layers of sandstone and conglomerate near San Juan Bautista. The sediments were deposited during late Cenozoic time and subsequently disrupted by tectonic activity along the nearby San Andreas fault. (Source: †Skapinsky, S. San José State University.)

Plate 3 Closeup view of Mesozoic plutonic rocks at Half Moon Meadow near Ten Lakes, Sierra Nevada. Bodies of darker, more mafic rock are enclosed within gray granitic rock formed from a more silica-rich magma. The white, quartz-rich veins cut through the both darker rock types, indicating that they crystallized last. (Source: Sarna-Wojcicki, A. U.S. Geological Survey.)

Plate 4 An outcrop of serpentinite along the beach south of the Golden Gate Bridge in San Francisco, showing both sheared and more massive bodies of serpentinite cut by light veins of dolomite. (Source: Sarna-Wojcicki, A. U.S. Geological Survey.)

†Deceased.

Plate 5

Plate 6

Plate 7

Fossils and Stratigraphy

Plate 5 The shell of a freshwater clam embedded in fine-grained lake deposits, Cronese Dry Lake, Mojave Desert. (Source: Wilshire, H. U.S. Geological Survey.)

Plate 6 Petrified redwood tree, Petrified Forest, Sonoma County. (Photo by author.)

Plate 7 A layer of volcanic ash within the sand and layers of the Rio Dell Formation at Centerville Beach, south of Eureka in Humboldt County, provides an age for the sediments. From its chemical fingerprint, the ash is identified as one erupted from the central Cascade Range about 1.2 million years ago. (Photo by author.)

Plate 8

Plate 9

Plate 10

Cascades and Volcanics

Plate 8 Lassen Peak viewed from the southwest. The light-colored area in the middle distance was covered by volcanic debris flows in 1915. (Source: Sarna-Wojcicki, A. U.S. Geological Survey.)

Plate 9 Late Cenozoic volcanic rocks of the Tuscan Formation at Long Gulch, along the northeastern edge of the Sacramento Valley. In this outcrop, conglomerate layers can be seen at the base, a relatively fine-grained volcanic debris flow in the center, and a very coarse volcanic debris flow at the top. (Source: Harwood, D.)

Plate 10 Columnar joints in volcanic rocks at Devil's Postpile. The columns result from weathering along the shrinkage cracks created as the lava cooled. (Source: Freeberg. U.S. Geological Survey.)

Plate 11

Plate 12

Plate 13

Plate 14

Desert Provinces

Plate 11 Tufa towers at Mono Lake. (Source: Sarna-Wojcicki, A. U.S. Geological Survey.)

Plate 12 View of the steep, faulted eastern front of the Sierra Nevada from New York Canyon in the southern Inyo Mountains. The Owens River can be seen in the middle distance, and the Alabama Hills at the base of the mountain front. (Photo by author.)

Plate 13 Pleistocene lake beds of Lake Tecopa, east of Death Valley. (Source: Sarna-Wojcicki, A. U.S. Geological Survey.)

Plate 14 Steeply dipping and folded Paleozoic sedimentary rocks in the northern Panamint Mountains. Light-colored dune sands of Death Valley's Racetrack can be seen in the valley beyond the Paleozoic rocks. (Source: McMackin, M. San José State University.)

Plate 15

Plate 16

Plate 17

Sierra Nevada

Plate 15 Yosemite's Half Dome viewed from Glacier Point. (Source: Sarna-Wojcicki, A. U.S. Geological Survey.)

Plate 16 A view of Hetch Hetchy reservoir showing the rounded domes typical of the glaciated granitic rocks of the High Sierra. Sarna-Wojcicki, A. U.S. Geological Survey.)

Plate 17 The Sevehah Cliffs, a roof pendant of folded Paleozoic marble near Convict Lake, eastern Sierra Nevada. (Stevens, C. San José State University.)

Plate 18

Plate 19

Plate 20

Plate 21

Coast Ranges

Plate 18 In this color infrared Landsat image of the San Francisco Bay area, the faults of the San Andreas system are clearly revealed as linear features on either side of the Bay. Irrigated fields of the Great Valley show as a distinct patchwork at the right edge of the photo, and Mt. Diablo is the teardrop-shaped mountain just south of the eastern part of the Bay.

Plate 19 The Golden Gate Bridge , looking northeast from Bakers Beach State Park. The Marin Headlands terrane of the Franciscan Complex forms the rugged mountains at the northern (left-hand) end of the Golden Gate Bridge. (Source: Sarna-Wojcicki, A. U.S. Geological Survey.)

Plate 20 A typical landscape formed on Franciscan rocks in the northern Coast Ranges. In this view, a small block of resistant blueschist forms an isolated knob on the hill. (Source: †Bailey, E. U.S. Geological Survey.)

Plate 21 A sample (about 20 cm across) of mercury ore from New Idria mine, San Benito County. This type of cinnabar is known as "strawberry ore" for its resemblance in color and shape to the fruit. (Source: †Bailey, E. U.S. Geological Survey.)

†Deceased.

Plate 22

Plate 23

Plate 24

Faults

Plate 22 Just south of the Cienega Valley winery (formerly the Almaden winery) in San Benito County, a drainage ditch crosses the creeping section of the San Andreas fault. In 1961, the ditch showed notable right-lateral offset. (Source: †Skapinsky, S. San José State University.)

Plate 23 In 1996, 35 years later, the amount of right-lateral offset has markedly increased. (Source: Sorenson, R.)

Plate 24 A thrust fault has placed red-toned, well-bedded sediments (left side) against more massive sediments on the right at the base of the outcrop. Layered sediments at the top of the exposure bury the fault. Kalorama Street, Ventura. (Source: Sarna-Wojcicki, A. U.S. Geological Survey.)

Plate 25

Plate 26

Plate 27

Plate 28

Beaches, Transverse Ranges

Plate 25 Typical features of an emergent coast, including sea stacks, steep cliffs, and coastal landslides, can be seen in this view along the Humboldt County Coast. Trinidad Head is the large point in the middle distance. The formation of sea stacks is enhanced by the nature of the Franciscan bedrock here, because a zone of melange containing resistant blocks parallels the coastline. (Photo by author.)

Plate 26 Natural arches formed by wave erosion at Natural Bridges State Park near Santa Cruz. In 1976, wave erosion collapsed the right-hand arch. The flat surface on top of the sandstone marks a marine terrace approximately 105,000 years old. (Sarna-Wojcicki, A. U.S. Geological Survey.)

Plate 27 The Ventura Avenue anticline, a recent fold formed in response to compression in the Transverse Ranges. Marine sediments formed in basins well below sea level as recently as a few hundreds of thousands of years ago. (Source: †Bradley, W.)

Plate 28 Hogbacks formed by resistant, dipping sandstone beds of the Cajon Formation near Cajon Pass, San Bernardino County. (Photo by author.)

PART

Basic Principles of Geology

The first four chapters of this book are designed to introduce readers with little or no geology background to some basic principles of geology. These principles are discussed with particular application to California geology, and geologic maps and cross sections are also introduced. Readers with a limited background in geology are also encouraged to supplement this information with additional readings from introductory geology textbooks as needed.

CHAPTER

1

Plate Tectonics and California

Most Californians know that earthquakes and the San Andreas fault have something to do with movement of the Earth's plates, but few residents realize the importance of ***plate tectonics*** in shaping California. According to the plate tectonics theory, which has been widely accepted for the past 25 years, the Earth's outer layers consist of a number of rigid plates (Fig. 1-1). These plates are mobile, and their encounters with each other cause many of Earth's dynamic events. Today plate tectonics is responsible for some obviously unwelcome aspects of our lives—earthquakes and volcanic eruptions—as well as for some of the scenery that attracts people to California: rugged mountain ranges and beautiful, rocky coastlines. The region we call California has been at an active plate boundary for the past 230 million years, and this has been the dominant factor in California's geologic history.

THE LAYERED EARTH

Although the deepest holes drilled into the Earth's interior have penetrated only 10,000 meters, scientists have long known from indirect evidence that the Earth is a layered sphere. The outermost thin skin of the Earth is the ***crust,*** a rigid layer of rocks rich in the elements silicon and oxygen (Fig. 1-2). The crust is only about 5 kilometers thick beneath the oceans and 50 or more kilometers thick beneath continental mountains like Sierra Nevada range (Fig. 1-3). Beneath the crust is the ***mantle,*** which extends inward about halfway to the center of the Earth and makes up more than 80 percent of the Earth's volume. Like the crust, the mantle is composed of rocks rich in silicon and oxygen, but its rocks also contain more iron and magnesium, making them denser than those found in the crust.

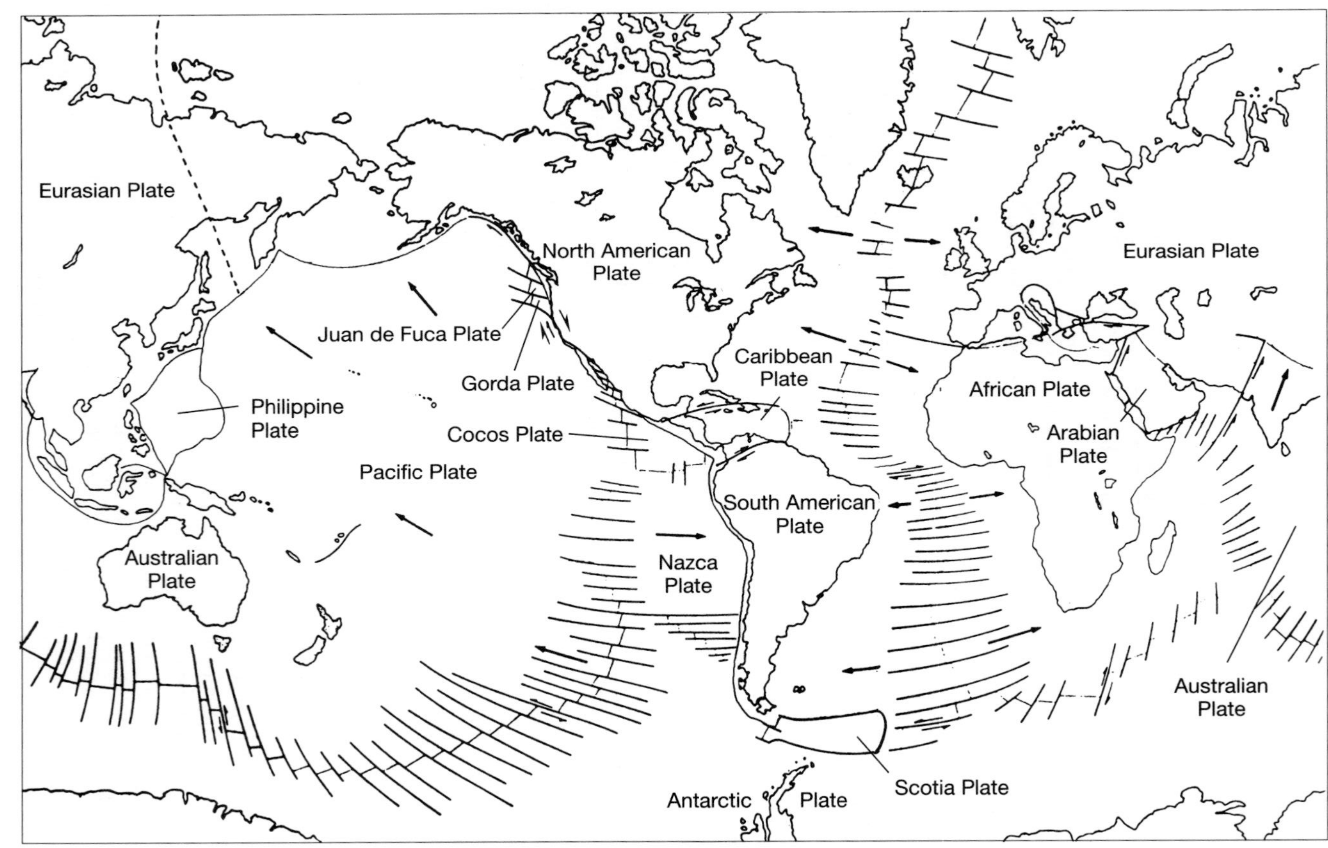

Fig. 1-1 Map showing the Earth's major plates. (Source: Modified from Coch, N., and Ludman, A. 1991. *Physical Geology.* New York: Macmillan Publishing Co., pp. 14-15.)

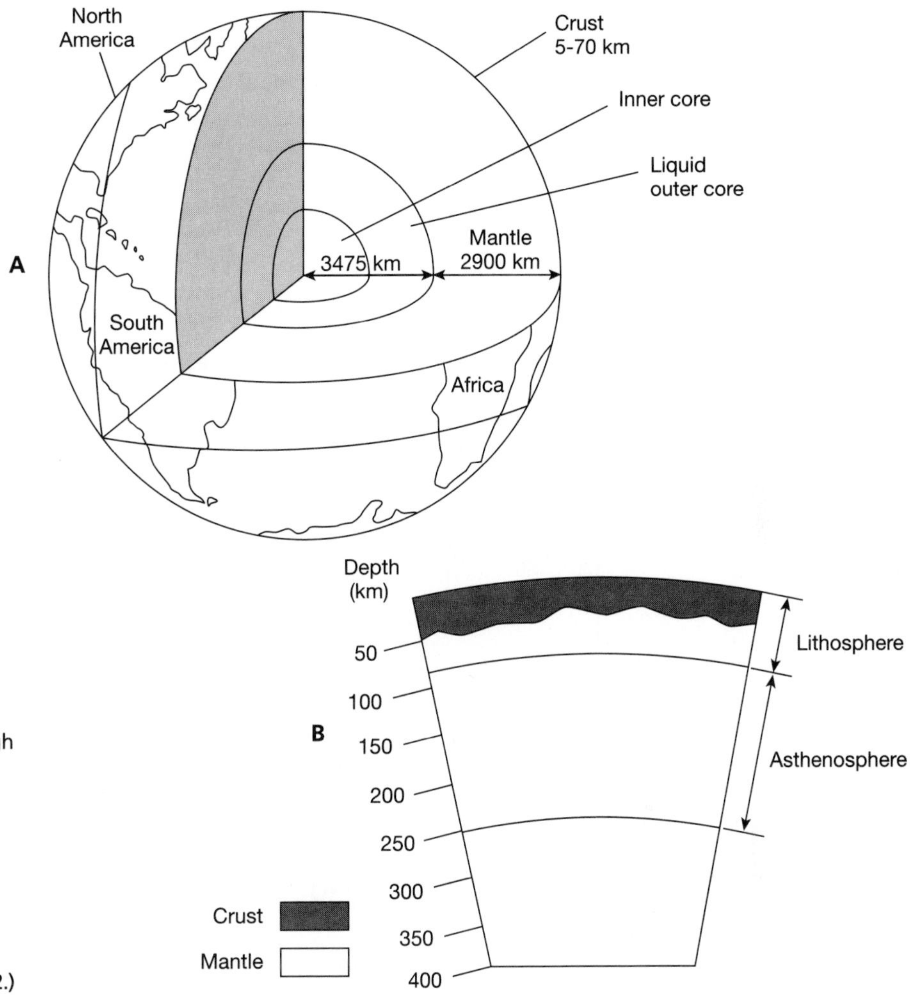

Fig. 1-2 **A,** Slice through the Earth, showing its layered interior. **B,** Enlargement of the outer 400 km. (Source: Modified from Coch, N, and Ludman, A. 1991. *Physical Geology.* New York: Macmillan Publishing Co., pp. 30, 32.)

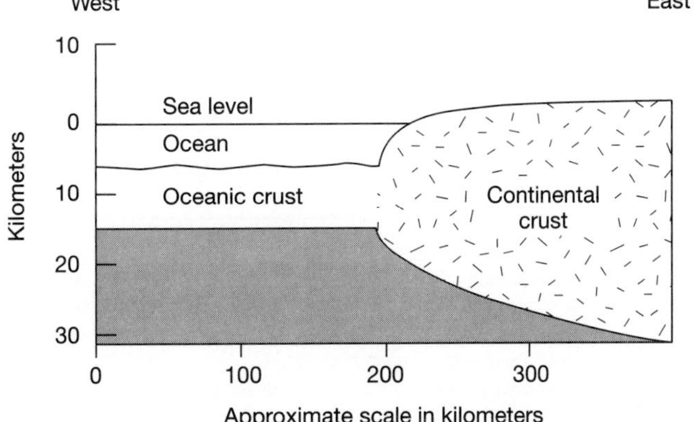

Fig. 1-3 A cross section beneath a continental margin showing the contrasts in the thickness of the crust. (Source: Modified from Skinner, B., and Porter, S. 1989. *The Dynamic Earth.* New York: John Wiley & Sons, p. 7.)

Rocks believed to originate in the mantle are found in several places in California. They were brought to the surface by a number of processes, including volcanic eruptions and plate collisions. For example, in the Mojave Desert, bits of mantle rocks have been brought to the Earth's surface in recently erupted lava (see Chapter 7). Indirect evidence about Earth's deeper interior is also provided by the behavior of seismic (earthquake) waves. Beneath the mantle is the ***core,*** which has a liquid outer part and a solid inner part. Both the inner and outer core are thought to be composed mostly of iron and nickel.

PLATE TECTONICS AND PLATE BOUNDARIES

Interactions between plates generate much of Earth's geologic activity. California owes its active seismicity and recent volcanic eruptions to the fact that both the Pacific Plate and the Gorda Plate (Fig. 1-4) meet the North American Plate in California. Most of the world's earthquakes occur at plate boundaries; in fact, the zones of concentrated seismicity are one of the main lines of evidence used to iden-

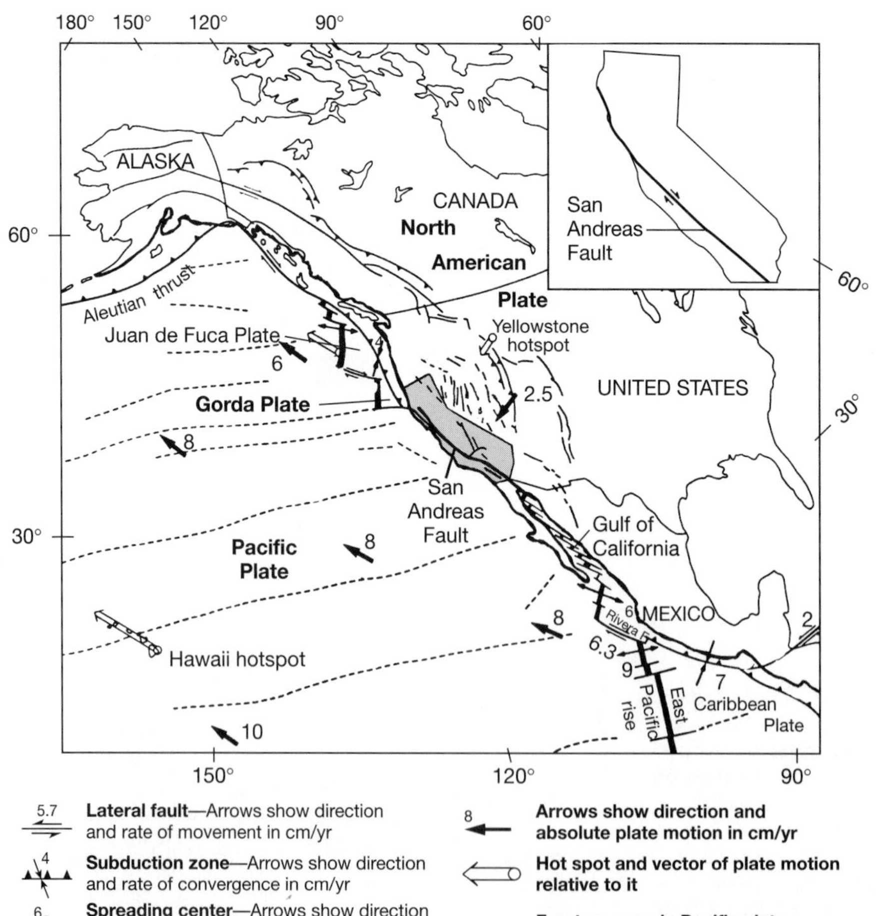

Fig. 1-4 Major plate tectonic features of western North America. (Source: Modified from Wallace, R. 1990. U.S. Geological Survey, Professional Paper 1515.)

tify plate boundaries. Along plate boundaries, the different motions of the plates create stress, and that stress is relieved during earthquakes (see Chapter 13).

Plates can interact in three ways. They may separate along ***divergent*** plate boundaries, where the crust is stretched and thinned (Fig. 1-5). At divergent boundaries, molten magma rises and erupts to form new oceanic crust. Today an active divergent boundary in the Gulf of California is causing Baja California, Mexico to break away from the Mexican mainland as new oceanic crust forms in the Gulf. This process is one example of ***seafloor spreading,*** the process that gave birth to the Atlantic Ocean as the Americas separated from Europe and Africa. Today satellite systems and laser-instrumented surveying devices allow scientists to measure the size of the opening of the Gulf of California from space. Based on the results of repeated surveys with this global positioning system (GPS), it appears that the area is spreading at the rate of about 5 centimeters per year.

Tectonic plates collide at ***convergent*** boundaries. At these zones, crust is either piled up into mountain ranges by buckling, faulting, and folding along the plate boundary or is shoved into the mantle during ***subduction.*** Subduction occurs if one or both of the converging plates is an oceanic plate. At the point of convergence, one plate—always the oceanic crust if the other plate is a continent—is overridden and driven beneath the other plate along a subduction zone. Along most subduction zones, a deep ocean ***trench*** marks the point where the plates converge (Fig. 1-6). As the plate descends into the mantle, earthquakes are generated. Subduction zones are the only tectonic settings at which earthquakes originate at great depths, and they also give rise to the largest-magnitude earthquakes. Earthquakes and their magnitudes are discussed in Chapters 13 and 14.

As subduction drives the dense, water-rich rocks of the oceanic plate deeper within the Earth's interior, water is driven out of the subducting rocks. This dewatering causes partial melting in the mantle above the subducted rocks, creating subduction zone volcanoes. As we will see in Chapter 5, these volcanoes are quite different from those found at divergent boundaries. Today along the northern margin of California, the Gorda Plate is being subducted beneath the North American Plate (see Fig. 1-4). As a result, northwestern California faces the threat of a great sub-

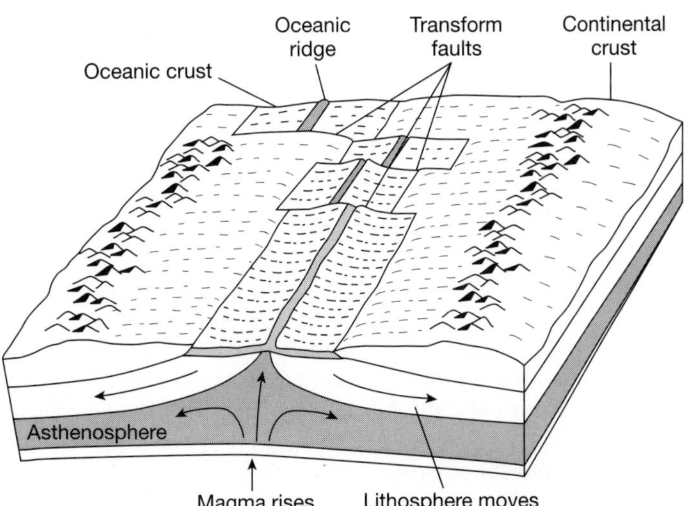

Fig. 1-5 Characteristic features of divergent plate boundaries. (Source: Skinner, B., and Porter, S. 1989. *The Dynamic Earth.* New York: John Wiley & Sons, p. 388.)

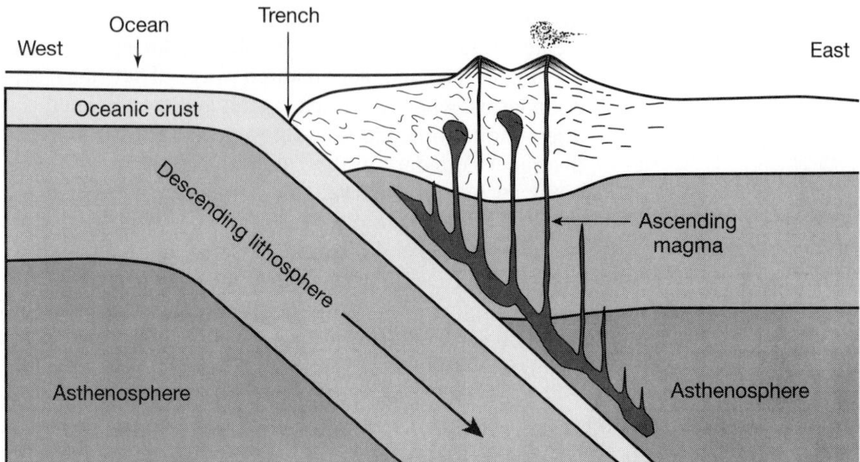

Fig. 1-6 Characteristic features of subduction zones. (Source: Skinner, B., and Porter, S. 1989. *The Dynamic Earth.* New York: John Wiley & Sons, p. 62.)

duction-zone earthquake, and areas further east must contend with the risk of major volcanic eruptions from Mt. Shasta and Lassen Peak.

During convergence and subduction, pieces of the subducted plate may also be scraped off along the plate margin, where they become attached to the other plate. These pieces are called ***accreted terranes.*** Throughout California, examples of ancient collisions are found in the form of oceanic fragments attached to the North American Plate (see Chapters 8, 9, and 12). In many accreted terranes, fragments of the mantle have been scraped up along with the oceanic crust, providing geologists with another opportunity to study rocks formed at great depths (see Chapter 9).

Plate boundaries where two plates are moving past each other without diverging on converging are known as ***transform*** plate boundaries. Along transform boundaries, earthquakes record the plate motions as they do along the other boundary types. However, volcanic eruptions are generally absent. The most famous transform plate boundary in the world is California's San Andreas fault system, which separates the Pacific and North American Plates (see Fig. 1-4 and Endpaper 2 inside the front cover). The faults of the San Andreas system show that the Pacific Plate is moving to the northwest relative to the North American Plate. The relative motion, or ***displacement,*** on these faults is in a right-lateral direction, as illustrated in Fig. 1-7.

The actual boundary between the Pacific and North American Plates is best thought of as a wide zone rather than as a line. The effects of plate motion, such as active transform faults and recently uplifted mountain ranges, are seen across the width of California and even further east (Fig. 1-8). The San Andreas fault is conventionally thought of as the plate boundary, but only about half to two thirds of the plate motion takes place within the San Andreas fault zone. In the San Francisco Bay area and south of the Transverse Ranges, plate motion takes place along other faults of the San Andreas system east and west of the San Andreas itself (see Fig. 1-7). Additional lateral motion takes place along the eastern edge of the Sierra Nevada, in the central Mojave Desert (see Chapter 14), and even east of California. The complexities of California geology are revealed when one realizes that even the question "Where is the exact boundary between the Pacific and North American Plates?" has no precise answer.

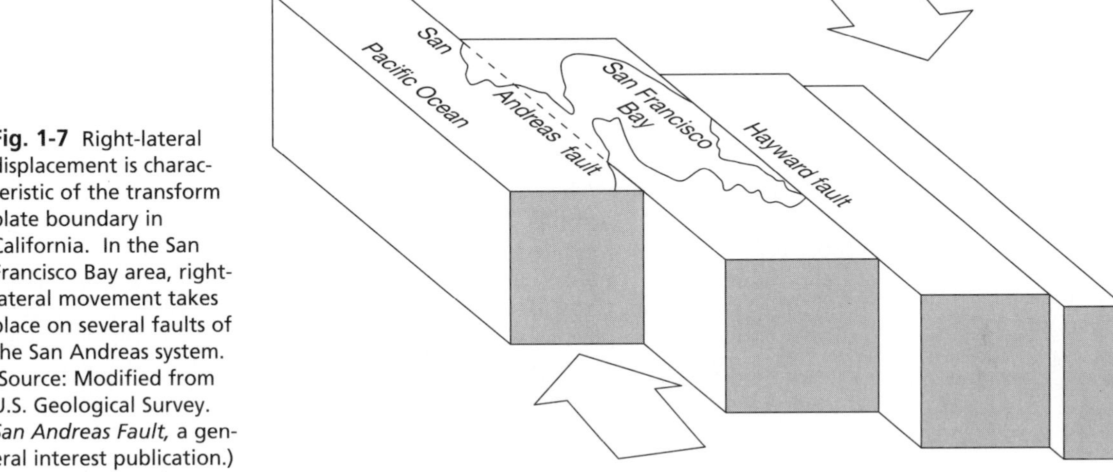

Fig. 1-7 Right-lateral displacement is characteristic of the transform plate boundary in California. In the San Francisco Bay area, right-lateral movement takes place on several faults of the San Andreas system. (Source: Modified from U.S. Geological Survey. *San Andreas Fault,* a general interest publication.)

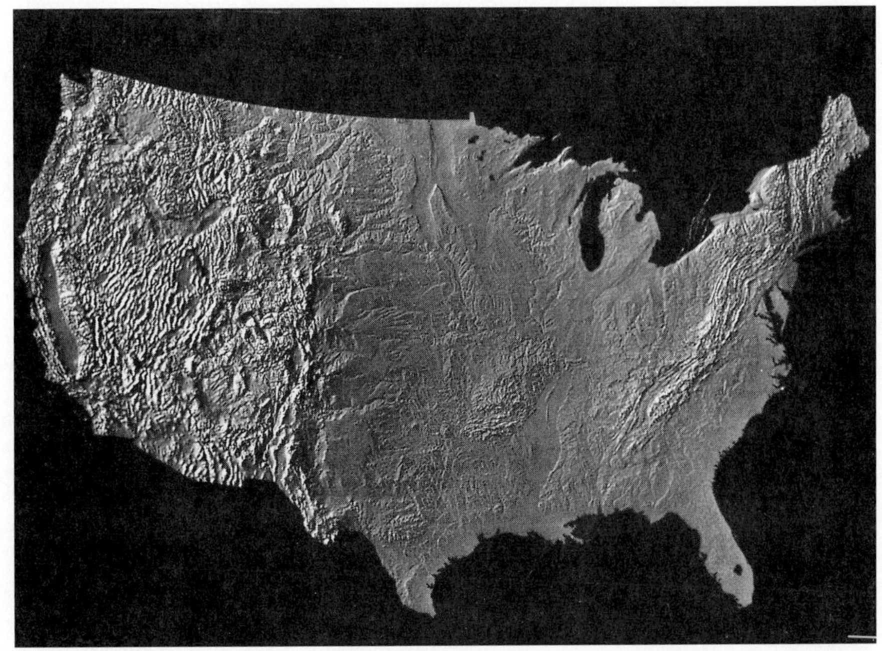

Fig. 1-8 Topography of the United States, illustrating rugged landscape of the western continent. The effects of recent tectonics are seen far east of the San Andreas fault. (Source: Thelein, G, and Pike, R. 1991. U.S. Geological Survey digital shaded-relief map I-2206 of the conterminous U.S.)

CAUSES OF PLATE MOTIONS

Within the Earth's upper mantle is a layer known as the ***asthenosphere,*** a weak, semiplastic layer that underlies the moving plates, or ***lithosphere.*** This layer is important in understanding plate tectonics, because it is the "conveyor belt" that move the plates. The boundary at the base of the plates, where relatively cool lithosphere meets hotter, more plastic asthenosphere, is about 100 to 250 kilometers beneath the Earth's surface. Scientists believe that they have located the top of the asthenosphere using seismic techniques, but they are less certain about the mechanisms that cause plates to move.

Until recently, geologists believed that ***convection currents*** in the asthenosphere caused the rigid plates above it to move. These convection currents cause hot mantle rock to rise, cool, and then sink again, creating circulation within the upper mantle. More recent ideas about plate tectonics focus on density and gravity differences within the lithosphere as the driving forces. After molten material rises to the surface and forms new oceanic crust at divergent boundaries, the plate cools and becomes denser. As a result, the plate sinks back into the asthenosphere at convergent boundaries (Fig. 1-9). Geologists hypothesize that this cooling and sinking of a plate could take as much as 60 to 70 million years.

HOTSPOTS AND PLATE MOTIONS

Because all of the Earth's plates are mobile, the plate boundaries provide clues only to the relative motions between plates. The actual movement directions and rates for individual plates (as they would be seen from space, for example) are measured in several ways. Global positioning satellites are one method for precisely determining plate motions or the vertical growth of mountain ranges. Geologists can also determine the long-term movement of plates at ***hotspots***. These are concentrated areas of very high heat flow deep within the mantle that causes rocks to melt above them. Because they are located below the plates, sci-

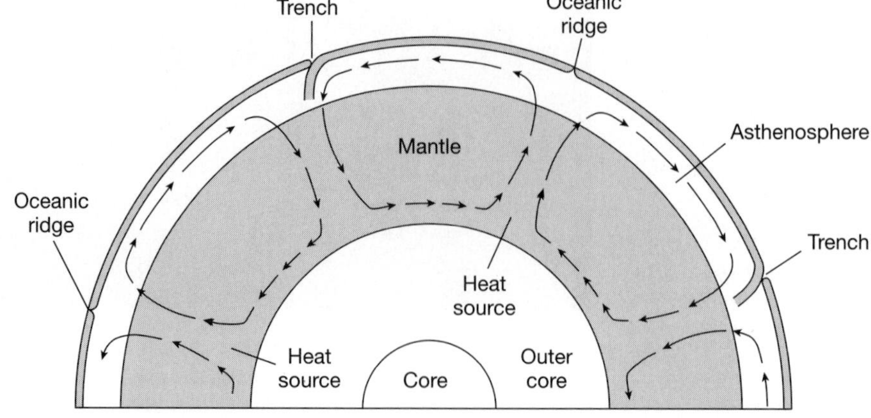

Fig. 1-9 Plate movement is thought to be related to convection in the Earth's mantle. (Source: Hamblin, W, and Christiansen, E. 1995. *Earth's Dynamic Systems,* 7th ed. Upper Saddle River, N.J.: Prentice Hall, p. 490.)

entists believe that hotspots remain fixed while the plates move over the hotspots. Directly above the hotspots, lava rises though the plate to form volcanoes. As the plate moves, new volcanoes appear in the part of the plate newly positioned over the hot spot. After millions of years of plate motion over a fixed hot spot, a chain of volcanoes and volcanic rocks is left as a marker of the direction and the long-term rate of plate motion. In this way, the Hawaiian islands record the movement of the Pacific Plate (Fig. 1-10). Similarly, the volcanic rocks of eastern Oregon, Idaho, and Wyoming record the movement of the North American Plate over the Yellowstone hotspot.

The absolute direction of movement of the Pacific Plate is northwestward, while that of the North American Plate is southwestward (see Fig. 1-4). Where the two plates meet along the northwest-oriented boundary, the relative motion between them results in the right-lateral motion that characterizes the San Andreas fault system (see Chapter 14). A small amount of this motion, about 10 percent, is convergent, resulting in the formation of mountains, as discussed next.

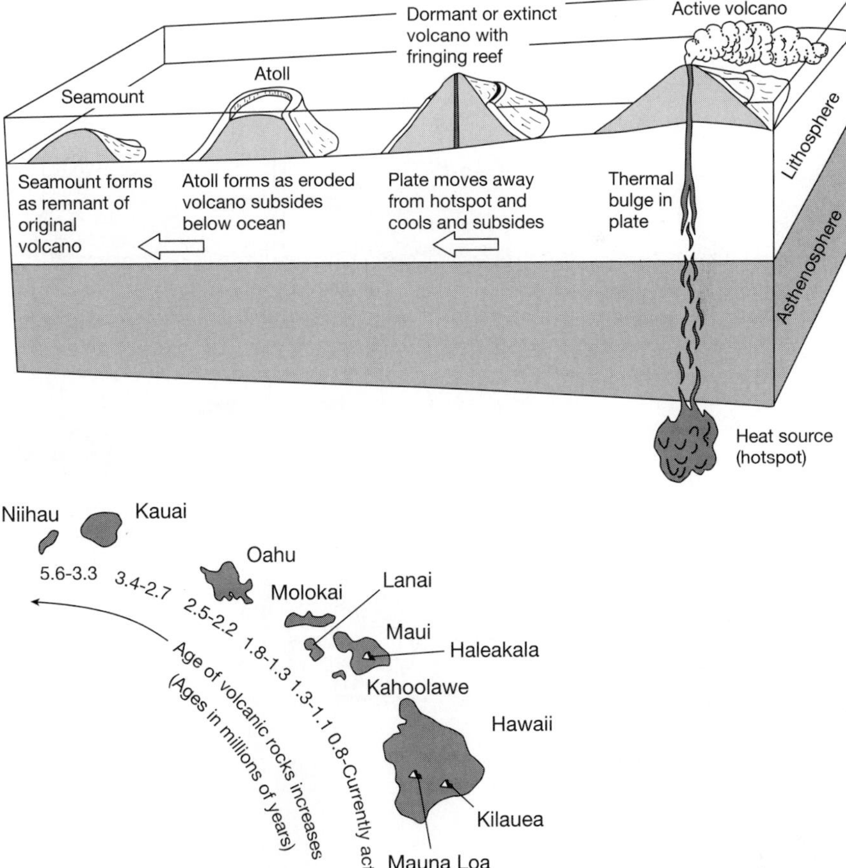

Fig. 1-10 The Hawaiian islands record the movement of the Pacific Plate during the past 5.6 million years. (Source: Modified from Coch, N., and Ludman, A.. 1991. *Physical Geology.* New York: Macmillan Publishing Co.)

DEFORMATION

The motions between plates create forces that act on the rocks near plate boundaries. In any given area, force applied to a particular body of rock, regardless of its size, creates ***stress*** on the rock. Tectonic stress causes rocks and landforms to be disrupted, creating features that geologists can use to interpret the types of forces that created them. By studying these patterns of ***deformation*** and analyzing their geometry, geologists can infer not only the modern stress conditions, but also the ancient forces recorded in the geologic features of an area. However, because plate boundaries are not simple planes between homogeneous blocks of rock, the patterns of deformation are often complex and difficult to interpret.

In general, convergent motion causes rocks to be ***compressed*** (Fig. 1-11). Compression causes rocks to be crumpled and buckled by folding and faulting, all of which act to shorten the earth's crust (Fig. 1-12). Divergent motion causes the crust to be ***extended.*** In this setting, as the crust is extended or stretched, blocks of rock may slide apart, creating depressions (Fig. 1-13). Further complexity may be added to the picture, as when a mountain range has been uplifted in response to compression, and the uplift itself may have produced extension along the crest.

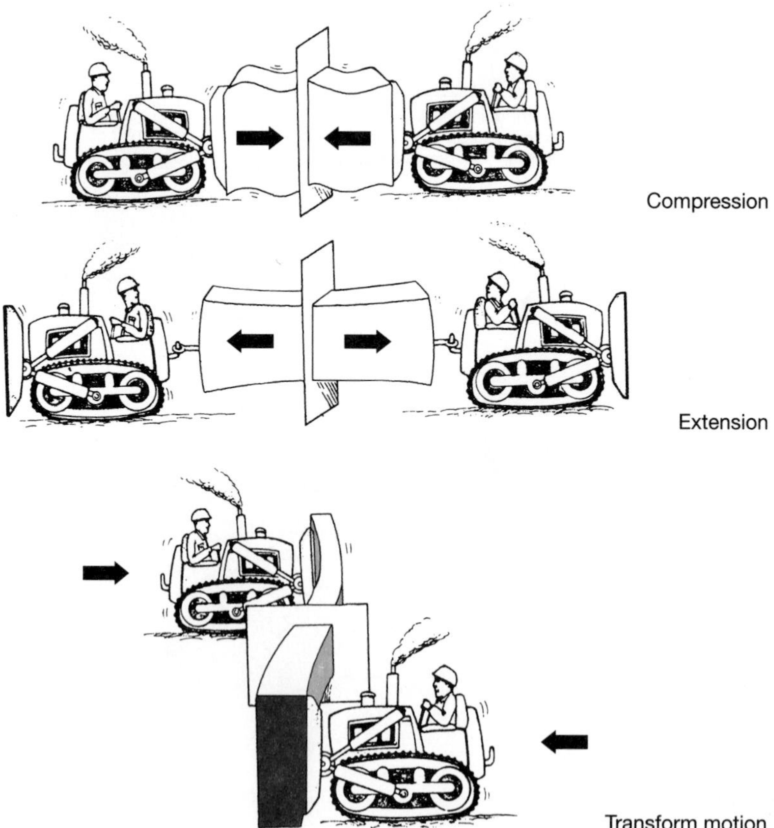

Fig. 1-11 This cartoon illustrates compression, extension, and transform motion. (Source: Coch, N., and Ludman, A. 1991. *Physical Geology.* New York: Macmillan Publishing Co., p. 466.)

An area affected by purely transform motion is subjected to ***shear*** stress (see Fig. 1-11). However, the forces operating along the San Andreas fault system are not created by simple transform motion. In many areas, convergent motion accompanies the transform motion along the plate boundary; geologists refer to this combination of forces as ***transpression*** (Fig. 1-14). As a result, vertical movements are recorded during earthquakes, and entire mountain ranges have been buckled and uplifted in the fault zone during the past few million years. California's Coast Ranges, Transverse Ranges, and even the Sierra Nevada are testimony to the significant, recent compression along the s boundary. During the geologically recent past, a sig-

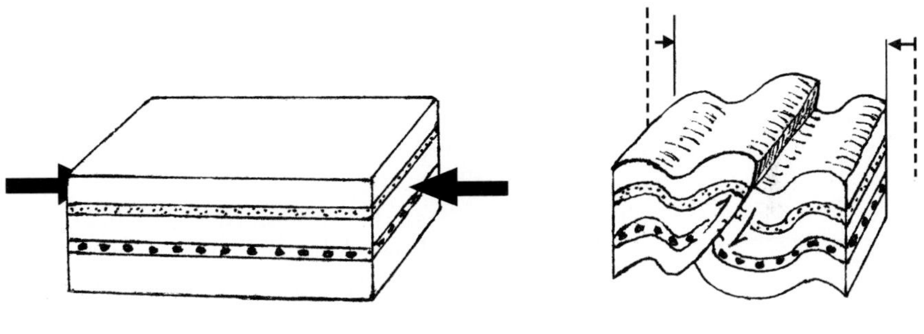

Fig. 1-12 Folded and faulted rocks record the shortening of the Earth's crust that results from compression. (Source: Creely, R.S. 1997.)

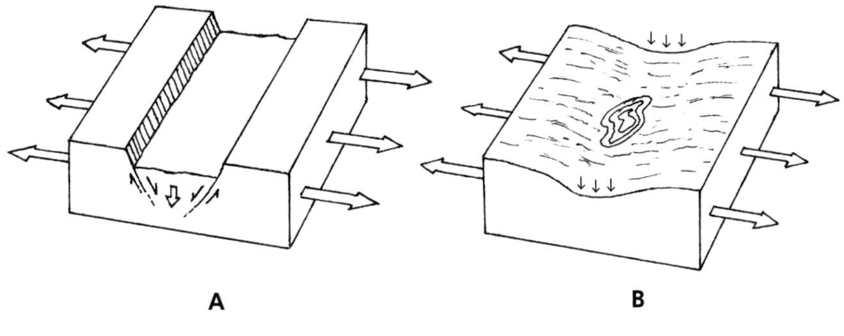

Fig. 1-13 Depressions formed as the crust is pulled apart. **A,** A fault-bounded valley, or ***graben,*** develops; **B,** a smaller and more localized depression contains a ***sag pond.*** (Source: Creely, R.S. 1997.)

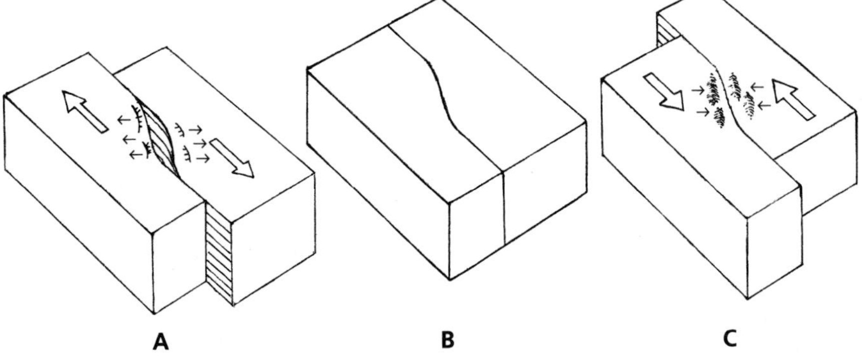

Fig. 1-14 Transpression or transtension may be created if compression or extension accompany transform motion. Bends in a fault **(B)** may create transtension **(A)** or transpression **(C)**, depending on the type of fault motion (shown by arrows). (Source: Creely, R.S. 1997.)

nificant component of extension accompanied transform motion along the San Andreas system, caused by ***transtension*** (see Fig. 1-14) along the plate boundary. This past extension is recorded in a series of fault-related marine basins and volcanism, as reconstructed from rocks of that period.

FURTHER READINGS

HOWELL, D.G., ed. 1985. *Tectonostratigraphic Terranes of the Circum-Pacific Region.* Houston: Circum-Pacific Council for Energy and Mineral Resources, Earth Science Series, No. 1, 581 pp.

MONROE, J.S., AND WICANDER, R. 1995. *Physical Geology: Exploring the Earth,* 2nd ed. St. Paul, Minn.: West Publishing, 639 pp.

MOORES, E.M., AND TWISS, R.J. 1995. *Tectonics.* New York: W.H. Freeman & Co., 415 pp.

TARBUCK, E.J., AND LUTGENS, F.K. 1996. *Earth,* 6th ed. Upper Saddle River, N.J.: Prentice Hall, Inc., 605 pp.

TARBUCK, E.J., TASA, D., AND MCNEILL, L. 1994. *The Theory of Plate Tectonics: Interactive CD-ROM.* Albuquerque, N.M.: Tasa Graphic Arts.

WYLLIE, P.J. 1976. *The Way the Earth Works.* New York: John Wiley & Sons, 289 pp.

CHAPTER

2

California's Rocks

Because of its long history as an active plate margin, California has an unusual assortment of rocks. Rocks that are common in many parts of the world—such as limestone—are uncommon in much of California, and rocks that are very rare in most places—such as ***serpentinite***—are fairly abundant. Although it is beyond the scope of this book to discuss the formation and classification of rocks in detail, this chapter presents a brief overview of California's rocks. Later chapters include more detailed discussions of some rock types, and many excellent sources are available for more comprehensive reading.

Even rocks that appear to be completely homogeneous are, in most cases, composed of a multitude of individual grains. These grains may be mineral crystals, pieces of rock or minerals, fragments of animals or plants, or a combination of two or more of these. The types of grains and their proportionate amounts define a rock's ***composition.*** The size and shape of the grains and the way the grains fit together define its ***texture.*** Both the composition and texture provide important clues about the origin of each rock, which is the reason that identification and classification of rocks are an important part of any geologist's work. Field study of rocks as they appear in the context of surrounding rocks and the landscape is an important part of the identification process (Fig. 2-1). Closer examination of hand samples with a magnifying lens may aid in the identification of minerals or fossils. Rocks composed of grains too small to be seen with a hand lens require examination with a microscope. (Figs. 2-2 and 2-3).

The minerals and textures of a given rock reflect its source, providing clues about either the preexisting rocks or the fluids from which it formed. For a particular mineral to form, the correct chemical building blocks must be present. For example, iron oxide cannot form in an environment in which oxygen is absent, and quartz cannot crystallize from pure water. Even if the correct ingredients are present, some minerals will form only if certain temperatures and pressures exist. After decades of detailed laboratory studies, geologists also know the temperature and pressure conditions favorable for the formation of most minerals. The presence of certain minerals in a rock reveals whether the rock formed at or near Earth's surface or thousands of meters beneath it.

Fig. 2-1 A geologist examines volcanic rocks and sediments near Mono Lake. (Source: Sarna-Wojcicki, A. 1995. U.S. Geological Survey.)

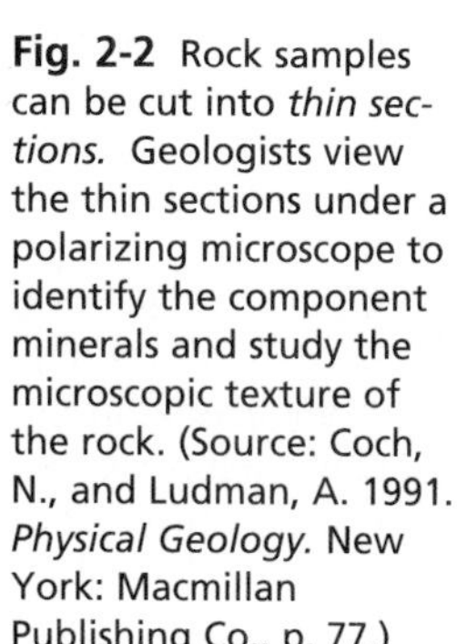

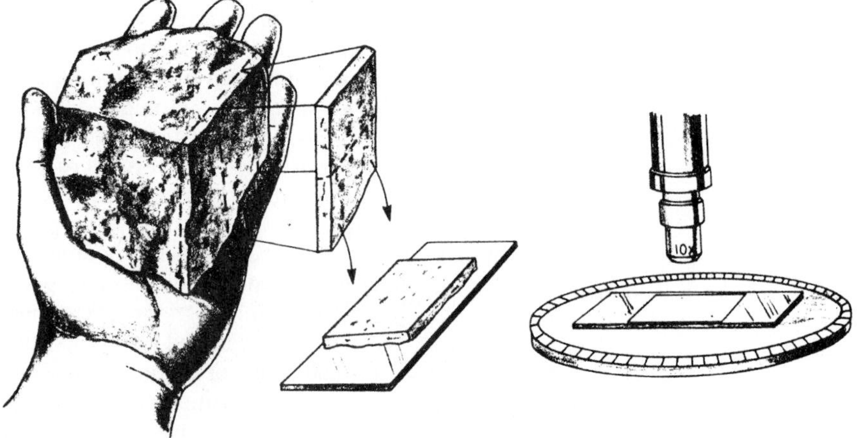

Fig. 2-2 Rock samples can be cut into *thin sections.* Geologists view the thin sections under a polarizing microscope to identify the component minerals and study the microscopic texture of the rock. (Source: Coch, N., and Ludman, A. 1991. *Physical Geology.* New York: Macmillan Publishing Co., p. 77.)

Fig. 2-3 Thin section of granite showing an interlocking network of silicate minerals.

MINERALS—THE BUILDING BLOCKS OF ROCKS

When confronted with the variety and complexity of California's rocks, a beginning student of geology has a difficult time believing that rocks are constructed in any systematic order. In fact, rocks are made of well-ordered building blocks, which are themselves composed of orderly materials. Rocks are composed of ***minerals,*** minerals are composed of one or more ***chemical elements,*** and the chemical elements are atoms, the basic building blocks of all matter. It is the variety of possible combinations of elements and minerals and the variety of physical and chemical conditions of formation that creates the variety in rock composition. On a larger scale, the hierarchy of Earth's building blocks continues: bodies of similar rock are grouped together in geologic ***formations,*** as discussed in Chapter 3. Rock formations may in turn constitute a distinctive geologic terrane. On a still grander scale, lithospheric plates—some of them aggregates of terranes—are the building blocks of the Earth's crust.

A mineral is a naturally occurring, inorganic, ***crystalline*** solid. A crystalline solid is a material whose atoms are arranged in an orderly, regularly repeating arrangement. Minerals may be composed of atoms of only one element, as is the case with gold nuggets. However, most of the nearly 3000 known minerals are made up of more than one element combined in fixed proportions in a regular crystalline arrangement. Even minerals that appear to be dirty, lumpy specks are in fact constructed with beautifully ordered atoms of their component elements. If minerals grow under ideal circumstances—with adequate space, time, and chemical conditions—the internal order may be reflected in the formation of crystals (Fig. 2-4, *A* and *B*).

Many people are familiar with the term *mineral* as it is sometimes used for nutritional supplements—such as iron, calcium, magnesium. Strictly speaking, these are chemical elements rather than minerals. Only 10 of the 105 known chemical elements make up 99.9% of the Earth's crust, and two, oxygen and silicon, alone account for almost 75%. It is therefore not surprising that the most common minerals contain oxygen and silicon. These minerals are known as the ***silicate minerals,*** and they are the building blocks of most rocks (Table 2-1). Although more than 600 minerals are found in California, and about 50 of these are found nowhere else, most of California's rocks are made up predominantly of the common rock-forming minerals.

Some of the rare elements in the Earth's crust are extremely valuable to society, and concentrations of these elements in rare minerals and rocks have long been sought after by miners. Geologic events have enriched California rocks with pockets of valuable commodities such as gold, mercury, lithium, chromium, boron, and other elements. As we will see in subsequent chapters, mineral resources have always played an important role in California's history.

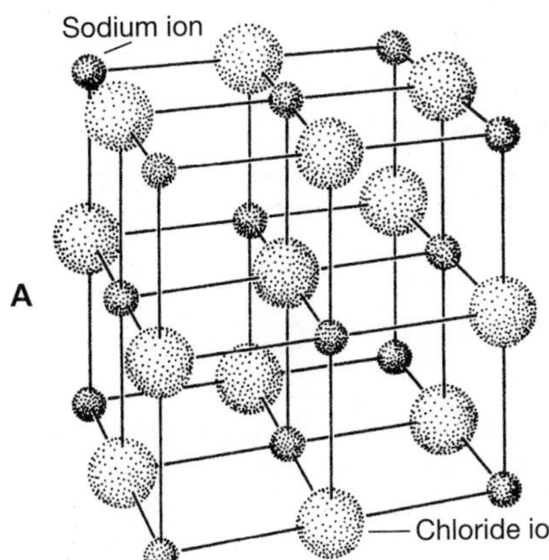

Fig. 2-4 A, The internal atomic structure of halite. **B,** Halite crystal. (Photo by author.)

TABLE 2-1 EXAMPLES OF COMMON MINERALS IN CALIFORNIA ROCKS

Mineral Group	Minerals	Component Elements	Occurrence
Silicates	Quartz	Silicon (Si), Oxygen (O)	All rock and sediment types
	Feldspars	Si, O, aluminum (Al), plus:	All rock and sediment types
	Potassium	Potassium (K)	
	Plagioclase	Calcium (Ca), Sodium (Na)	
	Clays	Si, O, Al, hydrogen (H), many others	Sediments, soils
	Olivine	Si, O, Magnesium (Mg), iron (Fe)	Igneous rocks,
	Serpentine	Si, O, Mg, Fe, O, H	Serpentinite
	Pyroxenes Augite	Si, O, Al, Ca, Mg, Fe	Igneous, metamorphic rocks
	Amphiboles Hornblende	Si, O, Al, Na, Ca, Mg, Fe, and others	Igneous, metamorphic rocks
	Biotite mica	Si, O, K, Mg, Fe, O, H	All rock types
	Garnet	Si, O, Al, Ca, Mg, Fe	Metamorphic rocks
Carbonates	Calcite	Carbon (C), O, Ca	Sedimentary rocks, sediments, marble
	Dolomite	C, O, Ca, Mg	Sedimentary rocks, sediments, marble
Oxides	Hematite (iron oxides)	Fe, O	Sediments, soils, sedimentary rocks, all weathered rocks
Sulfides	Pyrite (fool's gold)	Fe, sulfur (S)	All types of rocks
	Galena	Lead (Pb), S	Metamorphic rocks
	Cinnabar	Mercury (Hg), S	Metamorphic rocks

The texture of a rock also reflects its conditions of formation. In general, all rocks display one of two general textures. Rocks with a crystalline texture display a network of interlocking mineral grains, indicating that crystals grew together (Fig. 2-5). In contrast, rocks with ***clastic*** (from the Greek word meaning "broken") texture are composed of mineral and/or rock fragments held together by a cementing mineral (Fig. 2-6). Rocks with a clastic texture formed at least in part from pieces of older, preexisting rocks or from pieces of living organisms.

IGNEOUS, SEDIMENTARY, AND METAMORPHIC ROCKS

Rocks are classified into three major groups, based on the way in which they were formed. ***Igneous*** rocks form by the solidification of ***magma,*** molten silicate material formed by melting or partially melting rocks. Most magma is created at active plate boundaries or at hotspots (see Chapter 1). Igneous rocks are further broken down into two major subgroups: ***volcanic*** rocks that crystallize rapidly from ***lava,***

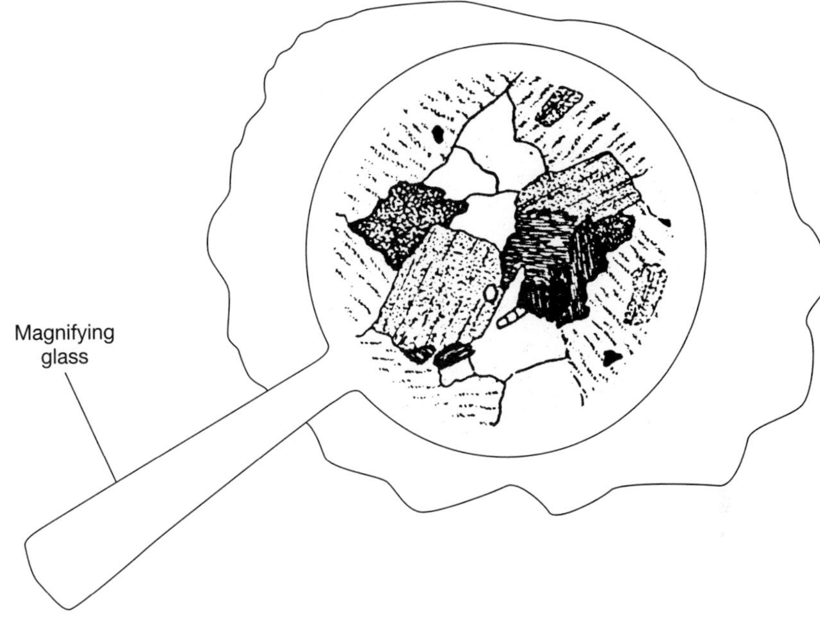

Fig. 2-5 A rock such as granite displays a crystalline texture. The silicate minerals are intergrown, indicating that they formed together. (Source: Sarna-Wojcicki, A. 1995. U.S. Geological Survey.)

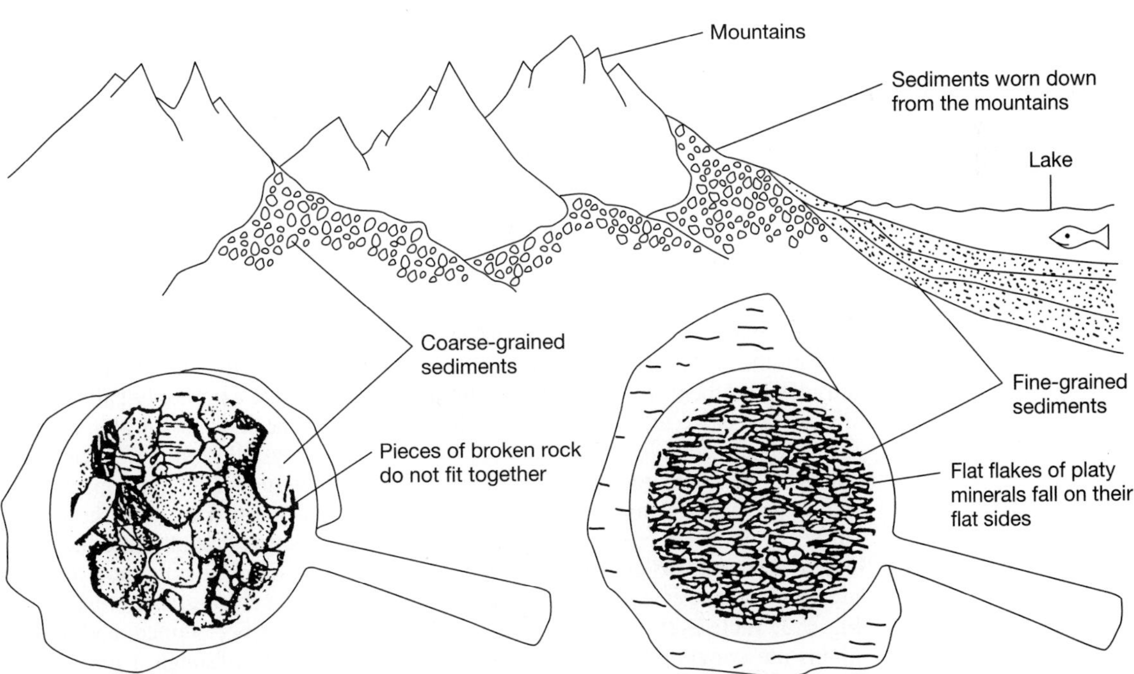

Fig. 2-6 Rocks composed of fragments of other rocks, such as the sediments shown here, display clastic texture. (Source: Sarna-Wojcicki, A. 1995. U.S. Geological Survey.)

TABLE 2-2 IGNEOUS ROCKS

Origins and Texture	Description of Texture	Compositional Category	Rock Color	Minerals Present	Rock Name
Plutonic: Coarse-grained	Visible, interlocking crystals	Silicic	Light; salt-pepper	Quartz, potassium feldspar, sodium feldspar, biotite and muscovite mica, amphibole	Granite
		Intermediate	Intermediate; pepper-salt	Sodium-calcium feldspar, amphibole, biotite mica	Diorite
		Mafic	Dark	Calcium feldspar, pyroxene	Gabbro
Volcanic: Fine-grained *or* porphyritic	Microscopic crystals or visible crystals in microscopic crystalline or glassy matrix	Silicic	Light; white or pink	Quartz, potassium feldspar, sodium feldspar, biotite and muscovite mica, amphibole; glass possibly also present	Rhyolite, rhyolite porphyry
		Intermediate	Intermediate gray *or* spotted-looking	Sodium-calcium feldspar, amphibole, biotite mica	Andesite, andesite porphyry
Volcanic: Fine-grained	Microscopic crystals	Mafic	Dark	Calcium feldspar, pyroxene	Basalt
Volcanic: Glassy or vesicular	No crystals present	Silicic	Black, red, *or* light	No minerals present	Obsidian, pumice

See Chapter 5 for further information about volcanic rock classification and Chapter 8 for further information about plutonic rocks.

which is magma that reaches the surface; and ***plutonic*** rocks that crystallize slowly beneath the surface. Beneath the surface, molten magma rises into solid rocks and intrudes into them: for this reason, plutonic rocks are also commonly called ***intrusive*** rocks. As we will see in Chapter 8, not all of the mechanisms of plutonic rock formation are completely understood. Plutonic rocks crystallize slowly because the overlying rocks are good heat insulators. A body of magma may require many thousands of years to completely solidify. In contrast, magma that reached the Earth's surface as lava will cool very rapidly because heat dissipates rapidly into the atmosphere or ocean. Because plutonic rocks have so much more time to cool, minerals grow to larger sizes, producing a rock with characteristic coarse-grained texture (see Fig. 2-3; Table 2-2). Volcanic rocks are characteristically finer-grained, and mineral grains are commonly so small that they can be seen only through a microscope. Igneous rocks displaying both visible and microscopic minerals—a ***porphyritic*** texture—have undergone slow cooling followed by rapid crystallization (Fig. 2-7). In all three of these cases, the igneous rocks display crystalline texture, indicating that crystals grew together. Some volcanic rocks produced during violent eruptions dis-

Fig. 2-7 A rock with porphyritic texture is evidence that slow early cooling was followed by rapid cooling.

play clastic texture, produced by broken fragments. Lava may also cool so rapidly that the atoms in the magma have no time to form orderly crystals. When this occurs, volcanic glass forms. Obsidian is the most common type of volcanic glass. Volcanic eruptions and the rocks they produce are discussed more completely in Chapter 5, where a classification table for volcanic rocks is included (Table 5-1).

Plutonic and volcanic rocks are further classified according to their mineral composition, which in turn reflects the composition of the magmas that generate them (Table 2-2). The relative abundance of silicon and oxygen (silica) and iron and magnesium are the key elements in the classification of igneous rock. Although there are dozens of volcanic and plutonic rock types, some of which are discussed in subsequent chapters, they can be grouped into general categories. ***Mafic*** (sometimes called ***basaltic***) rocks are relatively low in silicon and oxygen (silica) and high in iron and magnesium. It should be noted that mafic magma still contains about 50 percent silica. The term mafic denotes the high content of magnesium and iron, whose chemical symbol is Fe. About 90 percent of all of the lava erupted on Earth is mafic lava. Volcanic rock formed from mafic lava is called ***basalt,*** and its plutonic equivalent is ***gabbro.*** Both are dark-colored because of the high content of dark, mafic silicate minerals. ***Ultramafic*** rocks contain only about 40 percent silica. ***Silicic*** or ***felsic*** magma has a high silica content (about 70 percent) and is low in iron and magnesium. Light-colored minerals dominate in ***granite,*** the plutonic rock formed from silicic magma, and ***rhyolite,*** its volcanic equivalent.

Sedimentary rocks form at or near the Earth's surface as accumulations of mineral, rock, and/or plant and animal fragments. These sedimentary rock types display clastic texture (see Fig. 2-6). Other sedimentary rocks form by the crystallization of minerals from saline waters. These display crystalline texture, but in contrast to igneous rocks with crystalline texture, they contain minerals that form at near-surface temperatures and pressures. Other sedimentary rocks are organic, formed from the remains of living organisms. It is easier to conceptualize the formation of sedimentary rocks than the other rock types because the processes take place in surface environments, where they can be observed. In their daily lives people deal with many of the agents that transport ***sediment:*** rivers, waves, and wind. The processes that form marine sedimentary rocks are less familiar, but they too can be directly observed by oceanographers.

All rocks and minerals at the Earth's surface are affected by ***weathering.*** Chemical reactions between minerals and water may decompose rocks or cause new minerals to form. Minerals may also be chemically changed by interactions with the atmosphere and with water. One example of ***chemical weathering*** is the formation of iron oxide on the surface of rocks. The outside of a freshly broken piece of rock almost always appears more red or brown than the inside, because iron oxide has formed on the exposed surface (Fig. 2-8). Acting together with chemical weathering, physical processes cause rocks to be broken into smaller pieces. Freezing and thawing, crystallization of salt, or the growth of roots are examples of ***physical weathering*** processes that disintegrate rocks. Over geologic time, weathering weakens even the most solid of rocks. Massive rock disintegrates into small pieces available for transport by water, wind, or glaciers. Following decomposition and disintegration by weathering, rock materials are removed by ***erosion.***

One characteristic of sedimentary rocks is the fact that they are usually layered (Fig. 2-9). The layering, or ***bedding,*** results from the successive accumulation of sediments over time. Sedimentary beds, each of which generally represents a single depositional event, such as a windstorm or a flood, may be millimeters or meters thick. The bedding is visible because of differences between layers. An interval of no deposition or weathering between successive deposits will also produce layering. Volcanic rocks may also be layered, but these are easily distinguished from sedimentary layers because they are composed of volcanic materials.

When classifying sedimentary rocks, geologists are most interested in determining the environment where the sediments were originally deposited. In the case of clastic rocks, the nature of the sediment reflects the energy conditions in the environment where it accumulated. One important diagnostic characteristic of sediment and sedimentary rocks is the ***grain size*** of the particles in the rock (Table 2-3). For example, in high-energy environments like steep mountain streams, huge boulders may be part of the river channel. In contrast, in the almost flat channels of a delta, only the finest mud is found. Other important features of sediments and of clastic sedimentary rocks are their ***sorting*** and ***rounding.*** Sediments that are well sorted contain particles of about the same size and density, whereas those that are poorly sorted contain particles of many different sizes. The rounding of grains is another measure of how well-traveled and tumbled the grains are. Sand dunes and beaches are the two environments that produce highly rounded and well-sorted sediments (Fig. 2-10).

Fig. 2-8 The surface of this freshly broken rock along the Sonoma County coast is brown because chemical weathering has caused iron oxide to form. (Photo by author.)

Fig. 2-9 Tilted sedimentary rocks along Highway 99 near Fort Tejon. (Source: Sarna-Wojcicki, A. 1995. U.S. Geological Survey.)

TABLE 2-3 SEDIMENTS AND SEDIMENTARY ROCKS

CLASTIC SEDIMENTARY ROCKS			
Group/Texture	**Sediment Type and Size**	**Characteristics**	**Rock Name**
Clastic	Gravel, > 2 mm	Rounded fragments	Conglomerate
		Angular fragments	Breccia
	Sand, 1/16 to 2 mm	Mostly quartz	Quartz, sandstone
		Quartz and feldspar	Arkose sandstone
	Mud (silt and clay), < 2 mm	Usually gray	Shale or mudstone

CHEMICAL/BIOLOGIC SEDIMENTARY ROCKS			
Group	**Texture**	**Composition**	**Rock Name**
Inorganic	Clastic or crystalline	Calcite	Limestone
	Clastic or crystalline	Dolomite	Dolomite
	Crystalline	Halite (NaC1)	Rock salt
	Crystalline	Gypsum ($CaSO_4H_2O$)	Gypsum
Biologic	Clastic	Calcite	Limestone
	Clastic and crystalline	Microcrystalline quartz (radiolaria)	Chert
	Clastic	Silica (diatoms), altered plant remains	Diatomite, peat, coal

Fig. 2-10 Examples of well-sorted beach sand and poorly sorted sediment from a landslide. (Photo by author.)

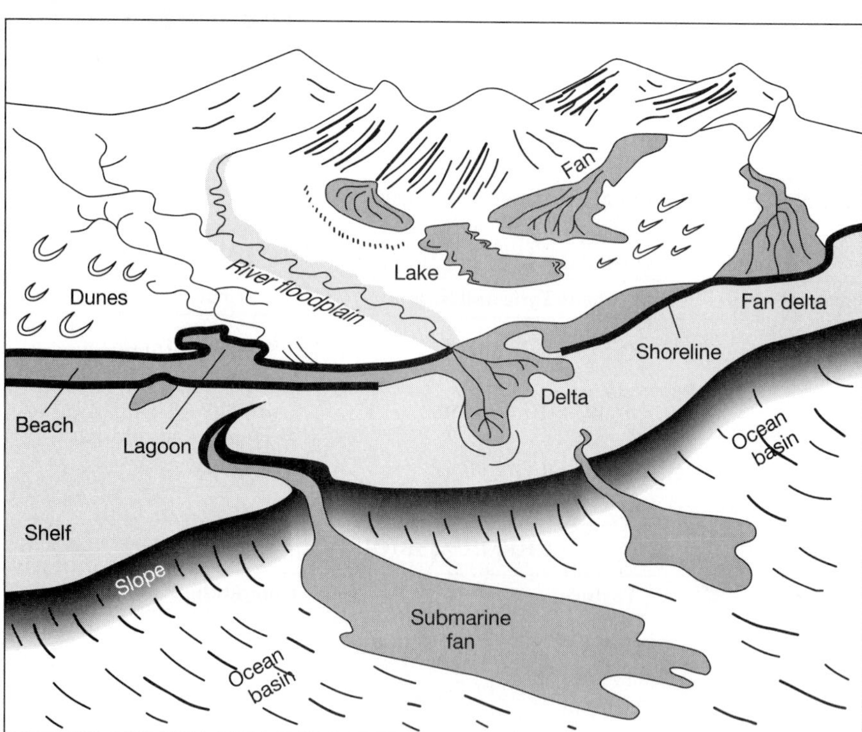

Fig. 2-11 Sedimentary environments near a coastline. (Source: Modified from Texas Bureau of Economic Geology.)

Using all of these characteristics, together with the composition of the grains and the type of fossils that may be present, geologists are able to determine with some certainty the environments where sedimentary rocks originally formed (Fig. 2-11). For example, a shale may originally have been deposited in a coastal lagoon. Even though that lagoon and its surrounding landscape have long since vanished, the rock preserves key evidence about the size, location, water chemistry, and surroundings of that environment.

Fig. 2-12 Mud cracks in Death Valley. The cracks form when the mud dries and shrinks. (Photo by author.)

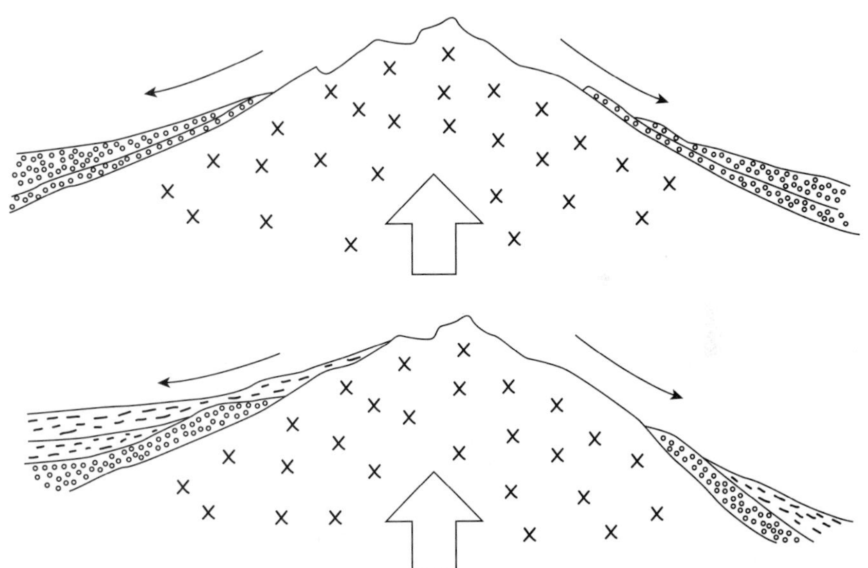

Fig. 2-13 As tectonic forces uplift mountains, erosion removes rock materials from high areas and deposits them in lower areas.

Other distinctive features may very precisely indicate a sedimentary rock's environment of deposition. For example, mid cracks (Fig. 2-12) are evidence that the surface of the sediment experienced episodic wetting and drying, indicating an environment such as a tidal flat or intermittently flooded lake bed. Other sedimentary features that aid interpretation of depositional environments include ripple marks, channels, ***cross-beds,*** mud lumps, and, of course, fossils. California fossils will be discussed further in Chapter 3.

Sedimentary rocks may also provide important information about the tectonic setting of a region. The sudden appearance of coarse-grained sediments in a section of rocks may indicate that uplift has occurred in nearby areas, causing an influx of sediment to be shed into nearby lowlands (Fig. 2-13). If the sediments contain distinctive rock or mineral types, it may be possible to identify their geologic source with some precision, even if the sediments have been displaced from their sources by faulting. Indicators of current directions in the sediments may identify the direction of sediment transport. Through careful study of sedimentary rocks, geologists are thus able to identify the precise location of newly uplifted mountain ranges long after those mountains have been eroded away. Reconstruction of the ancient geography or ***paleogeography*** of an area may in turn provide clues about past tectonic events.

Metamorphic rocks have been subjected to enough heat and pressure to cause grains to grow together, but not enough to melt them. During metamorphism, hot fluids containing reactive chemicals may also circulate through rocks and substantially change their composition. When classifying metamorphic rocks, geologists are particularly interested in interpreting the degree of metamorphism and the evidence of stresses during metamorphism. By reconstructing the temperature and pressure conditions that existed during metamorphism, geologists can estimate how deep beneath the Earth's surface the rocks were when they were metamorphosed (Fig. 2-14).

One type of metamorphism occurs when magma intrudes into a body of rock. The unmelted rocks surrounding the magma are heated. This type of metamorphism is known as ***contact metamorphism*** because it affects rocks that are in contact with igneous intrusions. Hot fluids may also be released into the rocks, reacting with them to form new mineral; this process is known as ***hydrothermal*** alteration. One important result of contact metamorphism is the concentration of valuable mineral deposits during hydrothermal alteration.

Most metamorphic rocks form during ***regional metamorphism.*** Large areas are subjected to high temperature and pressure when rocks are intensely deformed. The most obvious setting for regional metamorphism is along a convergent plate

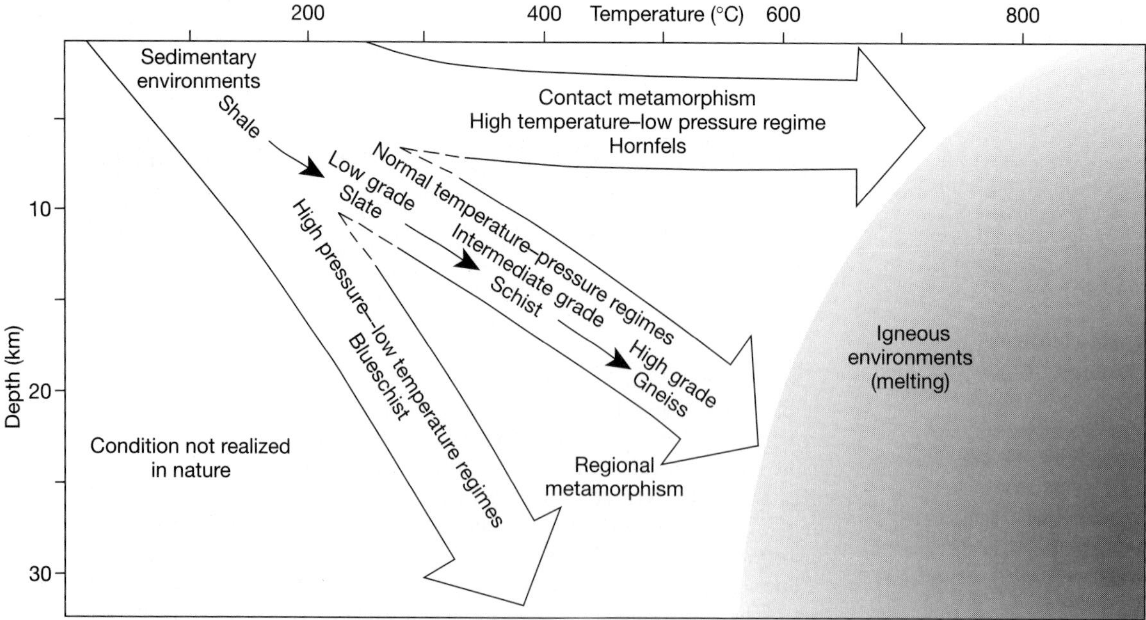

Fig. 2-14 Pressure and temperature increase with depth beneath the Earth's surface. (Source: Tarbuck, E., and Lutgens, F. 1993. *The Earth*, 4th ed. New York: Macmillan, p. 176.)

boundary, where rocks are buried to great depths in a subduction zone. An important characteristic of regional metamorphism is that rocks are subjected to stresses that are not equal in all directions. Rocks that are compressed by directed stress during metamorphism develop a ***foliated*** texture. Foliation is a distinctive layered texture that develops during metamorphism. If rocks are subjected to stress that is stronger in one direction, then new metamorphic minerals will crystallize in a preferred orientation (Fig. 2-15). It is the alignment of metamorphic minerals that produces foliation.

As the degree of metamorphism increases, rocks are subjected to higher temperatures and directed stresses. As a result, more highly metamorphosed rocks show more obvious foliation. The obvious layering of ***schist*** results from the growth of platy silicate minerals like mica, ***chlorite,*** or ***talc*** during metamorphism (Fig. 2-16). Less metamorphosed ***slate*** is also foliated, but the new metamorphic minerals are very small. Foliated metamorphic rocks may also contain other silicate minerals formed only by metamorphism, such as garnet, kyanite, or staurolite. ***Gneiss*** is a banded metamorphic rock composed mostly of mica, feldspar, and quartz that indicates a very high degree of metamorphism (Fig. 2-17).

Nonfoliated metamorphic rocks commonly display a grainy texture caused by the intergrowth of minerals with no preferred alignment. Metamorphism of lime-

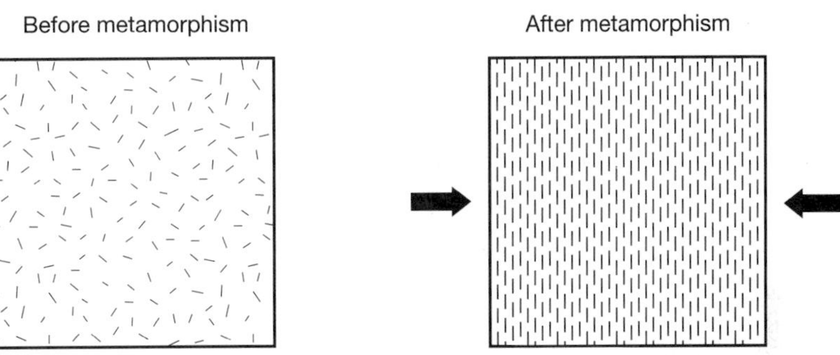

Fig. 2-15 When rocks are subjected to directed pressure during metamorphism, the mineral grains typically grow preferentially in the direction where stress is minimal. (Source: Tarbuck, E., and Lutgens, F. 1993. *The Earth,* 4th ed. New York: Macmillan.)

Fig. 2-16 Foliated schist from the Sierra Nevada foothills. (Source: Harwood, D., U.S. Geological Survey.)

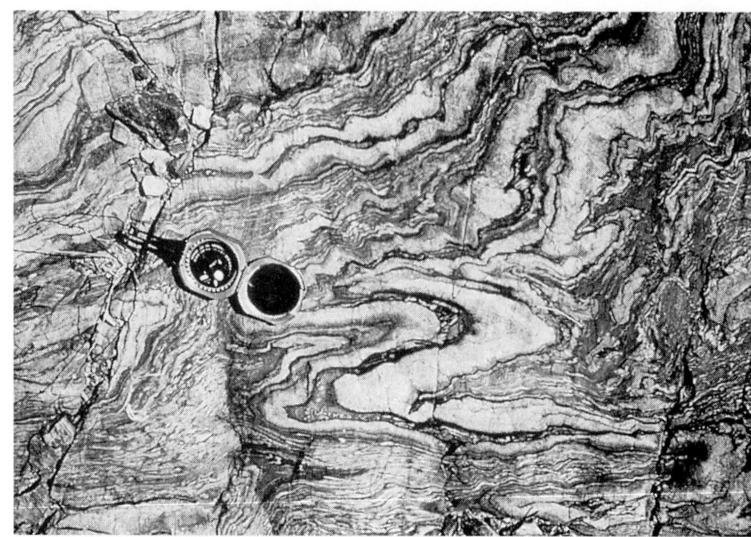

Fig. 2-17 Gneiss, Sierra Nevada. Note geologist's Brunton compass, shown for scale. (Source: Lockwood, J., U.S. Geological Survey.)

stone and dolomite produces larger crystals of calcite and dolomite with characteristically sugary texture—the metamorphic rock ***marble.*** Metamorphism of shale without directed stresses produces a hard, dark-colored, nonfoliated rock called ***hornfels.*** California's metamorphic rocks will be discussed in more detail in Chapter 8 and Chapter 12.

THE ROCK CYCLE

The rocks that we find at any given location have often experienced a complex history of transformations as they have been subjected to the forces that shape the Earth. Throughout all but its earliest history, when major extraterrestrial impacts were frequent, the planet has changed its mass very little. Only a tiny amount of new material has been added to the Earth by collision with meteors. The endless creation of new rocks and destruction of old rocks result from continuous recycling of rock material because of plate tectonics. The transformation from one rock type to another is referred to as the ***rock cycle***— the concept that a rock can be changed over geologic time. While it is useful to use an illustration to visualize the rock cycle (Fig. 2-18), any diagram fails to convey the almost endless possibilities created as rocks are subjected to geologic forces over millions, or even billions, of years.

Rocks are melted when they are subducted to depths at which temperatures are sufficiently great. Also because of plate tectonics or at isolated hotspots, magma rises and cools to form new igneous rocks. Rocks that are buried to great depths without melting can be changed to metamorphic rocks. As mountains are uplifted, they are worn down as rocks are weathered, eroded, transported, and deposited to form sediments. These sediments may then be gradually buried by younger sediments, eventually becoming sufficiently compacted and cemented to form sedimentary rocks (the process of ***lithification***).

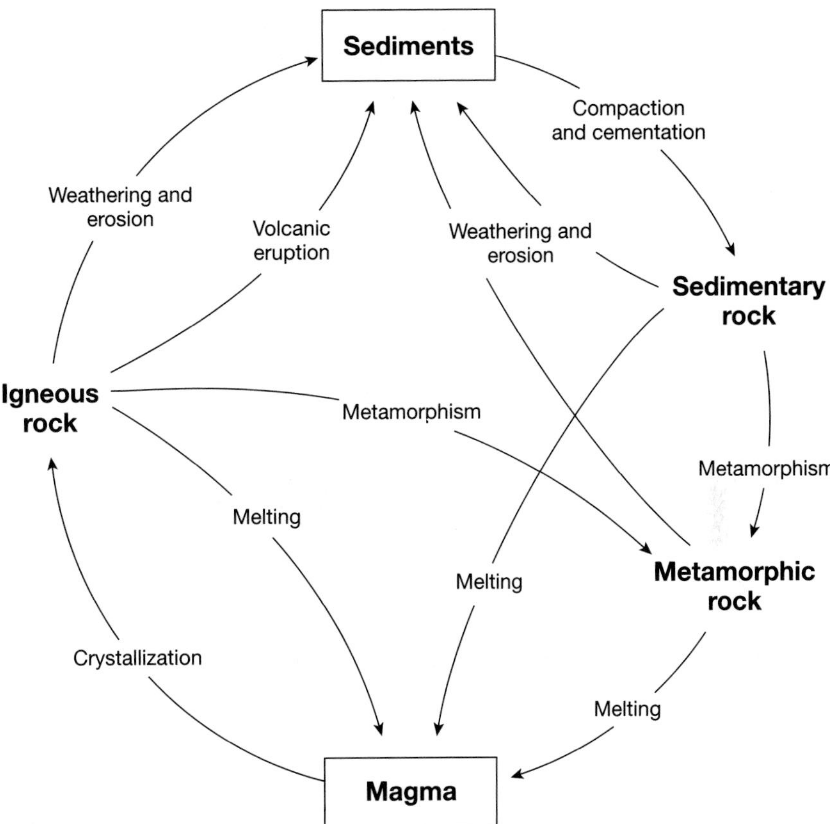

Fig. 2-18 The rock cycle.

By identifying the sediments that make up a sedimentary rock such as a sandstone, geologists may be able to decipher the sandstone's ancestor rock. For example, 140 million-year-old sandstone found today in the Great Valley contains abundant fragments of volcanic particles, suggesting that the sources of the sediments were volcanoes. The type of volcanic particles in the sandstone may even enable the geologist to infer what type of rocks were melted to generate the magma that in turn created the volcanoes. Using the principles of the rock cycle to trace the history of a rock back through geologic time is not unlike tracing a family's genealogy. In many cases, the "parents" of rocks can be identified, but their lineage becomes very sketchy beyond that—not surprising, considering the time spans involved!

It is important to realize that rocks can be transformed by many different pathways (see Fig. 2-18). For example, a volcanic rock can be transformed directly into a metamorphic rock—specifically a ***metavolcanic*** rock in this case—by subjecting it to sufficient heat and pressure. Belts of slate and schist in the western foothills of the Sierra Nevada and the Klamath Mountains were originally layers of volcanic rock formed from lava erupted from submarine volcanic chains. Any type of rock may be melted to form magma. As we will see in Chapter 5, lavas formed by melting oceanic rocks are recognizably different from lavas formed by melting continental rocks.

All of the rocks in California's landscape are subjected to erosion and weathering. In Death Valley, sand dunes contain quartz grains that were eroded from sedimentary rocks that formed hundreds of millions of years ago. Some of those ancestral rocks were themselves deposited in ancient sand dunes several hundred million years ago. The sediments carried by modern rivers from the high Sierra to San Francisco Bay may someday become sandstone. That sandstone will carry compositional clues that it was eroded from the Sierra Nevada—mountains of mostly ***granitic*** rock—even if the mountains themselves have been completely eroded away. For any rock, there is the almost endless possibility of becoming a very different type of rock in its geologic future. The only certainty is that if a rock melts to become magma, the magma can only be transformed to an igneous rock.

COMMON ROCKS IN CALIFORNIA

Igneous Rocks

Igneous rock types common in California include the following:

- *Granitic plutonic rocks* (shown in red on Endpaper 1, inside the front cover). Granitic rocks make up much of the Sierra Nevada Province. Smaller granitic bodies are found in the Klamath Mountains, in the Coast Ranges west of the San Andreas fault, in the Mojave Desert, and in the Peninsular Ranges of southern California. Types of granitic rocks and their origin will be discussed in Chapter 8.
- *Mafic and ultramafic plutonic rocks* (shown in purple on Endpaper 1). Serpentinite, which is actually a metamorphosed ultramafic rock, is included in this unit. Geologists believe that these rocks formed in the lower crust and upper mantle beneath oceanic plates. These rocks will be discussed further in Chapter 9.
- *Intermediate and silicic volcanic rocks.* Young volcanic rocks, shown in pink on the map, cover most of the Cascades Province, are abundant east of the Sierra Nevada, and are also found in the northern Coast Ranges. In these three areas, young volcanic rocks include large deposits of rocks with intermediate and high silica contents. California's volcanoes and volcanic rocks will be discussed further in Chapter 5.
- *Mafic volcanic rocks.* Recently erupted mafic volcanic rocks are also shown in pink on the geologic map. Young basalt is most common in California in the Modoc Plateau, the Mojave Desert, and in the Basin and Ranges of southeastern California.

  Basalt covers more of the Earth's surface than any other type of rock ; it forms the floor of all the ocean basins. In California, older volcanic rocks (included within the green and blue areas on Endpaper 1) occur in the Coast Ranges, where they have commonly been slightly metamorphosed, and in the Klamath Mountains and western Sierra Nevada, where they have been more highly metamorphosed. Even these metavolcanic rocks may show well-developed rounded forms called ***pillows,*** formed when lava entered the ocean and congealed into blobs. Studies have confirmed that many of California's metavolcanic rocks formed in ancient midocean ridges, seamounts, and island arcs and were then accreted to the North American continent. Mafic volcanic rocks will be discussed further in Chapter 5 and 12.

Sedimentary Rocks

On the California geologic map, sedimentary rocks are grouped according to their age. From youngest to oldest, the sedimentary rocks are colored tan, green, and blue. California's metamorphic rocks are grouped with the metasedimentary and metavolcanic rocks of Mesozoic (green) and Paleozoic (blue) age. Serpentinite is included with the ultramafic rocks (purple), and highly metamorphosed rocks of unknown or pre-Paleozoic age are shown in dark brown.

Sedimentary rocks common in California include the following:

- *Sandstone.* Sandstone is found in all provinces of California. The majority of California's sandstone was originally deposited in marine environments. Much of the sandstone in California is gray or brown and contains chips of shale and other rocks. Because of its location along a steep continental margin, California has produced these types of "dirty" sandstones rather than quartz-rich sandstones. The deposition of these sandstones will be discussed further in Chapter 12.
- *Shale.* Dark gray or black marine shale is also very common in California and usually is found together with sandstone (Chapter 12). Metamorphosed sandstone and shale examples of ***metasedimentary*** rocks, are common in the Sierra Nevada, Klamath Mountains, and other provinces.
- *Chert.* Chert is abundant in the Coast Ranges and can be black, red, green, brown or white. Much of California's chert formed by the slow settling of microorganisms called ***radiolaria*** onto the floors of the deep ocean basins. It is commonly found together with submarine volcanic rocks. Chert is also found in the Klamath Mountains, the Transverse Ranges, and the metamorphic rocks of the Sierra Nevada. Chert will be discussed further in Chapter 3 and 12.
- *Limestone and dolomite.* Carbonate rocks are common only in the mountain ranges of the Basin and Ranges and Mojave Desert provinces. Metamorphosed carbonate rocks (marble) are also common in the eastern Klamath Mountains and northwestern Sierra Nevada. Limestone also occurs as small blocks within the Coast Ranges and in the mountains of southern California. The formation of these rocks will be discussed further in Chapter 7.

Metamorphic Rocks

Metamorphic rocks common in California include the following:

- *Serpentinite.* This is California's state rock. It is commonly found along faults in the Coast Ranges, Klamath Mountains, and Sierra Nevada. Serpentinite forms when rocks originally formed as ultramafic rocks in the mantle are hydrated (see Chapter 9).
- *Slate.* Slate is abundant in the western Sierra Nevada foothills. It was originally mafic volcanic rock, shale, or sandstone and changed during metamorphism to slate (see Chapter 8).
- *Schist.* Schist occurs in the Santa Monica, San Bernardino, and San Gabriel Mountains in southern California, in the Sierra foothills, and in the Klamath Mountains. Like slate, many of the schists in California were originally sedimentary or volcanic rocks. A distinctive type of schist, ***blueschist,*** is found in the Coast Ranges. Metamorphic rocks and their significance will be discussed further in Chapters 8 (Sierra Nevada) and 12 (Coast Ranges).

- *Marble.* Marble, composed of metamorphosed limestone or dolomite, is relatively rare in California and occurs in blocks within the High Sierra Nevada, in the northwestern Sierra foothills, and in the Klamath Mountains.

Unconsolidated sediments are recent deposits that have not been buried, compacted, and cemented to form sedimentary rocks. These sediments, which cover virtually all of the low-lying areas of California, are colored yellow on the California geologic map. Most of these sediments are ***alluvium,*** sediments deposited by rivers. The processes that deposit nonmarine sediments will be discussed further in Chapters 6, 10, and 11, and the processes of marine and beach deposition along the California coast will be discussed in Chapters 15 and 16.

FURTHER READINGS

BARKER, R. 1990. *Collecting Rocks.* U.S. Geological Survey. General interest publication.

COCH, N, AND LUDMAN, A. 1991. *Physical Geology.* New York: Macmillan, 678 pp.

DIETRICH, R.V., AND SKINNER, B.J. 1979. *Rocks and Rock Minerals.* New York: John Wiley & Sons, 319 pp.

ESTAVILLO, W. 1992. *Gems and Minerals of California: A Guide to Localities.* Frederick, Colo.: Renaissance House, 48 pp.

Special California Issue. *Rocks & Minerals.* 69(6):360-430, 1994.

SYMES, R.F. 1988. *Rocks and Minerals.* New York: Knopf Eyewitness Books, 64 pp.

TOOKER, E.W., AND BEEBY, D.J. 1990. *Industrial Minerals in California: Economic Importance, Present Availability, and Future Development.* U.S. Geological Survey Bulletin No. 1958.

CHAPTER

3

Geologic Time, Dating Earth Materials, and California Fossils

Time is one factor that sets geology apart from many other sciences, and beginning geologists often have difficulty conceiving of the vastness of geologic time. Many of the processes that shape the Earth have little immediate effect on the overall landscape or distribution of materials. However, these processes operate over a seemingly endless time span when compared with the centuries or decades easily imagined by humans. Over tens of millions of years, entire oceans and continents can be created and destroyed. An understanding of how geologists determine rates and sequences of geologic events is important for deciphering California's complex geologic history.

RELATIVE DATING

Until the 1950s, it was not possible to determine the actual numerical age of a rock or sediment. Geologists relied on a scheme of ***relative age dating,*** which enabled them to order rock formations from oldest to youngest. Relative dating is still an important tool in many geologic investigations. Geologists use four fairly simple principles to determine the relative order of geologic events in an exposed section of rocks or sediments:

1. In an undisturbed sequence of layered materials, the layers at the bottom are oldest, and those at the top are youngest: the principle of ***superposition.*** Anyone who has worked his or her way through a stack of papers or dirty laundry is familiar with this principle.
2. Sediments are originally deposited in horizontal or nearly horizontal layers: the principle of ***original horizontality.*** Sediments accumulate in low-energy,

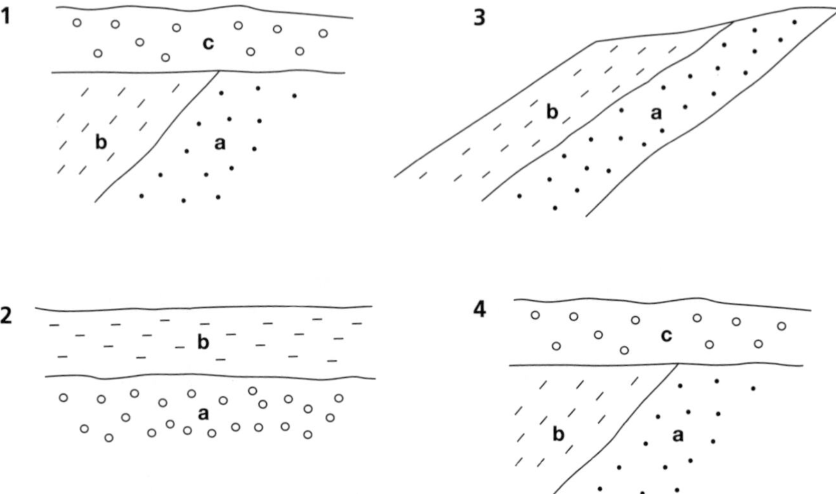

Fig. 3-1 The principles of relative age dating are used to determine a sequence of geologic events. The arrangement of rocks in diagram 1 could only result from the following sequence: layer *a* was deposited, followed by layer *b* (diagram 2). Both layers *a* and *b* were tilted (diagram 3). Following a period of erosion, layer *c* was deposited on the eroded surface (diagram 4).

gently sloping or flat environments. One can apply the first two principles to an exposure of folded or tilted rocks (Fig. 3-1) and conclude that tilting and folding must have occurred after the rocks were deposited.

3. A rock unit or feature such as a fault that cuts through a layer of rock must have formed after that rock layer was formed: the principle of ***cross-cutting relations.*** A vein of quartz that cuts through a granite body was intruded after the granite cooled. A fault that displaces rock layers moved after those layers formed (Fig. 3-2).
4. A rock or sediment containing fragments of other rocks must be younger than the rocks it contains. For example, lava may incorporate pieces of rock from the sides of a fissure as it erupts. The resulting volcanic rock will contain ***inclusions*** of the older, incorporated rock.

One important aspect of these principles is that the landscape is younger than any of the rock or sediment beneath it. Hills, valleys, and plains have been sculpted on the materials beneath them, and the forces of erosion and deposition that shape the landscape are constantly reshaping the landscape. If a fault causes displacement of a landscape feature such as a stream or a level field, it is most likely to be an active fault.

Fossils and the Geologic Time Scale

Many sedimentary and metasedimentary rocks contain ***fossils*** (remnants or impressions of plants and animals) that can aid geologists in assigning ages to the rocks. Fossils have been found from rocks formed in every conceivable environment that supports life, but the vast majority of fossils are the remains of marine life. Marine environments constitute more than 70 percent of the Earth's surface, and most marine environments are favorable for fossil preservation. Rocks formed in ancient terrestrial environments such as lake beds, peat bogs, and marshes also commonly contain fossils. Species with hard parts like shell, bone, and wood are more likely to

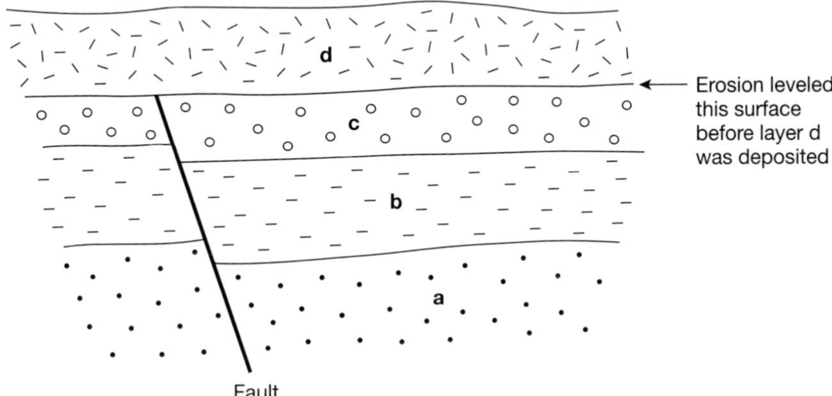

Fig. 3-2 Relative age dating is used to establish the time of movement along a fault. The fault moved after layers *a* through *c* were deposited, because these layers are broken (or offset) by the fault. The fault has not moved since layer *d* was deposited and the landscape developed, because layer *d* continues unbroken across the fault and the landscape.

be preserved than more delicate organisms. Environments conducive to rapid burial are particularly favorable for preservation of fossils, as well as indirect evidence of life such as worm trails, leaf impressions, and footprints.

The principles of ***paleontology,*** the study of ancient life, were developed during the nineteenth century, when paleontologists began to systematically collect and classify fossils. Evolutionary sequences of species were identified as paleontologists pieced together the successions of life forms represented in the fossil record. Although some species (for example, some ferns) have persisted through long stretches of geologic time without much change, others made only geologically brief appearances on Earth. The presence of these time-specific, characteristic ***index fossils*** is particularly useful. Their presence enables paleontologists to identify rocks formed during the same restricted time interval, even if the rocks are found in widely separated areas.

Long before methods of numeric dating of rock materials had been developed, paleontologists had established a geologic time scale based on the evolutionary sequence of species through time. Based on characteristic assemblages of species, geologic time was broken into major ***eras*** that encompassed several ***periods;*** periods were subdivided into ***epochs*** (Fig. 3-3). The boundaries between eras are marked by first appearances of groups of species and/or sudden, catastrophic ***extinctions*** of many life forms from the geologic record. Less dramatic changes in life forms mark the boundaries between periods and epochs. The seemingly odd names for some geologic periods arise from localities where the best examples of the fauna of that period are found, such as the Jura Mountains in Europe for the Jurassic period. Other periods are named for ancient peoples who inhabited parts of Britain where characteristic fauna occur, such as the Ordovician period, named for the Ordovice Celtic people of Wales.

ABSOLUTE DATING

Since the 1950s, geologists have relied on the Earth's natural radioactive "clocks" for determining the numerical ages of geologic materials. These are sometimes referred to as "absolute" ages. Some of the naturally occurring chemical elements have forms or ***isotopes*** that are unstable. These unstable forms account for a minute

Age	Eon	Era	Period	Epoch	Animals	Plants
	Phanerozoic	Cenozoic	Quaternary	Holocene Pleistocene		
66 MA			Tertiary	Pliocene Miocene Oligocene Eocene Paleocene	Humans	
		Mesozoic	Cretaceous	Late Early		
			Jurassic	Late Middle Early	Birds	Flowering plants
235 MA			Triassic	Late Early		
		Paleozoic	Permian	Late Early	Mammals	Gingkos
			Pennsylvanian	Late Middle Early		Pines
			Mississippian	Late Early	Reptiles	
			Devonian	Late Middle Early	Amphibians	Ferns
			Silurian	Late Middle Early		Club mosses Horsetail rushes
			Ordovician	Late Middle Early		
570 MA			Cambrian	Late Middle Early	Fishes	
	Proterozoic	Late Proterozoic Middle Proterozoic Early Proterozoic			Animals with shells	
	Archean	Late Archean Middle Archean Early Archean				
	Pre-Archean					

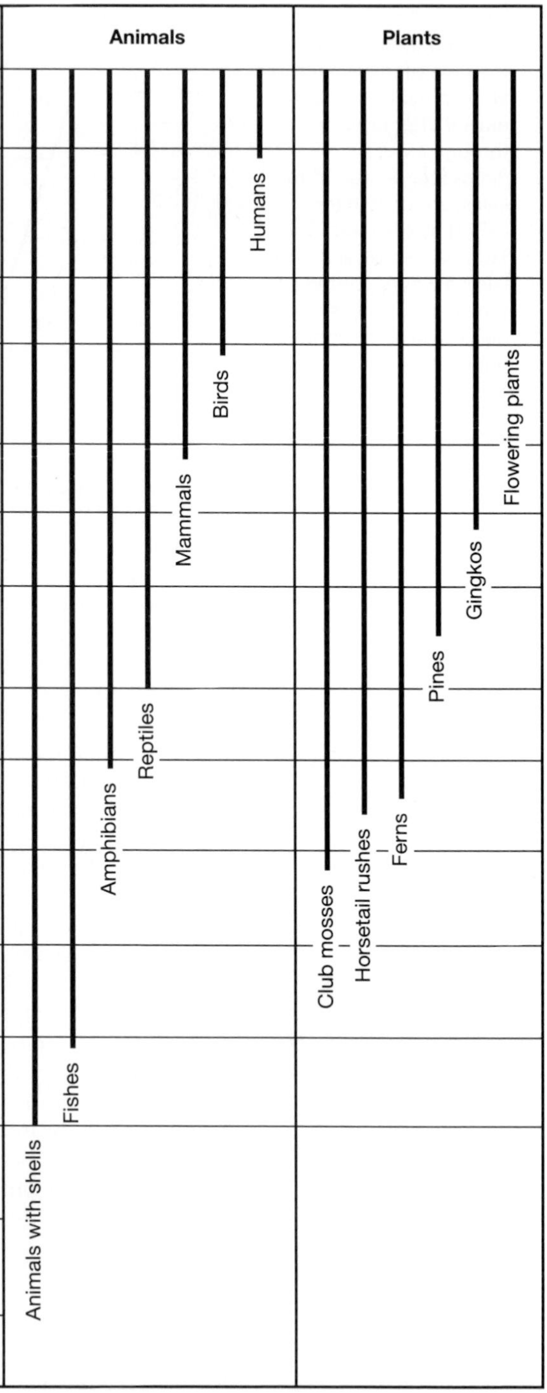

Fig. 3-3 The geologic time scale and the history of life on Earth. (MA = millions of years ago.)

fraction of the total amount of that element. They break down spontaneously to more stable atoms, and in the breakdown process they give off energy in the form of radiation. Unstable forms of elements are referred to as ***radioactive isotopes,*** and the breakdown process is known as ***radioactive decay.*** The decay of radioactive isotopes takes place at a specific and constant rate, as does the buildup of the products of decay. The time it takes for half of the initial amount of radioactive material to decay is the ***half-life*** of that particular isotope. Physicists have measured the half-lives of radioactive isotopes with great precision. They range from seconds to billions of years, but the half-life of each isotope is unique and unchanging.

The constant rate of decay of radioactive elements provides us with naturally occurring clocks that can be used to date Earth materials. Knowing the original amount of a radioactive isotope in a rock, sediment, or fossil, and knowing its half-life, one can calculate the amount of time that passed since the radioactive "clock" began ticking. The use of isotopes to date Earth materials is known as ***radiometric dating.***

A radiometric dating method commonly used for dating materials that are geologically young uses carbon-14, a naturally occurring radioactive isotope of carbon. Carbon-14 is produced when cosmic rays bombard nitrogen in the upper atmosphere, and a fixed amount of carbon-14 is present in all living tissues, including our own. When organisms die, they stop taking in carbon-14, and the carbon-14 in their bodies begins to decay (Fig. 3-4). Knowing that the half-life of carbon-14 is 5730 years, one can measure the amount of carbon-14 in formerly living material and calculate the time that has passed since it died. After about 50,000 years, the amount of carbon-14 remaining is so small that it cannot be measured accurately. For this reason, carbon-14 dating can be used to date only relatively recent materials and events. For fossils of wood, charcoal, bone, shell, and soil materials, carbon-14 dating is potentially useful. Geologists rely heavily on this technique to determine the ages of sediments formed by floods, landslides, earthquakes, and other potentially hazardous processes.

The "workhorse" of radiometric dating techniques for most rocks and geologic events is potassium-argon (K-Ar) dating. This method works best on igneous rocks and measures the time that has passed since the rock crystallized from magma or lava. If igneous or other rocks are metamorphosed, the K-Ar method measures the time that has passed since the rocks cooled after metamorphism. Because the decay of radioactive potassium-40 to argon-40 has a half-life of 1.3 billion years, the K-Ar technique can be used on all but the youngest geologic materials. The most commonly used technique for K-Ar dating involves measurement of the buildup of argon-40, a gas, in rocks.

By finding localities where sedimentary rocks containing key fossil assemblages are in contact with ancient lava flows, geologists have used K-Ar dating to add the numbers to the geologic time scale (see Fig. 3-3), as well as to date rocks from the moon. Other commonly used radiometric dating techniques use uranium-lead and rubidium-strontium isotopes.

CALIFORNIA'S FOSSILS

Because rocks of pre-Mesozoic age are relatively rare in California, fossils of the Paleozoic era are also restricted. Mesozoic rocks contain a wide variety of mostly marine fossils, because most of California's Mesozoic sedimentary and metasedimentary rocks were deposited in marine environments. The Cenozoic fossil record in California is rich and diverse. Marine fossils of many types are abundant, and

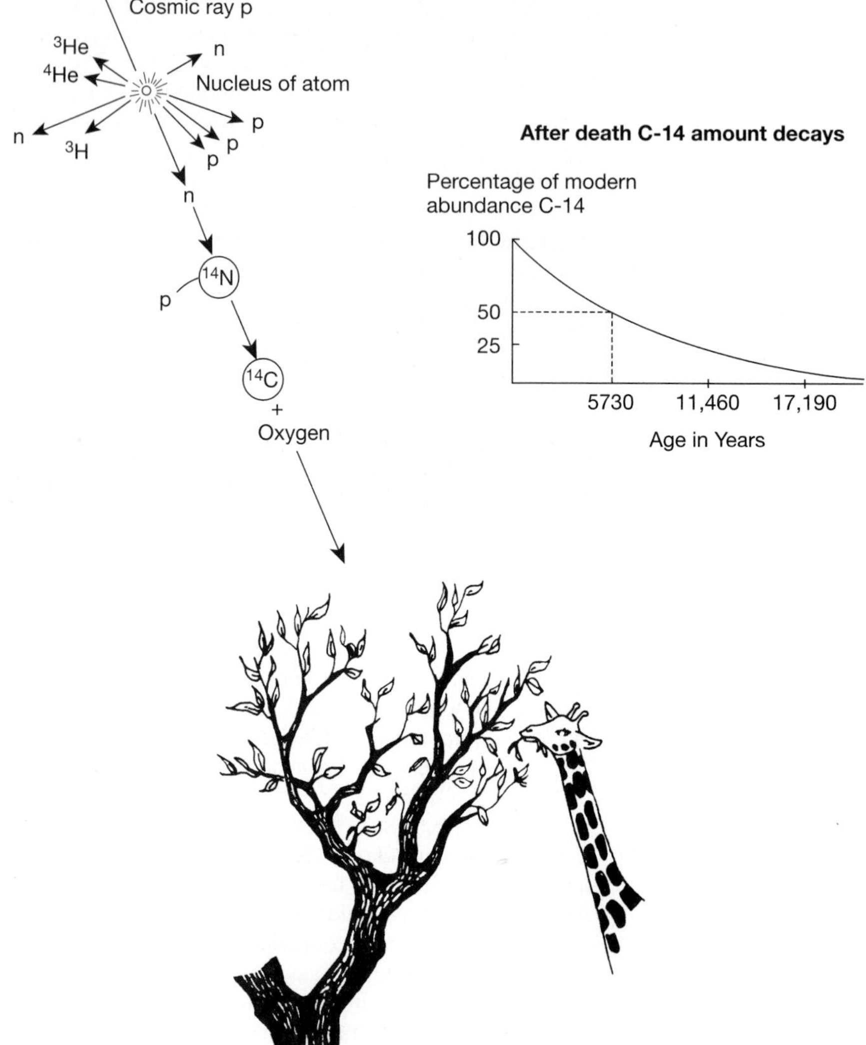

Fig. 3-4 Carbon-14 is formed at a nearly constant rate in the upper atmosphere when nitrogen atoms are bombarded by cosmic rays. All living organisms incorporate modern carbon-14 until they die, when radioactive decay begins. (Source: Levin, H. 1983. *The Earth Through Time,* 2nd ed. Philadelphia: W.B. Saunders.)

California is also home to some of North America's richest late Cenozoic terrestrial fossil finds. Although it is not possible to list all of California's fossil localities, it is worth describing some of the typical fossils of each era. Subsequent chapters include discussion of additional fossil localities in specific provinces.

Paleozoic Fossils

Paleontologists have identified at least 95 species of trilobites from early Paleozoic rocks in southeastern California. Trilobites were ocean-dwelling arthropods, ***invertebrate*** animals related to arachnids, insects, and crustaceans (Fig. 3-5). Because

Fig. 3-5 Cambrian trilobite fossil from the Marble Mountains, San Bernardino County.

many distinctive species developed during Paleozoic time, and because trilobites became extinct at the end of the Paleozoic era, trilobites are excellent index fossils for Paleozoic rocks. Trilobite fossils are present in the Cadiz Formation of Cambrian age in the Marble Mountains of San Bernardino County and in similar carbonate rocks in the southern Nopah Mountains near Death Valley.

Metasedimentary rocks found in the High Sierra west of Bishop contain graptolite fossils that place the formation of the original marine sedimentary rocks in Ordovician time. ***Graptolites,*** which persisted until Mississippian time, are very important index fossils for Ordovician and Silurian rocks. They were colonial organisms that secreted small skeletons; in rocks, graptolites look like small dashes or rods (Fig. 3-6). Fossil brachiopods are found in Silurian and Devonian rocks in the eastern Klamath Mountains.

Permian rocks are more widespread in California than rocks from earlier Paleozoic periods. Corals, including many forms that are extinct today, are found in Permian rocks near Death Valley and in the eastern Klamath Mountains. Fusilinids, large, single-celled foraminifera that are extinct today, are also found in Permian sedimentary rocks. ***Crinoids,*** particularly the cylindrical remnants of crinoids, and brachiopods are also well represented in fossiliferous rocks of Paleozoic age (Fig. 3-7).

Mesozoic Fossils

Several groups of marine invertebrate organisms are important fossils in California's Mesozoic rocks. Most of these are shells of mollusks, such as clams and snails. Ammonites, which are cephalopods that resemble the modern chambered nautilus, are found in Mesozoic rocks throughout California. In the Peninsular Ranges, Triassic ammonites are found in the Santa Ana Mountains. The age of the widespread Franciscan Assemblage in the Coast Ranges was originally determined from the presence of Jurassic and Cretaceous ammonites, and excellent Cretaceous specimens have

Fig. 3-6 **A,** Ordovician graptolite fossils from metasedimentary rocks in the Sierra Nevada. **B,** Pennsylvanian tabulate corals, Death Valley. (Source: McMackin, M. San José State University.)

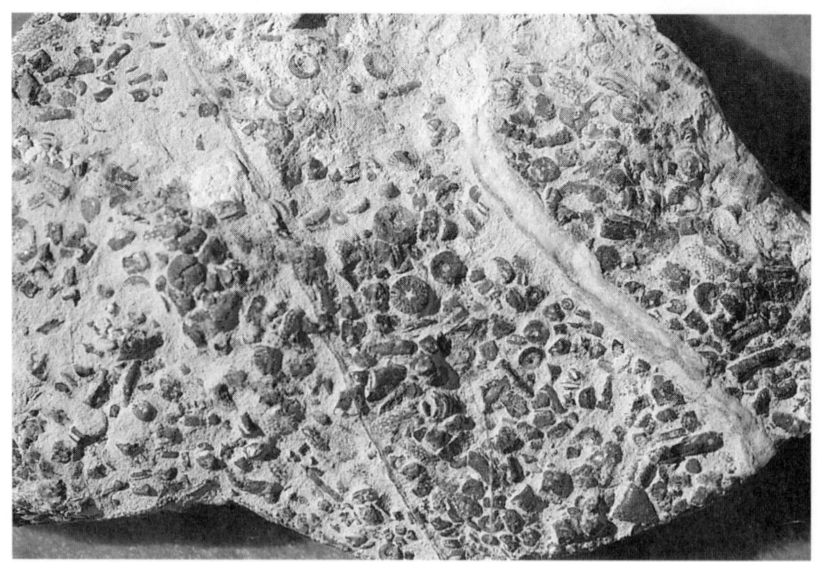

Fig. 3-7 Ordovician crinoid fossils from the Inyo Mountains.

Fig. 3-8 Examples of Cretaceous marine fossils (*Buchia*) from the western edge of the Sacramento Valley. Fossils are about 2 to 3 cm long.

been found along the western Great Valley (Fig. 3-8). Many other mollusks are abundant in California's Cretaceous rocks, including *Turritella,* a gastropod, and *Buchia* (Fig. 3-8), a distinctive pelecypod found in the Great Valley Sequence.

The Mesozoic era is famous for its dinosaurs, and in the 1990s, almost all Californians were introduced to the Jurassic period by the movie "Jurassic Park." Because little of California was dry land during Jurassic time, one must travel east to find fossils of the famed dinosaurs. However, a dinosaur footprint was recently discovered in the Jurassic Aztec sandstone in the easternmost Mojave Desert Province. Swimming dinosaurs have been found in Cretaceous rocks in California, including ichthyosaur remains found near Los Banos and Redding. A plesiosaur fossil and remains of a pteranodon were found in the Moreno Formation in western Fresno County.

Microscopic fossils of radiolaria have proved to be perhaps the most useful Mesozoic fossil for California geologists. Radiolaria are microscopic, single-celled protozoans that build their skeletons of silica. In the ocean basins, sedimentary layers may consist entirely of radiolaria, which, when lithified, form a sedimentary rock known as radiolarian chert (Fig. 3-9). Radiolaria may remain recognizable even after metamorphism, so they have proved to be very valuable to paleontologists and geologists interested in the origins of California's accreted oceanic terranes.

Diatoms are another microscopic organism composed of silica. Diatoms, a type of floating plant with a siliceous cell wall, live in fresh and marine water. Diatoms are found in California sedimentary rocks of late Cretaceous to modern age. Deposits composed almost entirely of diatoms (diatomite) are mined from several localities in California (see Chapter 16). Diatomite is used for soil amendments and filters. Diatoms are also important to California's geologic history because they are the likely source of much of California's petroleum (Chapter 16). Microscopic ***foraminifera,*** single-celled protozoans that construct their shells from calcium carbonate, are also important microfossils for rocks of Cretaceous and Cenozoic age.

Fig. 3-9 Photomicrograph of Jurassic radiolaria in chert from the central Coast Ranges. The white bar scale is 0.1 mm in length. (Source: Oscarson, R., and Murchey, B. U.S. Geological Survey.)

Cenozoic Fossils

During Cenozoic time, much of California emerged from the sea, creating a variety of land-based environments that supported an amazing assortment of organisms. The Cenozoic is the era of mammals, and many of California's best-known fossil sites are localities bearing the remains of mammals (Fig. 3-10). Many sites also preserve fossil plants. Near Elsinore, in the Peninsular Ranges, fossil leaves and ferns are found in rocks of Eocene age. At the Petrified Forest near Calistoga in the Coast Ranges of Sonoma County, redwood trees of Pliocene age were perfectly preserved when they were buried by a volcanic eruption from the nearby Sonoma volcanic field.

Fig. 3-10 Miocene mastodon tracks, Copper Canyon. (Source: McMackin, M. San José State University.)

The oldest known land animal fossils in northern California were found at Blackhawk Ranch on the southern side of Mt. Diablo, in the Diablo Range east of San Francisco Bay. At this site, fossils of rhinoceroses, antelope, camels, horses, rodents, oreodonts, and many others are found in rocks of Miocene and Pliocene age. Paleontologists continue to excavate fossils from this site, and many of the finds are on display at local museums. Just southeast of the Blackhawk site in the city of Fremont is a world-renowned mammal site. The community of land mammals excavated from the Irvington gravels there is the best early Pleistocene assembly found in North America. Fossils found at the quarries in Irvington include deer, camels, horses, saber-toothed tigers, a wolf, and the oldest mammoth elephant discovered in the western hemisphere. Many of the people living in the Irvington suburbs today have no idea that the area was home to such a collection of animals in the not-too-distant past.

California's fossil locality best known to nongeologists is the Rancho La Brea (Spanish for "tar") tar pools, which are now a part of Hancock Park in Los Angeles. The natural pools of tar trapped animals and plants of all kinds (Fig. 3-12, *A*), preserving them for excavation by twentieth century paleontologists. More than half a million specimens of mammals, birds, reptiles, insects, and plants—representing more than 200 species—have been excavated from Rancho La Brea. Extinct species of bison, mammoth, mastodon, and camel are among the mammals (Fig. 3-12, *B*). Also present are fossils of the extinct saber-toothed tiger, *Smilodon,* California's official state fossil. Many of the specimens from La Brea are on display at the Los Angeles County Museum of Natural History.

THE EXILED MAMMOTHS OF THE CHANNEL ISLANDS

The Channel Islands form an east-west chain that lies 20 to 50 kilometers off the coast of southern California near Santa Barbara. They are separated from the California mainland by water up to 200 meters deep. Today some of the Channel Islands are a popular recreational destination; others are wildlife sanctuaries. During the late Pleistocene they were apparently the home of an unusual group of land mammals.

In the late 1800s, the remains of extinct Pleistocene mammoths were first discovered on Santa Rosa Island. Since that time, geologists and paleontologists have found a number of additional specimens on Santa Cruz Island. The most recent discoveries were made on Santa Rosa Island in 1994 (Fig. 3-11). From what we know about mammoths, it is clear that they must have arrived in the Channel Islands during a time when the sea level was lower, a time when the islands were either separated by narrower bodies of water or connected by land bridges. Worldwide, the sea level was lower during several periods of Pleistocene glaciation (see Chapter 12), so it is likely that the original mammoths arrived in the Islands during one of these periods. The most recent period when sea levels were low was about 18,000 to 20,000 years ago.

What is most amazing about the fossil mammoths of the Channel Islands is that they are dwarf species of the more common fossil elephants, *Archidiskodon imperator* and *Parelephas columbi.* The dwarf species of the Channel Islands were only about 2 to 3 meters high at the shoulder (see Fig. 3-11), in comparison with mainland species, which stood about 4 meters at the shoulder. Paleontologists recognized the Channel Islands fossils as belonging to a separate species, *Mammuthus exilis,* or exiled mammoth. It appears that once the population was isolated from the mainland by rising sea levels, the smaller mammoths survived better because of limited food supplies. Over time, the process of natural selection favored the smaller animals. Fossils of dwarf elephants and other species have been discovered in other parts of the world on isolated islands.

Fig. 3-11 Fossil pygmy mammoth from Santa Rosa Island. (Source: Rockwell, T. San Diego State University.)

Fig. 3-12 **A,** Excavation of La Brea. **B,** Section of Rancho La Brea mural by Charles Knight, depicting Pleistocene life in southern California. (Source: **A,** U.S. Geological Survey; **B** reproduced with permission of the George C. Page Museum, Los Angeles.)

It is clear that fossils provide information far beyond establishing the ages of rocks. Fossils provide valuable information about past environments. Sensitive marine organisms like diatoms or foraminifera can give information about water temperature, salinity, depth, and amount of light (Fig. 3-13). Fossils with limited geographic distribution can provide information about the ancient latitudes of oceanic terranes. Fossils also give information about ecosystems and communities and their adaptations to changing habitats. By studying extinctions, we have learned about major crises in the history of life on Earth. By studying past changes in groups of organisms, we can better appreciate how organisms evolve and how their envi-

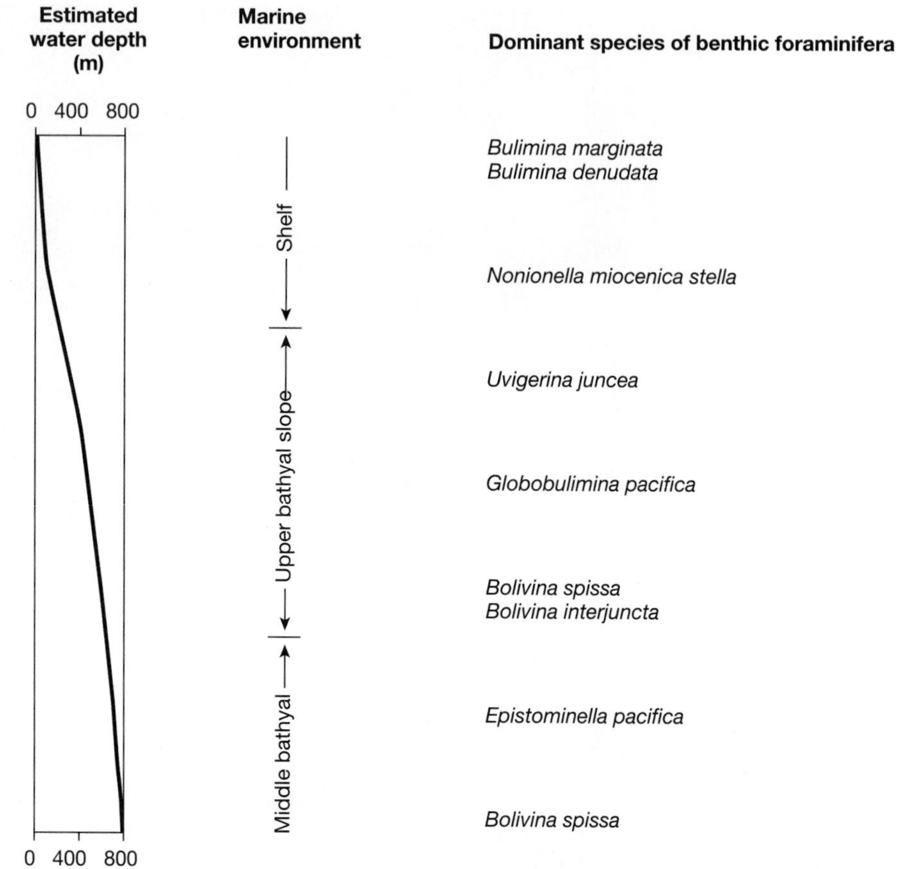

Fig. 3-13 Diagram showing changes in foraminifera species with different water depths (***bathyal*** = ocean depths, usually 180 to 1800 meters). (Source: Ingle, J. 1978. Neogene biostratigraphy and paleoenvironments of the western Ventura Basin with special reference to the Balcom Canyon section. U.S. Geologic Survey, Open-File Report 78-446.)

ronments change. From such studies we can make projections about future changes in Earth's environment and evaluate possible effects of humans on other organisms. Finally, fossils provide us with a broad perspective of our place on Earth and an appreciation of the expanse of geologic time.

FURTHER READINGS

BEDROSSIAN, T.L. 1971. Fossils: The living past. *California Geology* 24(12):227-239.

BEDROSSIAN, T.L. 1975. Vertebrate fossils and the history of animals with backbones. *California Geology* 28(11):243-259.

GOULD, S.J. 1987. *Time's Arrow, Time's Cycle.* Cambridge, Mass.: Harvard University Press, 222 pp.

LEVIN, H.L. 1996. *The Earth Through Time,* 5th ed. New York: W.B. Saunders Co., 700 pp.

CHAPTER

4

Geologic Maps and Sources of Information

GEOLOGIC MAPS AND CROSS SECTIONS

The purpose of most geologists' work is to understand how and where the rocks or sediments in an area were formed, what forces have affected them since they formed, and when these events occurred. To interpret geologic history, geologists must assemble a three-dimensional view of the rocks or sediments in an area. Understanding both the type and rates of changes in geologic conditions through time is also critical to solving many geologic problems. The three-dimensional picture of geology changes gradually, adding a fourth dimension—time—to the picture. Because this ever-changing geometric arrangement of geologic materials is so important for interpretation, geologists rely heavily on visual techniques for portraying information.

Maps are the fundamental tool for displaying geologic information. A geologic map may cover a large area, such as the entire state of California, or it may portray in detail the geology of a small area, such as a housing development. Regardless of scale, geologic maps show four fundamental types of information: (1) the types of rock or sediment at the surface; (2) the ages of the materials; (3) the orientation of the rocks or sediments, including the inclination or ***dip*** of layered rocks; and (4) the types of boundaries separating the different materials. On a generalized map such as Endpaper 1 (inside the front cover), only the major rock types and ages are portrayed. However, many geologic maps are very detailed, with rock and sediment types subdivided on the basis of physical characteristics, slight differences in composition, or environments of origin. Faults are typically portrayed by heavy black lines, differentiating them from other types of ***contacts*** between different rock or sediment units, which are shown by light black lines.

Map units are frequently shown by different colors and/or abbreviated symbols. The fundamental geologic map unit is a ***formation.*** A geologic formation is a distinctive, recognizable unit of rock or sediment. A formation is recognizable and mappable by its ***lithology***—rock or sediment type. On most geologic maps, the map symbols denote first the age of the unit, followed by the nature of the material or the formation name. For example, Quaternary alluvium (almost always shown in yellow on traditional geologic maps) is generally denoted by the abbreviated form "Qal." On many California geologic maps, "Tm" denotes the Tertiary Monterey Formation, "KJf" the Cretaceous and Jurassic Franciscan Assemblage, "Tv" Tertiary volcanic rocks, and so on. On many geologic maps, information about the landscape is also included because the geologic information is mapped on a ***topographic*** base map. A topographic map depicts the shape of the land surface with contour lines connecting lines of equal elevation.

A geologic ***cross section*** is a view of the geology beneath the surface to some depth, constructed as if one had sliced through the earth along a line. To construct a cross section, a geologist must understand the three-dimensional geometry of the map units and be able to project the contacts, including faults, from the surface. Although at least partly inferred, a completed cross section provides a clear view of the relative ages of map units, because one can use the principles of cross-cutting relationships and superposition (see Chapter 3). To a geologist, a cross section can provide an interpretive summary of geologic events in the area. Many published geologic maps show cross sections across one or more lines along the map area, with the symbols and colors on the cross section coordinated with those on the maps (Fig. 4-1).

In many parts of California, particularly in urbanized areas at risk from geologic hazards, Earth scientists have prepared specialized maps that are useful for planners and other government officials. Specialized maps may show flood-prone areas, potential and active landslides, active fault zones, or areas prone to liquefaction during earthquakes. The California fault map (Endpaper 2, inside front cover) is an example of such a map. Other types of hazard maps show the relative risk of certain events in an area, such as the probability of volcanic debris flows around the flanks of Mt. Shasta.

SOURCES OF GEOLOGIC MAPS

The California geologic map is available from the California Division of Mines and Geology (CDMG) in a larger format than the one included in this book (Endpaper 1). More detailed geologic maps of California, a series of 27 map sheets, at a scale of 1:250,000, are also available from the CDMG. Each map sheet is named for a city within the area; for example, the San Bernardino sheet covers an area including the city of San Bernardino. These regional maps are excellent references, and most include cross sections, fault maps, and source lists for additional information.

More detailed geologic maps are available for many areas of California. These include maps of individual topographic quadrangles or other restricted areas. Specialized geologic maps may emphasize materials of certain ages or types. Many of these maps are published by the U.S. Geological Survey or the CDMG. Other detailed or specialized maps are included in graduate theses; a yearly index of these is published in *California Geology,* a publication of the CDMG. City and county planning offices are an excellent source for obtaining hazard maps.

Maps showing earthquake epicenters and active faults are of particular interest to Californians. The California fault map, available from CDMG, shows all known and suspected active faults (Endpaper 2). This is the map used to define zones of special study for seismic hazards (see Chapter 13). The U.S. Geological Survey also has a series of maps showing earthquake epicenters plotted on raised-relief topographic

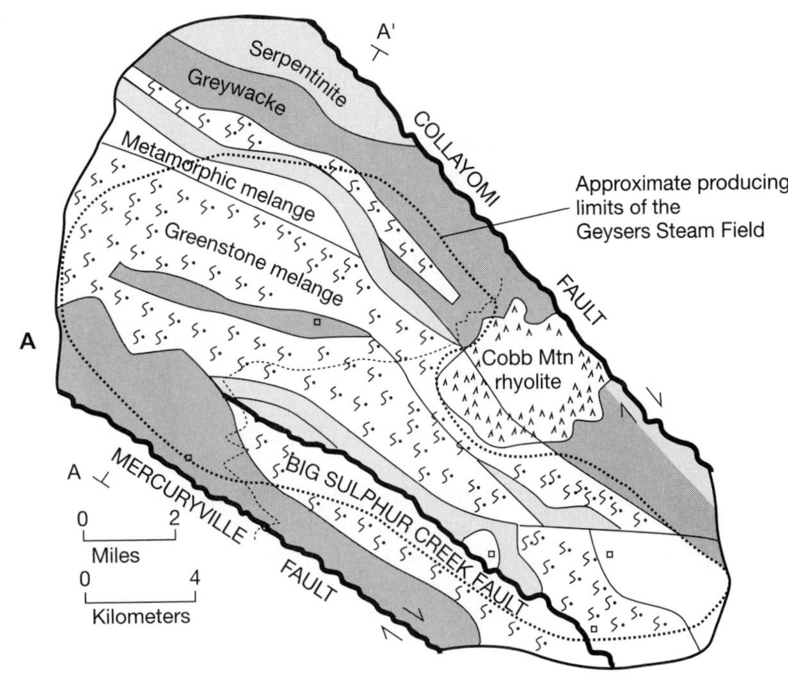

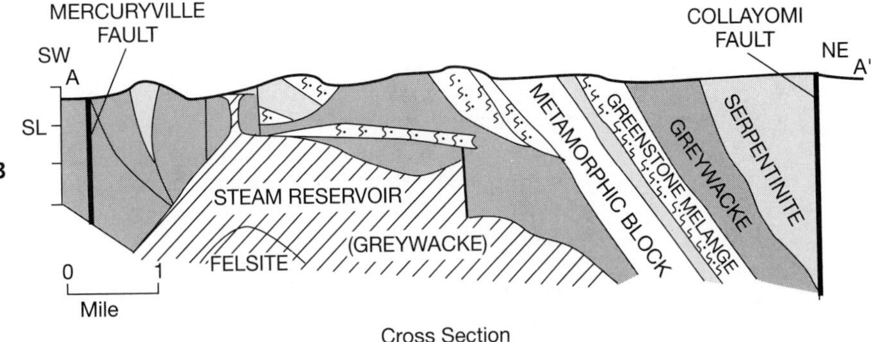

Fig. 4-1. **A,** Geologic map and **B,** a cross section of the Geysers geothermal field, Sonoma County. The cross section shows a slice through the area along the line A-A'. (Source: †Thomson, R.C. 1985. Unocal Geothermal Co.)

maps. Specialized hazard maps showing landslide areas, floodplains, liquefaction-prone areas, or volcanic hazards zones are published by the U.S. Geological Survey for areas where these hazards have been evaluated.

The California Water Map, published by the Water Education Foundation in Sacramento, shows the state's rivers, lakes, dams, aqueducts, and other water features (see Chapter 10). County soil surveys, published by the U.S. Department of Agriculture's Soil Conservation Service, are available for most counties. Topographic maps, which show the elevations of the ground surface with contour maps, are available from the U.S. Geological Survey and at many sporting and recreation stores.

GENERAL SOURCES FOR MAPS AND PUBLICATIONS

U.S. Geological Survey
Earth Science Information Service
345 Middlefield Road
Menlo Park, CA 94025

or

Publications Office
Denver Federal Center
Lakewood, CO 80225

California Division of Mines and Geology
Publication Sales
P.O. Box 2980
Sacramento, CA 95812
Free catalogs are available on request.

BOOKS AND PERIODICALS ABOUT CALIFORNIA GEOLOGY

This book is intended to provide a general overview of California's geology for those with little or no background in earth science. Because the emphasis is on geologic processes, the chapters only highlight particular aspects of the geology in each part of California and are in no way comprehensive. Many excellent sources of information are available to readers wishing more detailed treatment of particular areas or subjects. The materials described on pp. 50-51 should be considered a starting point for those who want to go beyond the material presented in this book.

ON-LINE RESOURCES

Many earth science institutions in California have information that can be accessed by computer networks. The U.S. Geological Survey (USGS), University of California's Paleontology Museum, Southern California Earthquake Center, and Monterey Bay Aquarium are examples of institutions with home pages on the World Wide Web. The list below provides net surfers a sampling of sites with information related to California geology.

Association of Bay Area Governments: *www.abag.ca.gov*

California Division of Mines and Geology: *www.consrv.ca.gov/dmg*

California Office of Emergency Services: *www.oes.ca.gov*

California Resources Agency: *www.ceres.ca.gov*

Monterey Bay Aquarium Research Institute: *www.mbari.org*

Museum of Paleontology, UC Berkeley: *ucmp1.berkeley.edu*

National Park Service: *www.nps.gov*

National Weather Service, California: *www.nws.mbay.net*

Northern California Earthquake Data Center: *quake.geo.berkeley.edu*

Scripps Institute of Oceanography: *sio.ucsd.edu*

Southern California Association of Governments: *www.sbag.ca.gov*

Southern California Earthquake Center: *scec.gps.caltech.edu*

U.S. Geological Survey: *www.usgs.gov*

FURTHER READINGS

ALT, D.D., AND HYNDMAN, D.W. 1975. *Roadside Geology of Northern California.* Missoula, Mont.: Mountain Press Publishing Co., 244 pp.

BAILEY, E.H., ed. 1966. *Geology of Northern California.* California Division of Mines and Geology, Bulletin 190.

JAHNS, R.H., ed. 1954. *Geology of Southern California.* California Division of Mines and Geology, Bulletin 170.

Both of these publications predate the acceptance of plate tectonics as the ruling theory of geology, and for that reason, they are somewhat out of date. However, both contain excellent summaries of different parts of California and discussion of important localities.

Burchfiel, B.C., Lipman, P.W., and Zoback, M.L., eds. 1992. *The Cordilleran Orogen: Conterminous U.S.* Vol. G-3, 724 pp.

This volume is an exhaustive summary of the geologic history of the western United States as understood in the early 1990s. This is a comprehensive and current treatment of the geology of the western United States, but any other than its intended audience of professional geologists may find it difficult reading.

California Division of Mines and Geology. *California Geology.*

This bimonthly periodical contains articles of interest to beginning geology students, and most are written at a level appropriate for beginners. In recent years the publication has included a "Teacher Feature" section for precollege teachers. Each December issue contains both a topical and a regional index.

Centennial Field Trip Guide: California, Oregon, and Washington. 1991. The Geology of North America Series, Decade of North American Geology (DNAG). Boulder, Colo.: Geological Society of America.

These field trip guides to important geologic localities are intended for geologists. However, many are appropriate for beginning geologists, and all include directions to the sites and reference lists.

Earth Magazine, Kalmbach Publishing.

This bimonthly periodical contains many articles about California and is written for a non-technical audience.

ERNST, W.G. 1981. *The Geotectonic Development of California.* Upper Saddle River, N.J.: Prentice Hall, Inc.

This book was the first to summarize California's plate tectonics history. Although many of the detailed studies presented are now somewhat out of date, the book is still a useful reference.

Geology of North America Series, *Decade of North American Geology* (DNAG). 1988 and later. Boulder, Colo.: Geological Society of America.

HUBER, N.K. *The Geologic Story of Yosemite National Park.* U.S. Geological Survey, Bulletin 1595.

MCPHEE, J. 1993. *Assembling California.* New York: Farrar, Straus, & Giroux.

This is an engaging account of the plate tectonics history of California. It also provides insight into the working life of field geologists.

NORRIS, R.M., AND WEBB, R.W. 1990. *California Geology,* 2nd ed. New York: John Wiley & Sons, 541 pp.

A comprehensive look at California geology, with detailed information about the rocks and history of its provinces.

OAKESHOTT, GORDON B. 1978. *California's Changing Landscape.* New York: McGraw-Hill Book Co., 379 pp.

A general reference with excellent photographs. Available by special order or at most libraries.

SHARP, R.P., AND GLAZNER, A.F. 1993. *Geology Underfoot in Southern California.* Missoula, Mont.: Mountain Press, 224 pp.

SHELTON, J.S. 1966. *Geology Illustrated.* San Francisco: W.H. Freeman & Co., 434 pp.

WALLACE, R.E., 1990. *The San Andreas Fault.* U.S. Geological Survey Professional Paper 1515, 283 pp.

PART

Geologic Highlights of California

Chapters 5 through 17 will introduce readers to the major geologic features found in different regions of California. The intent is not to give a comprehensive description of all geologic materials and events in each region, but rather to highlight the most important and interesting geologic features. Some features, like the California coastline or the San Andreas fault system, and some processes, like earthquakes, cross over region boundaries and are covered by individual chapters. The provinces are discussed in an order that presents less complicated areas and processes first, so the later chapters can logically build from them.

California's Geomorphic Provinces

Even a brief examination of California's geology, faults, and landscape features allows the reader to see that California has some distinct regional differences (see Endpapers 1-4 inside this book's covers). When these patterns are combined with California's patterns of regional climate, the state can be divided into distinct natural regions. Even before satellite imagery allowed geologists to view California's regions from space, these areas had been identified on the basis of their distinctive geology, climate, and landscape. Each of these regions, or geomorphic provinces (see the map on the next page), displays a distinctive landscape character. The amount of relief, the types of landforms, the orientation of valleys and mountains, and the type of vegetation in each province are elements of its geomorphology. Because geology is one of the controlling factors that shapes the landscape of each province, it is appropriate to discuss California's geology in the context of its geomorphic provinces.

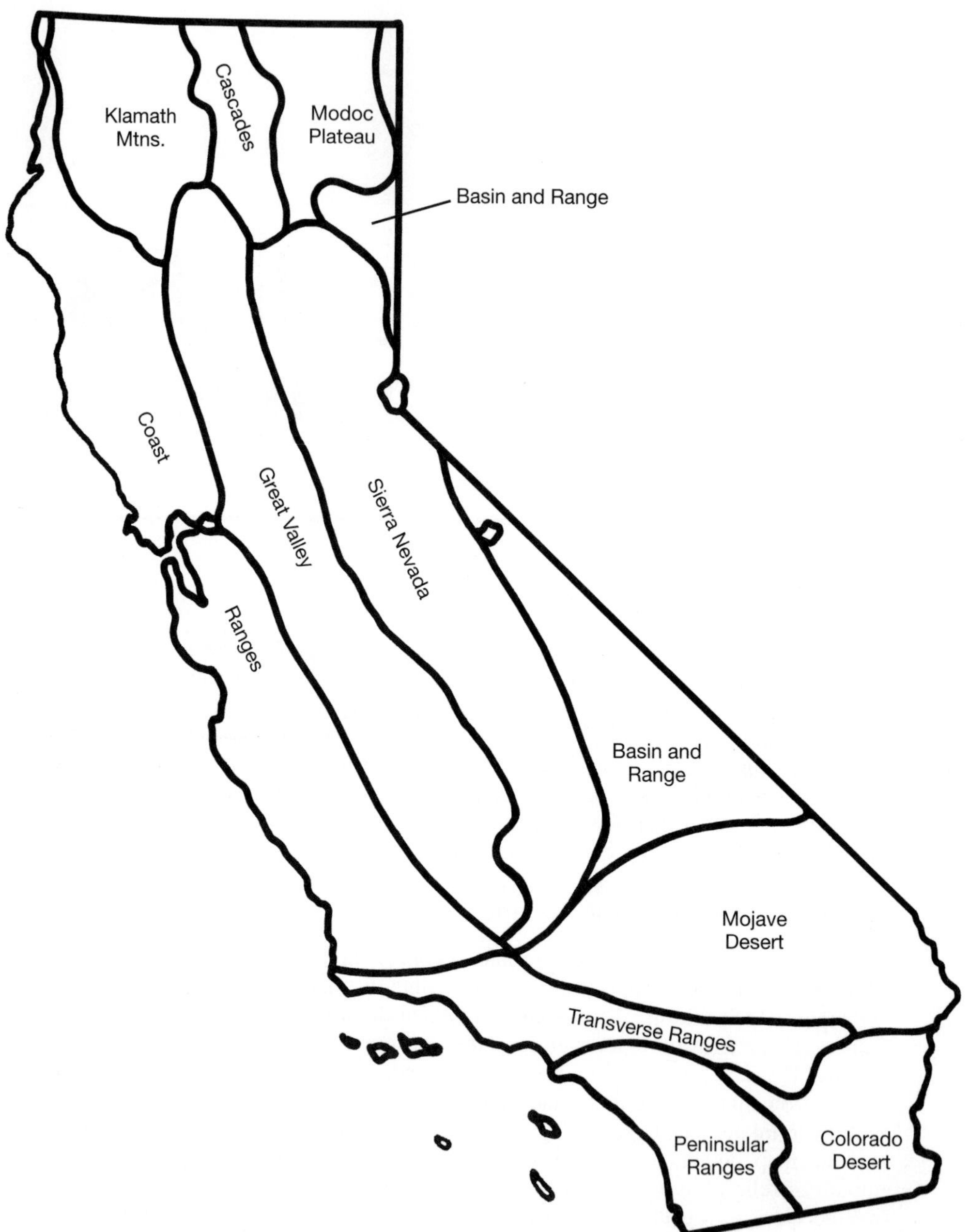

The geomorphic provinces of California.

CHAPTER

5

Young Volcanoes

The Cascades, the Modoc Plateau, and the Long Valley Caldera

GEOGRAPHIC HIGHLIGHTS

The Cascade Range is an 800-kilometer-long chain of young volcanic mountains that reaches from British Columbia to the northernmost part of California (Fig. 5-1). California's Cascades Province is the southern tip of this chain, and the geology of the Cascades Province is more similar to that of the Cascades of the Pacific Northwest than to other parts of California. Of the 12 major volcanic cones in the range, California's Mt. Shasta (elevation 4319 meters) is the second highest, behind Washington's Mt. Rainier, and the largest in volume. The towns of Weed, Mt. Shasta, and McCloud sit at the base of the volcano. The other large cone in California is Lassen Peak (elevation 3188 meters), which last erupted during the period from 1914 to 1921. The peaks of Mt. Shasta and Lassen Peak are spectacular landmarks that can be seen from as far as 150 kilometers away along Interstate 5 in the Sacramento Valley (Fig. 5-2 and Color Plate 8). The slopes of both mountains are forested, and the cones are surrounded by sparsely vegetated volcanic tablelands, while the summits are snow-covered bare rock.

The volcanoes that make up the Cascade Ranges are the result of active subduction of the Gorda and Juan de Fuca Plates beneath North America. The rocks formed by eruptions from this chain during the past 3 million years are known as the High Cascades Series. Late Cenozoic volcanic rocks that predate the present volcanic cones, known as the Western Cascades Series, are evidence that active volcanoes were present in the California Cascades 20 to 30 million years ago. Alluvium, lake deposits, and glacial deposits are associated with the young volcanic rocks of the Cascades Province. In some areas of the Cascades, older Mesozoic and Paleozoic rocks similar to those in the Sierra Nevada and Klamath Mountains Provinces can be found.

The Modoc Plateau lies east of California's Cascade Range in the northeastern corner of California. On the east, it is bordered by the Basin and Range Province, which will be discussed in Chapter 7. As its name implies, the Modoc Plateau is high, relatively flat country, although the province is cut by valleys (Fig. 5-3). This

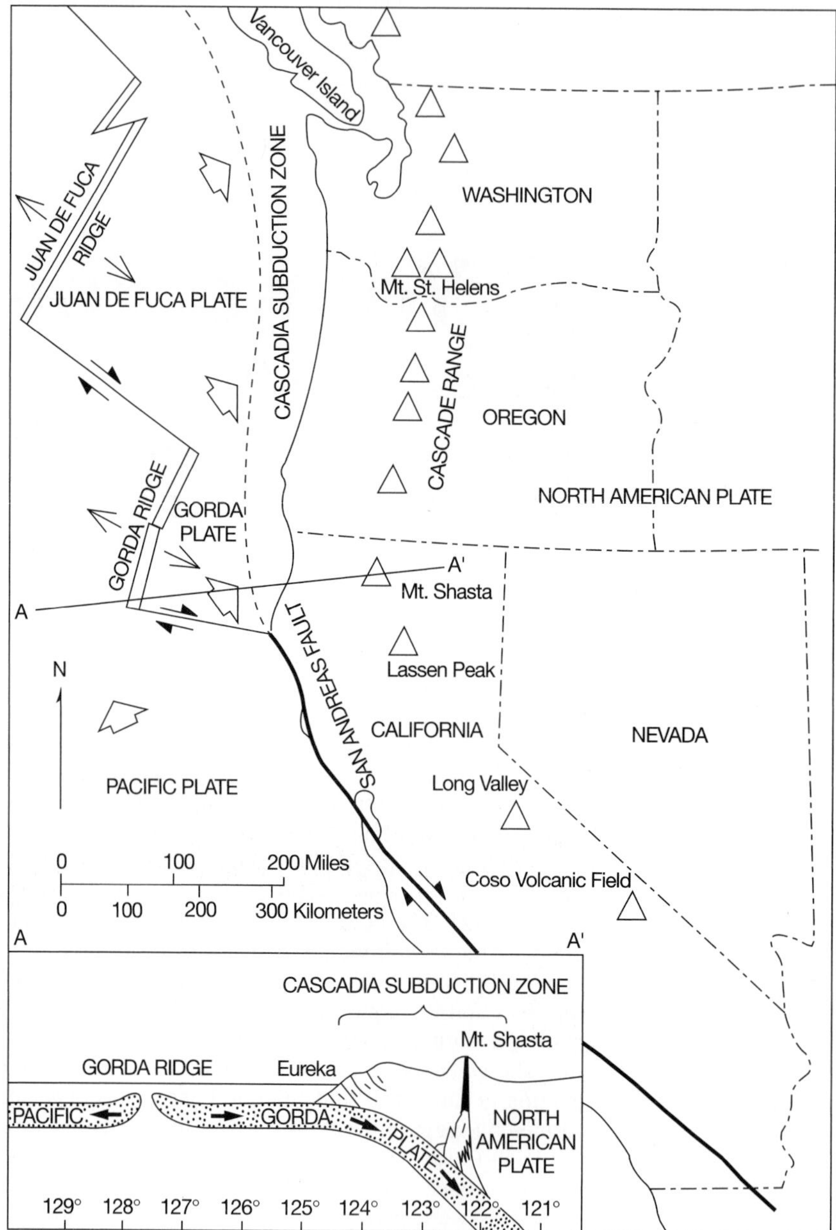

Fig. 5-1 Map and cross section showing the plate tectonic setting of active volcanoes of the Cascade Range. (Source: Dengler, L., and others. 1992. Sources of North Coast seismicity. *California Geology* 45:40.)

part of California is arid because the high ranges to the west block Pacific storms. Like the Cascades, the Modoc Plateau is made up of young volcanic rocks. However, we will see that the volcanic activity in the Modoc Plateau has been less explosive than that in the Cascades and has produced mainly lava flows like those seen in Lava Beds National Monument. On the border between the Modoc Plateau and the Cascade Range is Medicine Lake Volcano.

Fig. 5-2 Mt. Shasta. (Photo by author.)

Fig. 5-3 Volcanic plateaus of Lava Beds National Monument, showing typical dark basalt flows and sparse vegetation.

The contrast between the volcanic histories of the Cascades and Modoc Plateau Provinces provides an excellent opportunity to investigate the types and causes of volcanoes. A third area of recent major volcanic activity in California is the Long Valley area in the Basin and Range Province near Mammoth (Fig. 5-4). Although this area lies well beyond the Cascade Province, its geologic history is similar enough to that of Mt. Shasta and Lassen Peak to warrant its inclusion here. An evaluation of future volcanic hazards in California emerges from a review of the past volcanic events at these three major centers. The Coast Range Province and other parts

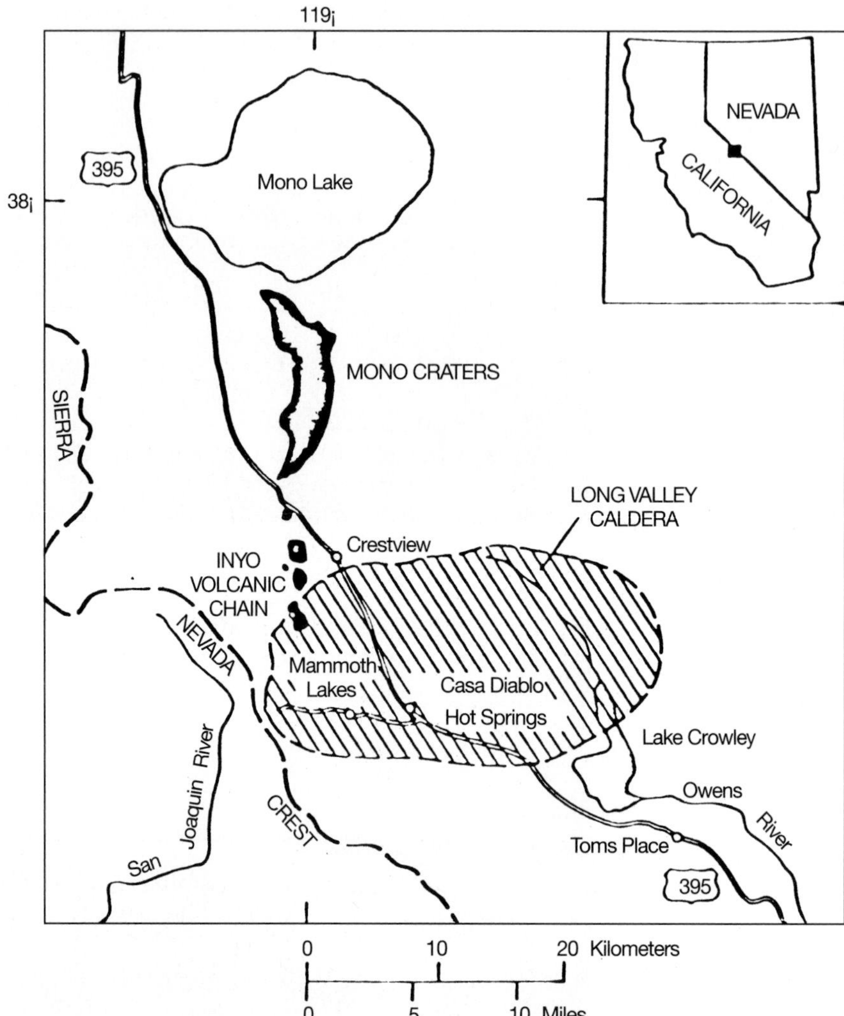

Fig. 5-4 Recently active volcanic centers in the Long Valley area east of the Sierra Nevada. (Source: Miller, C.D. 1984. *Friends of the Pleistocene Field Trip Guidebook.*)

of California's Basin and Ranges have also experienced geologically recent volcanic activity (see Chapters 7 and 12).

LAVA CHEMISTRY AND THE BEHAVIOR OF VOLCANOES

Visitors to the island of Hawaii often consider themselves lucky if an eruption takes place at Kilauea during their stay. Planes and helicopters fly over the lava flows and fountains, and bystanders take videos of lava as it knocks down trees and oozes down roads. At the Hawaiian Volcano Observatory, instruments predict eruptions several hours before lava breaks the surface, allowing geologists to travel to the likely site where lava will appear. Although thousands of homes have been inundated and burned by lava flows, the predictions allow orderly evacuations from threatened areas.

Contrast the behavior of Kilauea with the events of 1980 at Mt. St. Helens in Washington and those of 1991 at Mt. Pinatubo in the Philippines. The May 18, 1980 eruption of Mt. St. Helens killed 60 people, including a U.S. Geological Survey geologist who was observing the mountain from what was thought to be a safe site. If the area surrounding Mt. St. Helens had not been evacuated, the loss of life would have been far greater. A huge ***debris avalanche,*** traveling at speeds up to 300 kilometers per hour, filled the North Fork of the Toutle River, 22 kilometers away, with debris and ice up to 45 meters thick (Fig. 5-5). More than 200 houses and 300 kilometers of roads were destroyed, 4 billion board feet of timber damaged or destroyed, and 12 million salmon fingerlings killed (Fig. 5-6). Almost 2 million cubic meters (more than 2 million cubic yards) of ash had to be removed from Washington highways and runways. Air and highway traffic was disrupted for weeks. The estimated cost of the May 18 eruption was $1.1 billion, a sizable amount for what was considered by geologists to be a small eruption. At Mt. Pinatubo, the June, 1991 eruption produced about 5 to 7 cubic kilometers of debris, compared with about 1 cubic kilometer erupted from Mt. St. Helens in 1980. The Pinatubo eruptions killed 350 people, heavily damaged two U.S. air bases, and left thousands homeless.

What makes volcanoes like Mt. Pinatubo and Mt. St. Helens so deadly? Differences in the violence of volcanic eruptions can be explained by the chemistry of the magma feeding the volcano (Table 5-1). By sampling and analyzing molten lava and volcanic rocks, geologists can determine the chemical composition of

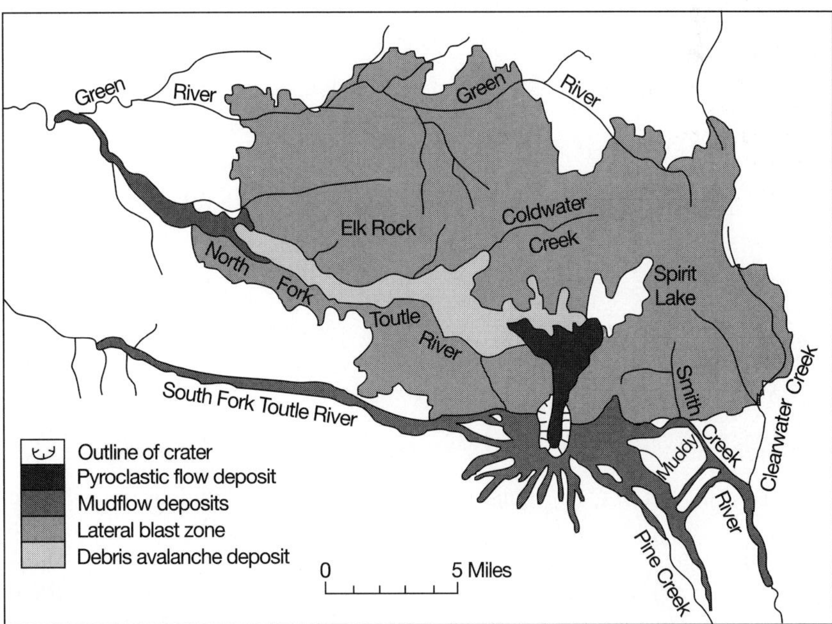

Fig. 5-5 Geologic map showing the volcanic deposits and the blast zone of the 1980 eruption of Mt. St. Helens. (Source: Tilling, R.I. *Eruptions of Mount St. Helens: Past, Present, and Future.* U.S. Geological Survey, Public Information Series.)

Fig. 5-6 Downed timber within the blast zone of the 1980 eruption of Mt. St. Helens. (Source: Keller, J.)

TABLE 5-1 EFFECTS OF MAGMA COMPOSITION ON VOLCANIC PROCESSES

Lava Type	Temperature of Lava	Fluidity of Lava	Gas Content	Events	Cone Type	California Examples	Rock Name(s)
Silicic							
High silica (~70% Si, 0) Low Fe, Mg	800° C	Viscous	High: explosive	Ash/steam; Pyroclastic flows; Gas clouds; Volcanic Debris flows; Glass flows	Dome	Mono Craters	Rhyolite Obsidian Pumice
Intermediate							
Intermediate silica (60%-65%)	1000° C	Intermediate			Strato-volcano (Caldera)	Mt. Shasta Mt. Lassen (Crater Lake, OR)	Andesite
Mafic							
Low silica (50%) High Fe, Mg	1200° C	Fluid	Lower: less explosive	Fluid lava flows; Cinder cones, spatter cones	Shield volcano; Fissure eruptions	Medicine Lake	Basalt

magmas. Although their chemistry is highly variable, lavas can be grouped into three broad categories. In all three, silica (see Chapter 2) is the predominant component.

- ***Mafic*** (sometimes called basaltic) lava contains relatively less silica (silicon and oxygen) and has high iron and magnesium. A mafic lava contains about 50 percent silica and 15 to 20 percent iron and magnesium. About 90 percent of all of the lava erupted worldwide is mafic or basaltic lava. Most volcanic rock formed from mafic lava is called ***basalt,*** which is dark-colored because of the high content of dark, mafic silicate minerals (Fig. 5-7).
- ***Intermediate*** or ***andesitic*** lava has a silica and iron-magnesium content that is intermediate between mafic and silicic lavas. Typical volcanic rock formed from intermediate lava is called ***andesite,*** and is typically gray or brown (see Fig. 5-7).
- ***Silicic*** or ***felsic*** lava has high silica content and is low in iron and magnesium. Silicic lava contains about 70 percent silica and only a few percent of iron and magnesium. Volcanic rock formed from silicic lava is called ***rhyolite,*** which is typically white or pink because of the absence of mafic minerals (see Fig. 5-7). Volcanic glass formed when silicic lava cools too quickly for minerals to crystallize is called ***obsidian*** (Fig. 5-8).

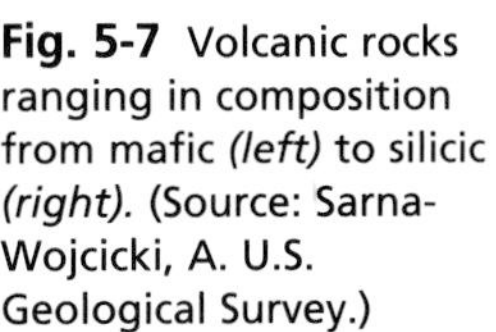

Fig. 5-7 Volcanic rocks ranging in composition from mafic *(left)* to silicic *(right).* (Source: Sarna-Wojcicki, A. U.S. Geological Survey.)

Fig. 5-8 Obsidian from Sonoma County *(left)* and from the Mono Craters area, Inyo County *(right).* Because of its superior cutting ability, obsidian had great value to early Native Americans in California *(top* and *center).*

The chemical differences between the major types of lava make them behave differently. In a silicic lava, the silica molecules tend to clump together, or *polymerize,* even though the lava is still a fluid. The clumping makes the lava ***viscous*** or sticky. In contrast, the iron and magnesium in a mafic lava act to keep the silica molecules separated, thus preventing the "clumping" action. As a result, mafic lava is very fluid and flows freely. In addition, silicic and intermediate lavas also contain more gas than mafic lavas. The gas is largely steam and carbon dioxide, but sulfur, nitrogen, and chlorine gases may also be present in small amounts.

An intermediate or high-silica lava moves sluggishly and may even become stuck in a volcanic vent because of its high viscosity. In contrast, mafic lavas are very *fluid,* sometimes behaving almost like water as they stream down valleys (Fig. 5-9). Extremely fluid basalt flows can move as fast as 16 kilometers per hour, and beneath the active volcanoes of Hawaii, basaltic magma can rise to the volcanic vents as fast as 1 kilometer per day.

The viscosity and the high gas content of intermediate and silicic lavas make eruptions explosive and potentially devastating. Because of the "sticky" movement within the volcanic vent, eruptions are also unpredictable. In extreme cases, the buildup of pressure is so great beneath the plug of lava that the eruption explodes with enough force to destroy an entire mountain. In contrast, eruptions of mafic lava low in gas produce relatively well-behaved lava flows and fountains. Because of its greater fluidity, mafic lava does not generally get stuck in the volcanic vents, eruptions are rarely explosive, and they occur with predictable regularity.

We can now see that whether a volcano will be a tourist attraction or a deadly menace can be explained by small differences in the chemical composition of silicate lavas. Mt. St. Helens and Mt. Pinatubo are both fed by intermediate to silicic magma, and both have had a long history of violent eruption.

Fig. 5-9 Fluid basalt flow, Hawaii. (Source: Griggs, J. U.S. Geological Survey.)

ORIGIN OF LAVA

Around the world today, chains of active volcanoes are concentrated near plate boundaries. As discussed in Chapter 1, volcanoes occur at both divergent boundaries and along subduction zones. Some volcanic centers, like the Hawaiian chain or the Yellowstone volcanic area, are found within plates; these reflect localized melting over hotspots in the mantle. The hundreds of volcanoes surrounding the Pacific Plate, often referred to as the "Ring of Fire," are mostly of andesitic or intermediate composition. In fact, andesite is named for one of these volcanic chains, the Andes Mountains. These volcanoes are associated with active subduction zones, some between and oceanic and a continental plate and others between two oceanic plates.

Geologists agree that water is a necessary ingredient for the production of andesite. Sediments that have accumulated in a deep-ocean trench or on the floor of the ocean basin are naturally saturated with the sea water that surrounds them. Water is also present within the crystalline structure of silicate minerals such as clays and mica. As a plate is pushed beneath the crust during subduction, it heats up and this water is released. The water acts like a catalyst, lowering the melting temperature of minerals to enable partial melting in the portion of the mantle that overlies the subducting plate.

Mafic magma rises at divergent boundaries and at zones where the crust is being extended within plates. At these localities, the magma is generated by partial melting of the mantle beneath the earth's crust. Magma rises relatively rapidly along pathways created as the crust is stretched. The magma is lower in silica, and because it is fluid and thus less dense than the surrounding rocks, it rises freely to the surface. Because it does not interact with crustal rocks on its journey, the magma retains its low-silica composition.

Volcanoes erupting only silicic lavas are found only within continental plates. These lavas are thought to originate from the partial melting of continental crustal rocks rich in silica as magma rises from beneath the crust. Because of its very high viscosity, high-silica magma tends to rise very slowly and cool beneath the surface to form plutonic rocks. As a result, silicic lavas are less common than either andesitic or basaltic lavas. However, an enormous eruption of silicic lava occurred in the Long Valley area in Inyo County 758,000 years ago.

Some volcanoes, including Medicine Lake in the Cascades Province and the Coso volcanic field east of the Sierra Nevada (discussed in Chapter 7), produce both mafic and silicic eruptions. In these areas, rising basaltic magma can heat adjacent crustal rocks and partially melt them to form silicic magma. At the same time, basalt can form directly from sources in the mantle. The result is a volcanic area that contains basalt and rhyolite of the same age, erupted from different source magmas. It is important to point out that an almost infinite variety of lavas is possible, considering the diverse types of rock that can be partially melted and the possibilities for mixing of different magmas.

Geologists can use the lava and young volcanic rocks found at a locality as a clue to the tectonic setting of the area. The Cascades Province, which is dominated by intermediate volcanism, shows the characteristics of a subduction zone setting. North of Cape Mendocino, Mt. Shasta and Lassen Peak are being fed by magma that is generated as the Gorda Plate is subducted beneath the North American Plate. The Modoc Plateau, where volcanic rocks are largely mafic, is an area where crustal extension is occurring. As the crust stretches, magma rises from beneath the North

American Plate to form the volcanic centers of the Basin and Range. Volcanic activity in the Long Valley area is also thought to be related to extension in the Basin and Range Province. In this area, however, silicic lavas are produced as basaltic magma rises into a more easily melted, thick section of Sierra Nevada granitic rocks, melting them and mixing to form the silicic magma beneath Long Valley.

VOLCANIC EVENTS AND FEATURES

Mafic

Lava flows are the most common product of basaltic volcanic eruptions. Lava may erupt from a central vent or from a ***fissure*** or crack along the flanks of the volcano (Fig. 5-10). Lava fountains or curtains of fire are common occurrences at Hawaiian volcanoes. Because of its fluidity, basaltic lava moves relatively freely down valleys, filling the low places in the landscape. Streams of lava may flow continuously in a channel for hours or days.

Although eruptions from low-silica volcanoes are less explosive than those from other volcanoes, some more explosive events can occur. The most common product of more explosive eruptions from basaltic cones is ***scoria*** or ***cinders,*** which are porous fragments, pebble-sized or larger, thrown from the volcano. The fragments typically have a frothy appearance because of the abundance of ***vesicles,*** or bubbles left by escaping gas.

Over a period of centuries or more, repeated eruptions of mafic lava from a central vent may create a ***shield volcano*** like Medicine Lake (Fig. 5-11). Shield volcanoes are broad and gently sloping, and they are often less dramatic landscape features than other volcanoes. Repeated eruptions along fissures may produce broad plateaus of lava, like those in the vicinity of Lava Beds National Monument, rather than a volcanic cone. ***Cinder cones*** are common features of all active volcanic terrains. They are usually fairly small cones, composed of fragments thrown from a

Fig. 5-10 Lava fountain from fissure eruption, Iceland. (Source: Atwater, B. U.S. Geological Survey.)

central vent. After the eruption, a funnel-shaped crater remains at the peak of the cone (Fig. 5-12). Small ***spatter cones,*** formed when blobs of lava are thrown from a vent and then fall in heaps, are also typical features of basaltic volcanic terrains.

Intermediate

During a large eruption from an andesitic volcano, huge volumes of gas and instantly chilled volcanic glass are released, along with chunks of rock from the volcano itself. An explosive eruption produces a wide variety of ***pyroclastic*** material; that is, fragmented debris produced by fire (from the Greek *pyr,* meaning "fire"). Great clouds of steam and other gases are released as the eruption releases the pressure within the

Fig. 5-11 Medicine Lake Volcano, northeastern California. Medicine Lake is a typical shield volcano formed by repeated eruptions of basaltic lava flows. (Source: Sarna-Wojcicki, A. U.S. Geological Survey.)

Fig. 5-12 Quarry excavated in a cinder cone near Manton, Lassen County. (Source: Sarna-Wojcicki, A. U.S. Geological Survey.)

Fig. 5-13 Petrified remains of victims of the 79 AD eruption of Mt. Vesuvius, Pompeii, Italy. (Source: Bradley, W.C.)

vent, in much the way that opening a shaken bottle of carbonated soda sends jets of carbon dioxide and soda from a bottle. Some of the gas clouds can be deadly: a few breaths of pure carbon dioxide is sufficient to asphyxiate a person. During the eruption of Mt.Vesuvius in 79 AD, many citizens of Pompeii dropped in their tracks as they inhaled the deadly gas streaming from the volcano (Fig. 5-13).

As the lava blows out from the vent, it is instantly chilled to form volcanic glass, and the mixture of debris and gas is sent high into the atmosphere. The pyroclastic material carried into the atmosphere is called ***tephra.*** Much of the tephra consists of very small particles known as ***volcanic ash,*** but some tephra fragments, called ***lapilli,*** are centimeters in size, and others, known as ***volcanic bombs,*** are a half-meter or more in diameter (Fig. 5-14). When gas trapped in the vent is violently separated from the lava, ***pumice*** is produced. Pumice is volcanic glass riddled with holes or vesicles. Because the vesicles make pumice fragments very light, they may be carried long distances from the source of the eruption.

Hot ash and other debris flow down the flanks of the erupting volcano, often releasing deadly gas as the lava travels. The flows may glow from incandescent lava fragments, and these deadly avalanches known as glowing clouds or ***nuee ardente*** can reach speeds as great as 300 kilometers per hour. In 1902, the glowing clouds produced by the eruption of Mt. Pelee in the Caribbean killed 30,000 people. ***Pyroclastic flows*** may be hot enough to partly weld the fragments within them (Fig. 5-15). When volcanic ash or pyroclastic flows are lithified to form rocks, they are referred to as ***tuff.***

Fig. 5-14 Large pumice blocks erupted from Mt. St. Helens in 1980. (Source: Sarna-Wojcicki, A. U.S. Geological Survey.)

Fig. 5-15 Welded tuff, Sonoma volcanic field. (Source: Sarna-Wojcicki, A. U.S. Geological Survey.)

The steep sides of an andesitic volcano are prime sites for debris avalanches, a type of landslide that can reach speeds of hundreds of kilometers per hour. A debris avalanche may occur in fairly dry materials, or it may involve some water. Huge debris avalanches may send debris tens of kilometers down the volcano's flanks.

Among the most deadly hazards during a large eruption of an andesitic volcano are ***volcanic debris flows.*** Consider a snow-capped volcanic peak, especially one with glaciers on its flanks, like Mt. Shasta. The flanks of an intermediate volcano are

Fig. 5-16 Effects of volcanic debris flows, Toutle River, Washington, 1980. (Source: Miller, C.D. U.S. Geological Survey.)

littered with volcanic ash, and early eruptions may produce a fresh covering of fine ash. A major eruption instantly melts the snow and ice, mixing in huge amounts of ash and debris, and sends the saturated mixture down the steep sides of the mountain at incredible speed. This is the recipe for disaster that has been repeated in the past 15 years. At Mt. St. Helens in 1980, debris avalanches became saturated debris flows that traveled about 35 kilometers from the mountains into the Cowlitz and Toutle Rivers, raising the river levels as much as 5 meters and filling the channels with debris (Fig. 5-16). The village of Armero, 30 kilometers from Nevada Ruiz in the Colombian Andes, was destroyed by debris flows in September 1985; 23,000 people were buried by the flows. At Mt. Pinatubo in the Philippines, debris flows continue to be a problem for villages in the valleys on the sides of the volcano; to date, hundreds of people have been killed by the flows.

Over a period of several thousand years, eruptions of intermediate lava may form a huge volcanic mountain known as a ***stratovolcano*** or ***composite cone.*** These are the "textbook" volcanoes, with summit craters and steep, symmetrical sides. The shape of a stratovolcano reflects the fact that it is composed of viscous pyroclastic material as well as lava flows.

A stratovolcano may be partly or completely destroyed by a catastrophic eruption. About 7000 years ago, an Oregon Cascades volcano, named Mt. Mazama by geologists, was destroyed by a series of huge explosions. After the magma was removed from the vent, the entire top of the mountain collapsed to form the depression known as Crater Lake. Crater Lake is an example of a ***caldera,*** a circular depression formed by the collapse of an area above a volcanic vent. Volcanic centers may undergo a cycle of volcano building and destruction, with new cones occupying the site of a destroyed mountain. For example, the north side of Mt. St. Helens was blown away in 1980, and since then, a new dome, called a ***resurgent dome,*** is being built in the summit crater.

Fig. 5-17 Glass Mountain obsidian flow, Medicine Lake. (Source: Sarna-Wojcicki, A. U.S. Geological Survey.)

Silicic/Felsic

Like the andesitic volcanoes, volcanic centers erupting high-silica lava can also produce explosive eruptions, creating many of the hazardous events typical of andesitic eruptions. Rhyolite pumice, tephra, and ash flows are common products of explosive eruptions from high-silica sources.

Eruptions of very high-silica lava also commonly produce ***obsidian domes*** and glass flows. Because high-silica lava is thick and pasty, it oozes sluggishly from vents and does not travel very far before it solidifies. Domes and flows formed from high-silica lava are generally steep-sided (Fig. 5-17). The glassy flows and domes near Medicine Lake volcano and in the Mono-Inyo volcanic chain, both discussed later in this chapter, are striking examples of high-silica volcanic features.

ERUPTIVE HISTORY OF CALIFORNIA'S ACTIVE VOLCANOES (Table 5-2)

Mt. Shasta

Mt. Shasta is a huge stratovolcano built by eruptions from four main volcanic vents (Fig. 5-18; see also Fig. 5-2). Geologists estimate that it has taken more than 100,000 years to construct the present volcano. During that time, eruptions have produced andesitic lava flows, pyroclastic flows, and debris flows, but few deposits of ash. Much of the volcano's construction has taken place during the past 10,000 years, when at least 13 eruptions have occurred. During the past 4500 years, the average interval between eruptions has been only 600 years. Today two small centers of hot springs remain near the summit.

Two of Mt. Shasta's eruptive centers have been very active during the past 10,000 years. Hotlum Cone at the summit has been constructed during the past 8000 years. During that time, it has erupted at least eight times, producing pyroclastic

TABLE 5-2 VOLCANIC ERUPTIONS IN CALIFORNIA

Date	Source Area	Volcanic Event(s)
1914-1922 AD	Mt. Lassen	Ash and steam clouds, debris flows
200-400 years BP	Paoha Island, Mono Lake	Most recent ash eruptions
200-300 years BP	Medicine Lake	Most recent lava flows, Glass Mountain area
Few centuries BP	Cima Dome, Mojave Desert	Youngest basalt flows and domes
1325-1365 AD (650 years BP)	Panum Crater, Mono Craters chain	Eruption of tephra ring. Obsidian dome in crater
9300-9700 years BP	Shasta area	Construction of Shastina Cone
40,000 years BP	Coso volcanic field	Youngest basalt flow
300,000-380,000 years BP	Shasta area	Giant debris flow, Shasta Valley
400,000 years BP	Lassen area	Rockland ash eruption
758,000 years BP	Long Valley	Caldera-forming eruption, Bishop ash and tuff, 500 km^3 ash erupted
3.4 million years BP	Lassen area	Nomlaki Tuff
3.4 million years BP (dated age), but older than Nomlaki, which is found above it in the Sacramento Valley	Sonoma (Coast Ranges)	Putah Tuff

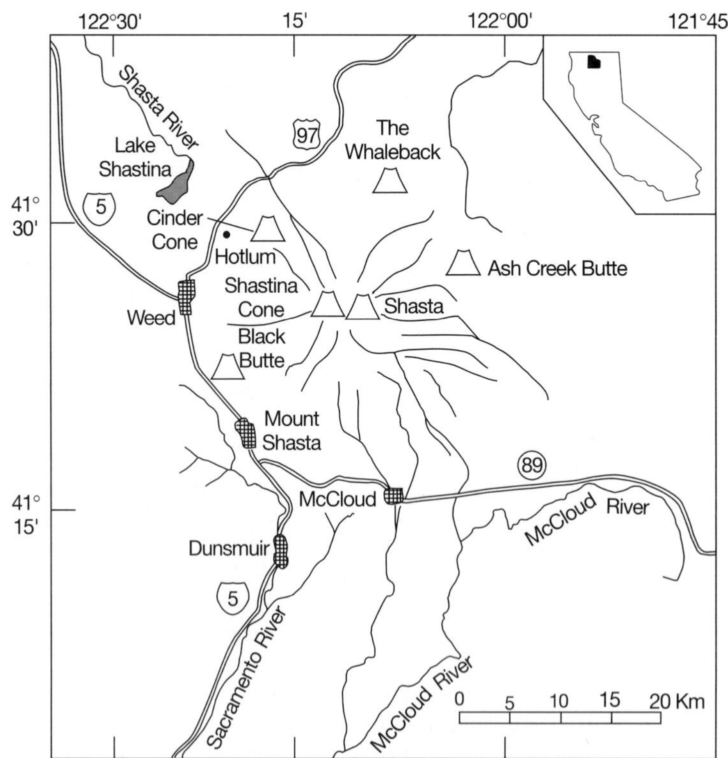

Fig. 5-18 Map showing eruptive centers at Mt. Shasta. (Source: Modified from Miller, C.D. 1980. U.S. Geological Survey Bulletin 1503, p. 2.)

Fig. 5-19 View of Black Butte, a young volcanic cone southeast of Redding. (Source: Harwood, D. U.S. Geological Survey.)

flows, andesitic lava flows, and debris flows that have traveled down the mountain's flanks. A more explosive eruption about 9600 to 9700 years ago also produced the Red Banks tephra, a layer of ash and pumice fragments found today north and east of the summit. About 200 years ago, several debris flows originated from the Hotlum area. These may have been produced by the eruption that was apparently observed in 1786 by explorer LaPerouse from a ship in the Pacific Ocean.

Shastina Cone, a subsidiary cone on the western flank of Mt. Shasta, was built by eruptions between 9300 and 9700 years ago. These eruptions produced andesitic lava flows and pyroclastic flows; the town of Weed is built on one pyroclastic flow erupted from Shastina about 9500 years ago. Shastina's summit is composed of several ***dacite*** domes (a material with a composition between andesite and rhyolite) built at the end of its eruptive period. A subsidiary center of recent volcanic activity is Black Butte, a group of dacite domes about 13 kilometers west of Mt. Shasta. At both Shastina and Black Butte, the collapse of volcanic domes resulted in pyroclastic flows about 9500 years ago.

Before the cone-building activity occurred at Shastina and Hotlum Cones, volcanic centers at Misery Hill, just south of the summit, and Sargeants Ridge, on the south flank of the mountain, were responsible for most of the activity. Geologists can identify older eruptive centers by tracing volcanic deposits back to their source, where they are thickest. By dating the deposits, geologists can determine the time period when that center was active. At Mt. Shasta, the two older cones have been eroded by the glaciers that were present until about 15,000 years ago. In contrast, Shastina and Hotlum Cones show no evidence of glaciation.

MEDICINE LAKE HIGHLANDS AND LAVA BEDS NATIONAL MONUMENT

Within the uplifted plateau known as Medicine Lake Highlands lies Medicine Lake volcano, a recently active shield volcano on the border between the Cascades and Modoc Plateau provinces about 50 kilometers northeast of Mt. Shasta (see Fig. 5-11).

A 300,000-YEAR-OLD CATASTROPHE

The western two thirds of the valley of the Shasta River, on the northwest side of Mount Shasta, is a maze of mounds, hills, and ridges (Fig. 5-20) that have puzzled geologists for more than 80 years. The blocks of volcanic rock that form the hills are up to hundreds of meters long, and they are strewn chaotically about the valley. Although the volcanic rock is similar to the andesite found in the modern Mt. Shasta cones, a variety of volcanic rock types are found in the hills and mounds. One of the original interpretations of the deposits of Shasta Valley was that they are deposits left by glaciers or by streams washing out from the glaciers that once occupied Mt. Shasta.

After a careful study of Shasta Valley, U.S. Geological Survey geologist Dwight Crandell came up with a startlingly different conclusion: the deposits are the remains of a giant debris avalanche from the sides of an ancestor of Mt. Shasta. Based on the location of volcanic blocks, Crandell determined that the debris avalanche traveled 49 kilometers from its source and covered 675 square kilometers of the Shasta Valley. As estimated from records of water wells drilled in the valley, the debris avalanche filled the valley to a depth of 50 meters. The estimated volume of 45 cubic kilometers makes it the largest volcanic debris avalanche measured by geologists.

The age of the debris avalanche can be bracketed fairly closely. Some of the blocks within the deposit contain the 400,000-year-old Rockland Ash, indicating that the debris avalanche postdated that eruption. Also contained within the deposits are blocks of andesite about 360,000 to 380,000 years old. Overlying the debris avalanche material at some sites is a lava flow dated at about 300,000 years old. Using the principles of relative dating outlined in Chapter 3, these dates bracket the age of the debris avalanche between 300,000 and 380,000 years ago. Considering that modern Mt. Shasta's four cones are all younger, the debris avalanche must have come from the sides of an ancestral cone.

A

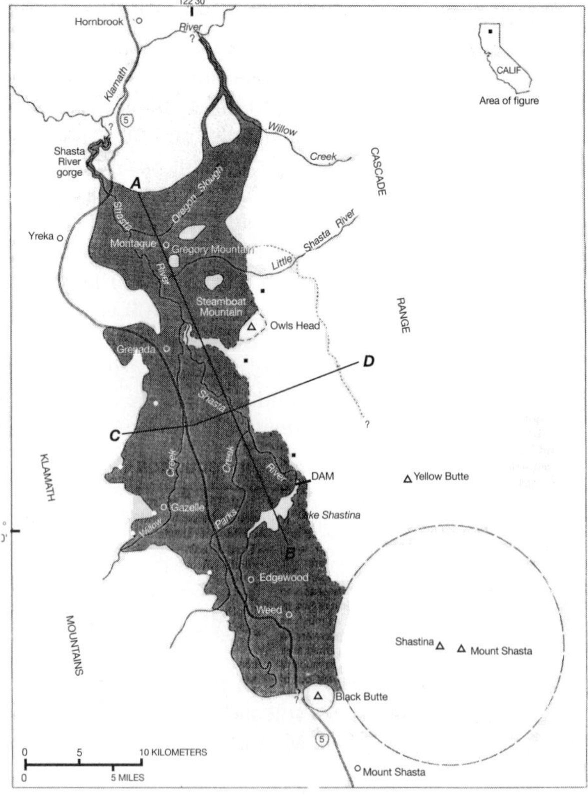

B

Fig. 5-20 **A,** Photo and **B,** map of the remains of a 350,000-year-old debris avalanche from Mt. Shasta. (Source: **A,** Hotz, P. U.S. Geological Survey; **B,** Crandell, D.R.. 1989. U.S.Geological Survey, Bulletin 1861, p. 2.)

Volcanic activity in the Medicine Lake area has been varied, and it is an excellent place to view the results of basaltic to silicic volcanic eruptions. Many of the volcanic features formed within the past few thousand years, and the youngest flows took place no more than a few hundred years ago. During the summer of 1989, a flurry of tens of thousands of earthquakes indicated that a new eruption might occur, but none materialized and the seismic activity ceased.

The earliest volcanic activity at Medicine Lake, about 700,000 years ago, was the eruption of basalt that can be seen today at the base of the Highlands. The eruptions were similar to those that formed other parts of the Modoc Plateau, including Lava Beds National Monument on its northern flank (Fig. 5-21). Cinder cones, explosion pits, and spatter cones are found near the vents. Higher on the slopes of the volcano are younger and more silica-rich volcanic rocks. These include Glass Mountain and Little Glass Mountain, produced by rhyolite lava flows erupted from a number of vents. Some of these eruptions occurred only 200 to 300 years ago. Associated with the flows are domes of obsidian, pumice, and rhyolite. The obsidian of Glass Mountain and Little Glass Mountain was the source for tools and trade materials used by the Modoc Indians. On the north side of the volcano is the Callahan andesite flow, dated at about only 1110 years old. Medicine Lake itself sits within a crater at the summit of the Highlands. The entire shield volcano complex, which appears gentle and nonthreatening, actually has a greater volume than towering Mt. Shasta.

To the north of Medicine Lake is the bare, flat, but rough-surfaced country of Lava Beds National Monument. This is a land of dark basalt flows, sparsely vegetated because of the desert climate. Only ranches and the town of Tulelake break up the open spaces of this part of California. At the northeastern end of Lava Beds is Tule Lake itself, a large lake that straddles the California-Oregon border.

Lava Beds National Monument is famous for its network of tunnels and caves (Fig. 5-22). The tunnels are ***lava tubes,*** some of which carried mafic lava 25 to 35 kilometers from vents on the north side of Medicine Lake. As fluid lava streams along a channel, a solid crust or roof may form on the surface and sides of these channels, because the outer surfaces cool more rapidly. Lava may continue to flow within the channels, just as water can flow beneath an ice-covered stream. The lava may drain out to leave an open lava tube cave, or the tunnel may be partly or completely filled by later lava flows. At Lava Beds National Monument, parts of the roofs of many of the tubes have collapsed, providing access to the tunnels (Fig. 5-23).

The last major war between the United States government and Native American tribes was fought in the Modoc Plateau in 1872-1873. Led by Captain Jack, a band of about 150 members of the Modoc tribe managed to elude as many as 650 Army troops for 7 months by taking advantage of the rough volcanic terrain of the Modoc Plateau. The blocky lava, lava tubes, collapse pits, and low areas behind lava flows provided shelter and escape routes for the outnumbered Modocs. They were captured only after being betrayed by fellow Modocs.

Mt. Lassen

It was 5 PM on Memorial Day, May 30, 1914, and Ike McKenzie of Chester happened to be gazing toward Lassen Peak. To his astonishment a huge cloud of steam rose from the center of the mountain, signaling the beginning of the first volcanic eruption in the United States since the 48 states had been united (Fig. 5-24). Few people lived in the area around Mt. Lassen in 1914, but many local residents made

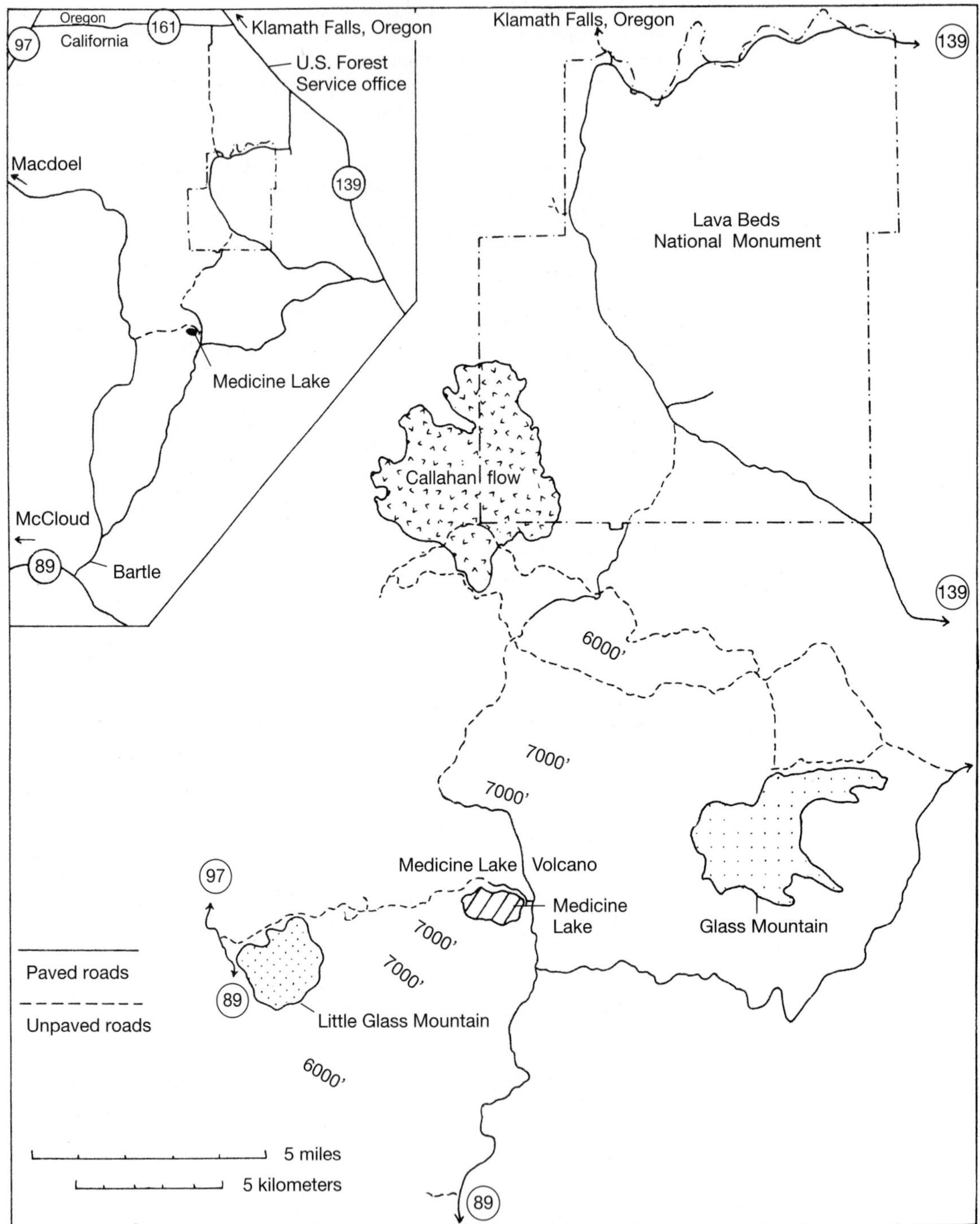

Fig. 5-21 Map showing recent volcanic flows and domes at Medicine Lake volcano. (Source: Donnelly-Nolan, J. 1987. *Geological Society of America Centennial Field Trip Guide No. 66,* p. 290.)

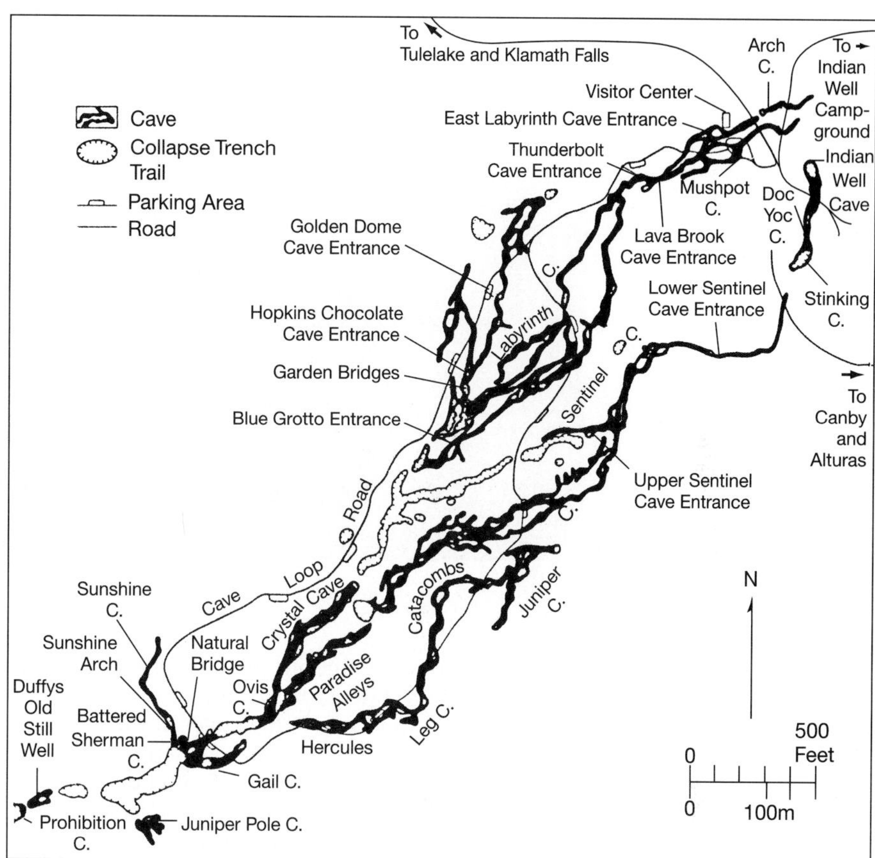

Fig. 5-22 Caves and lava tubes at Lava Beds National Monument. (C = Cave.) (Source: Waters, A., and others. 1990. U.S. Geological Survey, Bulletin 1673.)

Fig. 5-23 Entrance to lava tube, Lava Beds National Monument. (Source: †Skapinsky, S. San José State University.)

Fig. 5-24 Ash eruption from Mt. Lassen, photographed by B.F. Loomis in 1914. (Source: National Park Service, Loomis Museum.)

careful notes of the eruptions that took place between 1914 and 1921. During the first year, 170 eruptions of steam and ash were recorded. Many local residents made the climb to look into the crater of the erupting volcano, where most of the fragments ejected were lapilli and small rocks. This foolish behavior accounted for the only injury during the 7 years of eruption. Lance Graham and two friends were standing on the crater when the volcano blew out a shower of ash and rock. One rock hit Graham, knocking him unconscious and breaking his collarbone.

In the Spring of 1915, Lassen picked up its activity. "Fire," actually glowing lava, could be seen coming from the crater on the night of May 19; the next day, viewers could see the dark mass of a new lava flow extending 2000 feet down the west side of the peak. On the eastern slope, a massive debris flow swept down Lost Creek, picking up enough speed to jump over a divide into Hat Creek. The flow washed away buildings and timber and carried 20-foot boulders along its path. Barking dogs had alerted residents to the oncoming hazard, allowing them to escape. On May 22, a glowing cloud leveled and scorched trees in what is now known as the Devastated Area (Fig. 5-25). Today in and near Lassen National Park, hot springs and ***fumaroles,*** areas where volcanic gas and steam boil from vents, remind visitors of the recency of volcanic activity at Lassen Peak (Fig. 5-26).

Until the eruption of Mt. St. Helens in 1980, the Lassen eruptions of 1914 through 1921 were the only eruptions of the twentieth century in the lower 48 states. By studying the deposits of past eruptions from Lassen, as well as comparing them with eruptions from other stratovolcanoes, geologists know that the eruptions of 1914-1921 were minor events: the geology of the Lassen area reveals a history of repeated catastrophic eruptions during the past few million years.

Lassen Peak itself is a plug dome of dacite that rose on the eastern flank of a huge hole created by a very large blast about 200,000 years ago. That eruption, centered west of today's Lassen Peak, almost totally destroyed a large stratovolcano, which

Fig. 5-25 The Devastated Area, Lassen National Park, 1915. (Source: Model Studio, Anderson. National Park Service, Loomis Museum.)

Fig. 5-26 Active fumaroles, Bumpass Hell area, Lassen National Park. (Source: Sedlock, R. San José State University.)

geologists have named Mt. Tehama (Fig. 5-27). Brokeoff Mountain, appropriately named, marks one edge of the hole. An earlier explosion took place in the same general area 400,000 years ago. This event sent ash, named the Rockland Ash by geologists, downwind for at least 800 kilometers, covering much of northern California and western Nevada. Near the source, the eruption produced pyroclastic breccia containing pumice fragments as well as fine ash. Here the pyroclastic flows and ash were up to 12 meters thick. As far away as Humboldt County 23 centimeters of ash fell, and ash fell in San Francisco and in the Santa Clara Valley. The eruption was at least as large as, and perhaps even greater than, the eruption of Mt. Mazama, which produced Crater Lake about 7000 years ago. By comparison, the 1980 eruption of Mt. St. Helens produced about 1/50 to 1/70 the volume of ash and pumice (Fig. 5-28).

Projected former height of Brokeoff Mountain (Mt. Tehama)
1915 Lava
LASSEN PEAK

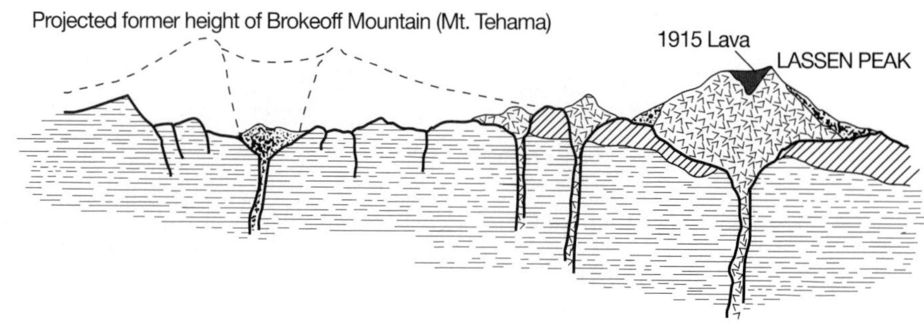

Fig. 5-27 Cross section of Lassen Peak and the remains of Mt. Tehama, which was destroyed by a major eruption about 400,000 years ago. (Source: Harris, S. 1988. *Fire Mountains of the West.* Missoula, Mont.: Mountain Press.)

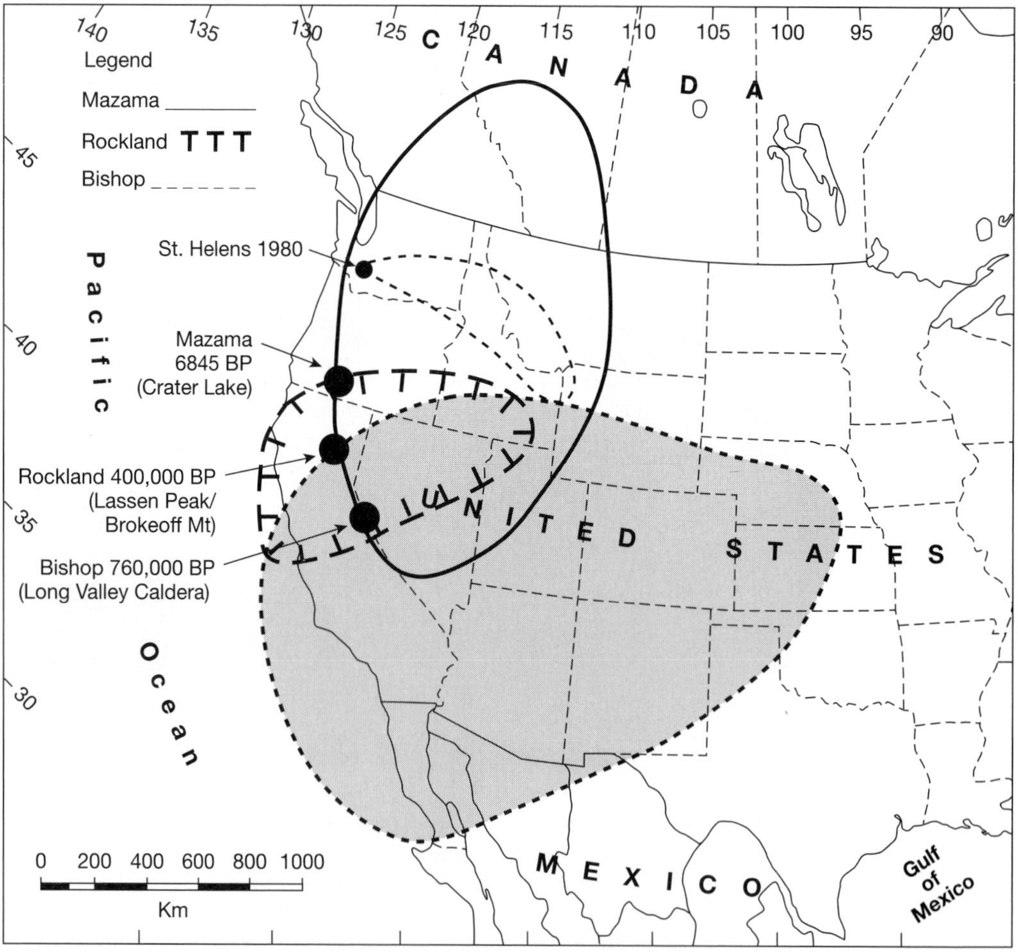

140
135
130
125
120
115
110
105
100
95
90
45
40
35
30
Legend
Mazama
Rockland TTT
Bishop
C A N A D A
St. Helens 1980
Mazama
6845 BP
(Crater Lake)
Rockland 400,000 BP
(Lassen Peak/
Brokeoff Mt)
Bishop 760,000 BP
(Long Valley Caldera)
Pacific
Ocean
UNITED STATES
MEXICO
Gulf of Mexico
0 200 400 600 800 1000
Km

Fig. 5-28 Comparative distribution of some volcanic ashes erupted in the western United States. (BP = years before present.) (Source: Sarna-Wojcicki, A. U.S. Geological Survey.)

Fig. 5-29 Sediments of the Tuscan Formation, interlayered with the Nomlaki Tuff, which was erupted from the Lassen area about 3.4 million years ago. (Source: Harwood, D. U.S. Geological Survey.)

The Mt. Lassen area has been the source of at least one other major explosion in the recent geologic past. Approximately 3.4 million years ago, an even earlier stratovolcano experienced a catastrophic eruption that sent ash flows down the Sacramento Valley for at least 130 kilometers. The rhyolite flows, which contain large pieces of pumice carried by the ancient Sacramento River, reached as far south as what is now the town of Willows. Today they are found as a volcanic tuff, named the Nomlaki Tuff, which is up to 90 meters thick at some localities east of Redding. Along the east side of the Sacramento Valley, the Nomlaki Tuff is found interlayered with volcanic flows and alluvial deposits of the Tuscan Formation (Fig. 5-29). On the western side of the Sacramento Valley, the Nomlaki Tuff is found within alluvial deposits of the Tehama Formation. Much further south, pumice and ash from the eruption are found in marine sedimentary deposits of the San Joaquin Formation in the Kettleman Hills near Coalinga (see Chapter 12).

The Long Valley Caldera

The Long Valley area is located on the western edge of the Basin and Range Province along the base of the Sierra Nevada Range (Fig. 5-32, *A* and *B;* p. 82). Volcanic activity began there about 3.6 million years ago, when mafic and andesitic lavas were erupted over a fairly wide area. These flows were followed by the eruption of more silicic domes and flows near the northern edge of the present caldera. About 2 million years ago, a series of high-silica eruptions began near Mono Glass Mountain, resulting in the formation of a series of domes and flows surrounded by ash and ash flows. The change in the lava chemistry occurred as a large, fairly shallow magma chamber developed beneath Long Valley.

About 760,000 years ago, the roof of the Long Valley magma chamber ruptured, causing a catastrophic eruption of 600 cubic kilometers of rhyolite magma.

VOLCANIC ASHES RECORD A MOMENT IN GEOLOGIC TIME

As we have seen, the chemistry of lavas and the composition of the volcanic rocks formed from them are highly variable. Although all volcanic ashes erupted from the Cascades volcanic chain show the general characteristics of subduction-zone volcanoes —they typically have intermediate amounts of silica and mafic materials—not all Cascades lavas are exactly the same. Each eruptive source has a distinctive chemistry, and ashes erupted from each source show slight variations in their composition. For example, volcanic ashes erupted from Mt. St. Helens have a different composition from those erupted from Long Valley (Fig. 5-30), and even though all volcanic ash erupted from Mt. St. Helens is generally similar, each individual eruption is slightly different. When an eruption occurs, the composition of the lava remaining in the magma chamber changes slightly. Over time, the magma chemistry may also change because new molten material is produced.

Because it is instantaneously cooled lava, volcanic glass gives geologists a particularly accurate look at the chemistry of the lava at the time of an eruption (Fig. 5-31). Crystals of silicate minerals formed during the slow cooling of the magma before a volcanic eruption may also be present in the ash. Because they may have formed long before the eruption, these crystals are a less instantaneous record of the lava chemistry. The volcanic glass in each ash has a unique chemical "fingerprint" and can be identified even when the source volcano has long since been destroyed by eruption or erosion.

Geologists analyze the chemistry of volcanic ash using a variety of techniques. Using a microscope, geologists can observe the shapes of the glass fragments and their vesicles and the types of minerals within the glass. An electron microprobe can be used to measure the abundance of some of the chemical elements in the glass. For very detailed comparisons, neutron activation analysis, which involves use of a nuclear reactor, can be used to measure rare elements present in concentrations of only a few parts per billion.

What can be learned from "fingerprinted" volcanic ashes? Geologists can trace the source of each chemically distinct ash by comparing its thickness at many places and pinpointing the area in which it is thickest. By mapping out its general distribution and thickness, geologists can also reconstruct the size of an ancient eruption. The size estimates are only a minimum, because areas distant from the source would receive only a fine covering of ash that probably would not be preserved in the geologic record. Frequently, the volcanic rocks near the source can be dated using radioactive isotopes (see Chapter 3). For each source area, then, geologists can estimate the number and size of eruptions over a time period of thousands of years. Volcanoes with a history of frequent, major eruptions, like Mt. St. Helens, can be identified. The characteristic frequency of major eruptions, or the ***recurrence interval,*** can be estimated from each active volcano. This information allows a much more accurate assessment of volcanic hazards that the picture obtained by the historic record of eruptions over the past two centuries.

Another important use of volcanic ash is for dating and correlating the sediments interlayered with the ash. Many sedimentary deposits lack fossils or materials that can be dated using radiometric methods. Because an ash layer represents a moment in geologic time, it can provide an excellent ***marker bed*** that can be used to correlate sediments from different areas. For example, the 400,000-year-old Rockland Ash discussed above is found in marine sediments off Cape Mendocino, in alluvial deposits in the Santa Clara Valley, and in lake and alluvial fan deposits in eastern California and western Nevada.

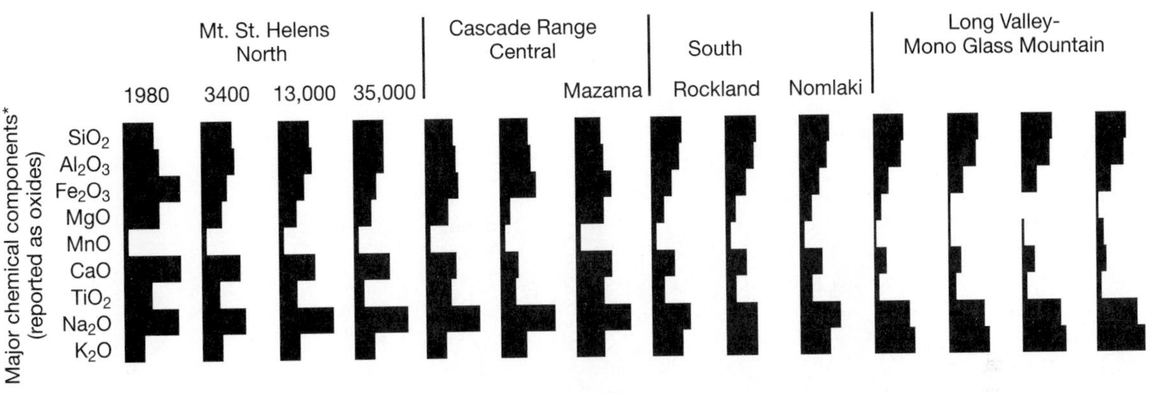

Fig. 5-30 Chemical comparison of volcanic ash erupted from various volcanoes. The length of the bars for each component shows its relative abundance. Note the general similarity among glasses from the same source area. (Source: Sarna-Wojcicki, A. U.S. Geological Survey.)

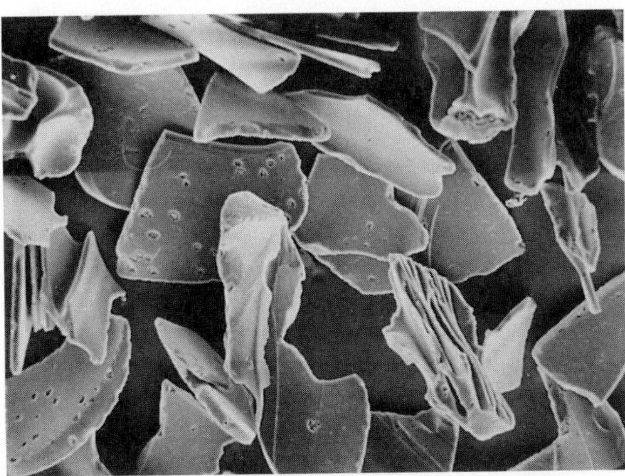

Fig. 5-31 Scanning electron micrograph of volcanic glass. (Source: Sarna-Wojcicki, A. U.S. Geological Survey.)

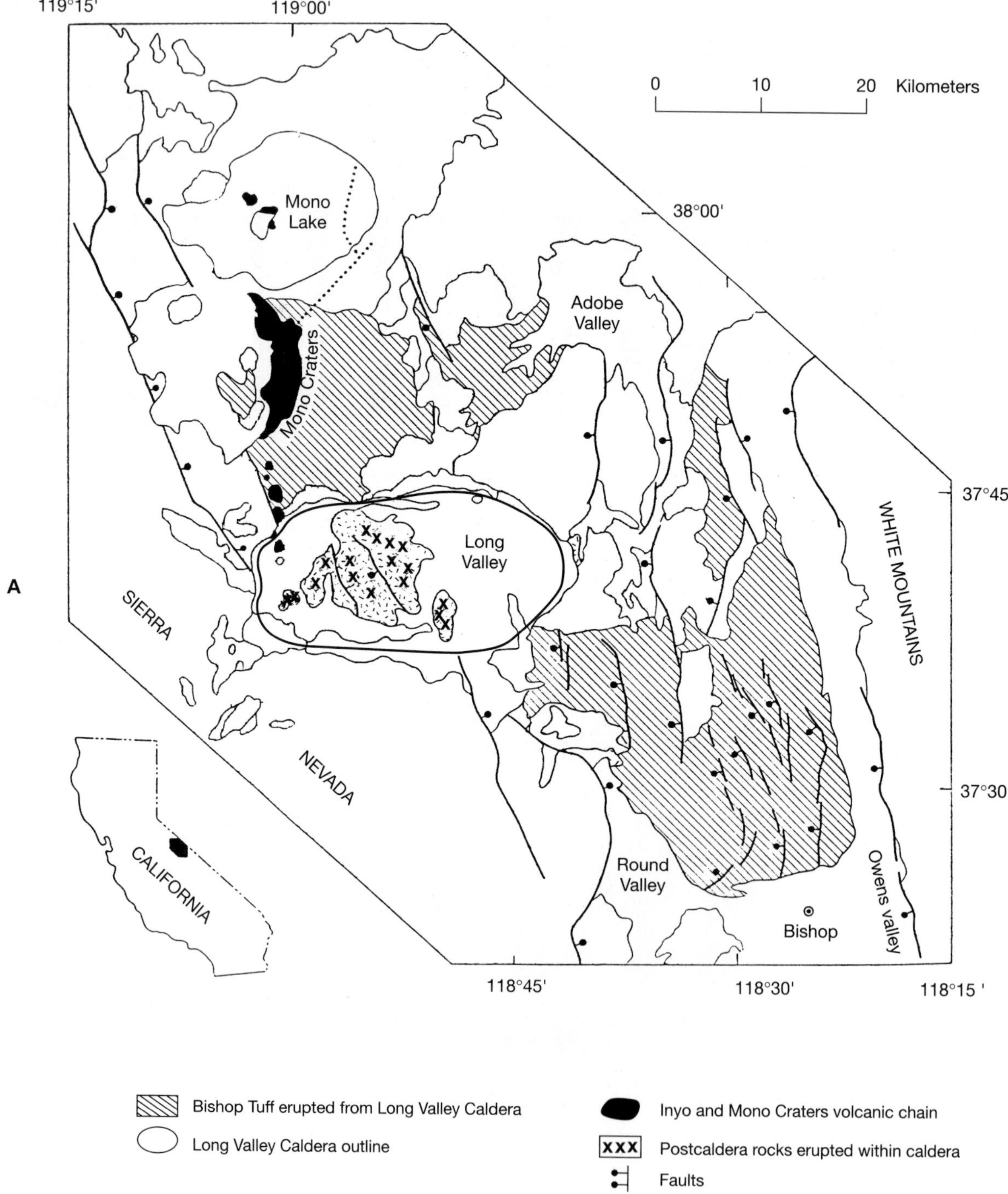

Fig. 5-32 A, Map showing the Long Valley Caldera and other volcanic centers on the east side of the Sierra Nevada.

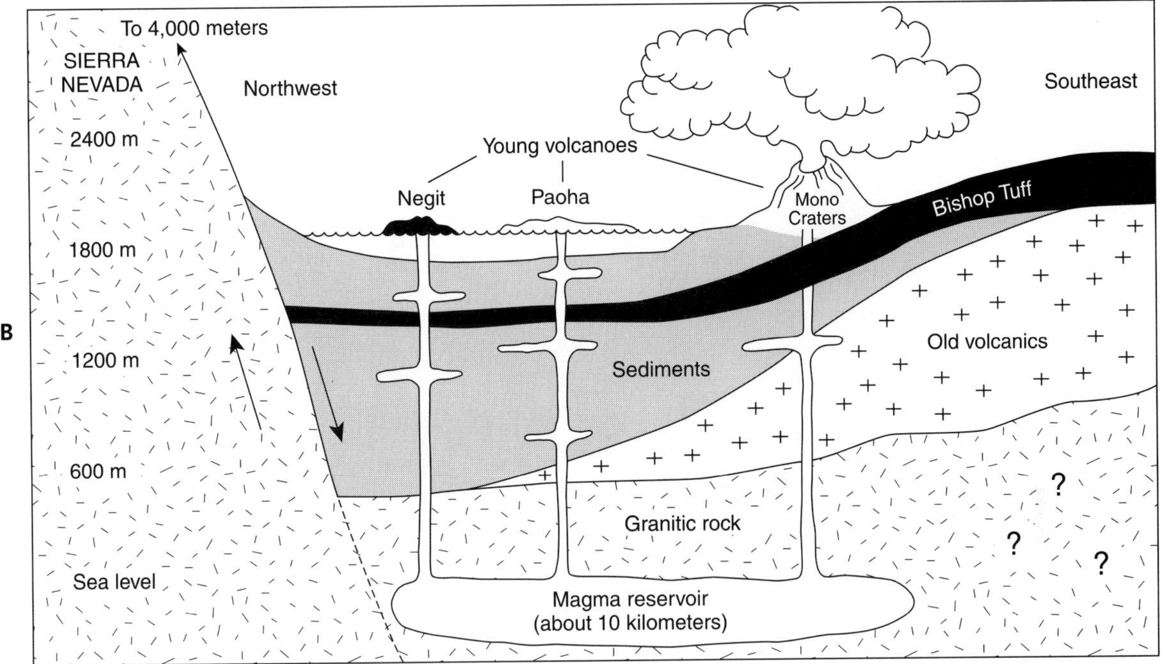

Fig. 5-32—cont'd **B,** Cross section beneath the Mono Lake area. (Source: **A,** Bailey, R. 1976. *Journal of Geophysical Research;* **B,** Shearin, R. 1989. *Mono Lake Guidebook.* Lee Vining, Calif.: Kutsavi Books.)

Most of the lava formed hot, glowing ash flows that covered 1500 square kilometers of the surrounding area to a depth of up to 200 meters. Ash was also blown high into the atmosphere and carried long distances by the prevailing winds. Today airfall ash from this explosion is found in sediments as far east as Missouri and off the coast of California (see Fig. 5-28). The deposits from the eruption are known as the Bishop Tuff, named for the town of Bishop in Inyo County (Fig. 5-33). The sudden emptying of the magma chamber caused the ground above it to collapse, forming an oval depression 2 to 3 kilometers deep known as the Long Valley Caldera, which measures about 17 by 32 kilometers. The resort area of Mammoth Lakes lies within its rim, and U.S. Highway 395 crosses through the caldera.

After the roof of the magma chamber collapsed, eruptions continued within the caldera. Pyroclastic flows and obsidian flows erupted onto the caldera floor. A resurgent dome then formed in the center of the caldera floor, accompanied by obsidian flows and rhyolite lava flows. Following a relatively quiet period, eruptions began again about 500,000 years ago and continued at roughly 200,000 year intervals. No eruptions have occurred in the past 100,000 years, but magma appears to remain beneath the central dome at fairly shallow depths. The presence of this magma 5 to 15 kilometers beneath the surface is suggested by seismic surveys and by unusual earthquake activity. Between 1980 and 1986, swarms of thousands of earthquakes occurred in the area, alarming residents and geologists alike. The activity appears to have been the result of magma move-

Fig. 5-33 Air fall ash (lower cliffs) and ash flow (upper cliffs) units of the Bishop Tuff, north of Bishop. (Source: Sarna-Wojcicki, A. U.S. Geological Survey.)

ment within the chamber. The activity has decreased since 1986, but the area continues to show signs of unrest. The resurgent dome in the center of the caldera is rising at a rate of 2 to 3 centimeters per year and has risen a total of 60 centimeters since the current activity began. In 1995, carbon dioxide gas leaked from the southwestern part of the caldera, killing large stands of trees. Numerous hot springs in the Long Valley Caldera also suggest that magma is present beneath the surface.

Mono-Inyo Volcanic Chain

Another center of recent volcanic activity on the east side of the Sierra Nevada near the Long Valley Caldera is the Mono-Inyo volcanic chain. The Mono-Inyo chain is a system of vents 45 kilometers long that overlaps with the Long Valley Caldera at the southern end and extends north to Mono Lake (Chapter 6). The chain of volcanic vents is relatively linear, and the volcanic activity may be controlled by a fault in the underlying rocks. Volcanic activity at the Mono-Inyo chain is more recent than most of the Long Valley eruptions.

The Mono Craters chain of at least 30 rhyolite and obsidian domes, flows, and craters began to form only 40,000 years ago. The highly viscous lava from these vents flowed very slowly and maintained very steep front edges; these obsidian flows are striking features of the Mono-Inyo chain (Fig. 5-34). The most recent eruption at Mono Craters took place about 650 years ago, between 1325 and 1365 AD. The Inyo Craters chain, also obsidian and rhyolite domes, flows, and craters, extends south of the Mono Craters chain. The youngest eruptions from the Inyo Craters chain occurred as recently as 520 years ago in about 1472 AD. Within Mono Lake, several cinder cones and rhyolite flows form Negit Island and part of Paoha Island. The most recent explosions from these centers may have taken place as recently as 150 years ago.

Fig. 5-34 View domes and craters of the Mono Craters chain; Sierra Nevada in the distance. (Source: Miller, C.D. U.S. Geological Survey.)

FUTURE ERUPTIONS? VOLCANIC HAZARDS IN CALIFORNIA

In 1980, the newspapers in California came alive with news of earthquake swarms near the resort town of Mammoth on the eastern side of the Sierra Nevada. The thousands of small to moderate earthquakes were occurring at fairly regular intervals, a pattern known as ***harmonic earthquakes.*** Believing the earthquakes to be a signal that magma was rising beneath the Long Valley Caldera, seismologists at the U.S. Geological Survey issued an alert for the area. Leveling instruments detected a bulging of the earth's surface on the floor of the caldera, giving strength to the idea that magma was on the move. However, the flurry of activity at Long Valley calmed down in 1986, causing geologists to conclude that magma movement has ceased for the time being. Since 1989, renewed seismicity and ground movement (Fig. 5-35) have been closely watched by geologists and seismologists.

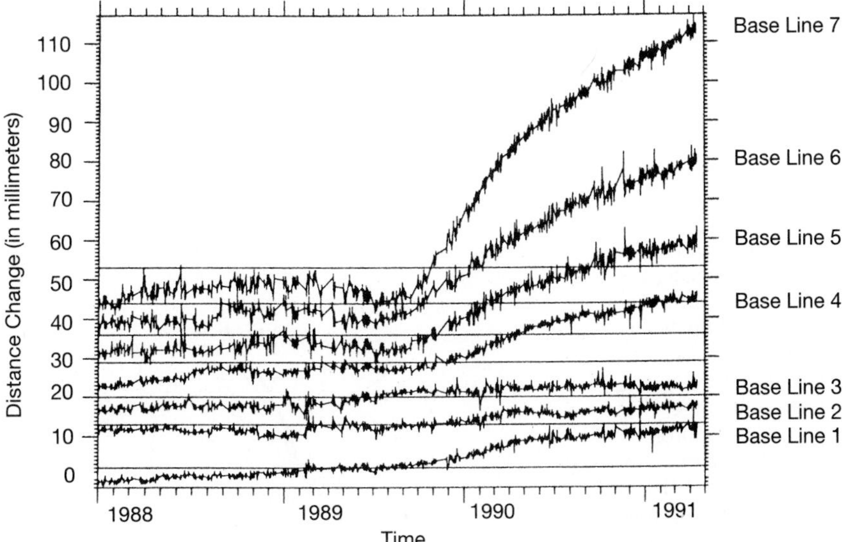

Fig. 5-35 Repeated laser surveys show changes in the ground surface at Long Valley; the changes may reflect upward movement of magma beneath the caldera. (Source: U.S. Geological Survey.)

At Long Valley, Mt. Shasta, Lassen Peak, and Medicine Lake, networks of seismographs and ground-leveling stations are monitored as part of the U.S. Geological Survey's volcanic hazards program. The purpose of the monitoring programs is to detect precursors of impending volcanic eruptions, enabling government agencies to plan for evacuations if the signals indicate that eruption is imminent. However, because of the unpredictable nature of eruptions from the Cascade volcanoes and the Long Valley area, precise warnings of eruptions may not be possible.

After detailed studies of each of the known active volcanoes in California, geologists are able to piece together a record of the characteristic eruptive history for each volcanic center. By examining and dating the volcanic deposits formed around each volcano, geologists predict the type, size, and frequency of likely future events. Based on the geologic information, ***volcanic hazards maps*** can be prepared for each area. These maps outline the areas likely to be affected by future volcanic events (Figs. 5-36 and 5-37). The maps can then be used by government agencies as they debate the advisability of future developments in the vicinity of known active volcanoes. In addition, evacuation routes and special high-risk activities and structures can be pinpointed (Fig. 5-38).

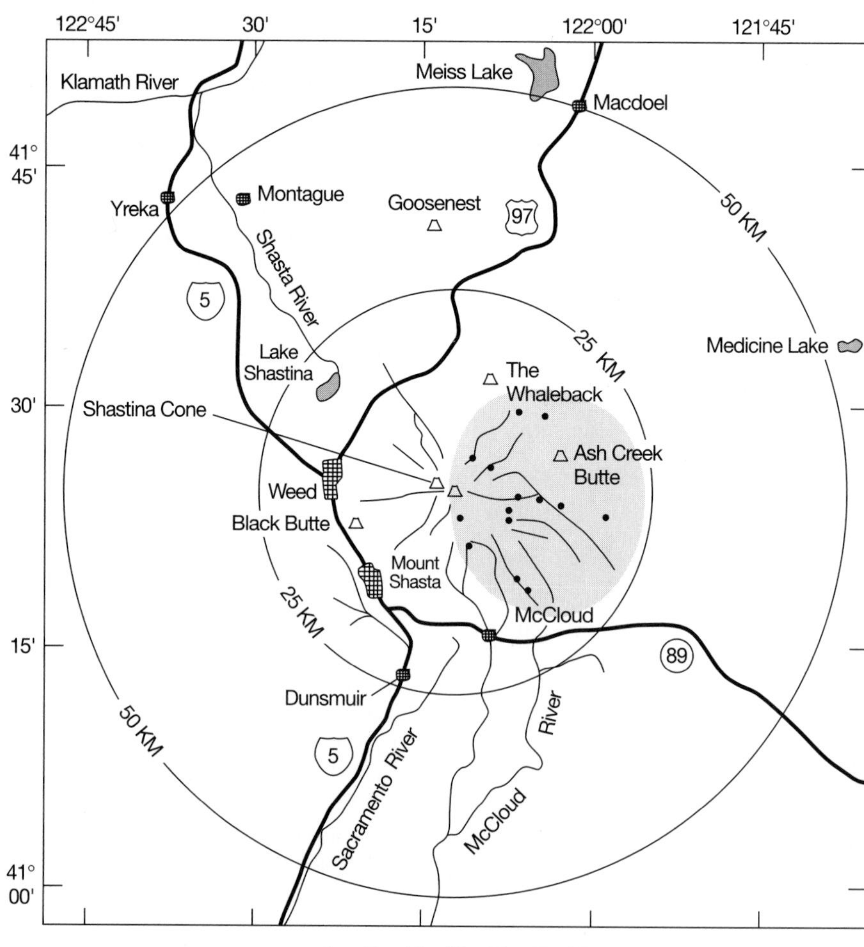

Fig. 5-36 Volcanic hazards map for Mt. Shasta, showing the areas likely to be affected by eruptions of volcanic ash. The outline of the hazardous area was based on the distribution of the 9600-year-old Red Banks tephra, found today at localities within the shaded circle. (Source: Crandell, D.R., and Nichols, D.R. 1989. U.S. Geological Survey, public information pamphlet.)

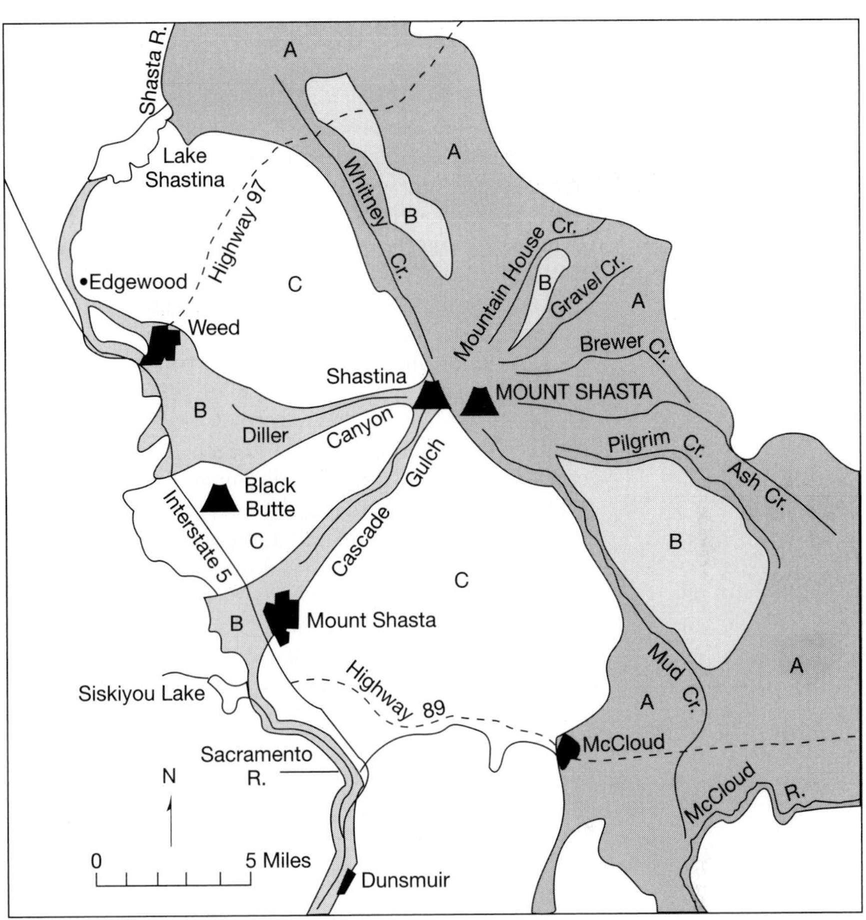

Fig. 5-37 Volcanic hazards map for Mt. Shasta, showing the areas likely to be affected by volcanic debris flows. Zones designated by letters show relatative likelihood of being affected by debris flows: zone A is most likely and zone C is least likely to be affected. No debris flow hazard exists on high areas within or beyond the zones. The hazard decreases everywhere within the zones with greater height above stream channels and greater distance from Mt. Shasta. (Cr. = Creek; R. = River.) (Source: Crandell, D.R., and Nichols, D.R. 1989. U.S. Geological Survey, public information pamphlet.)

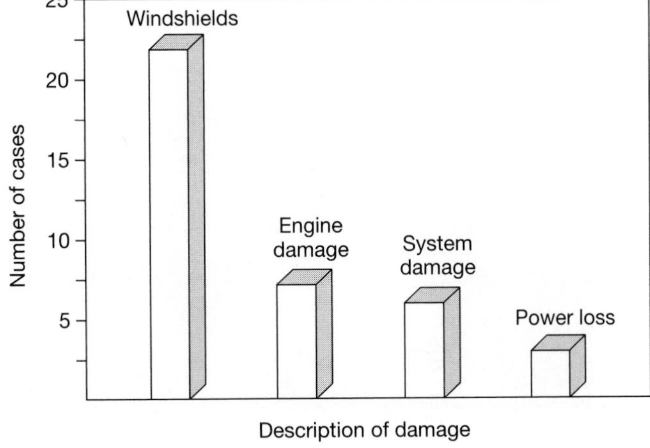

Fig. 5-38 Reported encounters of aircraft with erupting volcanic ash. (Source: Wright, T.L. and Pierson, T.C. 1992. *Living With Volcanoes.* U.S. Geological Survey Circular 1073, p. 27.)

COMMERCIAL USES OF VOLCANIC ROCKS

Anyone who has purchased decorative red rock at the local building supply center to use in a planter box or the yard has contributed to California's volcanic rock industry. Late Cenozoic volcanic eruptions have supplied California with abundant cinders, pumice, ash, and volcanic glass that are used in a variety of industries. These materials have been mined at quarries in the Cascade and Modoc Plateau Provinces, as well in the northern Coast Ranges and in southeastern California. In 1990, \$5 million worth of pumice alone was mined at quarries in the Cascade and Modoc Plateau Provinces, as well as from cinder cones in southeastern California.

Volcanic cinders are reddened fragments (see Fig. 5-12) of basalt thrown from a volcanic vent and often piled up into cinder cones. They are used as base material for roads and railroad beds, as decorative rocks, and to make lightweight concrete blocks. Pumice is used as an aggregate in lightweight concrete and in acoustical insulating materials. It has long been known to be an effective abrasive for cleaning everything from oven surfaces to skin (Fig. 5-39). Pumice can be added to soil as a conditioner and is used in some absorbent materials. Volcanic ash is used as a base material in agricultural pesticides and insecticides as well as in some ceramics.

Throughout California's early history, volcanic glass was a valuable commodity. Discoveries of obsidian tools found at archaeologic sites hundreds of kilometers from their volcanic sources prove that Native Americans traded obsidian across much of the western United States. Even today, some surgeons prefer obsidian blades to steel scalpels because of obsidian's sharpness. Hydrated volcanic glass known as ***perlite*** is used in aggregates for plaster and fire-resistant wallboard. It is also used in insulating materials.

Fig. 5-39 Products made from pumice.

FURTHER READINGS

BAILEY, R.A. 1989. *Geologic Map of the Long Valley Caldera, Mono-Inyo Craters Volcanic Chain and Vicinity, Eastern California.* U.S. Geological Survey, Map I-1933, scale 1:62,500.

HARRIS, S.L. 1988. *Fire Mountains of the West: The Cascade and Mono Lake volcanoes.* Missoula, Mont.: Mountain Press, 379 pp.

HILL, M.R. 1970. Mount Lassen is in eruption and there is no mistake about that. *California Geology,* November 1970, pp. 211-227.

MILLER, C.D. 1980. *Potential Hazards from Future Eruptions in the Vicinity of Mount Shasta Volcano, Northern California.* U.S. Geological Survey Bulletin 1503.

WATERS, A.C. 1992. Captain Jack's stronghold: The geologic events that created a natural fortress. *California Geology,* pp. 135-144.

WATERS, A.C., DONNELLY-NOLAN, J.M., AND ROGERS, B.W. 1990. *Selected Caves and Lava-tube Systems in and near Lava Beds National Monument, California.* U.S. Geological Survey Bulletin 1673, 102 pp.

WRIGHT, T.L., AND PIERSON, T.C. 1992. *Living with Volcanoes.* U.S. Geological Survey, Circular 1073, 57 pp.

CHAPTER

6

California's Deserts

Climate, Changing Environments, and Resources

THE DESERT ENVIRONMENT

Covering more than 25 million acres, southeastern California's deserts encompass as much land as the entire state of Ohio. Together, the deserts of the southern Basin and Range Province, the Mojave Desert, and the Colorado Desert account for about one fifth of California's landscape (Fig. 6-1). Almost 20 million people live within a few hours' drive of the California desert, but the population there remains relatively low because of the lack of available water. However, Californians have made extensive use of the desert, removing billions of dollars' worth of mineral wealth since the late 1800s. The clear skies, open landscape, and low population have made the desert ideal for military training facilities since World War II. Perhaps the open spaces of the desert will find their most important role as the growing urban population looks to these areas for recreation.

Technically, a desert is an area where average evaporation is high relative to precipitation, with the result that vegetation is scarce and the population is low. California's deserts are the result of topography. The winter storms that deliver most of California's water originate in the Pacific Ocean near Alaska and approach the coast from the west. Moist air masses generally travel eastward across the state, cooling as they rise over the Coast Ranges, the Sierra Nevada, the San Bernardino Mountains, and the ranges to the east. Because the cooler air cannot hold as much moisture, condensation occurs on the western slopes of the mountains, relieving the air masses of much of their moisture. This pattern, which results from the topography of the ranges, is known as an ***orographic effect,*** and it creates a ***rain shadow*** on the eastern side (Fig. 6-2). The rain shadow is most pronounced on the eastern side of the Sierra Nevada, because the high range is a major barrier. However, a less pronounced rain shadow exists in the western Great Valley east of the Coast Ranges.

The high mountains of the Sierra Nevada and, further south, the San Gabriel and San Bernardino Mountains are so effective in blocking moist Pacific air that

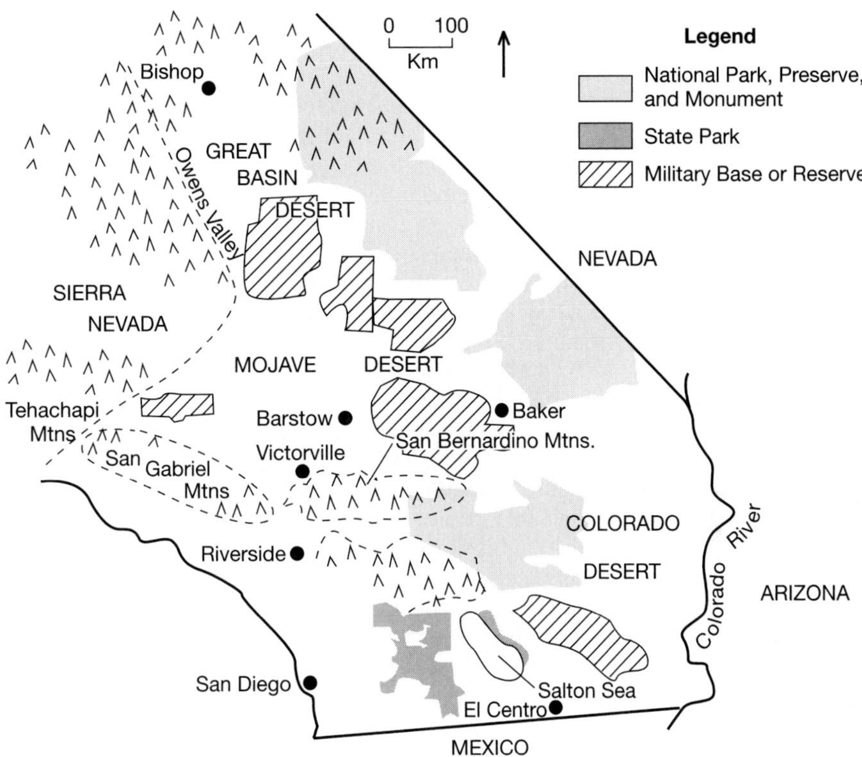

Fig. 6-1 Map showing California's deserts, including parks, monuments, and other areas included in the 1994 California Desert Protection Act.

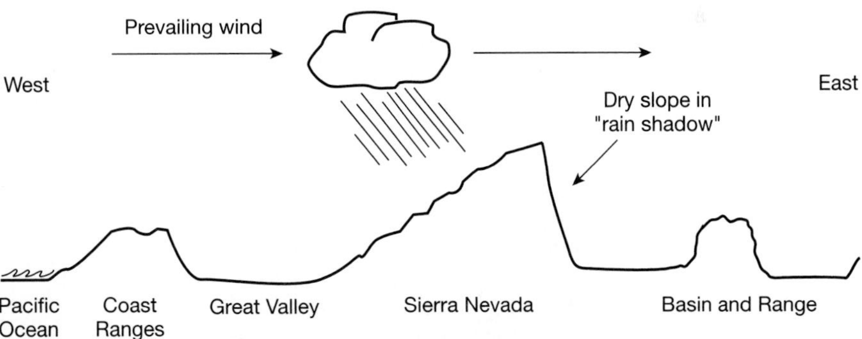

Fig. 6-2 The *orographic effect* of a mountain range on precipitation. In most of California, moist air masses travel eastward from the Pacific Ocean, rising first over the Coast Ranges and then over the higher Sierra Nevada, where the strongest orographic effect, or rain shadow, is created.

they have created the desert environment of southeastern California. In the Basin and Range Province, east-traveling air masses must cross three additional ranges to reach Death Valley, by which time they are usually stripped of almost all moisture (Endpaper 4). The average annual precipitation on Death Valley's floor is about 3 centimeters, but some years pass with no precipitation. The town of Bagdad in the Mojave Desert experienced 767 consecutive days from 1909 through 1912 without a single drop of precipitation. Summer temperatures are fierce throughout California's deserts, particularly in the low-lying areas. The hottest temperature ever recorded in the Western Hemisphere was 134° F (57° C) at Furnace Creek in Death Valley on July 10, 1913.

One of the distinguishing features of a desert is the scarcity of vegetation. The plant species that can survive are well adapted to the dry environment. Many spectacular desert plants—the Joshua tree, the palo verde, the smoke tree, and various cacti—are found nowhere else in California (Fig. 6-3). Near springs, streams, and lakes and on the higher slopes of some of the ranges, the ground surface is vegetated. However, throughout most of the California desert, the rock and soil at the surface are directly exposed to the forces of weathering and erosion.

Chemical weathering of rocks proceeds slowly in the desert because of the lack of water and vegetation. Because decayed plant material is generally absent, organic acids, which are highly effective chemical weathering agents, are also absent. Without sufficient water to dissolve them, soluble minerals such as salt (halite), gypsum, and calcite persist at the ground surface and in desert soils. One major effect of weathering is the formation of ***desert varnish*** (also called rock varnish), a black iron-oxide and manganese-oxide coating on the rocks. Over tens of thousands of years, the varnish slowly turns the rock surfaces dark (Fig. 6-4). Early desert dwellers chipped through the varnish coatings to carve intricate ***petroglyphs,*** rock drawings that remain visible today in several places in the California desert (Fig. 6-5).

Fig. 6-3 Joshua trees in the Mojave Desert. (Source: Nakata, J. U.S. Geological Survey.)

Fig. 6-4 Desert varnish and desert pavement have formed on the surface of an alluvial fan near Las Vegas, Nevada.

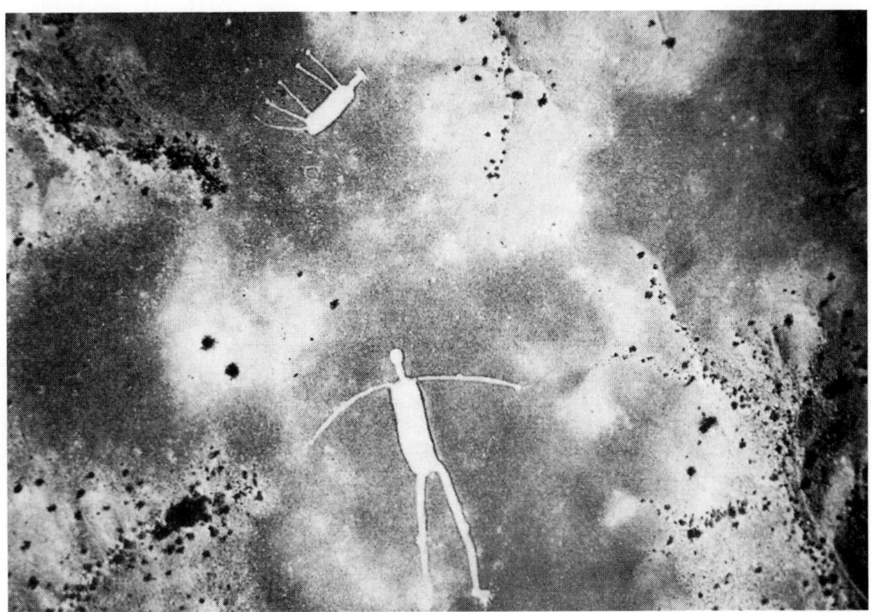

Fig. 6-5 Native American intaglios near Blythe, Riverside County. (Source: Wilshire, H. U.S. Geological Survey.)

WIND AND SAND DUNES

Wind is an important geologic agent in the desert regions of California. Strong winds can be generated by local temperature variations between the mountains and basin floors of the desert, as well as by more regional storms. Enormous dust storms occur periodically in the Mojave Desert when the Santa Ana winds blow from the desert toward the coast. During these storms, sand particles can be carried as high as 2 meters above the ground surface, sandblasting the windshields of travelers unlucky enough to be caught in the storm. Finer dust is carried higher and can be transported across the Los Angeles Basin as far as the Channel Islands west of Santa Barbara (Fig. 6-6).

The evidence of wind erosion is abundant in the California desert. One of the features produced by the removal of fine sediment by wind (a process called ***deflation)*** is ***desert pavement.*** Removal and downward settling of fine particles leaves behind the larger rock fragments. The larger rocks then become coated with desert varnish, and the ground surface becomes armored with the desert pavement. The surface is then protected from further wind erosion, unless the pavement is disturbed by erosion or human impact (see Fig. 6-4).

Sand dunes come to most people's minds when they think of a desert, but dunes make up a very small part of even the sandiest desert. In Death Valley, sand dunes cover only 2 percent of the total area. Within the California deserts, however, are some spectacular dune fields. The Algodones Dunes, one of the largest dune fields in the United States, stretch for 70 kilometers along the eastern edge of the Salton Sea near the Mexican Border, reaching heights of 100 meters in some places (Fig. 6-7). In the Mojave Desert south of Soda Lake, Kelso Dunes are as high as 200 meters. Large dune fields can also be found in Death Valley at Stovepipe Wells (Fig. 6-8), in the Eureka Valley, and at Edwards Air Force Base east of the Rogers Lake Bed. Cat Mountain in

Fig. 6-6 A dust storm in the Mojave Desert. (Source: Nakata, J. U.S. Geological Survey.)

Fig. 6-7 Algodones Dunes, east of El Centro, Imperial County. Note vehicle tracks in the foreground for scale. The rock in the distance approximately marks the Mexican border. (Photo by author.)

Fig. 6-8 Dumont Dunes, Death Vallley National Park. (Source: McMackin, M. San José State University.)

Valley, and at Edwards Air Force Base east of the Rogers Lake Bed. Cat Mountain in San Bernardino County is named for a large isolated dune shaped like a cat.

Three conditions must be met for a sand dune to accumulate. First, wind speed in an area must be adequate to lift and transport sand grains. Second, the area must have little or no vegetation; plant roots are effective anchors that stabilize the sand. Third, sand-sized sediment (1/16 to 2 millimeters) must be available for transport. Smooth, polished rock surfaces will not give rise to sand dunes, even in the fiercest winds. The valleys of southeastern California are ideal places for dune formation because they are generally unvegetated and windy. The sandy alluvium in dry washes and alluvial fans are ideal sources of sand. Sand is picked up by the prevailing winds, which typically blow from west to east across the valleys. As air masses reach the front of a mountain range, the wind speed decreases and the sand piles up against the mountain front. For example, sand is carried eastward across the Panamint Valley, through Emigrant Pass, and deposited at Stovepipe Wells at the mouth of Death Valley's Emigrant Canyon (Fig. 6-9). At Algodones Dunes, sand has been transported from the unvegetated sandy delta of the Colorado River southeast of the dunes.

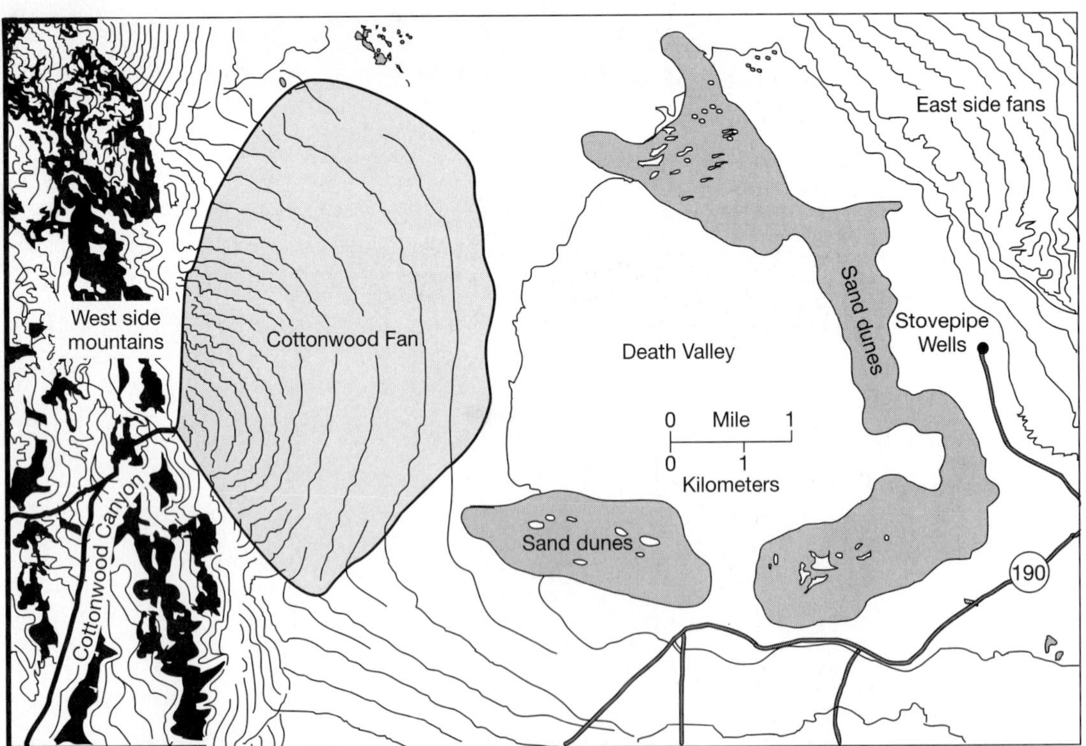

Fig. 6-9 The topographic setting that favors dune formation at Stovepipe Wells in Death Valley. Prevailing winds from the west pick up sand particles from the Cottonwood fan, carrying them eastward until they encounter the mountainous barrier along the eastern edge of the valley. (Source: U.S. Geological Survey.)

RUNNING WATER

Even in the driest desert, running water is the primary force that shapes the landscape. In the California desert, infrequent rainstorms can be intense, particularly during the summer. Storms moving north from the Gulf of California may bring late summer precipitation, and thunderstorms can produce locally intense rainfall in the mountains. Several centimeters of rain may fall within an hour during these storms. Because the mountain slopes are steep and unvegetated, little of the rainfall infiltrates into the ground. Water quickly collects in steep mountain channels, resulting in ***flash floods*** in desert canyons. In larger canyon systems, floodwaters may travel so quickly downstream that streams begin to flow without warning. When intense rains result from thunderstorms in the mountains, the skies in downstream areas may even be cloudless. Flash floods are a significant hazard in the desert. In Clark County, Nevada, near the California border, nine people were killed by a destructive flash flood in 1974. Desert hikers learn to take precautions, checking the weather continuously while exploring steep canyons (Fig. 6-10) and never leaving their vehicles in a dry wash.

When floodwaters are carried from the steep, confined canyons or ***arroyos*** (water-carved channels) of the mountains onto the basin floors, the water spreads out. As the velocity decreases, the sediment carried by the stream, usually a mixture of sand and gravel, is deposited at the mouth of the canyon. The channel shifts back and forth at the canyon mouth, and after repeated floods, a fan-shaped deposit of alluvium accumulates. Throughout the California deserts and along the fronts of other ranges in semiarid environments, these ***alluvial fans*** are prominent features of the landscape (Fig. 6-11). The apex of the fan is at the canyon mouth, and the fan slopes toward the basin floor (Fig. 6-12). In

Fig. 6-10 Flash floods in steep-walled desert canyons can pose a serious threat to unsuspecting hikers. Note the flood-deposited boulders at the mouth of this canyon, as well as the light-colored high water line on the cliff to the hiker's left and right. Jumpup Canyon, southern Utah. (Source: Burke, D.B.)

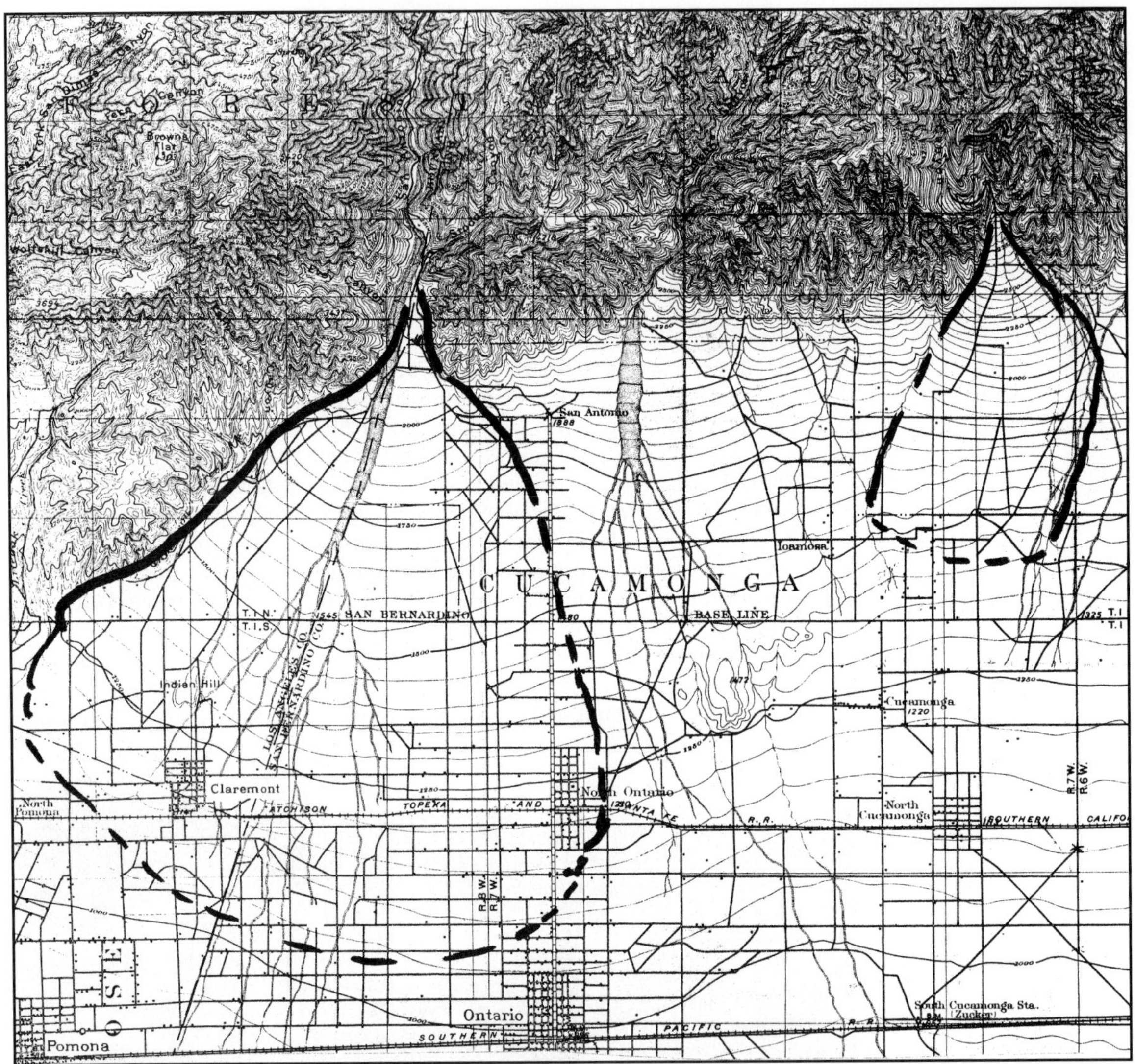

Fig. 6-11 A portion of the U.S. Geological Survey Cucamonga topographic map, showing the outlines of two alluvial fans along the margins of the San Bernardino Mountains near Ontario. This map was published in 1903, long before the area was extensively developed.

general, the size of an alluvial fan is proportional to the size of the drainage network that feeds it. Along the front of some ranges in the California desert, the alluvial fans coalesce along the range front to form a broad, sloping surface known as a ***bajada.***

Alluvial fans may be built by both flash floods and debris flows or mudflows. The shifting of channels within the fan may occur suddenly as a channel becomes blocked during a flood. Because all active surfaces of alluvial fans are vulnerable to

Fig. 6-12 Alluvial fan, Death Valley. (Source: Bradley, W.C.)

Fig. 6-13 Large boulder in alluvial fan gravel along the San Andreas fault near Palm Springs. (Photo by author.)

periodic floods, construction on fans is risky. Floods may occur only once in a century, or even less frequently, on average. However, the potential magnitude of those floods may be very large, as indicated by the size of boulders left behind during past events (Fig. 6-13). (See Chapter 10 for more information on California floods.)

Because of the high temperature and relatively permeable soil along the mountain fronts, floodwaters from the desert mountains quickly evaporate or sink into the desert floor. Even during floods, most streams lack the water to flow beyond the confines of the basin floor. A system of unconnected streams flows into each basin. In many areas, the streams are isolated by the ranges that separate the basins. This pattern is typical of most of the California desert under today's climatic conditions and is known as ***interior drainage.***

PLAYAS AND ANCIENT LAKES

During major flash floods, the water that runs off from the canyons may cover the basin floor. Because most of the basins have no outlet, the water collects in a shallow lake on the basin floor. The flat basin floors are called ***playas,*** from the Spanish word for "beach" or "shore," and the lakes that periodically form in the basins are known as playa lakes (Fig. 6-14). A series of unexpected storms can flood roads for weeks, leaving unprepared tourists stranded until the lakes recede. During times without runoff, the water in the lakes evaporates, leaving behind fine-grained lake sediment and minerals crystallized from the evaporating lake water. Because many of the ***evaporite*** minerals ("salts") are light-colored, the dry surfaces of the playas are white and highly reflective. When viewed from a nearby mountain summit, the basins of the California desert appear flat and white. These white flats shimmer in the glare of the hot summer sun as light reflects off the lake beds. From some angles, the flats reflect the blue of the clear desert sky, creating a mirage of clear blue water. Tired and thirsty pioneers making the journey across the desert on foot commonly saw sailing ships, green forests, and flocks of birds floating on these blue seas. However, as they came closer, the mirage faded into the reality of rocky outcrops along the dry valley floors.

A few of the basins in the California desert contain permanent lakes today. In the Basin and Range Province of northeastern California are Goose Lake, along the California-Oregon border, and Honey and Eagle Lake in Lassen County. Mono Lake, in Inyo County, is fed by Sierra Nevada snowmelt (Fig. 6-15 and Color Plate 11), as was Owens Lake until 1913, when the Owens River was diverted into the Los Angeles Aqueduct (see Chapter 10). At that time, Owens Lake was as much as 10 meters deep and covered 275 square kilometers, but today it is dry except during periods of high runoff.

Fig. 6-14 Shallow lake, Racetrack playa, Death Valley. (Source: Bradley, W.C.)

Another example of desert streams is the Mojave River, the single stream that flows through the Mojave Desert. Most of the time, the floor of the Mojave River is dry. However, every few years, intense rainfall brings enough water to the river that it flows into Soda Lake, near the town of Baker in San Bernardino County. Further south, the Salton Sea also contains water year-round. (See Chapter 10 for a discussion of the Salton Sea's origin.)

Throughout California's deserts and across the Basin and Ranges of Nevada, Utah, and Arizona, geologists have found conclusive evidence of landscapes dotted

Fig. 6-15 Mono Lake and tufa towers along the shore. The steep eastern front of the Sierra Nevada can be seen at the edge of the lake. (Source: Bradley, W.C.)

MARK TWAIN AND MONO LAKE

Not all visitors to California's deserts are captivated by their beauty, as is evident in this excerpt from Mark Twain's *Roughing It.* However, Twain did observe features such as the Mono Craters volcanic chain, as well as exaggerate the good use made of the alkaline lake water:

> Mono Lake lies in a lifeless, treeless, hideous desert, 8000 feet above the level of the sea, and is guarded by mountains 2000 feet higher, whose summits are always clothed in clouds. This solemn, silent, sailless sea—this lonely tenant of the loneliest spot on earth—is little graced with the picturesque. It is an unpretending expanse of graying water, about a hundred miles in circumference, with two islands in its center, mere upheavals of rent and scorched and blistered lava, snowed over with gray banks and drifts of pumice stone and ashes, the winding sheet of the dead volcano, whose vast crater the lake has seized upon and occupied.
>
> The lake is 200 feet deep, and its sluggish waters are so strong with alkali that if you only dip the most hopelessly soiled garment into them once or twice, and wring it out, it will be found as clean as if it had been through the ablest of washerwomen's hands. While we camped there our laundry work was easy. We tied the week's washing astern of our boat, and sailed a quarter of a mile, and the job was complete, all but the wringing out. If we threw the water on our heads and gave them a rub or so, the white lather would pile up 3 inches high.

Source: Twain, M. 1872. *Roughing It,* Chapter 38. Hartford, Conn.: American Publishing Co.

Fig. 6-16 Shorelines marking former positions of Mono Lake near Lee Vining. (Photo by author.)

with lakes as recently as 11,000 years ago. In the basins containing modern lakes, shorelines can be seen as high as hundreds of feet above the modern lake levels (Fig. 6-16). In many dry basins, some of which have no history of even rare flooding, shorelines are preserved like bathtub rings after the water has been drained. Some of the ancient lakes were so large that they spilled over to connect basins that are separated today. Throughout the desert basins, lake beds interbedded with layers of evaporite minerals and volcanic ashes are found, all associated with ancient shorelines, and in areas where shorelines have been erased by later erosion or deposition. When the ancient lake deposits and the shorelines are used to reconstruct the area in California covered by these lakes, the picture is an amazing contrast to today's deserts (Fig. 6-17). From Utah's Great Salt Lake, which pioneering geologist Grove Karl Gilbert recognized as a small remnant of ancient Lake Bonneville, to California's Mono Lake, whose larger predecessor is Lake Russell (named for geologist I.C. Russell), water covered almost all of the low parts of the landscape. This would have been a considerably more hospitable environment than today's barren deserts for the Native Americans and grazing animals who lived along the lake shores.

Geologists have established the ages of the ancient lakes by carbon-14 dating of scarce wood fragments, shells, and tufa deposits, and by fingerprinting volcanic ash layers interbedded with lake beds. (Fig. 6-20). Although the exact ages of the ancient lakes are not yet known, it appears that lakes were periodically extensive during Pleistocene time, until about 11,000 to 13,000 years ago throughout the southwestern United States. The lakes began to shrink as the global climate warmed following the most recent glaciation (see Chapter 8). High lake levels occurred during or at the end of the last glacial period, an observation made by Gilbert more than a century ago. The ancient shorelines and lake beds record high lake levels during a time when temperatures were cooler and evaporation rates much lower than present. The lakes in eastern California and western Nevada were fed principally by precipitation and by meltwater from the Sierra Nevada.

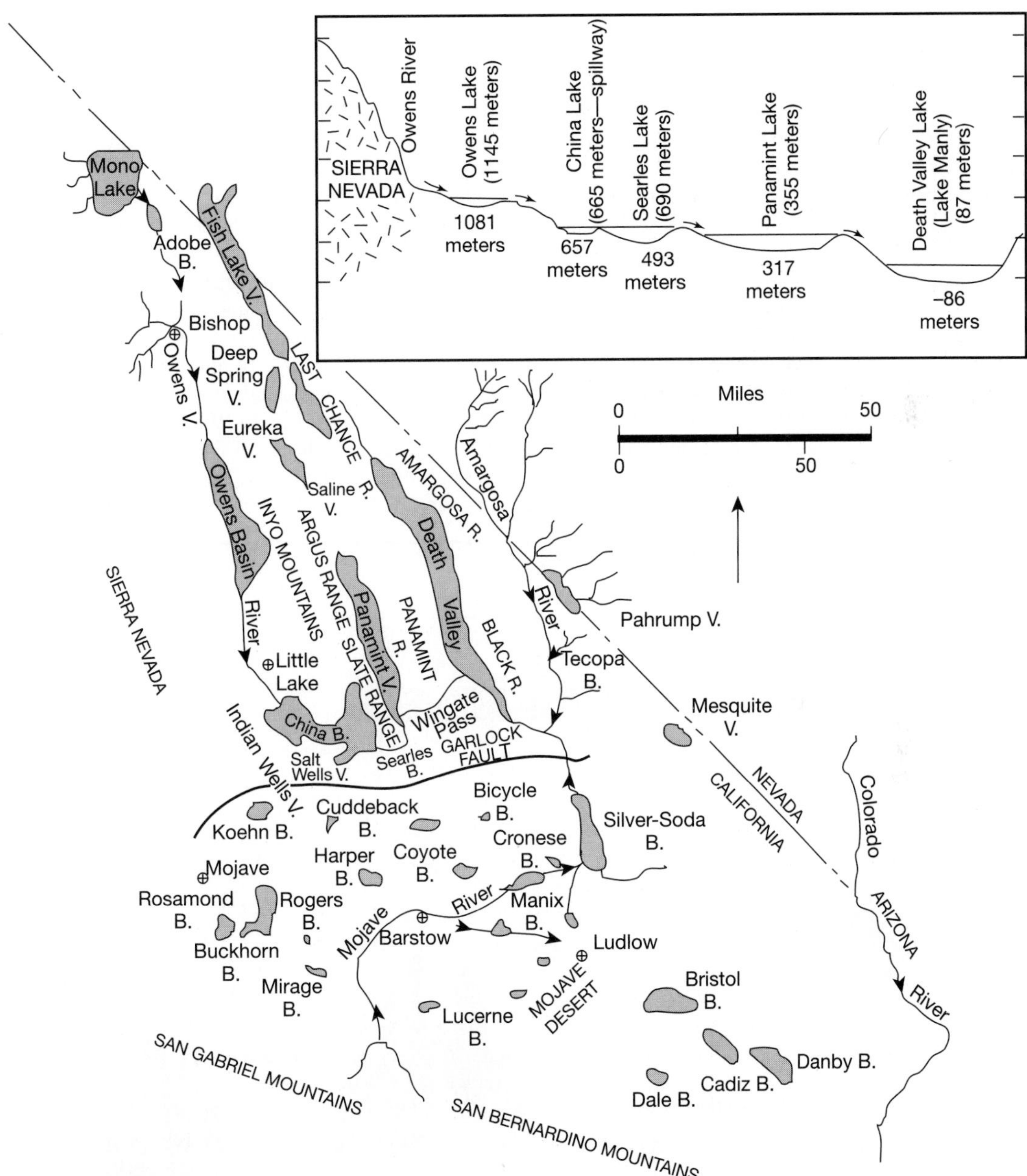

Fig. 6-17 Basins filled by Pleistocene lakes in southeastern California. *Inset* shows the connected system during times of overflowing lake levels. Today only Mono Lake remains. (Source: Used with permission of California Department of Conservation, Division of Mines and Geology. In Blackwelder, E. 1954. Bulletin 170.)

TUFA

Scattered along the shores of Mono Lake are clusters of rounded white towers that look like the pieces of a giant chess game or, as described in a Mono Lake guidebook, "giant towers of cemented cauliflower" (see Fig. 6-15 and Color Plate 11). The towers are a form of calcium carbonate known as ***tufa.*** Tufa forms in zones where water from springs that discharge into the lake mixes with the lake water. The two waters differ in their chemistry: the springs carry dissolved calcium from nearby rocks, and the alkaline lake water contains abundant carbonate. The mixed water is saturated with calcium and carbonate, leading to the crystallization of the mineral calcite ($CaCO_3$). Some tufa contains cells of algae, indicating that the process may be aided by the algae (Fig. 6-18). Tufa towers form at or near the shore of the lake, while tufa-cemented sand forms beneath the lake.

Tufa towers have been left standing along the shoreline long after the lake level has dropped. Along Highway 167, ancient tufa towers, turned gray by weathering and covered with orange lichens, can be seen several miles north of the present lake. These towers mark the level of Pleistocene Lake Russell approximately 16,000 years ago. For geologists studying the fluctuations of Pleistocene lakes, the tufa provides an important marker because it can be dated using radiocarbon methods.

The Pinnacles near Trona, along the southwestern shore of Searles Lake, are another example of tufa in California. The towers stand more than 30 meters above the now-dry basin (Fig. 6-19). Tufa can also be found along the ancient shores of Pleistocene lakes at Honey Lake, Indian Wells Valley, Panamint Valley, the Salton Sea, and in Nevada.

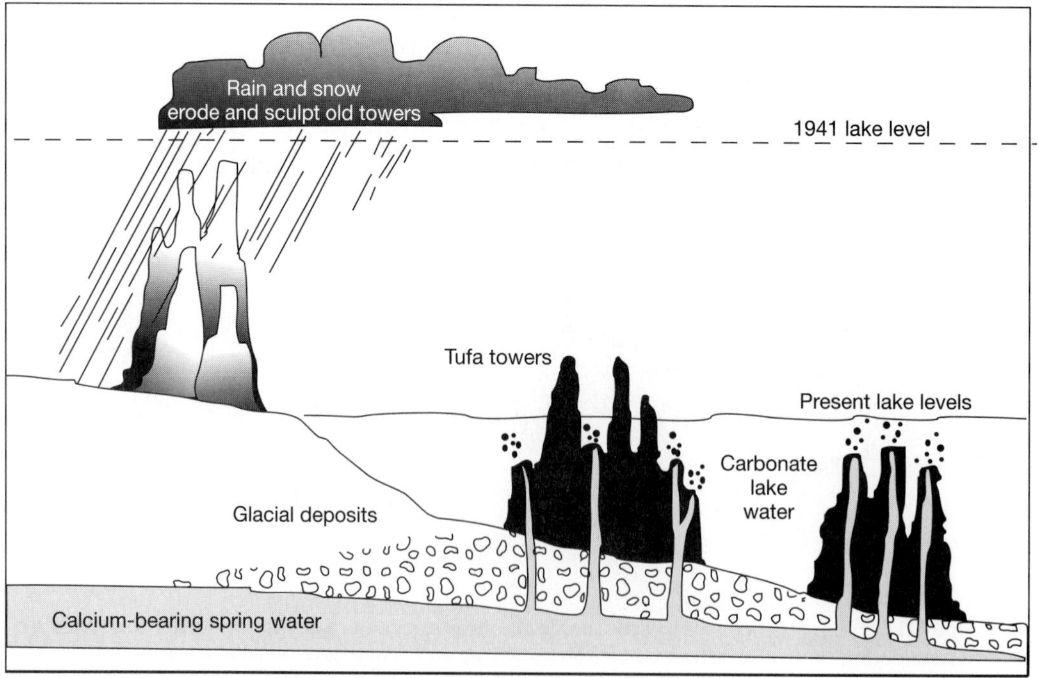

Fig. 6-18 The formation of tufa towers at Mono Lake. (Source: California Department of Parks and Recreation. In *California Geology,* 1992.)

Fig. 6-19 Tufa towers along the edge of Pleistocene Searles Lake, near Trona. (Source: Bradley, W.C.)

Fig. 6-20 Volcanic ash and pumice erupted from Mono Craters, interbedded with lake sediments of Pleistocene Lake Russell, Mono Lake. (Source: Sarna-Wojcicki, A. U.S. Geological Survey.)

In basins where lake history has been studied in detail, it appears that lakes have grown and shrunk episodically, reflecting major changes in the climate during the past 1 to 3 million years. Mono Lake did not overflow during its most recent high level 13,000 years ago or during its maximum level during the most recent glacial episode between about 100,000 and 18,000 years ago (Fig. 6-21). However, at some time during the previous glacial period, Mono Lake spilled eastward into Adobe Valley, then into Owens Lake. Water from Owens Lake spilled into the China Lake Basin, Searles Lake, the Panamint Valley, and into Death Valley (see Fig. 6-17). The lake that occupied Death Valley has been named Lake Manly for an early explorer of the region. At its height, Lake Manly was up to 200 meters deep. Today, the ancient shorelines and lake beds of Lake Manly, including ancient beaches and gravel bars, can be seen at Shoreline Butte, along the Black Mountains, and along the highway between Death Valley and Beatty, Nevada.

In the Mojave Desert, a much more vigorous Mojave River once flowed beyond its present mouth at Soda Lake, perhaps to the Colorado River, passing through a chain of overflowing lakes on its way (see Fig. 6-17). Today these lakes, Bristol, Cadiz, Danby, and Chuckwalla, are dry playas. The original watercourse has been broken up by recent faulting and lava flows as well as by the drying up of the river and lakes. Isolated lakes were present throughout other parts of California's deserts and in the southern San Joaquin Valley, as indicated by shorelines and lake beds preserved there. In the Imperial Valley, Pleistocene Lake Cahuilla, periodically filled by floods from the Colorado River, occupied the basin now filled by the Salton Sea (Fig. 6-22).

Because the desert lakes are excellent recorders of past climatic conditions, they are being intensively studied today as researchers attempt to better understand the earth's global climate system. Deep cores have been sampled from several of California's desert lakes in order to determine the history of lake fluctuations (see Fig. 6-21). Because they were originally horizontal, ancient shorelines can also be used like long-term level lines to determine the amount of faulting and tilting that has occurred in a basin since the shoreline was cut by the lake.

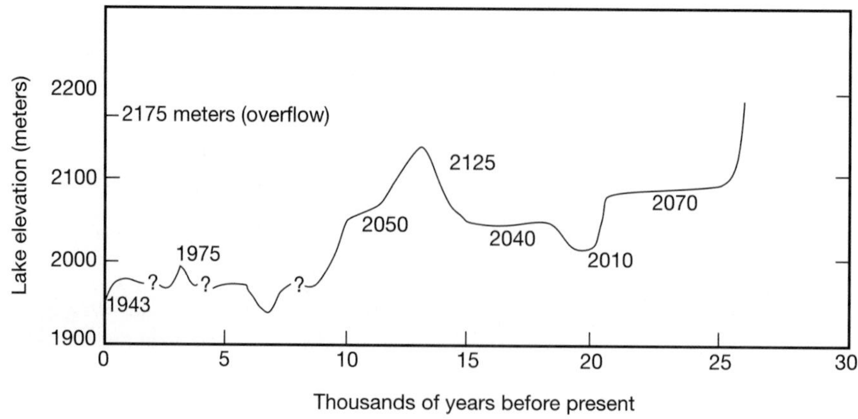

Fig. 6-21 Graph showing fluctuations in the level of Mono Lake during the past 30,000 years, based on the elevations of ancient shorelines, lake beds, and tufa. The elevation in 1975, at the 0 point on the graph, was 1943 meters. (Source: LaJoie, K.R. 1984. *Friends of the Pleistocene Field Trip Guidebook*.)

Fig. 6-22 Lake beds of Pleistocene Lake Cahuilla, near today's Salton Sea, Imperial County. (Photo by author.)

RICHES IN THE DESERT LAKE BEDS

By the 1880s, gold and silver miners had become disillusioned with the mountains of the California desert. Under the harshest of conditions, fortunes had been made, and many more lost, extracting silver and gold from the area. The hopeful turned their sights to the common white coverings on the dry lake beds. Two minerals discovered in the playas in 1873 were sodium borates (Fig. 6-23): ulexite (chemical formula $NaCaB_5O_9 \cdot 8H_2O$), also known as "cottonball," and tincal or borax ($Na_2B_4O_7 \cdot 10H_2O$). A large deposit of a third borate mineral, colemanite ($Ca_2B_6O_{11} \cdot 5H_2O$), was discovered in 1880 in ancient lake bed deposits east of Death Valley's floor. These discoveries brought prosperity to the desert, and borate minerals were mined from Death Valley from 1882 until 1928.

At the time of the Death Valley discoveries, borate of soda was used by glassmakers and metalsmiths as alloys, by potters to give ceramics a lustrous finish, and by meatpackers as a preservative. Used in enamel, borax created the smooth surfaces on the newly fashionable bathtubs, toilets, and sinks of the period. Soap companies had discovered that borax was an excellent water softener and buffer. In the late 1800s, borax was sold in apothecaries for as much as $.50 per ounce. The world's supplies of borax came originally from Tibet, and by the 1800s, from Tuscany in Italy, from Turkey, Chile, Argentina, and near Clear Lake in the northern Coast Ranges.

The greatest problem in the mining operation was transporting the borax across 275 kilometers of dry, steep mountains and salt marshes between Death Valley and the nearest railroad station at Mojave. Many stretches of the journey were entirely without water, and there were only three springs over 225 kilometers of the distance. To make the mining of borax profitable, huge weights had to be hauled in each trip.

Fig. 6-23 Borate mineral specimens collected from evaporite deposits in southeastern California. (Photo by author.)

However, the weight of the loads created problems both climbing and descending the steep ranges (see the box on p. 110). Crossing the thin salt crust over the playa was treacherous: even a hiker choosing the wrong route could fall through it. One route with a firm surface was found, but it was covered with jagged salt crystals. Building a road through this terrain, chopping through the maze of salt crystals by hand in temperatures greater than 110° F (43.3° C), was a difficult job accomplished by Chinese laborers. Over this road passed enormous wagons designed to carry the borax. The wagons stood almost two stories high, and each could carry half a freight car's worth of borax. The wagons were hitched in trains of two wagons and a water tank, and each train was pulled by 20 mules—the famous 20 mule teams whose picture appeared for years on borax boxes in American households (Fig. 6-24).

In 1913, a homesteader named John Suckow, while drilling a water well 60 kilometers east of Mojave, discovered colemanite in the cuttings brought up by the drill bit. After a series of exploratory wells were drilled, mining engineers concluded that a major borate deposit lay beneath a cover of younger stream gravels. The area is known as the Kramer Borate District, and the town of Boron grew up around the mining operations. Along with tincal, ulexite, and colemanite, a new mineral was discovered: kernite ($Na_2B_4O_7 \cdot 4H_2O$), also referred to as rasorite. In 1928, the Pacific Coast Borax Company shut down operations at Death Valley and began production from the Kramer deposits.

The Kramer borate deposit is found within lake beds deposited in a basin during Miocene time, long before development of the present topography and the development of the Pleistocene lakes (these older basins will be discussed in Chapter 7). The Miocene lake beds, including the borate deposits, were deposited on a series of basalt flows. Geologists believe that the borate originated from the volcanic vents in the area and then migrated to the lakes in the circulating groundwater. The borate deposits formed when the lake water evaporated to the point that it became super-

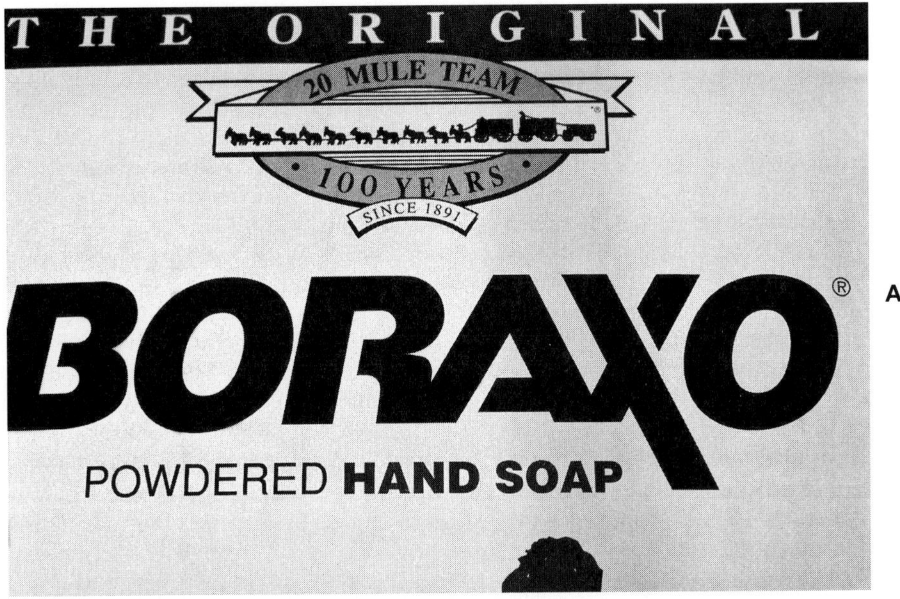

A

B

Fig. 6-24 **A,** Box of Boraxo hand soap showing the 20 mule team trademark. **B,** The 20 mule team wagons, Death Valley National Park. (Source: **A,** Photo by author. **B,** McMackin, M. San José State University.)

DRIVING THE MULE TRAINS OF DEATH VALLEY

The grade down the mountain shoulder near Granite Spring was the most hazardous stretch. It was straight; it was steep; it was miles long. Here was no sand to drag at the wheels; the surface was hard packed and the great wheels rolled down it all too easily. The difficulty in maneuvering that grade was in braking the wheels. And many times even the mighty brake blocks were not heavy enough to cope with 30 tons of borax, wagon, and water pushing from behind. As the wagons gained momentum the skinner threw his weight on the front brake; the swamper did the same on the trailer. Groaning, screeching and sliding, the cumbrous monsters lumbered on faster and faster.

The race with death began when the brakes no longer held and the wagons were out of control. Then the full dexterity of the skinner was called into play. There was no provision in the harness hitch for holdbacks; the animals simply had to be kept ahead of the load to avoid being run down. Their quickstep became a trot, the trot a kind of canter, the canter a dead run, a lope, and a gallop.

All precaution was thrown to the winds. Like a succession of pistol shots, the skinner's whip cracked over the heads of the frantic mules. Bowling behind, the wagons thundered their threat. They groaned under their weight, bounced and careened.

...Often they made it to the bottom of the grade. Sometimes not. And a wagon would come along 2 days, 4 days, a week later, to shovel away the drifts of spilled borax from the tangle of wreckage.

Source: Excerpted from Lee, W.S. 1963. *The Great California Deserts.* New York: Putnam & Sons.

saturated with borate, causing the minerals to crystallize.

The process of evaporation and precipitation of minerals is responsible for producing evaporite deposits of several types. The minerals that crystallize are saline minerals (salts), including sodium chloride, or common table salt. In the California deserts, salts have been mined from lake beds and ancient lake deposits in several areas. These include borates produced at Death Valley, Kramer, and Owens Lake; potassium, bromine, and lithium salts at Searles Lake; and sodium and calcium chloride at Bristol Lake near Amboy. Most of the world's supply of boron comes from the Kramer deposits and Searles Lake. In 1990, the value of California's boron minerals was more than $400 million. More than $2 billion worth of minerals have been extracted from the beds of Searles Lake alone.

THE FUTURE OF THE CALIFORNIA DESERT

As the population of southern California increases, the open spaces of the nearby California desert are increasingly appealing as a place for development and recreation. Today, the once-remote deserts are criss-crossed by roads and trails that bring one out of every three vacationing southern Californians to the desert for recreation and solitude. Conservationists have long been concerned about the impact of humans on the land surface, ecology, water resources, and scenery of the desert. One impact of concern is the disturbance of fragile vegetation and desert pavement by off-road vehicles. Studies of areas used for training tank operators during World War II have shown that vehicle tracks persist on the desert floor for over 50 years. While areas have been set aside for operation of off-road vehicles, much damage can be done by those few who stray from designated trails (Fig. 6-25). Controversy also exists about the extent of mining that

Fig. 6-25 Tracks left by off-road vehicles on the Soda Lake playa, Mojave Desert. (Source: Wilshire, H. U.S. Geological Survey.)

should be permitted on desert lands and about the reclamation of existing mined lands. Excessive withdrawal of groundwater from desert basins for agriculture has caused loss of plants, salt damage to soils and water, and increased erosion in some areas.

Recognizing the unique value of the California desert, Congress designated 25 million acres as the California Desert Conservation Area (CDCA) in 1976 and authorized development of a comprehensive land use plan for the area. About 72 percent of this land is owned and managed by the Federal Government, including 3 million acres in military reservations, 2.5 million acres in Death Valley and Joshua Tree National Monuments, and 12.5 million acres managed by the Bureau of Land Management. In 1994, Congress passed the comprehensive California Desert Protection Act, which increased to 9.2 acres the amount of land set aside in federal park and wilderness areas. The bill also converted Death Valley and Joshua Tree from national monuments to national parks and created the Mojave National Preserve.

FURTHER READINGS

BLACKWELDER, E. Pleistocene lakes and drainage in the Mojave Region, southern California. 1954. In California Division of Mines and Geology. *Geology of Southern California.* Part V, pp. 35-44.

GAINES, D. 1989. *Mono Lake Guidebook.* Lee Vining, Calif.: Kutsavi Books, 104 pp.

HILL, M. 1984. *California Landscape, Originated Evolution. California Natural History Guide 48.* Berkeley, Calif.: University of California Press, 262 pp.

JAEGER, E. 1965. *The California deserts.* Stanford, Calif.: Stanford University Press,

208 pp.

LEE, W.S. 1963. *The Great California Deserts.* New York: Putnam & Sons, 306 pp.

MUMFORD, R. 1954. Deposits of saline minerals in southern California. In California Division of Mines and Geology. *Geology of Southern California.* Chapter VIII, pp. 15-22.

RAE, C. 1992. *East Mojave Desert: A visitor's guide.* Santa Barbara, Calif.: Olympus Press, 206 pp.

RIEGER, T. Calcareous tufa formations, Searles Lake and Mono Lake. *California Geology,* 45:99-109, 1992.

VER PLANCK, W.E., Salines in southern California. 1954. In California Division of Mines and Geology. *Geology of Southern California.* Chapter VIII, pp. 5-14.

WALKER, A.S. 1992. *Deserts: Geology and Resources.* U.S. Geological Survey General Interest Publication, 60 pp.

CHAPTER

7

The Basin and Range and Mojave Desert

Old Rocks and Young Faults

THE BASIN AND RANGE LANDSCAPE

Anyone who has traveled across Nevada from east to west, or from west to east, has encountered the major topographic characteristic of the Basin and Range Province. The climb to the summit of each mountain is followed by a rapid descent onto the flat desert surface (Fig. 7-1). From each summit, you can view the highway stretching across a dry valley and up the next range, which may be more than 30 kilometers away. As the hours pass, you repeat this experience many times, traveling west until you reach the Sierra Nevada, or east until you reach Salt Lake City. A relief map tells the same story: a series of ranges running north-south, each between 75 and 250 kilometers long, separated by parallel valleys. Most of the region is fairly high, with the valley floors standing over a thousand meters above sea level, and the highest peaks exceeding 4000 meters.

The Basin and Range Province extends from eastern California to central Utah, from southernmost Oregon and Idaho on the north to southern Arizona and southwestern New Mexico and further south into the Sonoran State of Mexico (Fig. 7-2). The California portion of the province includes a small area in the northeastern corner of the state and a much larger area bordered by the Sierra Nevada on the west and the Mojave Desert on the east (see the map on p. 54). Death Valley, perhaps the most famous of the basins in the entire Basin and Range Province, contains the lowest point in the U.S., 86 meters below sea level. The highest point, east of Owens Valley in the White Mountains, stands 4341 meters above sea level. Prominent California ranges include the Warner Range, the White and Inyo Mountains, the Panamint Range, and the Funeral Mountains. The major valleys of California's Basin and Range Province are Owens Valley, along the eastern edge of the Sierra Nevada, and Saline, Panamint, and Death Valleys. Like all of California's deserts, the Basin and Range is an area of low rainfall and scarce vegetation. Only the higher parts of some ranges support a forest cover.

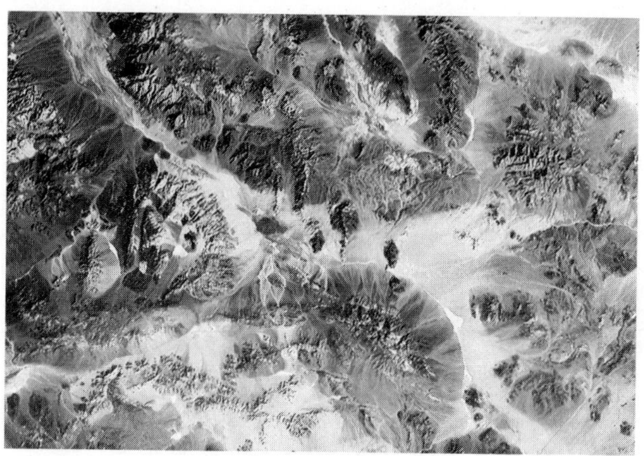

Fig. 7-1 Landsat satellite image showing the southern portion of Death Valley, the northwest-trending linear zone in the upper left. The Garlock fault, which separates the Basin and Range and Mojave Desert Provinces, crosses the image roughly through the center, from left to right.

Fig. 7-2 The Basin and Range Province in western North America. (Source: Stewart, J. 1978. Geological Society of America, Memoir 152.)

Fig. 7-3 View of Clark Mountain from the Kingston Range, looking to the south at a typical landscape of the eastern Mojave Desert. (Source: McMackin, M. San José State University.)

THE MOJAVE DESERT

South of California's Basin and Range Province, the Mojave Desert is a large triangle bounded on the east by the Colorado River and the California-Nevada border, on the north by the Garlock fault (see Fig. 7-1 and Chapter 14), and on the southwest by San Gabriel and San Bernardino Mountains and the San Andreas fault.

A comparison of the topography in the Mojave Desert with the Basin and Range (Endpaper 4) shows some general similarity and some distinct differences. Basins and ranges are characteristic of both provinces, but the landscape is more subdued in the Mojave Desert. Most ranges of the Mojave Desert are shorter and lower than those in the Basin and Range (Fig. 7-3). The basins are broader, particularly in the western Mojave Desert, which is a fairly flat plain. The basins and ranges in most of the Mojave Desert are also less aligned than those of the Basin and Range to the north. Both the similarities and the differences in the landscapes of the two provinces can be explained by their recent geologic history.

EARTHQUAKES, RECENT FAULTING, AND BASIN-RANGE TOPOGRAPHY

On March 26, 1872, a major earthquake rumbled through Owens Valley, leveling the entire town of Lone Pine and killing 23 of its 250 to 300 inhabitants. Most of the adobe and brick buildings in the region were completely destroyed—leading the local newspaper to point out that the deaths would not have occurred if the buildings had been constructed of wood. The earthquake, with an estimated magnitude of 7.8, was one of the largest in the history of the United States (Chapter 13). The quake triggered huge rockfalls in Yosemite Valley, one of which was witnessed by famous naturalist John Muir. Shaking was felt as far away as Salt Lake City and San Diego. The earth-

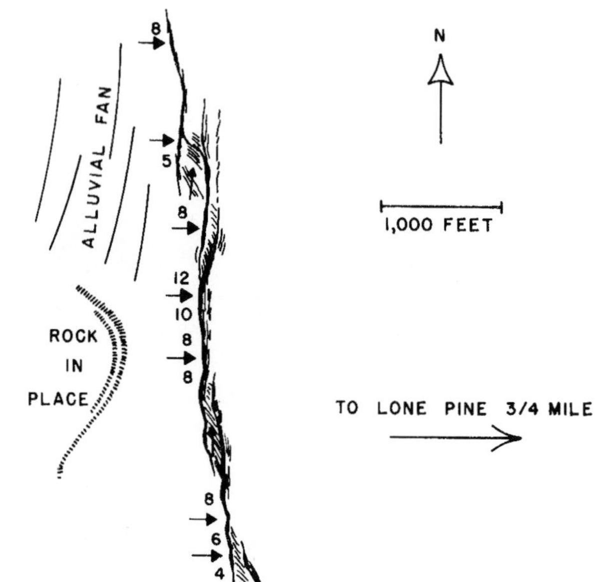

Fig. 7-4 Map showing the surface ruptures produced during the 1872 Owens Valley earthquake, about 1.3 kilometers west of the town of Lone Pine. The original maps were drawn by W.D. Johnson in 1907. At that time, the fault scarps measured as much as 12 feet high, as indicated by the numbers in the sketch (given in feet). (Source: *California Geology,* March 1972. Used with permission of the California Department of Conservation, Division of Mines and Geology.)

quake broke the ground along a 100-kilometer stretch of the Owens Valley fault system and separated the ground surface several meters in some places (Fig. 7-4).

Today, some of the evidence of the 1872 quake is still well preserved. A prominent ***fault scarp,*** a break in the topography caused by surface faulting, can be seen west of the town of Lone Pine along the Lone Pine Fault. A fault scarp forms when an earthquake ruptures the ground surface, displacing the landscape along a sharp break (Fig. 7-5). Following the earthquake, the scarp fades away as erosion and deposition smooth out the land surface. By careful study of the fault scarp and the young sediments broken by the fault near Lone Pine, geologists have estimated that the 1872 earthquake shifted the land surface about 1 to 2 meters vertically and as much as 6 meters horizontally. The fault scarp west of Lone Pine was built by three separate earthquakes during the past 10,000 years, each one moving the western side of the fault upward and to the northwest relative to the eastern side. Over geologic time, repeated earthquakes along the faults of the Owens Valley system have gradually elevated the Sierra Nevada 3000 meters (9000 feet) above the floor of Owens Valley. The occurrence of the 1872 earthquake, as well as smaller quakes west of Mono Lake in 1991, are dramatic proof that this process continues.

Throughout the Basin and Range province, ***range-front faults*** can be found along the base of the ranges. Some of these faults are clearly as active as the Owens Valley fault. Moderate earthquakes are relatively common along some faults of the Basin and Range Province. Two large earthquakes in 1954 in Dixie Valley, central Nevada, produced fault scarps as high as 5 meters, which are well preserved today in the desert climate.

Sharp, well-preserved fault scarps along the base of many ranges in southeastern California indicate that earthquakes have occurred there during the past few thousand years. Along other range fronts, fault scarps are subtle or absent, indicating that a long period of erosion and deposition has passed since the last ground-

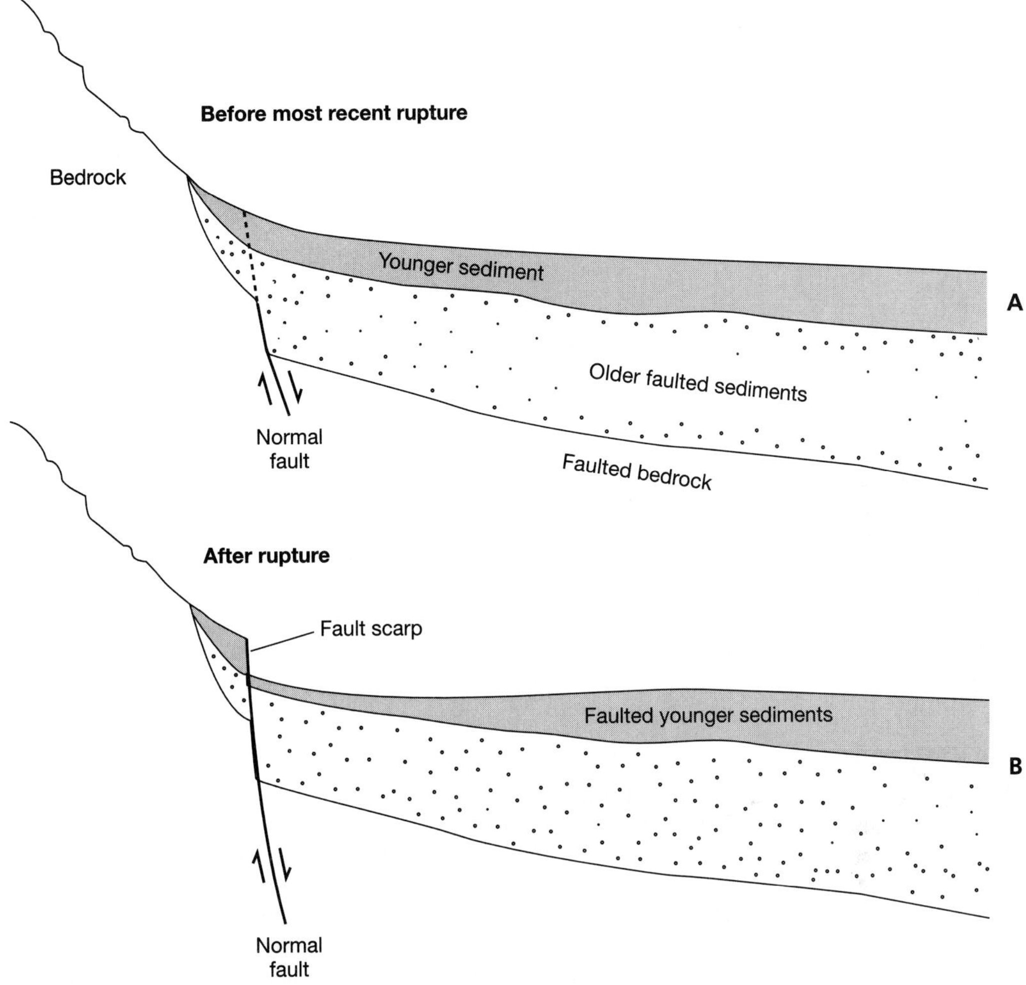

Fig. 7-5 Diagram showing the displacement of rock and sediment layers and the formation of a fault scarp along a normal fault.

breaking earthquake. Geologists can use the freshness of scarps to estimate the age of the most recent earthquake on a fault. This technique has been particularly successful in the Basin and Range and Mojave Desert, where the type of faulting, the rock types, and the erosion rate are similar in many ranges. When detailed investigations are carried out along range-front fault scarps, the deposits of alluvial fans can provide important information about the timing of recent faulting. Many active range-front faults have broken through recent alluvial fans, while older fault scarps have been buried by young alluvial fans.

Along the faults of the Basin and Range Province, recent movement has elevated the ranges relative to the basins. Most of the range-front faults are ***normal faults.*** On a normal fault, movement drops one side along the fault surface in the direction that one would expect by gravity, given the geometry of the fault—hence

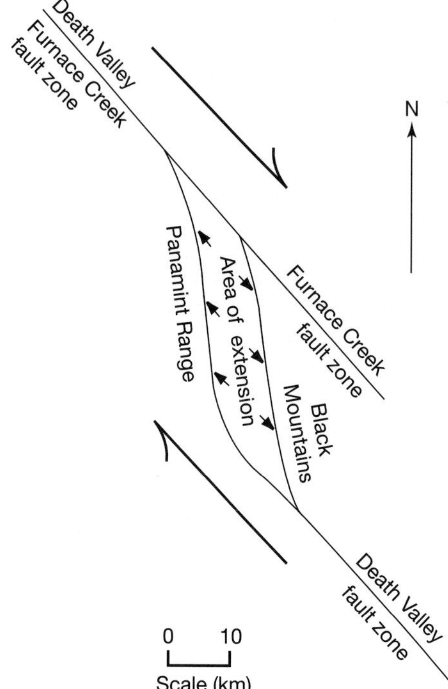

Fig. 7-6 Map showing recent faults in the Death Valley area. (Source: Burchfiel, P., and Stewart, J. 1966. Geological Society of America Bulletin, Vol. 77 [4].)

the term "normal" (see Fig. 7-5). It is repeated vertical movements along normal fault systems that produce the characteristic basin and range topography.

The overall shape of the mountain fronts and the sharpness of the basins and ranges also provide clues about an area's faulting history. The steep, straight mountain fronts of the Basin and Range province reflect the recent movement along normal faults. In contrast, the more subdued basins and ranges in most of the Mojave Desert are a clue that normal faulting is less recent there than in the Basin and Range. After normal faulting ceased in the Mojave Desert, erosion of the ranges and deposition in the basins have softened the stark contrast between mountains and valleys. Over time, the straight mountain fronts have also become more ***sinuous*** as erosion wears back the ranges.

Many of the faults in southeastern California show lateral displacement (Chapter 1), as well as normal movement. In the Basin and Range Province, the Owens Valley, Panamint Valley, Death Valley, and Furnace Creek faults all show lateral motion (Fig. 7-6 and Endpaper 2). Recent lateral faulting has also occurred in three zones in the Mojave Desert: the San Andreas system along its southwestern edge, the Garlock fault along the northern boundary, and the eastern California shear zone in the central Mojave desert (Chapter 14). As discussed in Chapter 1, lateral motion on faults east of the San Andreas system accommodates some of the relative motion between the Pacific and North American Plates that is not taken up by the San Andreas itself. The eastern California shear zone and the 1992 Landers earthquake, which provided important new information about southern California faulting, are discussed further in Chapter 14.

DETACHMENT FAULTS

Recent studies have demonstrated that many of southeastern California's normal faults are steep at the surface but dip at a more shallow angle at depth. In some areas, multiple normal faults converge or are cut off along a single, gently dipping fault surface. Rocks on the upper surface move over those on the lower plate along these master faults, which geologists refer to as ***detachment faults*** (Fig. 7-7). The detachment faults may extend below the earth's surface to great depths, perhaps more than 10 kilometers.

At depths where temperature and pressure are great enough, the rocks along a detachment fault may be greatly altered. The shearing motion of the fault blocks subjects the rocks in fault zone to great stresses, and a characteristic metamorphic rock known as ***mylonite*** may form (Fig. 7-8). Mylonite has a smeared texture, indicating that the rocks deformed in a plastic manner, rather than being crushed or broken. The rocks must therefore have been buried to great depths, where temperatures and pressures are great enough to make them deformable. In areas where mylonite is now seen at the Earth's surface along detachment faults, geologists can conclude that 8 to 15 kilometers of overlying rock must have been removed (Fig. 7-9).

As the crust in the Basin and Range and Mojave Desert regions stretched, the rocks at the surface responded by normal faulting (Fig. 7-10). Extension thinned the earth's crust, and some areas became so extended that surface rocks no longer covered the extended area. In these regions, not surprisingly referred to as "highly extended terranes," the mylonite and other metamorphic rocks from the lower plate beneath the detachment fault can be seen at the earth's surface.

Detachment faulting caused by Cenozoic extension time may explain unusual surfaces along the eastern side of Death Valley. These features were named ***turtlebacks*** for their curved shapes (Fig. 7-11), and geologists have long been puzzled about their origin. One model proposes that they are the surfaces of detachment faults and that the sheared rocks along their surfaces are mylonite produced by the fault movement. The turtleback surfaces indicate that the region was pulled in a

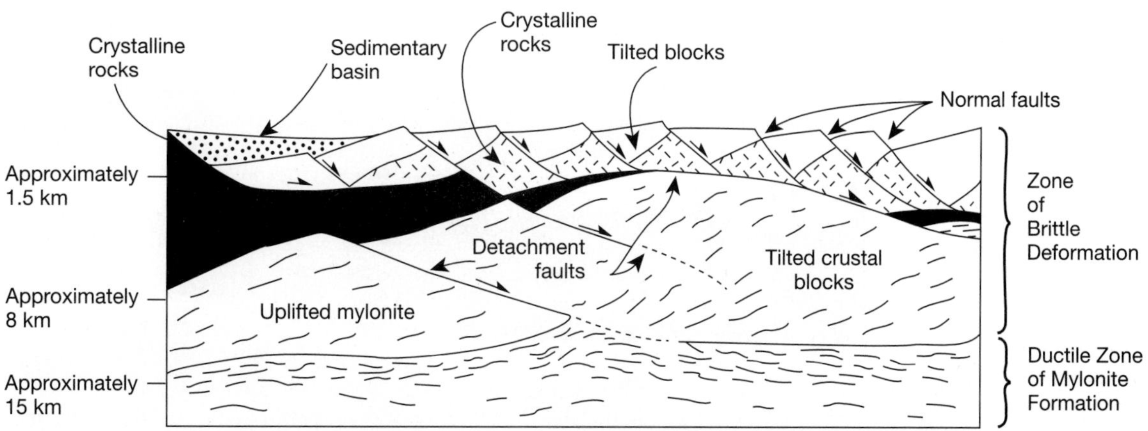

Fig. 7-7 Diagram showing the formation of detachment faults. (Source: Modified from Pridmore, C., and Frost, E. Detachment faults. *California Geology* 45, 1992. Used with permission of the California Department of Conservation, Division of Mines and Geology.)

Fig. 7-8 Mylonite formed by ductile deformation along a detachment fault in the Basin and Range, Redtail Canyon. (Source: McMackin, M. San José State University.)

Fig. 7-9 View of a detachment fault, northern Mesquite Mountains (view to south). Dark craggy areas at the top of the hills are Cambrian rocks of the upper plate. Proterozoic lower plate metamorphic rocks are snow-covered. (Source: McMackin, M. San José State University.)

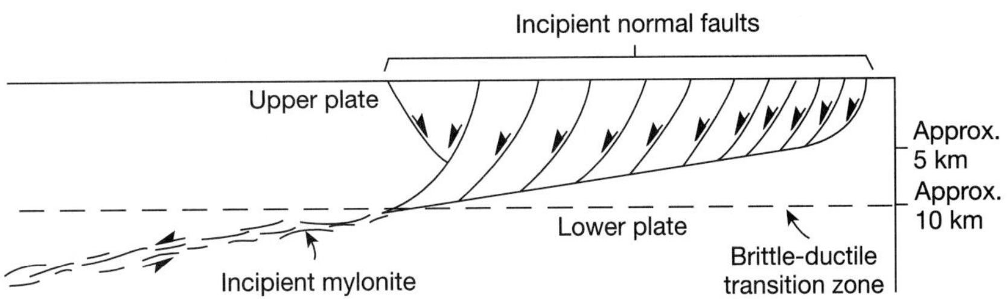

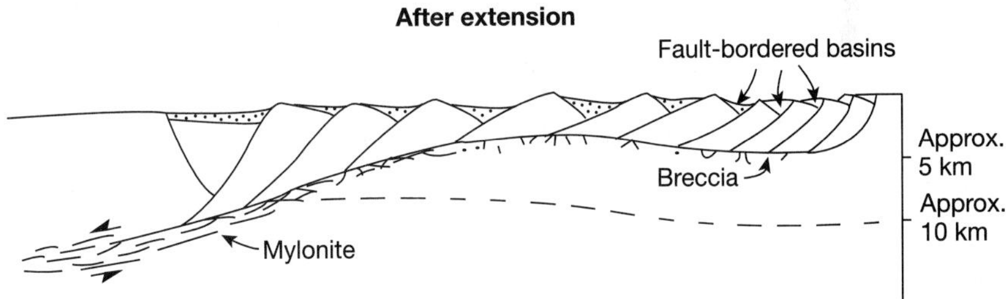

Fig. 7-10 The development of detachment faults in response to crustal extension. (Source: Modified from Pridmore, C., and Frost, E. 1992. Detachment faults. *California Geology* 45, 1992. Used with permission of the California Department of Conservation, Division of Mines and Geology.)

Fig. 7-11 A turtleback along the eastern side of Death Valley. (Photo by author.)

northwest/southeast direction (see Fig. 7-6). The stretching caused the valley to open up, with younger sedimentary units moving downward. As extension progressed, the detachment faults, or the turtlebacks, became exposed.

Geologists have hypothesized that the Paleozoic rocks of the Panamint Range west of central Death Valley were originally located above or next to the pre-Cambrian rocks of the Black Mountains. Extension and detachment faulting carried them 80 kilometers to the northwest, exposing the older rocks in the Black Mountains. Another complex fault zone in the southern Black Mountains known as the Amargosa Chaos is now thought to have resulted from detachment faulting. Gently dipping normal faults that are likely detachment faults have also been recognized at several localities between the Panamint Range and the California-Nevada border, as well as along the southern boundary of the Basin and Range Province north of the Garlock fault.

Detachment Faults in the Mojave Desert

In many areas of the Mojave Desert, the upper crust moved along detachment faults sufficiently to expose the rocks beneath the detachments. In the eastern Mojave Desert, detachment faults are exposed in the Old Woman Mountains, Chemehuevi Mountains, Whipple Mountains, and other ranges. In the central Mojave Desert, detachment faults can be seen in the Bullion Mountains, the Newberry Mountains, and the Waterman Hills (Fig. 7-12). In many of these areas, the detachment faults have uncovered metamorphic rocks that were 8 to 15 kilometers below the surface before the detachments formed.

EXTENSION AND VOLCANIC ACTIVITY IN SOUTHEASTERN CALIFORNIA

Beginning about 16 to 19 million years ago in Miocene time, lava erupted from several sources scattered throughout California's Basin and Range Province. Like detachment faulting, the volcanic activity is a result of extension of the crust (see Chapter 5). Recent volcanic eruptions have left young cinder cones, craters, and lava flows. These include Ubehebe Craters in Death Valley, a series of cones and flows in the Owens Valley near Big Pine, as well as the Bishop Tuff and recent volcanic features in the Modoc Plateau discussed in Chapter 5.

The Coso volcanic field is a center of recent volcanic activity along the southern edge of Owens Lake, at the western edge of the Basin and Range Province. Lava has erupted intermittently in this area during the past 6 million years. During the past million years, both high-silica lavas (Fig. 7-13) and mafic lavas have been erupted (Fig.7-14) a pattern known by geologists as ***bimodal volcanism.*** Geologists believe that the mafic lavas are erupted directly from sources in the mantle. The high-silica lavas form when rising magma causes crustal rocks to melt. Eruptions from both mantle and crustal sources in the same area results in both basalt and rhyolite being formed in the same area. In the Coso volcanic field, 38 rhyolite flows and domes formed in the Coso volcanic field during the past million years. During the same time period, basaltic lava erupted from at least 14 volcanic vents. The youngest dated feature in the Coso field is a basalt flow erupted about 40,000 years ago.

An interesting feature of the eruptive history of the Coso field is the very regular pattern of eruptions during the past 500,000 years. By dating the rocks formed during individual eruptions and measuring the volume of erupted materials, geolo-

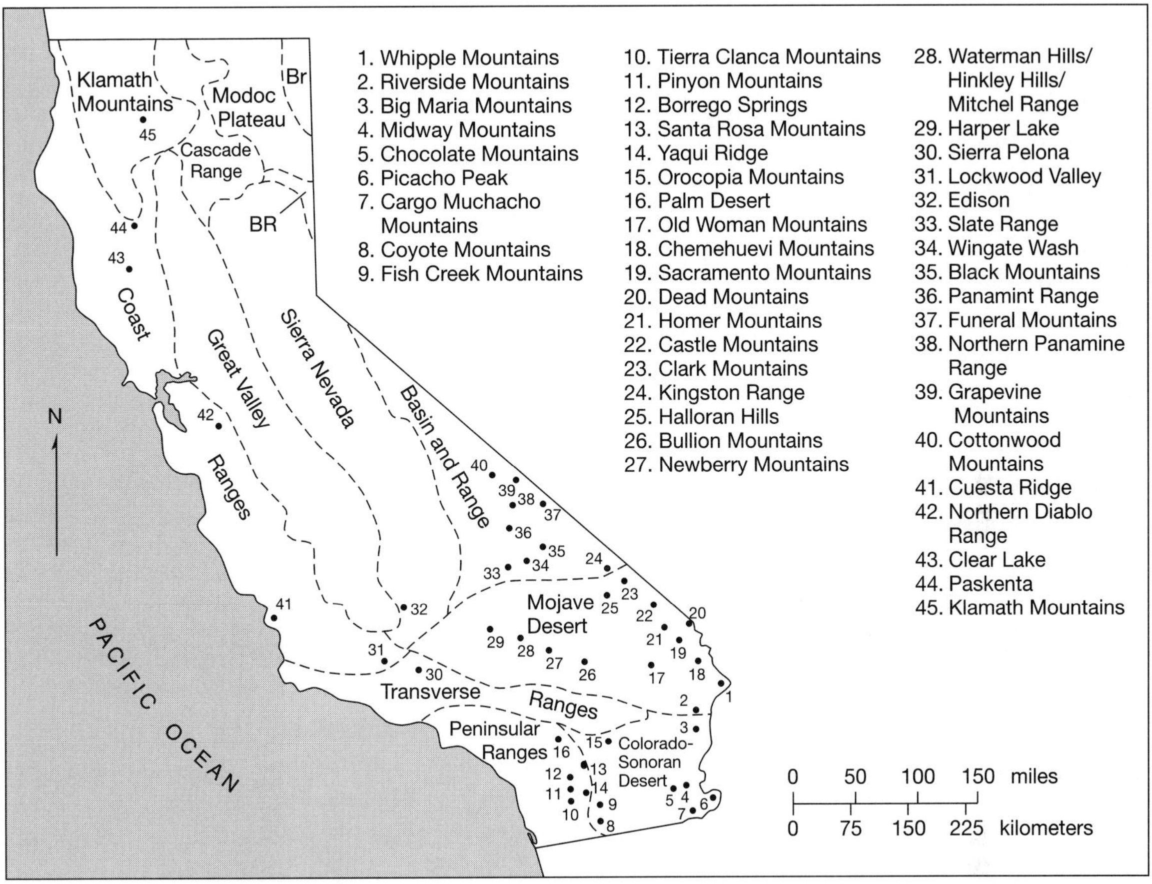
Klamath Mountains
Modoc Plateau
Br
Cascade Range
BR
Coast Ranges
Great Valley
Sierra Nevada
Basin and Range
Mojave Desert
Transverse Ranges
Peninsular Ranges
Colorado-Sonoran Desert
PACIFIC OCEAN
N
0 50 100 150 miles
0 75 150 225 kilometers
1. Whipple Mountains
2. Riverside Mountains
3. Big Maria Mountains
4. Midway Mountains
5. Chocolate Mountains
6. Picacho Peak
7. Cargo Muchacho Mountains
8. Coyote Mountains
9. Fish Creek Mountains
10. Tierra Clanca Mountains
11. Pinyon Mountains
12. Borrego Springs
13. Santa Rosa Mountains
14. Yaqui Ridge
15. Orocopia Mountains
16. Palm Desert
17. Old Woman Mountains
18. Chemehuevi Mountains
19. Sacramento Mountains
20. Dead Mountains
21. Homer Mountains
22. Castle Mountains
23. Clark Mountains
24. Kingston Range
25. Halloran Hills
26. Bullion Mountains
27. Newberry Mountains
28. Waterman Hills/ Hinkley Hills/ Mitchel Range
29. Harper Lake
30. Sierra Pelona
31. Lockwood Valley
32. Edison
33. Slate Range
34. Wingate Wash
35. Black Mountains
36. Panamint Range
37. Funeral Mountains
38. Northern Panamine Range
39. Grapevine Mountains
40. Cottonwood Mountains
41. Cuesta Ridge
42. Northern Diablo Range
43. Clear Lake
44. Paskenta
45. Klamath Mountains

Fig. 7-12 Map showing identified detachment faults in California. (Source: Pridmore, C., and Frost, E. Detachment faults. *California Geology*, 45, 1992.)

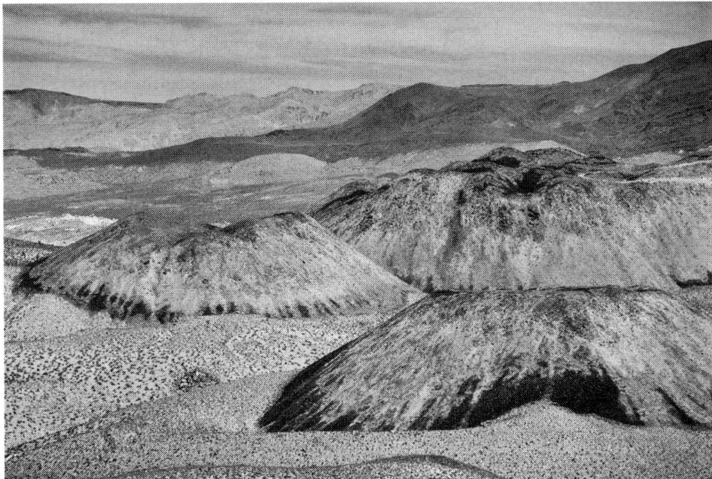

Fig. 7-13 Rhyolite domes of the Coso volcanic field, viewed from Sugarloaf Mountain. (Source: Hill, M. U.S. Geological Survey.)

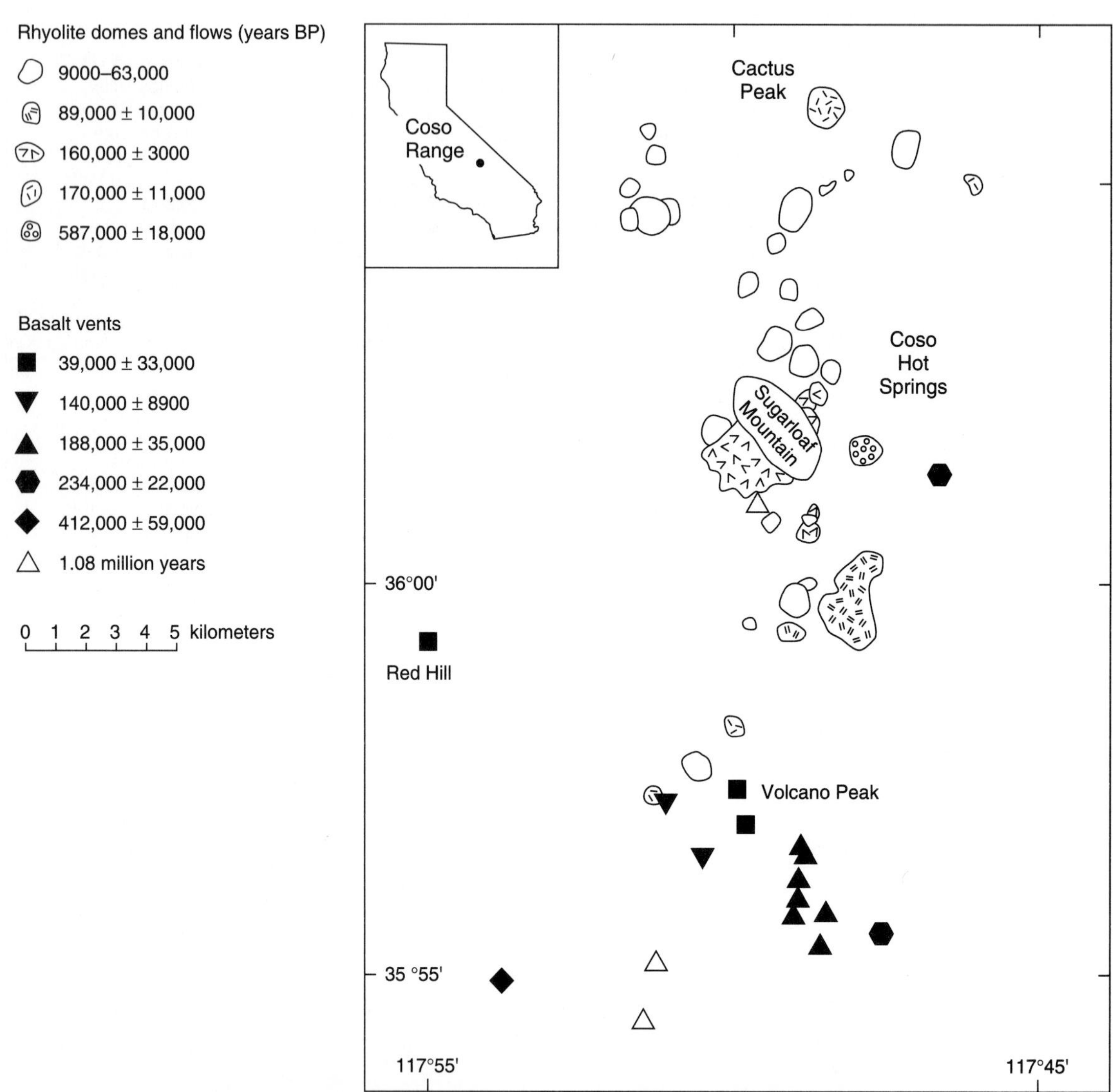

Fig. 7-14 Map showing volcanic eruptions in the Coso volcanic field during the past million years. (Source: Bacon, C. *Geology* 10:66, 1982.)

gists discovered a pattern. For both basalt and rhyolite, the volume of erupted material could be predicted by calculating the time that had passed since the previous eruption: more lava was erupted if more time had passed since the previous eruption. From the pattern of volcanic eruptions in the Coso volcanic field, geologists concluded that the eruptions were controlled by the opening of ***feeder dikes*** that carried magma to the volcanic vents. Feeder dikes form when magma is injected into linear conduits. Geologists believe that extension of the crust created these pathways for the Coso eruptions.

RECENT VOLCANIC ACTIVITY IN THE MOJAVE DESERT

A string of young volcanic flows, cones, and craters stands out along a northwest-oriented zone between Barstow and Bristol Lake (Fig. 7-15). Volcanic eruptions have occurred along a linear depression in the Mojave landscape, and both the depression and the volcanic activity are thought to be caused by continuing extension. Basalt has been the dominant rock produced in the zone, as would be expected in an extensional environment (see Chapter 5). Many of the volcanic features appear unweathered, evidence of both recent activity and excellent preservation in the desert environment.

The Cima volcanic field, located about 20 kilometers north of Kelso, contains 30 cinder cones and associated flows. Cones and flows in the northern part of the field are as old as about 10 million years, whereas the southern part of the field is much younger. There some of the eruptions are thought to have occurred only a few centuries ago, based on their excellent preservation (Fig. 7-16) and on radiocarbon dating of charcoal found beneath one of the flows. At the Aiken Mine, the red cinders have been mined for gardens, paths, and other decorative uses. Cinders are also mined from Pisgah Crater, a 250-foot-high cinder cone just south of Interstate 40 about 35 miles east of Barstow. Recent lava flows that covered part of Lavic Lake are associated with the crater. Three craters known as Sunshine Peak Craters, and their associated lava flows, are a few miles south of Pisgah.

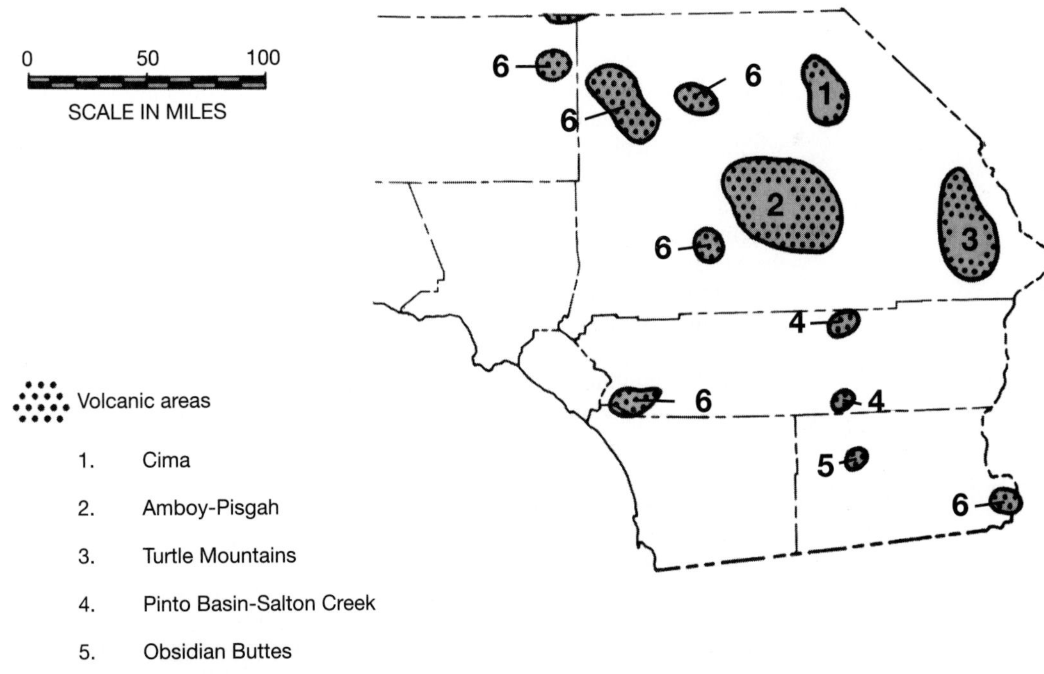

Fig. 7-15 Map showing the distribution of young volcanic features in the Mojave desert. (Source: Used with permission of the California Department of Conservation, Division of Mines and Geology. 1954. Bulletin 170.)

Fig. 7-16 Aerial view of recent basalt flows and cones, Cima volcanic field. (Source: Nakata, J. U.S. Geological Survey.)

Dish Hill contains two cinder cones erupted about 2 million years ago. Both at Dish Hill and in the Cima volcanic field, the volcanic eruptions have brought pieces of the lower crust and the mantle to the earth's surface in the form of ***xenoliths*** found within the volcanic rock. These fragments or ***inclusions*** of already-crystallized rock are important because they provide samples of rocks from as deep as 70 to 80 kilometers below the surface, far beyond the reach of even the deepest drilling. Fragments of gabbro (mafic plutonic rock) suggest that the lower part of the crust in the Mojave desert is composed of gabbro. Presumably, the underlying mantle comprises bits of rarely seen ultramafic plutonic rocks.

Amboy Crater is a young cinder cone that rises about 75 meters above the surrounding terrain. Lava from Amboy Crater flowed onto the surface of Bristol Lake, probably within the past several thousand years. Volcanic features such as collapsed lava tubes, pits, and lava blisters can still be seen on the surface of the flow.

SEDIMENTARY BASINS AND EXTENSION

Throughout southeastern California, extension resulted in the formation of down-dropped basins between the ranges. Sediments eroded from newly raised mountains accumulated in those basins (see Chapter 2). By identifying and dating the oldest sedimentary rocks that were deposited in the basins, geologists can determine when crustal extension began. Throughout southeastern California, sedimentary rocks of this type began to accumulate during the Miocene period, between about 22 and 16 million years ago. Not surprisingly, the timing coincided with early volcanic eruptions caused by crustal extension in southeastern California.

Fig. 7-17, A, A generalized stratigraphic column showing the members of the Miocence Artist Drive Formation on the left and their inferred environments of formation on the right. **B,** The Artist Drive Formation at Zabriske Point, Death Valley. (Source: **A,** adapted from South Coast Geological Society Field Trip Guidebook; **B,** †Skapinsky, S. San José State University.)

As topographic basins were formed by normal faulting, they created a trap for alluvium eroded from nearby mountains and for sediments deposited in lakes within the basins. Later faulting has uplifted Miocene sedimentary units to expose them in several areas of southeastern California. The Miocene sedimentary units include alluvial fan gravels carried from the Miocene ranges into the ancient basins. The gravels contain recognizable pieces of older rocks recycled into Miocene sediments (see Chapter 2). One example of Miocene alluvial fan deposits is the Artist Drive Formation, 14 to 6 million years old, in Death Valley. As the Miocene basin developed, younger lake sediments and intermixed volcanic layers, the Furnace Creek Formation, accumulated about the fan gravels (Fig. 7-17).

BARSTOW BEDS

The Pleistocene lakes discussed in Chapter 6 demonstrate that the past environment in the California desert was vastly different from today's. Another contrast with the present can be seen in the sedimentary layers preserved about 18 kilometers north of Barstow in an area known as the Rainbow Basin. A section of sedimentary rocks up to 650 meters thick records the history of deposition in and around a lake during Miocene time. Miocene extension created the down-dropped basin in which the sediments accumulated. The rocks, which have been named the Barstow Formation, are a mixture of mudstone, sandstone, and conglomerate, with some interbedded ash layers. Based on the ages of volcanic rocks above and below the sediments, the age of the Barstow Formation is placed between 16 and 13 million years ago.

The most interesting feature of the Barstow Formation is that it contains a diverse assortment of fossils. Since 1919, when the fossils were first described, paleontologists have identified fossilized remains of mammoths, dogs, bears, cats, horses, camels, antelope, bison, sheep, turtles, shellfish, flamingos, and palm trees.

Fig. 7-18 Red Rock Canyon, Highway 14, Kern County. (Source: Nakata, J. U.S. Geological Survey.)

RED ROCK CANYON

Travelers on State Highway 14 pass through Red Rock Canyon about 40 kilometers north of the town of Mojave. Even though many do not stop at Red Rock Canyon State Park, they still admire the beautiful red, orange, and pink rocks as they pass by (Fig. 7-18). Those who stop for a closer look will find interlayered volcanic and sedimentary rocks. The sedimentary rocks—orange conglomerate and sandstone, and white mudstone—were deposited in ancient rivers and lakes. Periodic volcanic eruptions added layers of white ash, pink pyroclastic flows, and dark andesite and basalt flows. Within the sedimentary beds are fossils of the same diverse animals that inhabited the Rainbow basin.

The rocks at Red Rock Canyon belong to the Ricardo Group, deposited between 19 and 8 million years ago in a basin similar to the basin where the Barstow Formation accumulated. Like the Barstow Formation, the Ricardo Group records the extension of the crust in the Mojave Desert during Miocene time.

EXTENSION OF THE CRUST IN SOUTHEASTERN CALIFORNIA

Geologists and geophysicists working in the Basin and Range in the western United States agree that ***extension*** of the earth's crust began about 40 million years ago and continues today. In the California portion of the Basin and Range, extension began about 16 million years ago. Some areas, including the Death Valley region, may have been stretched by 100 percent, and geophysicists estimate that the entire Basin and Range province has been lengthened by about 250 kilometers during the past 40 million years. The 10-hour drive from San Francisco to Salt Lake City along U.S.

Fig. 7-23 Preserved algal mounds in the Noonday dolomite, Do Not Pass Hills, southeastern Kingston Range. (Source: McMackin, M. San José State University.)

Fig. 7-24 A reconstruction of a Middle Devonian reef. (Source: Smithsonian Institution.)

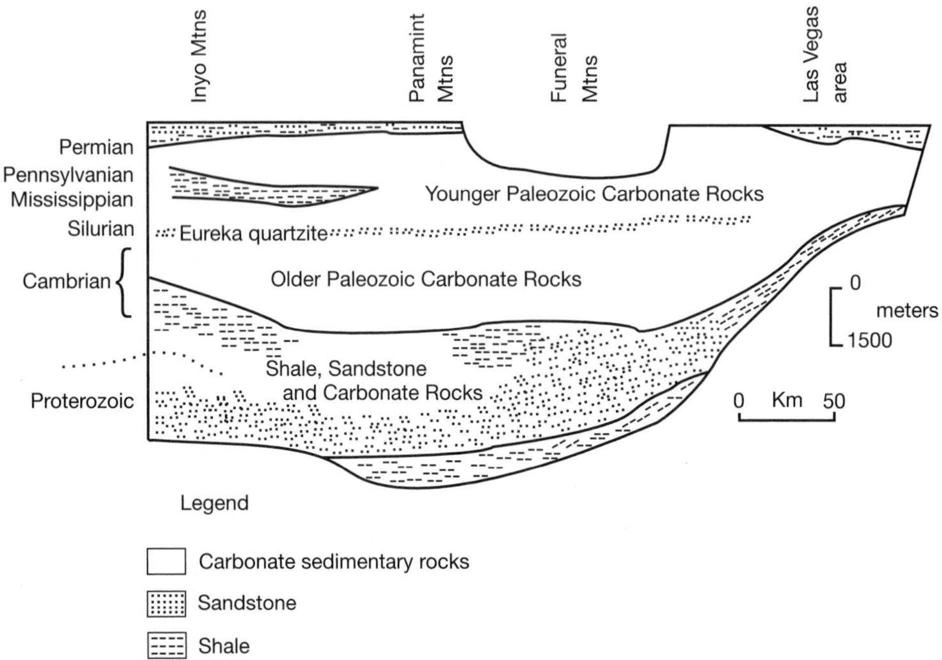

Fig. 7-25 A generalized stratigraphic cross section across the Basin and Range between Mono Lake and Las Vegas, Nevada, showing the general extent of the late Proterozoic and Paleozoic carbonate platform. *PM* = Permian; *PE* = Pennsylvanian; *M* = Mississippian; *S* = Silurian; *O* = Ordovician; *C* = Cambrian; *PR* = Proterozoic; *PC* = Precambrian. (Source: Modified from Burchfiel, P., and Davis, G. 1981. *Geotectonic Development of California.* Upper Saddle River, N.J.: Prentice Hall.)

tions represent more than 7000 meters of deposition (Fig. 7-25). Some individual formations can be found across most of the entire Basin and Range Province—for example, the Eureka Quartzite, a quartz sandstone named for the old mining town of Eureka in eastern Nevada. Fossils found within the formations indicate that the rocks formed from latest Proterozoic time through the entire Paleozoic era, a period of more than 350 million years between about 600 and 250 million years ago (refer to Fig. 3-3, p. 36).

Because many formations can be traced for such great distances to the east, the ancient marine environments where they accumulated must have been fairly uniform over a large area. Geologists reason that the slope of the continental shelf was broad and very gently sloping, so that water depth increased only slightly westward. Any slight rise or fall in sea level would cause the shoreline to shift many miles east or west (Fig. 7-26).

A sample section of rocks in the southern Inyo Mountains records such changes in amazing detail. The type of sedimentary rock and the fossils provide the clues to each formation's environment of deposition. The Al Rose Formation contains intermixed mudstone and carbonate. As indicated by the presence of sedimentary features formed in the tidal zone, part of this unit must have formed in fairly shallow water. Above the Al Rose Formation is the Badger Flat Formation, dolomite con-

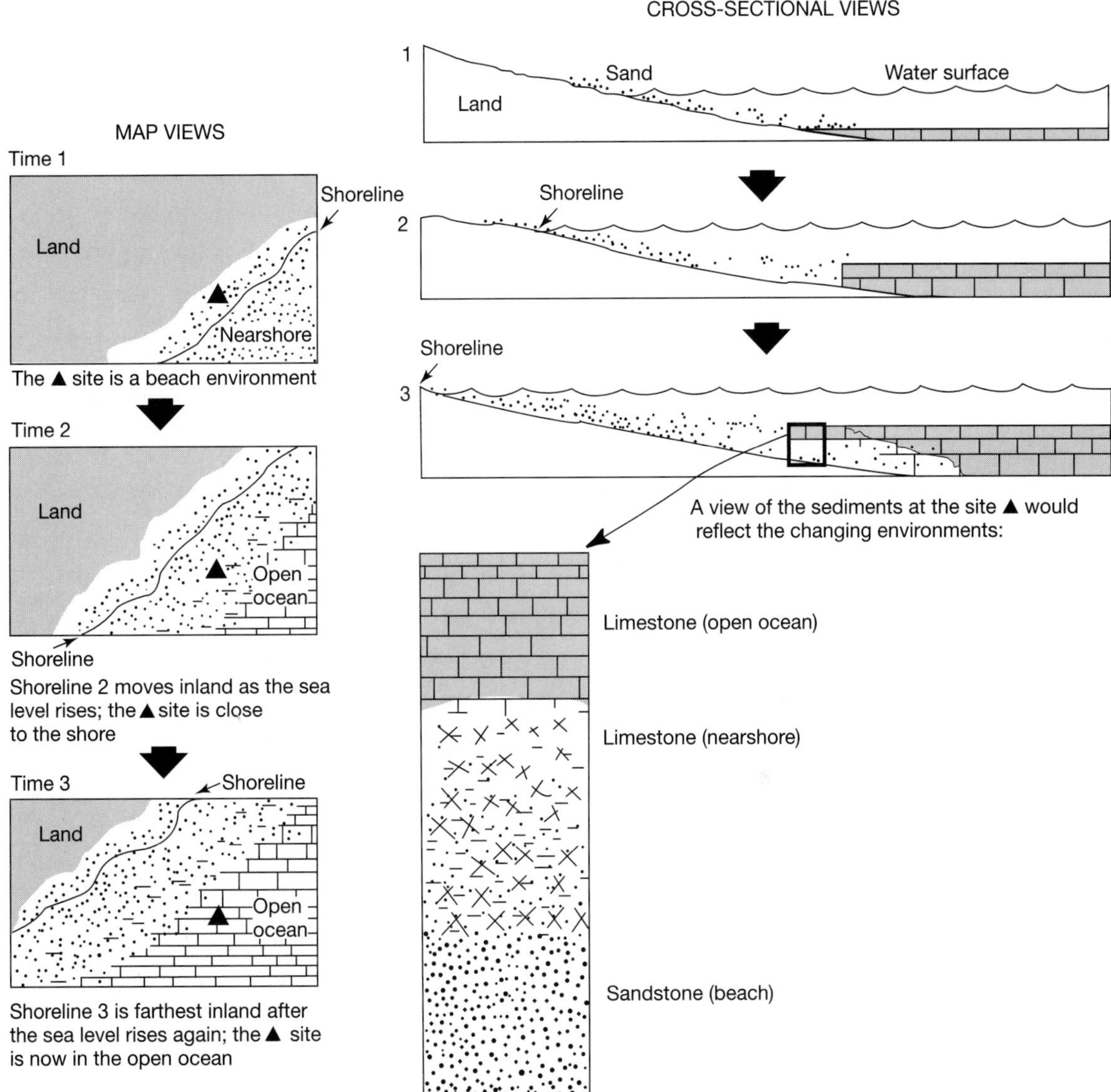

Fig. 7-26 Diagrams illustrating the effects of sea-level change on the position of a shoreline and on sedimentary environments. (Source: Modified from Montgomery, C. *Physical Geology.* Dubuque, Iowa: Wm. C. Brown.)

taining huge snail fossils, formed in shallow marine water (Fig. 7-27; see also Fig. 7-21). By the time the next unit formed, the marine basin had filled in enough with sediment to expose this part of the continental margin above sea level. The sedimentary unit representing these conditions is the Eureka Quartzite, originally deposited on a beach or a sand dune. Finally, the sea rose again to create a shallow marine environment that produced the Ely Springs Dolomite, and then a deeper envi-

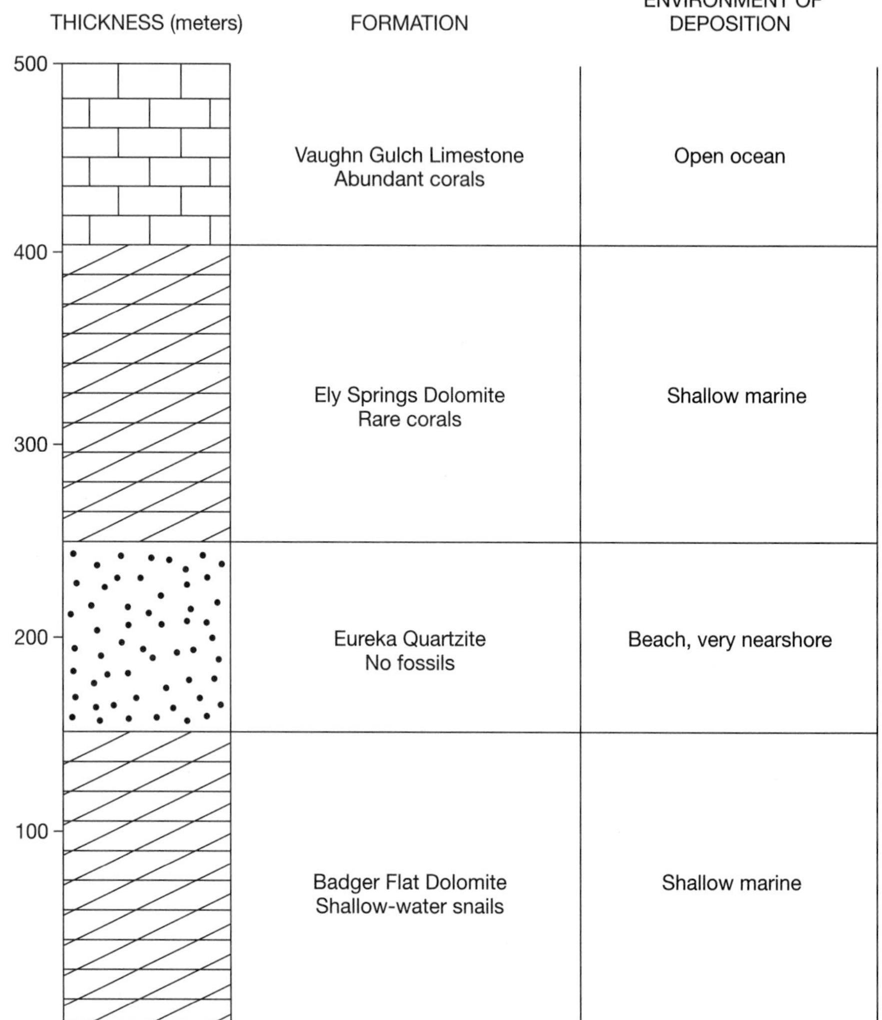

Fig. 7-27 Schematic column illustrating sedimentary formations, fossils, and ancient environments represented by older Paleozoic rocks in the southern Inyo Mountains. (Source: Stevens, C. San José State University.)

ronment that produced the Vaughn Gulch Limestone. Thus, the rising and falling of sea level hundreds of millions of years ago can be reconstructed for a single location. By piecing together this kind of geologic information across the entire Basin and Range, geologists have come to understand some of the workings of an ocean that disappeared from southeastern California more than 200 million years ago.

THE PASSIVE CONTINENTAL MARGIN

West of ancient North America, the edge of the continental margin began to ***subside*** or sink about 600 million years ago, in late Proterozoic time. Geologists believe that the sinking resulted from rifting within the ancient North American

Fig. 7-28 Proterozoic rocks in the Black Mountains, east of southern Death Valley, looking east. (Source: Bradley, W.C.)

continent. As the edge of the continent subsided, sediments gradually accumulated on what was then the continental shelf. The beginning of this 350-million-year-long period of subsidence and sedimentation is represented today by formations composed of quartz sandstone, siltstone, and shale. The early sediments washed onto the continental shelf from the ancient land mass, carried by rivers that flowed from the coastal plain. In the Death Valley Area, the Johnnie Formation, Wood Canyon Formation, and the Zabriskie Quartzite are some of the recorders of these events (Fig. 7-28).

By earliest Paleozoic time, little sediment from the ancient continent to the east was reaching the area that is now easternmost California. As a result, sediments accumulating on the continental margin were composed mostly of carbonate produced by organisms, with only minor amounts of sandstone and siltstone. Today these sediments are preserved as the thousands of meters of Paleozoic carbonates seen in the ranges of the Basin and Range Province and the Mojave Desert. These rocks record the history of the ***stable carbonate platform*** that existed along the western edge of North America for 350 million years. During this time, no dramatic tectonic events occurred in the eastern part of California's Basin and Range or Mojave Provinces.

Active plate boundaries were apparently distant from this ***passive margin*** until the very end of the Paleozoic era. Oceanic plates were in motion somewhere offshore to the west of this margin during Paleozoic time, creating trenches, volcanic arcs, and subduction zones. As we will see in Chapters 8 and 9, these plates interacted with the North American continent in areas that are now part of the Klamath Mountains and northern Sierra Nevada.

It has taken scores of geologists and paleontologists a century to piece together the Paleozoic history of the Basin and Range and Mojave Desert. As the geology of each mountain range is better understood and compared with that of neighboring ranges, a coherent picture has evolved. For about 350 million years, easternmost California was a tectonically quiet, stable continental shelf. At no time before or since that period was California geology so uneventful, and in no other part of California was deposition of sediments less affected by active tectonics. However, when a visitor admires the broken and swirled rocks of the desert ranges, this picture is anything but obvious. The Paleozoic rocks have been greatly folded, faulted, and eroded during Mesozoic and Cenozoic time, and the imprint of this disruption on the rock outcrops is very strong.

BEFORE THE PASSIVE MARGIN: OLDER ROCKS IN SOUTHEASTERN CALIFORNIA

Less is known about the geologic environment of the southeastern California before the development of the passive continental margin, because rocks older than latest Proterozoic age are found only in a few places. The rocks of the Pahrump Group in Death Valley, dated between about 1.2 billion and 700 million years old, appear to document the continental rifting that immediately preceded formation of the passive continental margin. Geologists believe that the rifting occurred as an ancient supercontinent broke apart during this time period (see Chapter 18). In Death Valley, the Kingston Peak Formation (Fig. 7-29), which is about 750 million years old, contains the types of coarse sediments produced along faults or in rift valleys. Abrupt changes in the types and thickness of the sediments indicate that faults were active at that time. Geologists reason that these rocks record some of the continental rifting that gave rise to the passive continental margin.

Rocks older than abut 1.2 billion years old are also exposed in some of the ranges of southeastern California. These rocks are significant because they represent California's oldest known rocks, and they are confined to a very small por-

Fig. 7-29 Proterozoic rocks of the Kingston Peak Formation, eastern Kingston Range. (Source: McMackin, M. San José State University.)

MOUNTAIN PASS MINE

At the Mountain Pass Mine in San Bernardino County, some of California's oldest rocks provide elements necessary for some of our most advanced technologies. Located between Barstow and Las Vegas along Interstate 15, at the southern end of the Clark Range, the Mountain Pass area contains some very unusual plutonic rocks. One of these rocks is ***carbonatite,*** an igneous plutonic rock composed predominantly of carbonate minerals. As outlined in Chapter 2, most igneous rocks are composed of silicate minerals, and carbonate minerals are typically formed in sedimentary environments like those discussed earlier in this chapter. Because carbonatite magma is highly unusual, carbonatites are very rare. Geologists believe that carbonatite magmas formed very deep in the Earth's mantle. The carbonatite at Mountain Pass crystallized about 1.4 billion years ago, when magma intruded even older rocks.

About 9 percent of the carbonatite is the mineral ***bastnasite,*** a rare carbonate mineral. Bastnasite—chemical formula (Ce, La)(CO_3)F—contains varying amounts of the 15 chemical elements known as the ***rare earth*** or ***lanthanide elements.*** The primary rare earth elements in the bastnasite at Mountain Pass are cerium (Ce) and lanthanum (La). Many of the rare earth elements have important commercial uses (Table 7-1), and the carbonatite at Mountain Pass mine is the largest concentration of rare earth elements in the Western Hemisphere. About 2000 tons of rock are mined daily at Mountain Pass, and mining engineers estimate that enough bastnasite remains for another 50 to 60 years of production at the current rate.

Table 7-1 COMMERCIAL USES OF RARE EARTH ELEMENTS MINED AT MOUNTAIN PASS

Element	Uses
Cerium	Polishing compounds for optical glass, camera lenses and TV face plates, decolorizer in glass, catalytic converter ceramics
Lanthanum	Petroleum refining, manufacture of optical fibers, camera lenses, x-ray intensification screens
Neodymium	Magnets, ceramic capacitors, enhancement of brightness and contrast in color television sets
Praseodymium	Magnets, yellow glaze for tile
Europium, yttrium	Red phosphor in color television, fluorescent lighting

tion of the state. Ancient rocks are found in the San Bernardino and San Gabriel Mountains (see the box), several ranges in the Mojave Desert, and in the Panamint Range near Death Valley. They include ancient metamorphic rocks that were metamorphosed about 1.7 billion years ago. Plutonic rocks in the same areas are dated at about 1.4 billion years old.

ACTIVE TECTONICS IN THE MESOZOIC

The tectonic quiet of the Paleozoic ceased in southeastern California at the beginning of Mesozoic time, about 245 million years ago. Throughout California, the passive margin gave way to an active boundary between the North American continent and plates of the Mesozoic ocean to the west. The evidence for this great change can be seen in the rocks of the Basin and Ranges and the Mojave Desert. As we will see in subsequent chapters, the development of an active plate boundary along the western margin of North America has probably been the dominant force shaping California's geology.

Granitic plutonic rocks and volcanic rocks similar to those found along modern subduction zones occur throughout the ranges of southeastern California. Most of the granitic plutonic rocks in the Basin and Range and Mojave Desert have been dated between 200 and 100 million years old. Most of the Mesozoic plutonic rocks are parts of the Sierran Batholith, which is discussed in Chapter 8. In eastern California, granitic magma intruded into the older sedimentary rocks and metamorphosed the rocks along the contacts with the plutons. In many areas, valuable mineral deposits of silver, gold, talc, and other commodities were formed in these contact zones and later extracted by miners.

Evidence for widespread compression during Mesozoic time is also preserved in southeastern California. The older sedimentary formations are greatly disturbed by folds and thrust faults (see Fig. 7-20). These structures can be dated as Mesozoic, because younger rocks are not affected by the folds and faults. In addition, the folds and thrust faults are themselves displaced by the normal faults discussed earlier in this chapter, as are the older rocks. Following the principles of cross-cutting relations outlined in Chapter 3, the normal faults must be younger than the structures created by Mesozoic compression. The Mesozoic igneous activity and the faults and folds produced by compression are typical events that occur along a subduction zone boundary.

FURTHER READINGS

COLLIER, M. 1990. *An Introduction to the Geology of Death Valley.* Death Valley, Calif.: Death Valley Natural History Association, 60 pp.

DOKKA, R., AND OTHERS. 1991. Aspects of the Mesozoic and Cenozoic evolution of the Mojave Desert. In *Geological Excursions in Southern California and Mexico.* Geological Society of America Annual Meeting Guidebook, pp. 1-43.

FIERO, B. 1986. *Geology of the Great Basin.* Reno, Nev.: University of Nevada Press.

HILL, M. 1972. The great Owens Valley earthquake of 1872. *California Geology* 25(3):51-61.

HUNT, C.B. 1975. *Death Valley.* Berkeley: University of California Press, 234 pp.

OAKESHOTT, G.B., Greesfelder, R.W., and Kahle, J.E. One hundred years later. *California Geology* 25(3):51-61.

PRIDMORE, C., AND FROST, E.G. 1992. Detachment faults. *California Geology,* 45(1):3-17.

REYNOLDS, R. 1991. *Crossing the borders: Quaternary studies in eastern California and southwestern Nevada.* Redlands, Calif.: Mojave Desert Quaternary Research Center, Special Publication.

TROXEL, B.W., AND WRIGHT, L.A. 1976. *Geologic Features of Death Valley, California.* California Division of Mines and Geology, Special Report 106, 72 pp.

CHAPTER

8

The Sierra Nevada
Granite, Gold, and Glaciers

Stretching for 600 kilometers along much of California's eastern border, the Sierra Nevada has been the geographic key to California's history. The mountain range itself was a barrier to those seeking California, but the gold of the western foothills brought immigrants by the thousands. The range is responsible for the rain and snow dumped by Pacific storms (see Chapter 6), and without the Sierra Nevada's water, California could not support even a fraction of its 31 million people (see Chapter 10). The Sierra Nevada is also home to some of the world's most spectacular scenery, including Yosemite, Sequoia, and Kings Canyon National Parks.

The Sierra Nevada Province includes the Sierra Nevada range and a broad belt of foothills on its western flank. From its southern end east of Bakersfield to its northern end south of Lassen Peak, the Sierra Nevada trends north-northwest, with a slightly more north-south trend than the Coast Ranges (see cover image and Endpaper 4, inside the back cover). The range is between 65 and 130 kilometers wide, with a broad region of foothills along the western slope.

In 1776, Father Pedro Font, a Spanish missionary and a member of the Anza expedition, which had journeyed from Mexico to settle along San Francisco Bay, made a sketch from a hill near the mouth of the Sacramento River. About 200 kilometers to the northeast he had seen *"una gran sierra nevada"*—a great snowy range—running from south-southeast to north-northwest. Font was not the first European to view the Sierra Nevada, which had been observed by Father Crespi 4 years earlier, but he is credited with naming California's most famous mountain range. Trapper Jedediah Smith became the first non–Native American to cross the range in 1827. Earlier that year, Smith and 17 others had crossed the Mojave Desert and arrived at the San Gabriel mission, where they were promptly invited to leave by the local authorities. After traveling north along the San Joaquin Valley, Smith and two companions struggled eastward through deep snow across what is today Ebbetts Pass.

The box on p. 142 provides some helpful information about mountain ranges.

PEAKS AND PASSES

For those unfamiliar with mountainous terrain, it is useful to review the geographic elements of a mountain range. A ***mountain range*** is a chain or group of mountains consisting of a number of individual ***peaks*** or mountains. In the Sierra Nevada, individual mountain peaks stand above the surrounding terrain—for example, Mt. Whitney (Fig. 8-1) or Mt. Coness. The ***summit,*** or highest point, of Mt. Whitney stands 4418 meters (14,495 feet) above sea level. Some peaks in the Sierra Nevada stand out less obviously because the surrounding areas are of similar elevation.

An imaginary line connecting the summits and highest parts of the ridges along a mountain range is the ***crest*** of the range (Fig. 8-2). The crest of the Sierra Nevada, which includes its most spectacular summits, is highest in the southern part of the range near Mt. Whitney, where 12 peaks reach elevations greater than 4267 meters (14,000 feet). At the lowest northern end of the range near the Feather River, the crest is between 1800 and 2450 meters. Along a 30-kilometer-wide band from north of Yosemite National Park to south of Mt. Whitney lies the part of the Sierra Nevada known as the High Sierra. Because the crest of a mountain range marks its highest points, streams flow from the crest to lower elevations on both sides of the crest. The crest of the Sierra Nevada is therefore also the ***divide*** between the watersheds of the streams that flow east and west from the crest (see Chapter 10 and Fig. 10-5 for further information about watersheds). The Sierra Nevada's crest was also chosen as the boundary between many of California's mountain counties, so it is easily traced on even a simple road map.

Fortunately for those traveling across the Sierra Nevada, the crest of the range has points of lower elevation, even in the higher southern part of the range. These low notches in the crest are the ***passes*** across the range. The major passes of the Sierra Nevada crest are at the heads of the larger river valleys, most of which have been shaped by glaciers. Some of them, including Mono Pass near Yosemite, were well traveled by Native Americans seeking obsidian at Mono Craters, trading with groups in the opposite side of the range, or escaping the summer heat. During the 1800s the passes became the key element of trails used by the pioneers headed for California. Because of the greater height of the southern crest, the trails were limited to the northern Sierra Nevada, and even today no roads cross the highest part of the range. The most heavily traveled route today is Interstate 80, which crosses the Sierra Nevada at Donner Pass (elevation 2175 meters), one of the lower passes through the range.

Fig. 8-1 View of the southern Sierra Nevada, looking westward. Mt. Whitney, California's highest peak, appears in the center of the image beneath the small cloud and behind the range of mountains in the foreground. (Photo by author.)

Fig. 8-2 Oblique aerial photograph of the Kaweah peaks area, southern Sierra Nevada, looking westward. The steep mountains in the foreground show prominent peaks and a sharp ridge crest. Note also the bowl-shaped, snow-filled glacial cirques in the foreground. (Source: U.S. Geological Survey.)

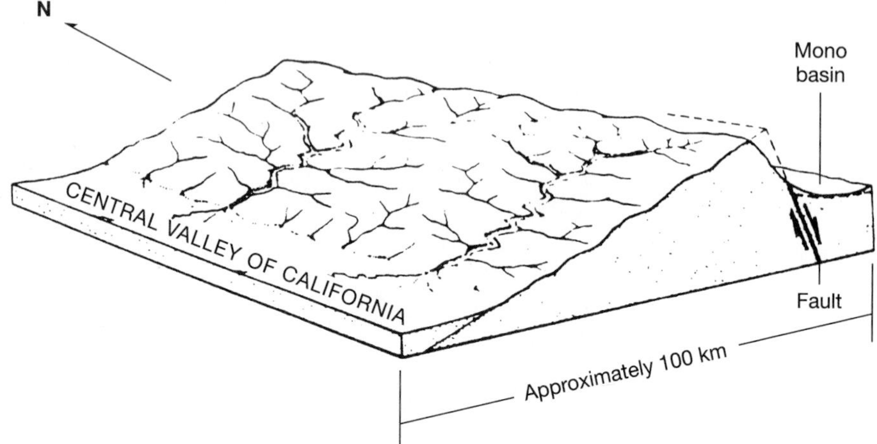

Fig. 8-3 Uplift and westward tilting cause the asymmetry of the Sierra Nevada. (Source: Huber, N.K. 1987. *The Geologic Story of Yosemite National Park.* U.S. Geological Survey, Bulletin 1595.)

FORMATION OF THE SIERRA NEVADA RANGE

Any driver who crosses the Sierra Nevada going east on Interstate 80 learns firsthand that the mountain range is not symmetrical. From the base of the mountains near Roseville, the drive up the western slope to Donner Pass is about 130 kilometers, but the drive from Donner Pass down to Reno is only about 55 kilometers. Along the entire range north of the Kings River, the western slope is more gradual than the steep eastern front (Fig. 8-3). Active normal faults that mark the western edge of the Basin and Range Province account for the steepness of the eastern Sierra Nevada front (Chapter 7). Uplift of the range along the eastern faults has been accompanied by westward tilting so that the entire Sierra Nevada has been tipped westward, producing the gently sloping western slope. However, faults are present along the western edge of the range as well, and some may be active. For example, a 1975 earthquake at Oroville along the western fault system was apparently triggered by the weight of the water filling the reservoir behind the Oroville dam.

Through a series of fortuitous events, the uplift and tilting of the Sierra Nevada have been preserved in the geologic record with enough detail that geologists are able to reconstruct when the present mountains formed. Evidence suggests that the Sierra Nevada mountains are very young by geologic standards: much of the uplift and tilting of the modern range has taken place in the past 5 million years. Before 9 to 10 million years ago, rivers flowed westward into the Central Valley across what is now the crest of the Sierra Nevada. Sediments from those ancient rivers were then buried by huge lava flows, ash flows, and volcanic mud flows about 9 to 10 millions years ago. Sources of some of the volcanic eruptions were east of the present Sierra crest—one east of Sonora Pass and one northeast of Mono Lake. Volcanic flows traveled down stream channels into the ancient Stanislaus, Tuolumne, and San Joaquin rivers (Endpaper 3, inside the back cover). Clearly, no large mountain range blocked their path.

Fig. 8-4 A sketch of Table Mountain near Sonora, taken from an 1886 guidebook to Yosemite Valley. Table Mountain is the flat-topped mountain in the center.

Geologists can reconstruct the general topography of the Sierra Nevada using the 9- to 10-million-year-old volcanic flows and the alluvial sediments they buried. Because only fragments of the original flows and sediments have been preserved, the remnants must be correlated to reconstruct the ancient valleys. By estimating the original slopes of the 10-million-year-old rivers, it is possible to determine the amount of uplift and tilting that have occurred since that time. The estimated uplift of the Sierra Nevada since the flows were deposited varies from about 1830 meters at Tioga Pass to 2150 meters at Deadman Pass at the head of the San Joaquin River.

The best known of the late Cenozoic volcanic flows in the Sierra Nevada forms Table Mountain (Fig. 8-4), which can be seen west of Sonora along Highway 108. Viewed from the air or on a geologic map (Fig. 8-5), Table Mountain looks strikingly similar to a sinuous river, with one major difference. Rather than winding along a valley bottom, it forms a flat-topped mountain. Table Mountain is capped by 90 meters of volcanic rock, and underneath are sediments of the river channel that was filled with lava 9 million years ago. At the time of eruption, Table Mountain was a low-lying river valley, but once it filled with volcanic rock, it became resistant to erosion. The areas surrounding Table Mountain have been eroded at a faster rate, so that they are now lower than the lava-filled valley. As a result, the topography has become reversed or ***inverted,*** with former valley bottoms now the highest part of the landscape (Fig. 8-6). Other table mountains formed in the same way are found along the western flank of the Sierra Nevada; for example, San Joaquin Table Mountain. Table mountains are excellent examples of the process of ***differential erosion,*** which is important in shaping all landscapes. Over time, areas underlain by weak rocks are eroded more than areas underlain by resistant rocks.

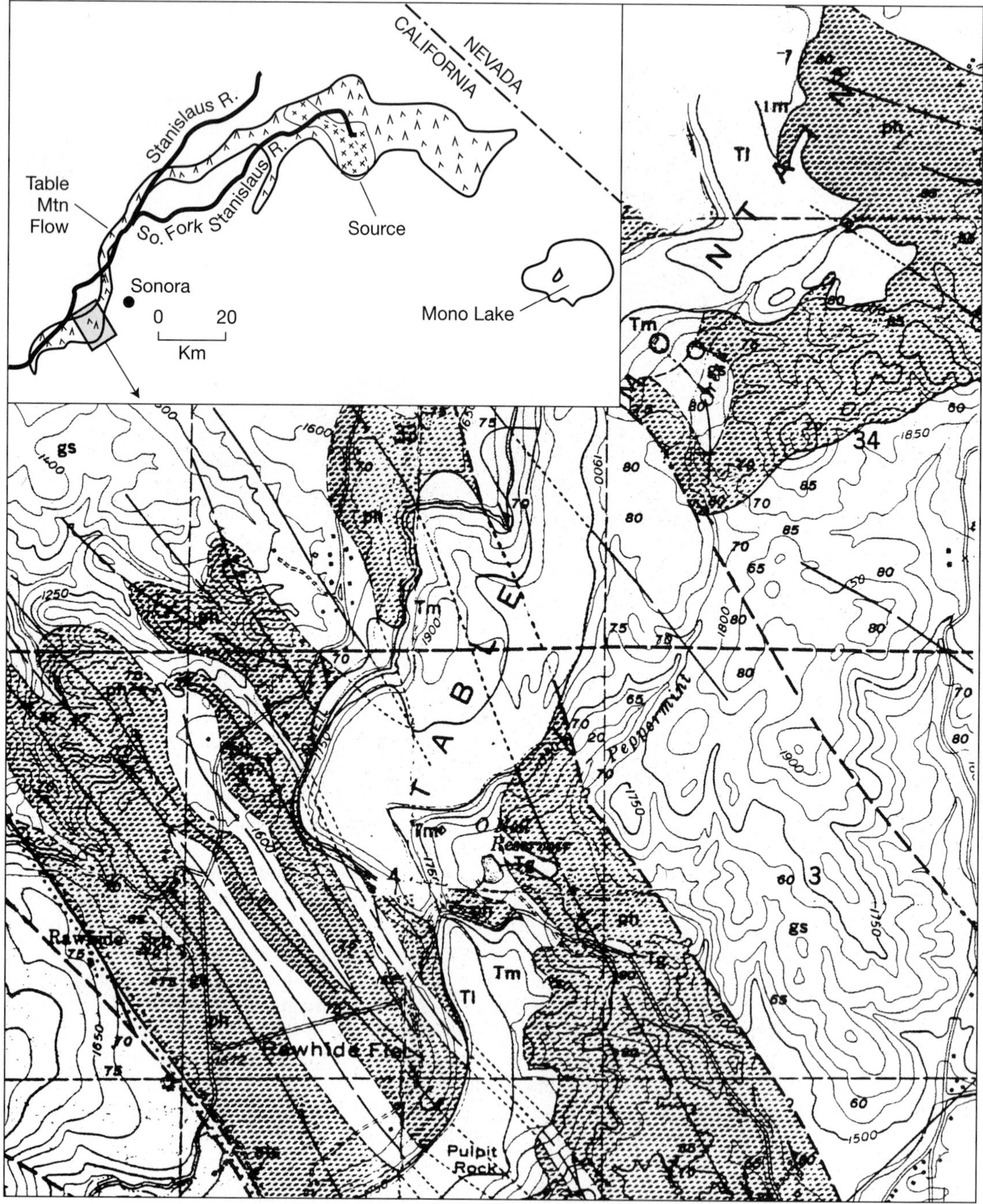

Fig. 8-5 A portion of the Sonora geologic map, showing the sinuous pattern of the Table Mountain volcanic flow. Areas covered by the flow are labeled *Tm* and *Tl.* On the location map (*inset*), locations of recent volcanic flows are shown in the V-pattern area. (Source: Adapted from Heyl, G., and others. 1955. *Geologic Map of the Sonora Quadrangle.* California Special Report 41.)

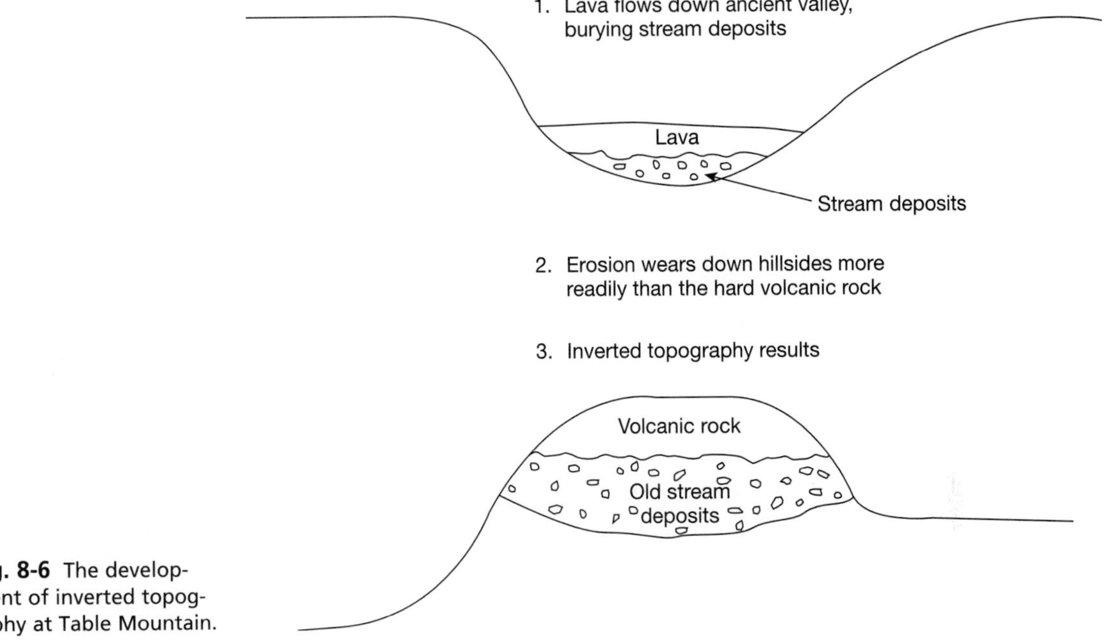

Fig. 8-6 The development of inverted topography at Table Mountain.

THE SIERRAN BATHOLITH: ROOTS OF AN ANCIENT VOLCANIC ARC

Today's Sierra Nevada mountains are youthful by geologic standards, but most of the rocks of the Sierra Nevada are much older. Visitors to the High Sierra can see vast expanses of bare rock surfaces, recently scraped clean by glaciers and exposed at high elevations too cold for most vegetation (Fig. 8-7). These rocks provide evidence that, during Mesozoic time, an older mountain range must have stood in about the same position as today's mountains. The older mountains and the rocks beneath them were a direct result of the tectonic upheaval that began when the passive margin gave away to subduction along the western margin of North America. Although the Mesozoic mountains have long been eroded away, the rocks of the Sierra Nevada remain as testaments to their existence.

Today we have a first-hand look at the working of the ancient magma chambers that were the roots of the Mesozoic mountains. Within the space of a 10-minute stroll, we can observe rocks that originally cooled and solidified 10 or more kilometers below the Earth's surface, at temperatures of about 750° to 800° C (1800° to 2000° F). This amazing view has been provided by the tremendous uplift of the Sierra Nevada and the erosion of thousands of meters of sediment and rock from above the crystallized magma. In fact, without the workings of the rock cycle (Chapter 2) and the dynamics of plate tectonics, plutonic rocks would never be seen at the Earth's surface.

Even a very quick look at the California geologic map (Endpaper 1) reveals that Mesozoic plutonic rocks make up most of the Sierra Nevada—they constitute the Sierran ***batholith,*** a word combined from the Greek words *bathos* (deep) and

Fig. 8-7 El Capitan, Yosemite Valley. (Source: Sarna-Wojcicki, A. U.S. Geological Survey.)

Fig. 8-8 Closeup view of the Cathedral Peak granodiorite near Saddlebag Lake, showing large rectangular feldspar crystals. (Photo by author.)

lithos (rock). The plutonic rocks that make up the Sierran batholith crystallized from many different batches of magma to form hundreds of recognizable ***plutons*** that geologists have distinguished by their mineral components and ages. The magma emplacement and crystallization were hardly instantaneous, even by geologic standards. The oldest plutonic rocks are about 210 million years old, and the youngest plutons crystallized about 80 million years ago. A single body of magma might crystallize over thousands of years, allowing large crystals to grow slowly to several centimeters in length (Fig. 8-8).

Almost all of the rocks in the batholith are granitic in their composition; that is, relatively high in silica. Because of the high silica content of the magmas from which they crystallized, granitic rocks contain abundant crystals of light-colored quartz and feldspar, minerals rich in silica. Set within the light minerals are randomly distributed dark silicate minerals, most commonly hornblende and biotite mica. lending an overall salt-and-pepper look—in more silicic rocks, mostly "salt," but in intermediate and mafic rocks, a more even mixture of light and dark (Figs. 8-9 and 8-10).

Because they crystallized from magmas of somewhat different compositions, granitic rocks contain variable proportions of silicate minerals. Depending on the

Fig. 8-9 Closeup view of Half Dome granodiorite showing large crystals of hornblende. (Source: Huber, N.K. U.S. Geological Survey.)

Fig. 8-10 Three distinct episodes of magma intrusion can be seen in this single rock along Highway 120 in Yosemite National Park. Following the principles of cross-cutting relations outlined in Chapter 3, we can see that the light-colored quartz vein was the last to form, because it cuts through both of the other plutonic rocks. (Source: Sarna-Wojcicki, A. U.S. Geological Survey.)

relative amounts of quartz and the two feldspar minerals, the most common granitic rocks of the Sierra Nevada are classified as granite, granodiorite, or tonalite; granodiorite is the most abundant (Fig. 8-11). Plutonic rocks such as diorite, with more dark minerals and little or no quartz, are also present in the Sierra Nevada, but are much less abundant than granitic rocks.

Many decades of study in the Sierran batholith have led geologists to abandon the idea that it is a simple and homogenous body of granitic rock. Today, geologists continue to recognize distinct plutons of variable composition, structure and age. The differences seen within the rocks of the batholith reflect the complexities of magma generation. Important factors that influence the composition and texture of rocks within the batholith include the origin of the magmas, the changing composition of the magmas as they partially crystallized, and the variable depth of solidification. By examining the plutonic rocks of the batholith within the period of greatest magma generation, about 100 million years ago, it is possible to "see" several levels of the batholith exposed in different parts of the Sierra Nevada today (Fig. 8-12).

Many lines of geologic evidence indicate that the Cretaceous plutons in the central and northern Sierra Nevada, as well as those in the Peninsular Ranges (Chapter 17), appear to have been emplaced in the crust at depths of less than 10 kilometers. One key used to infer the emplacements depth of plutonic rocks is the metamorphic minerals created in the surrounding rock by the magma's intrusion. The temperature and pressure conditions for the formation of some of these mineral are so restricted that geologists use these minerals as ***geobarometers*** and ***geothermometers.*** The presence of one or more of these metamorphic minerals–for example, andalusite–can be a reliable indicator of the depth at which magmas were emplaced. On a larger scale, the type of deformation of the rocks can indicate whether they were brittle or plastic when the magma was emplaced, as can the sharpness of contacts between plutons and the shapes of the plutons themselves. These physical properties also provide evidence of the depth of magma emplacement.

Plutons of Cretaceous age in the northern and central Sierra Nevada are clearly separable from each other; that is, contacts between different bodies of magma are sharp. Geologists are able to distinguish multiple plutons, indicating that discrete batches of magma were injected from below. Because most of the plutonic rocks are rich in silica, geologists infer that magmas rose through a thick section of silica-rich continental crust, partially melting it, mixing with the melt, and thus enriching the magma in silica. Many of the intruding magmas ballooned outward as they rose within the crust, because the pressure of the overlying rocks decreases at higher levels. The rocks at the margins of the plutons were plastically deformed by the heat. Evidence of this deformation can be seen in the rocks surrounding the Tuolumne series in Yosemite Valley.

The plutonic rocks along the southern "tail" of the Sierra Nevada, where it is abruptly cut off by the San Andreas fault, are surprisingly different from those seen to the north. Here, in the Tehachapi and San Emigdio Mountains, is a belt of mafic plutonic rocks and metamorphic rocks which appear to have formed at very high temperatures and pressures (Fig. 8-13). The most abundant rock in the area is mafic ***gneiss,*** a banded rock composed mainly of hornblende and plagioclase feldspar, sometimes containing red garnets as large as 10 cm in diameter. Based on its composition, the most likely source rock for the gneiss was material much lower in silica than continental crust. Compared with the central and

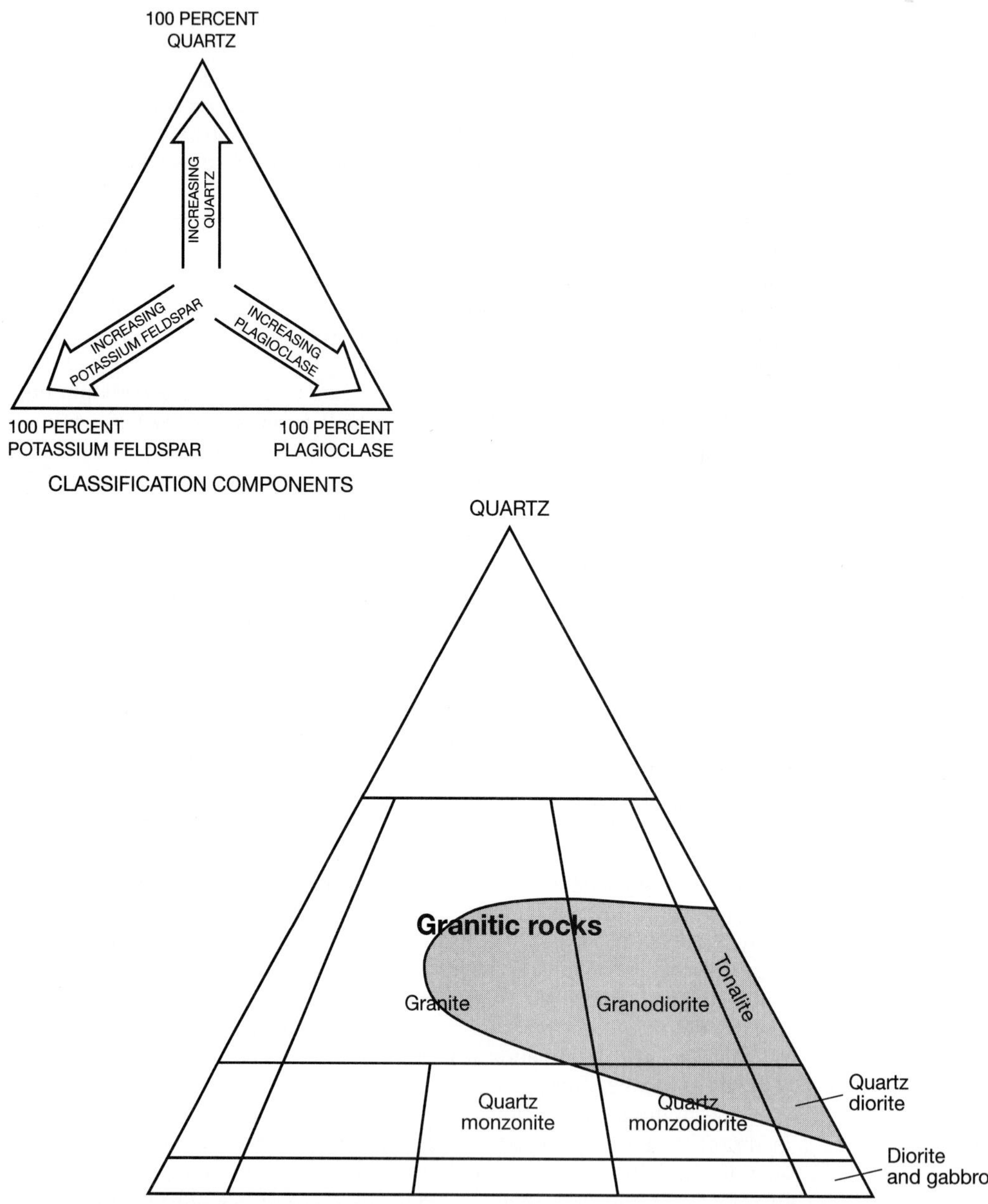

Fig. 8-11 Proportional content of quartz and the two feldspars used to classify common plutonic rocks. Most Sierran plutonic rocks fall within the shaded area. (Source: Huber, N.K. 1987. *The Geologic Story of Yosemite National Park.* U.S. Geological Survey, Bulletin 1595.)

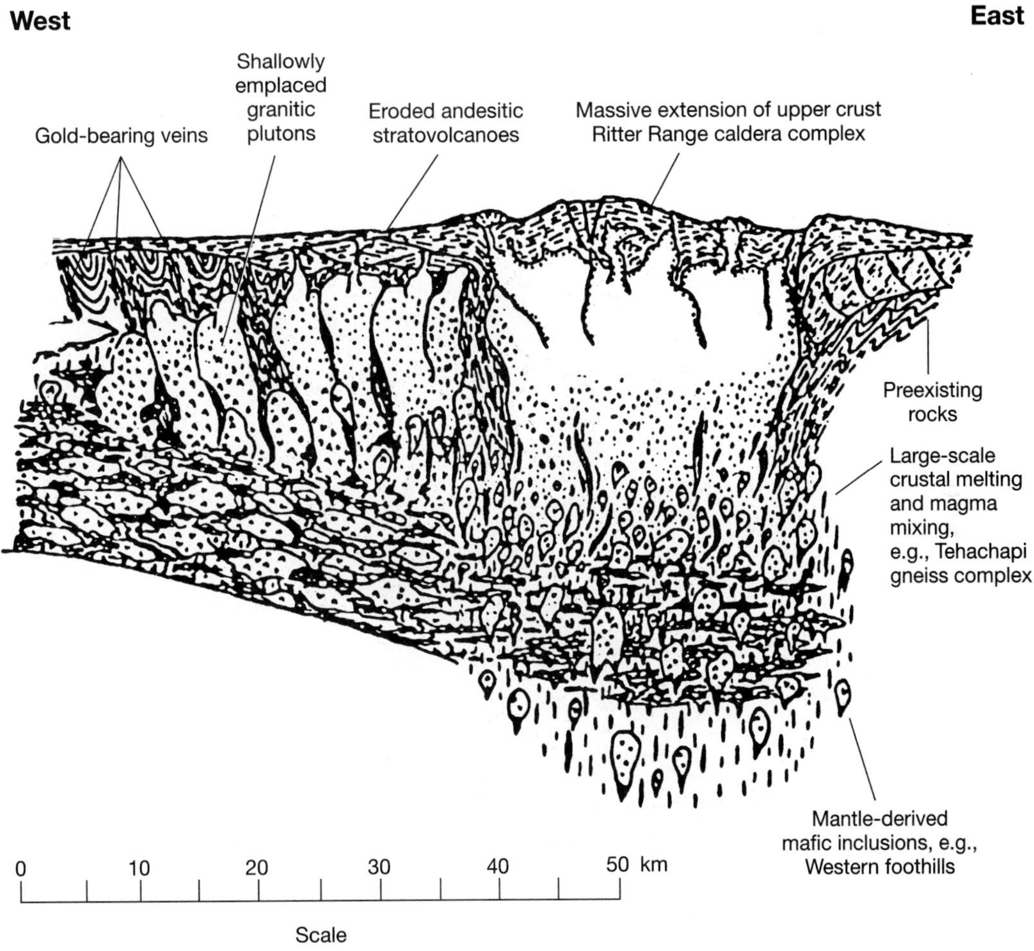

Fig. 8-12 Schematic cross section of the plutonic rocks in the southern Sierra Nevada and the different levels of the Mesozoic magma chambers they appear to represent. (Source: Modified from Saleeby, J. 1990. *Exposed Cross-Sections of the Continental Crust.* Norwell, Mass.: Kluwer Academic Publishers, p. 143.)

northern Sierra Nevada, the contacts between plutons are hard to define in this part of the batholith. In many areas, contact zones are gradational and the metamorphic rocks are ***ultrametamorphosed***—that is, they are difficult to distinguish from igneous rocks.

All of the characteristics of the southern Sierra Nevada batholith indicate that the plutonic rocks crystallized at a much greater depth than those seen further north. The mafic gneisses and mafic plutons in the southwestern Sierran foothills indicate that magma was emplaced in the middle crust at about 30 kilometers depth. Here, then, are the "roots" of the batholith (see Fig. 8-12); that is, these rocks have been brought to the surface from the deepest levels of magma emplacement. Further north, small chunks of unmelted rock similar to the mafic rocks of the southern Sierra, known as ***inclusions,*** are found in the more typical granitic plutons. It is

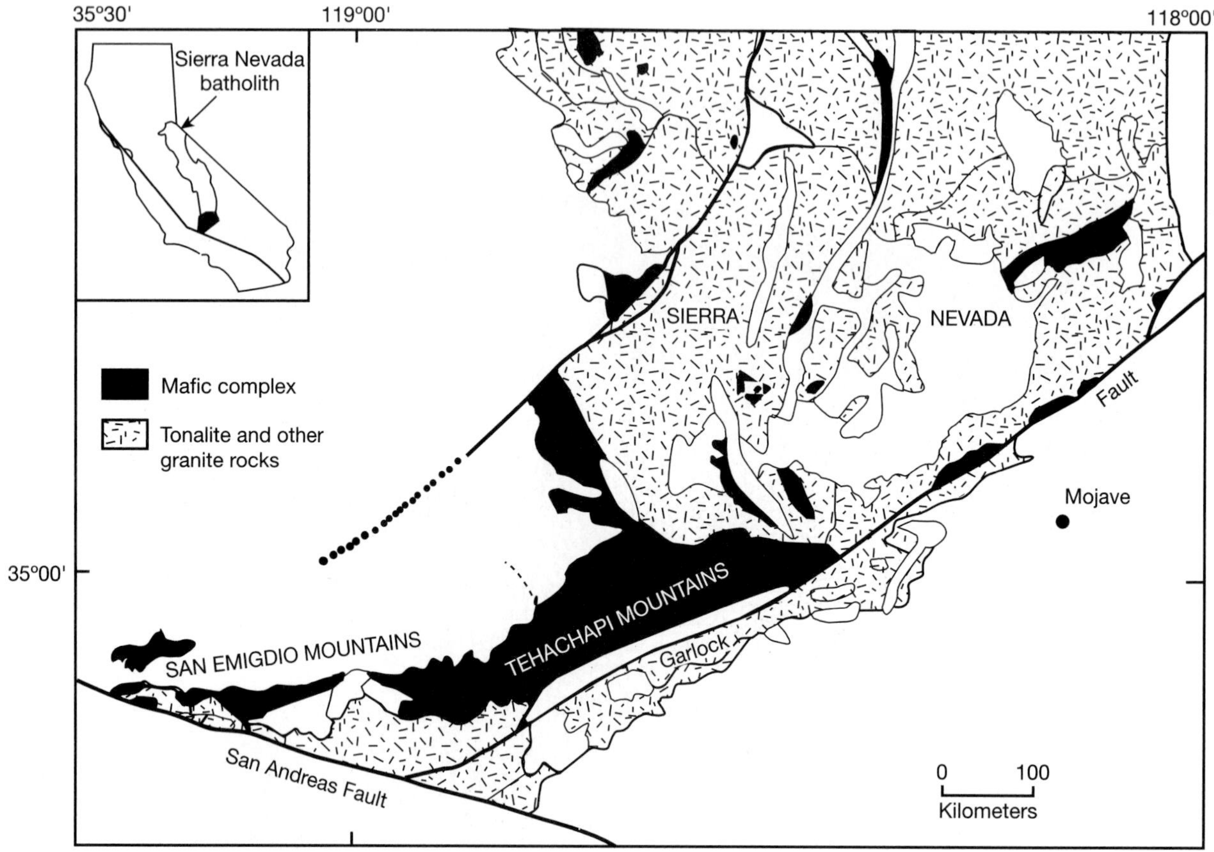

Fig. 8-13 Simplified geologic map of the southernmost Sierra Nevada. Mafic plutonic rocks and gneiss are found in dark-shaded areas, and granitic rocks in the patterned areas. (Source: Ross, D. *Geology* 13[4]:289, 1992.)

likely that rocks similar to those seen at the batholith's tail underlie the granitic plutons further north.

About 280 kilometers to the north-northeast of the batholithic roots are the peaks of the Minarets and the Ritter Range (Fig. 8-14). These dark, jagged pinnacles stand in stark contrast to the smooth, rounded granitic slopes for good geologic reason: they are composed of an entirely different rock type. Naturalist John Muir called them black slate during his climbing adventures of 1873. Geologists working in the area today have recognized them as metamorphosed volcanic rocks that were originally erupted about 100 million years ago. They have thus discovered parts of the Mesozoic volcanoes that must have sat above the Sierran batholith in the same fashion as the Andres Mountains today lie above a huge batholith. In the area near the Minarets, a large caldera covered about the same-sized area as the more modern Long Valley Caldera (see Chapter 5). Pyroclastic flows, unusual deposits formed by the collapse of huge chucks of the caldera walls, and even part of the ancient caldera rim have been recognized by careful mapping and study of the composition of these rocks (Fig. 8-15). The

Fig. 8-14 Granitic and metavolcanic rocks near Mt. Humphreys, Sierra Nevada. The granitic rocks in the foreground are lighter colored and smoother, contrasting with the dark, jagged metamorphic rocks behind them. (Source: †Cooper, A.)

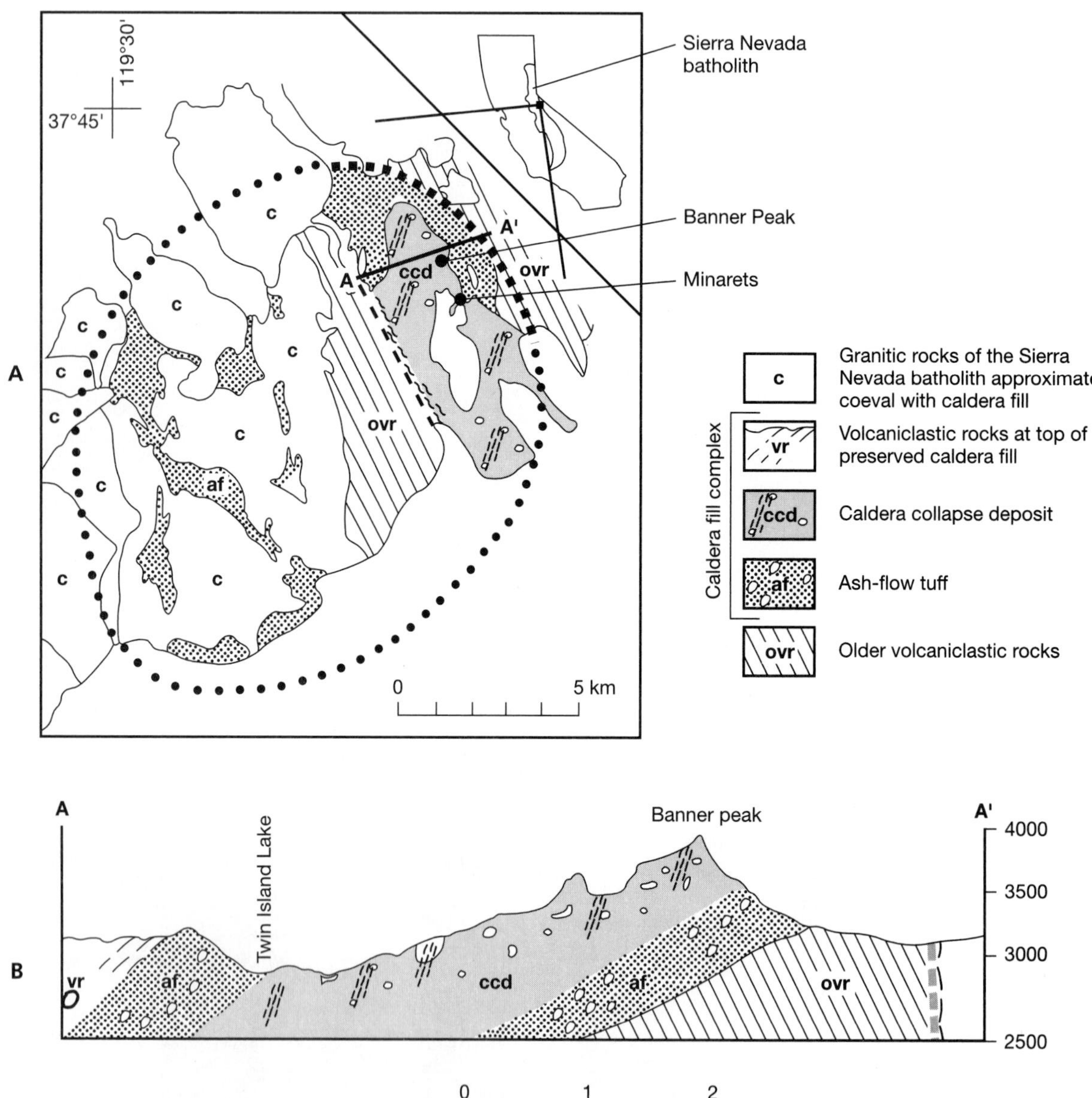

Fig. 8-15 **A,** Map and **B,** cross section showing the extent of the 100-million-year-old Minarets Caldera, reconstructed from the metavolcanic and plutonic rocks found in the area today. Source: Fiske, R., and Tobisch, O. 1994. Geological Society of America Bulletin, Vol. 106[5].)

recognition of the Minarets Caldera has come about only recently, because it was necessary to "see through" all of the later metamorphism, intrusion of magma, and faulting that have affected the rocks in the Sierra Nevada.

ORIGIN OF THE BATHOLITH

Geologists agree that batholiths throughout the world are formed at plate margins where oceanic and continental plates converge. Today the Andean volcanic arc is the surface expression of an immense quantity of magma being emplaced in the South American Plate as the Nazca Plate descends beneath it to the west (see Figs. 1-1 and 1-7). During Mesozoic time, the Sierran batholith was presumably overlain by an arc of active volcanoes similar to today's Andes Mountains (Fig. 8-16). In fact, volcanic

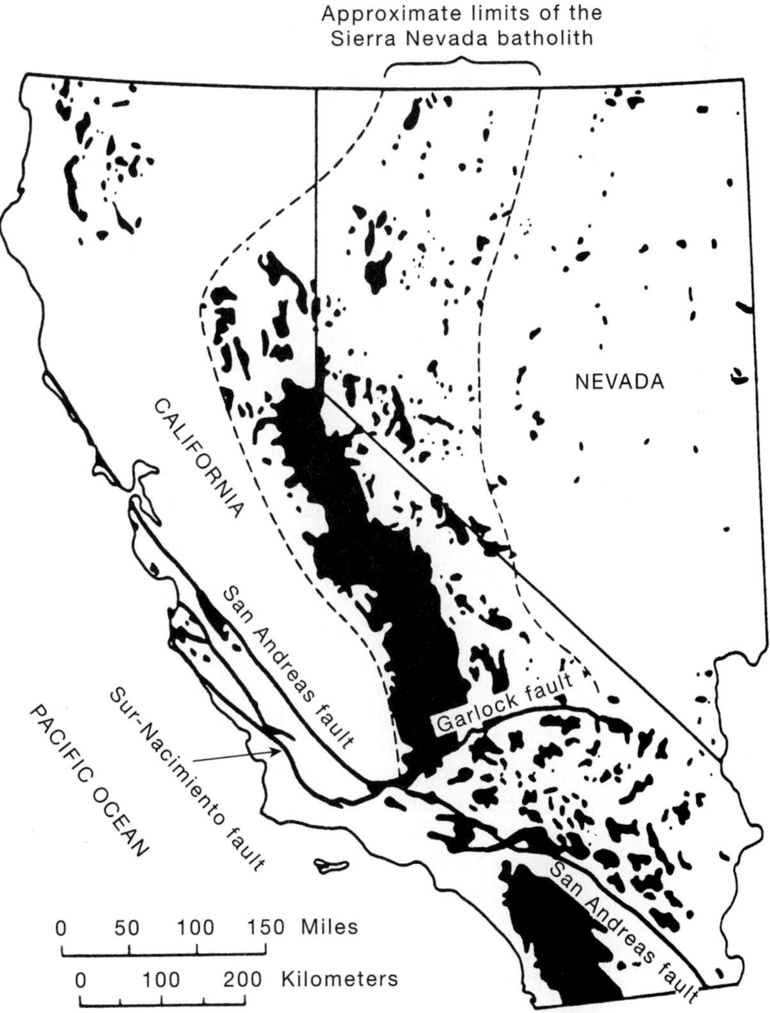

Fig. 8-16 Map showing the extent of the Sierra Nevada batholith. (Source: Norris, R., and Webb, R. 1990. *Geology of California.* New York: John Wiley & Sons. From Geological Society of America.)

ashes found within Cretaceous sedimentary rocks in Colorado are believed to record some of the eruptions from those now-gone volcanoes.

The generation of batholiths has long been a mystery to geologists. It is currently thought that the formation of granitic batholiths is fueled by water. Water trapped in the subducting oceanic crust is driven off as the plate descends. This water "leaks" into the mantle above it and lowers the melting point of the surrounding rocks, causing them to partially melt. Laboratory experiments have demonstrated that water effectively lowers the melting temperature of many silicate minerals. As heated basaltic magma rises from the mantle into the lower crust, some of it mixes with large amounts of silica-rich magma, generating the batholith-forming granitic magma. Some magma rises to the Earth's surface was less mixing, producing the andesitic lavas of volcanoes like the Andes or the Cascades.

One unresolved problem of batholiths is a question of space. How can such a huge volume of magma be emplaced in the crust, and, what happens to the materials that were already there? One obvious way to make room for the batholith is to incorporate crustal materials into the magma, but this does not in itself create enough room. Other mechanisms include the uplift and erosion of materials above the batholith and the removal of material by volcanic eruptions. Another mechanism favored by some geologists is the creation of space by extension of the crust in a direction perpendicular to the axis of the batholith. Extension or transform faulting would favor the continued upwelling of magma into the extended crust.

PLATE MOTIONS AND THE BATHOLITH

Plutons in the central Sierra Nevada do show some regular patterns of ages and types of rocks. The oldest plutonic rocks, about 210 million years old, are found along the eastern edge of the Sierra Nevada west of Mono Lake. Slightly younger plutons of Jurassic age are found on both the western and eastern edges of the batholith. The most voluminous plutons, accounting for more than 90 percent of the batholith, are those emplaced between about 120 and 80 million years ago during the Cretaceous period. These form the center of the batholith and include the Half Dome granodiorite, the Cathedral Peak granodiorite, and the other famous granitic rocks of the Yosemite area. During this latest phase of activity in Cretaceous time, plutonic activity moved systematically eastward (Fig. 8-17). The emplacement of magma was episodic during this period, so that today we find north-trending groups of older Cretaceous plutons in the western Sierra Nevada and younger Cretaceous plutons in the eastern Sierra.

The history of the Sierran batholith is an important record of the subduction history along the western margin of North America. As geologists learn more about the exact timing and location of magma emplacement, they can track the motions of the oceanic plates that collided with North America during Mesozoic time. For example, the fact that the greatest volume of plutonic rock is between 150 and 80 million years old tells us that the subduction ultimately responsible for the magmas was most vigorous at that time. The episodic emplacement of magmas, which can be deduced from the clustering of ages of various plutons, probably reflects episodic periods of more rapid subduction.

Geologists speculate that about 80 million years ago, the rate of subduction of the oceanic plate to the west, known as the Farallon Plate, increased, and the angle of subduction flattened out. As a resült of this change in geometry, the zone of magma generation shifted to the east, "shutting off" the Sierran batholith and triggering magmatic activity to the east (Fig. 8-18).

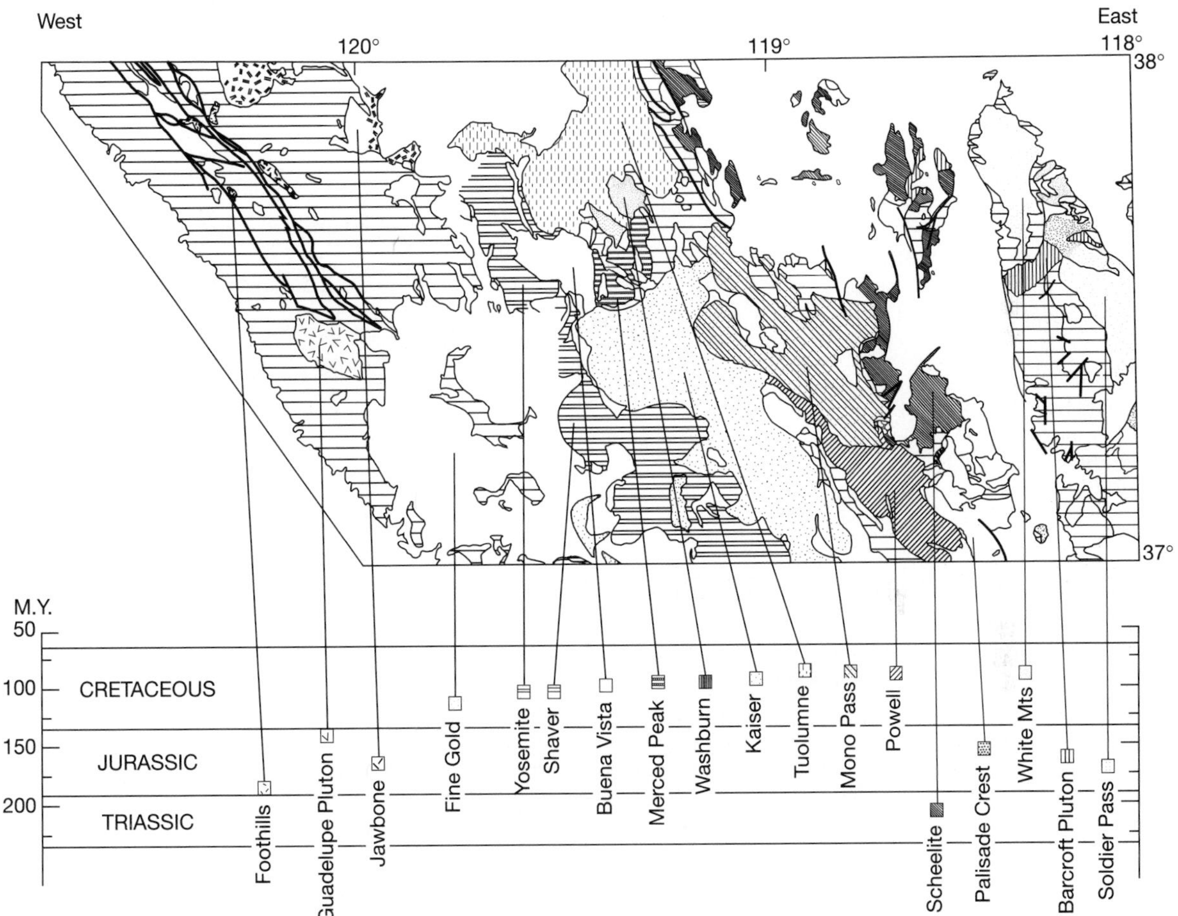

Fig. 8-17 Diagram showing the ages of plutonic rocks in the central Sierra Nevada, demonstrating that plutonic activity migrated eastward during Cretaceous time, between crystallization of the *Fine Gold* and *Powell* Plutons. (Source: Bateman, P. 1983, Geological Society of America, Memoir 159.)

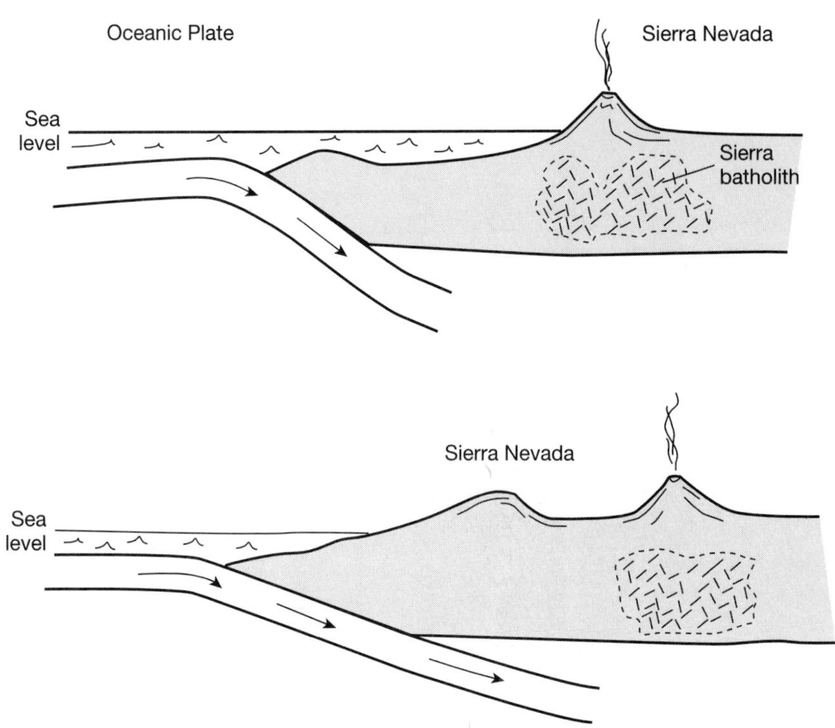

Fig. 8-18 Diagram showing how magmatic activity might shift in response to flattening of a subducting plate.

METAMORPHIC ROCKS OF THE SIERRA NEVADA

Although Mesozoic granitic rocks make up most of the Sierra Nevada province, metamorphic rocks are scattered throughout much of the batholith. Blocks of metamorphic rocks in the High Sierra are called ***roof pendants*** (Fig. 8-19; see also Fig. 8-21)—where they appear to represent the rocks that were originally the top or roof of the magma chamber—or as ***septa***—where they form screens or dividers at the edges of plutons. A more extensive zone of metamorphic rocks, known as the western metamorphic belt, occurs along the northwestern foothills of the Sierra Nevada. The rocks here have been of particular interest to geologists because they are the host rocks for California's gold deposits.

The majority of the metamorphic rocks in the Sierra Nevada have been only moderately metamorphosed. Highly metamorphosed rocks such as gneiss (see Chapter 2) are relatively rare and are concentrated in the southern "tail" of the batholith. Many of the metamorphic rocks are foliated; that is, they show evidence of metamorphic layering caused by the growth of metamorphic minerals in preferred orientations (see Chapter 2; Fig 8-20). Some foliation is inherited from original sedimentary layering and enhanced during the process of metamorphism. In other rocks, the foliation crosses the original sedimentary layers. In some rocks, geologists can detect several directions of foliation in a metamorphic rock, suggesting that the rock was subjected to several different episodes of metamorphism. Common foliated rocks found in the Sierra Nevada include slate and schist; these have wide variety of compositions, reflecting their diverse origins.

Fig. 8-19 Mount Morrison roof pendant, looking north. (Source: Stevens, C. San José State University.)

Fig. 8-20 Foliated metavolcanic rocks in the western Sierra foothills, south of Folsom. Because the rocks weather along the foliation planes, they form aligned outcrops commonly called "tombstone rocks" for their resemblance to grave markers.

The most common nonfoliated metamorphic rocks are hornfels, marble, ***quartzite,*** and ***metachert*** (see Chapter 2). The presence of certain metamorphic minerals or assemblages of minerals in a metamorphic rock provides geologists with clues about the identify of the premetamorphic rock, called the ***protolith.*** As discussed previously, metamorphic mineral assemblages also act as geobarometers or geothermometers to record the pressure and temperature conditions of metamorphism. Other features, including layering and even fossils, may survive metamorphism and provide additional clues about the nature of the "parents" of slates and schists.

An understanding of the metamorphic rocks of the Sierra Nevada is particularly critical to the interpretation of California's tectonic history during Mesozoic time. Unfortunately, this understanding continues to present great difficulties because of the very nature of the metamorphic rocks. It is particularly challenging to identify and correlate the rocks as they had been assembled before metamorphism, because many of the rocks have been subjected to multiple episodes of deformation. West of the batholith, oceanic rocks were buried in subduction zones, chemically altered by hot fluids during subduction, and subjected to great stress as they were accreted to the North American margin. The arrival of each new accreted terrane generated faulting, folding, and shearing in the adjacent belts of rocks. For over 130 million years, rocks surrounding the Sierran batholith were subjected to episodes of great heat and pressure as magma was emplaced. At depth, the intrusion of magma squeezed the rocks along the pluton walls, causing them to be folded and sheared. The heat from plutonic intrusions mobilized fluids in the adjacent rocks, causing chemical reactions and concentrating certain elements—including gold—in veins and pockets.

PIECES OF THE PUZZLE: METAMORPHIC ASSEMBLAGES OF THE SIERRA NEVADA

Fragments of Paleozoic North America

As outlined in the previous chapter, a stable, shallowly sloping marine platform lay west of the margin of the North America continent before the development of an active subduction zone. Deeper marine environments would have existed further west of that margin—the area now the location of the Sierra Nevada—during Paleozoic time. It is logical to expect that Paleozoic sedimentary rocks would have accumulated in that area; those rocks would reflect their deposition on the stable carbonate platform. With the transformation of the continental margin to a subduction zone in early Mesozoic time, the Sierra Nevada region become the host for the Sierran batholith. However, bits of the former stable continental margin have survived as roof pendants.

In the eastern part High Sierra, roof pendants contain marble, quartzite, slate, and hornfels. The sedimentary parents of these rocks—limestone, dolomite, quartz sandstone, and shale—accumulated offshore of the stable western margin of North America (see Fig. 8-24). In some localities, fossils of early Paleozoic age have also survived metamorphism. Some of the rocks have been correlated to their unmetamorphosed equivalents further east, in the Inyo and White Mountains, but some geologists believe that the metamorphic fragments may correlate to sedimentary rocks farther south in the Mojave desert. Areas where these remnants of the passive margin are present include the eastern parts of the Mt. Morrison and Saddlebag Lake roof pendants, and the Bishop Creek roof pendant near Convict Lake.

Paleozoic and Mesozoic Volcanic Rocks in the High Sierra

Some of the Sierran roof pendants are hornfels and schist whose protoliths were clearly of volcanic origin. Those that are of the same age as the Sierran batholith presumably represent volcanic flows and ash erupted from the same large magma bodies that formed the batholith. Rocks of the Minarets Caldera (discussed on p. 154) are included in this category. Volcanic rocks in the Mount Dana roof pendant may represent an older Mesozoic caldera.

Some volcanic roof pendants in the High Sierra predate the Sierran plutonic rocks. These volcanic rocks were deformed and intruded by younger Mesozoic intrusions of magma. Geologists hypothesize that the volcanic rocks, which are found along a narrow northwest-trending zone in the High Sierra, represent one or more earlier volcanic arcs. One example of prebatholith volcanic rocks is found in the Saddlebag Lake pendant just north of Tioga Pass, where metavolcanic rocks are found with slate and schist that were formerly marine sedimentary rocks (Figs. 8-21 and 8-23). Geologists have speculated that some of the prebatholith volcanic rocks were moved from their Paleozoic volcanic centers to their present locations along later lateral faults (see Chapter 18).

Accreted Oceanic Terranes of the Sierran Foothills.

During Paleozoic time, the western margin of North America stood roughly where the Mojave Desert and Basin and Range Provinces of California are today. As outlined in Chapter 7, this was a time of tectonic quiet for most of California; in fact, much of what is now California did not exist. When the active convergent plate boundary developed along western North America, the relative quiet ended. As discussed earlier in this chapter, the Sierran batholith is one striking proof of tectonic upheaval during Mesozoic time. The metamorphic rocks of the northwestern Sierra Nevada provide additional evidence of plate collisions and terrane accretion during Mesozoic time. The metamorphic rocks of the western foothills also document earlier collisions between the North American continent and oceanic terranes that lay to the west.

On the California geologic map (Endpaper 1, inside front cover), the western metamorphic belt of the Sierra Nevada appears as two extensive bands of metamorphic rock on the northwestern side of the Sierra Nevada province. On the map, Mesozoic and Paleozoic units are separated by an extensive fault system, and the faults are further accentuated by the large slices of serpentine. Although the geology of the western foothills appears relatively simple on the state map, each unit is actually very complex. Geologists now believe that each unit contains several individual slivers of oceanic crust that were accreted to the western margin of North America during subduction (Fig. 8-22). It appears that some slivers are terranes that formed near the continental margin, while others formed at greater distances from North America and are more "exotic" to the Sierran region.

Because all of the rocks have been metamorphosed, faulted, and folded, they have been extremely difficult to correlate and identify as separate crustal fragments. At the present time, work on the metamorphic rocks in the northwestern Sierra Nevada is allowing geologists to make revisions and additions to California's tectonic history with each new study.

By about 140 million years ago, in the mid-Jurassic period, the oceanic terranes in the western foothills had been accreted, and the subduction zone moved west of the modern Great Valley. The largest plutons of the batholith crystallized as oceanic crust was accreted to North America along a subduction zone that ran

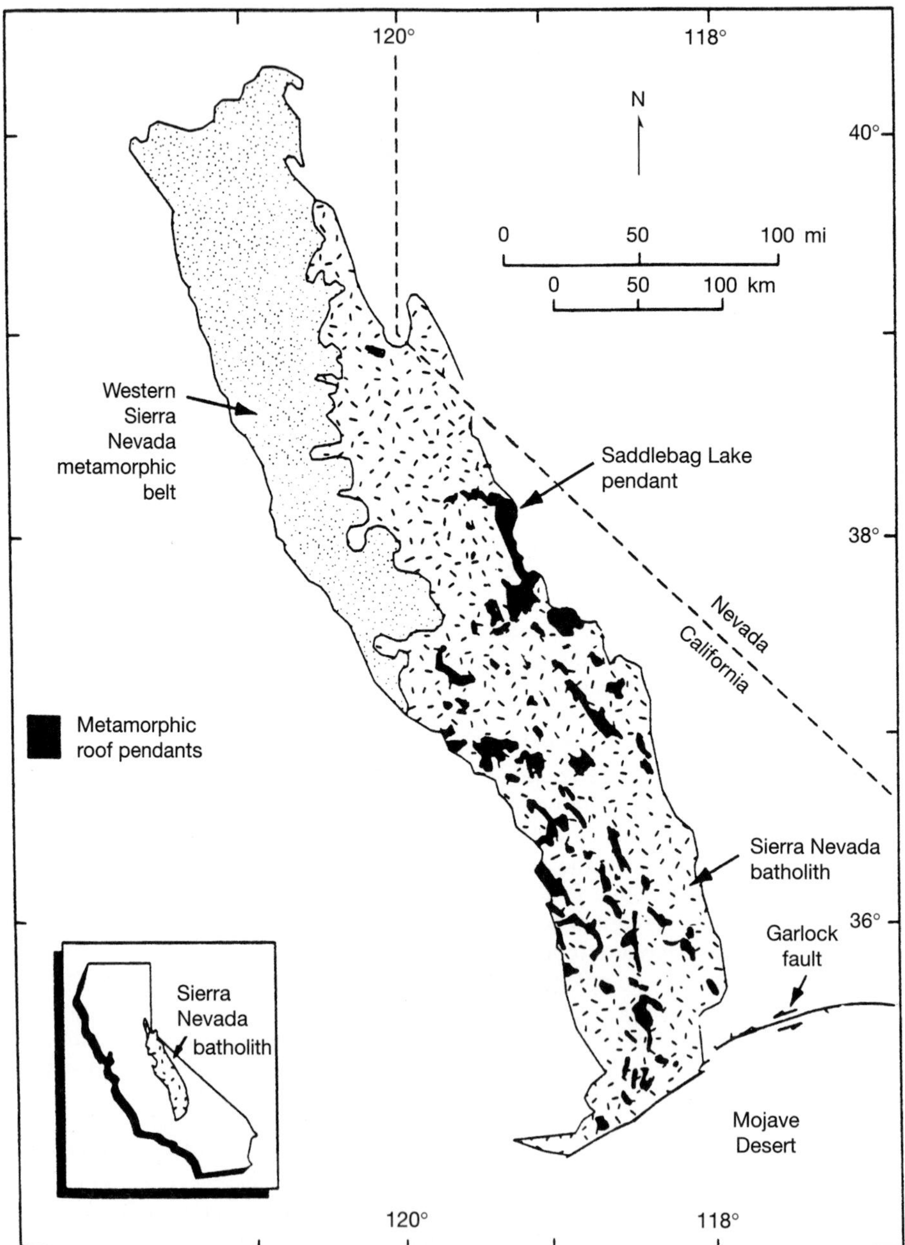

Fig. 8-21 Map showing roof pendants in the Sierra Nevada. (Source: Modified from Lahren, M., and Schwiekert, R. 1994. Geological Society of America Bulletin.)

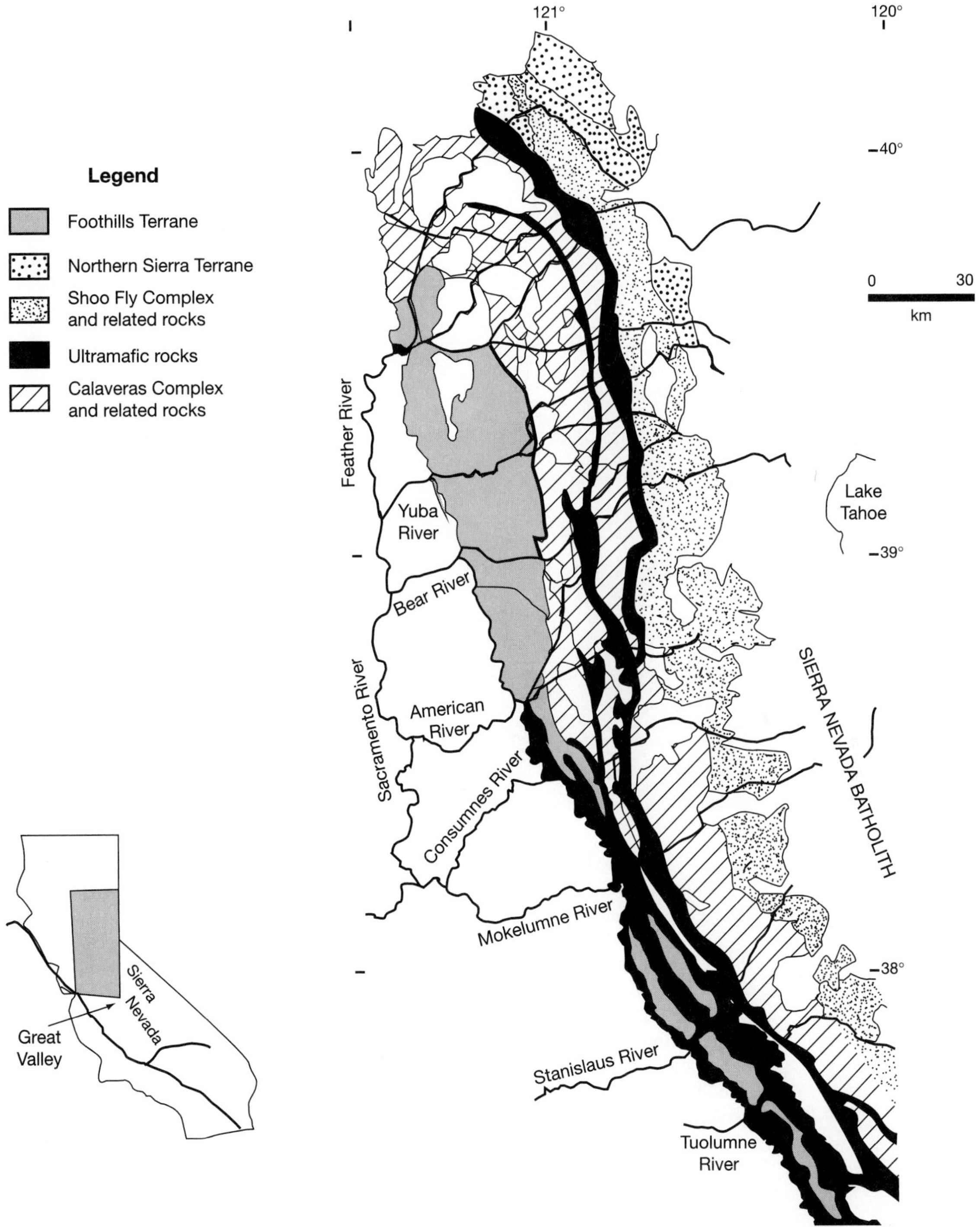

Fig. 8-22 Map showing the accreted terranes of the western metamorphic belt, as interpreted from studies of the metamorphic rocks. (Source: Modified from Fagan, T., and others. 1995. *Northern Sierra Nevada Field Trip: Tectonic Belts of the Sierra Nevada.* Northern California Geological Society.)

Fig. 8-23
Metasedimentary and metavolcanic rocks of the Saddlebag Lake pendant, north of Saddlebag Lake. (Photo by author.)

the length of western North America in the approximate location of today's Coast Ranges (see Chapter 12).

Paleozoic Metamorphic Belts. The oldest rocks in the northern Sierra Nevada, which are as much as 500 million years old, make up the Shoo Fly Complex. The Shoo Fly Complex, named in the early 1900s for a small settlement north of Quincy, is found in the northwestern Sierra Nevada. It contains a variety of metamorphic rocks that are of oceanic origin, including both metasedimentary (Fig. 8-25) and metavolcanic rocks. The Shoo Fly Complex has been subdivided by geologists into several units, and some geologists believe that the units represent different oceanic terranes. About 375 million years ago, during late Devonian time, a period of uplift and erosion is marked by an unconformity between Shoo Fly Complex and all younger Paleozoic rocks. During this period of deformation, rocks of the Shoo Fly Complex were faulted and folded.

East of the Shoo Fly Complex in the northern Sierra Nevada is a younger belt of Paleozoic rocks termed the Northern Sierra Terrane (see Fig. 8-25). Rocks in this belt include chert and black shale, both containing radiolaria, formed in deep marine water during the Mississippian and Permian periods. During Permian time, about 250 mil-

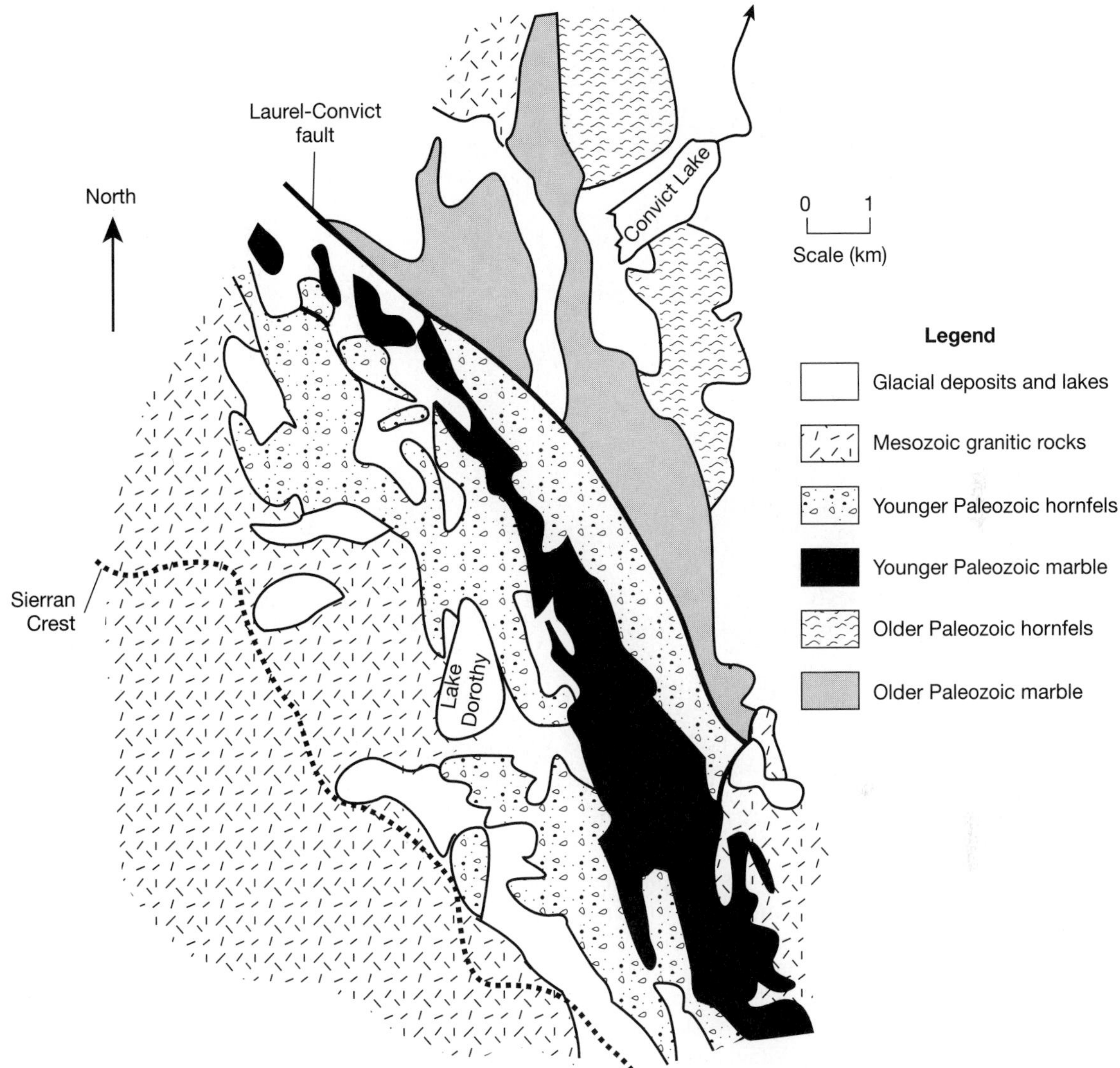

Fig. 8-24 Map showing the roof pendants of the Mount Morrison area. (Source: Modified from Reinhart, C.D., and Ross, D. 1964. U.S. Geological Survey, Professional Paper 385.)

lion years ago, large volumes of andesite erupted in the vicinity, as evidenced by the presence of Permian metavolcanic rocks associated with the sedimentary formations. The source of the eruptions was a volcanic island arc. Associated with the volcanic rocks are limestones that once formed reefs that fringed the islands. Some of the metamorphosed limestone contains fossils that are distinct and unique to the ancient reefs of the Klamath Mountains and northern Sierra Nevada (see Chapter 9). In fact, the geologic history of the northern Sierra Nevada during Paleozoic time appears to match well with that of the eastern Klamath Mountains. Geologists believe that the

Fig. 8-25 A rare block of dolomite marble in the Shoo Fly complex, along the North Fork of the American River. (Source: Harwood, D. U.S. Geological Survey.)

two areas were continuous and were part of a series of island arcs that lay west of the North American continental margin during Paleozoic time (see Chapter 9).

Late Paleozoic and Mesozoic Subduction Complexes and Accreted Terranes.

West of the rocks of the northern Sierran terrane—and separated by a major fault system—are rocks which geologists interpret as belonging to an ancient subduction zone. Within this zone are complex assemblages of rocks representing many elements of a ***subduction complex:*** submarine volcanic rocks erupted at spreading centers; sedimentary rocks deposit in deep-ocean trenches; and metamorphic and plutonic rocks formed at great depths within and beneath the crust. This mixture of rocks forms a discontinuous belt stretching into British Colombia and including the Central metamorphic terrane of the Klamath Mountains (see Chapter 9). In the Sierran foothills, rocks within this belt have been named the Calaveras Complex. The types of rock found in a subduction complex are discussed in more detail in Chapters 9 and 12.

The westernmost belt of metamorphic rocks in the Sierra Nevada is referred to as the Foothills terrane (see Fig. 8-22). Rocks found here are of middle to late Jurassic age, about 160 million years old, thus representing a younger period of subduction and accretion than the Paleozoic rocks discussed above. Along the central part of the Sierran foothills, the Foothills terrane is separated from the rocks of the Calaveras Complex by the Melones fault, the easternmost branch of the Foothills fault system.

Volcanic rocks from submarine spreading centers, now metamorphosed, are a major component of this unit. Unweathered metavolcanic rocks with high iron content are commonly green, and in some areas are called ***greenstone***. Metamorphosed marine sandstone and shale formed along the convergent margin are also found as slate and schist within the Foothills terrane. One of the more extensive formations of this unit is the Mariposa Formation. Gabbro that was formed at greater depths within the oceanic plate and ultramafic rocks thought to have formed in the mantle are also present in the Foothills terrane.

One of the most unusual and interesting metamorphic rocks in the western Sierran foothills is mariposite schist, a banded green and white rock containing the bright apple-green mineral ***mariposite.*** Mariposite, named for Mariposa County, where it was first discovered in 1868, is a form of mica that gets its green color from small amounts of chromium. Mariposite forms by the alteration of serpentine by hydrothermal fluids, and it is most common along the Melones fault zone, where it is associated with massive white quartz veins. Mariposite schist may also contain gold-bearing veins. Mariposite schist has long been valued as a decorative stone. A huge boulder from Placerville was selected in 1987 to represent California in the Constitution Monument built in Philadelphia for the two hundredth anniversary of the Constitution and the Bill of Rights.

The box on p. 170 describes the discovery of gold in California.

Formation of Sierran Gold Deposits

Most of the gold in the Sierra Nevada foothills is pure or "native" gold, rather than gold combined with other elements to form goldbearing minerals. The gold forms in veins of quartz 30 centimeters to 30 meters thick (Fig. 8-26). Other minerals, including calcite and pyrite or "fool's gold," occur in some veins. Rare gold-bearing minerals such as ***calaverite,*** gold ***telluride*** ($AuTe_2$) are also present in some veins. Calaverite was named for Calaveras County, the county where it was first found. Within some veins are rich pockets of gold known as ***ore shoots.***

The quartz veins were emplaced along shear zones and faults, which provided pathways for the migration of hydrothermal fluids along the weaker rocks in those zones. Metal deposits of this type are referred to as ***lode deposits.*** In some areas, particularly in the southern part of the foothills, more diffuse areas of metamorphic rock are also mineralized with gold. Although most of the gold has been mined from quartz veins in metamorphic rocks, veins in granodiorite have accounted for California's richest gold district, the Grass Valley mines.

A continuous zone of gold-bearing quartz veins and mineralized rock, known as the ***Mother Lode,*** runs for 195 kilometers along the Melones fault. A trip along California's historic Highway 49 through the Gold Rush towns and mines is also a trip along the tectonic suture zone represented by the Melones fault (see Fig. 8-28). Hundreds of mines were excavated to follow the quartz veins within this 1.5-kilometer-wide belt. Because most of the veins are dipping, the lode deposits may extend to thousands of meters below the surface.

The Grass Valley mining district, in western Nevada County, was the richest and most famous of California's gold mining districts. At least $300 million in gold was produced from its lode mines between 1852 and 1956. The Grass Valley district lies north of the Mother Lode, but along the same general trend. Quartz veins there are found in granodiorite, rather than in metamorphic rocks. The Empire Mine, richest producer of all California mines, alone produced $130 mil-

GOLD

In January 1848, James Marshall was attempting to correct a defect in his sawmill. He had convinced his boss, Captain John Sutter, to build a sawmill on the American River at Coloma and float the much-needed lumber downstream to Sutter's Fort, located near Sacramento. The tailrace—the stretch of channel downstream from the mill—had not been dig deep enough, and water was backing up to the wheel and preventing it from turning. Marshall blasted a deeper channel, hoping that additional sediment would be removed by the stream. On January 24, Marshall closed the gate that stopped the water flowing through the tailrace and walked down it to see whether enough sand and gravel had been removed.

> . . .He stood for a moment examining the mass of debris that had been washed down, and his eye caught the glitter of something that lay, lodged in a crevice, on a riffle of soft granite, some six inches under water. He picked up the substance. It was heavy, of a peculiar color, and unlike anything he had seen in the stream before. . . .The weight assured him that it could not be mica. Could it be sulphurets of copper? He remembered that the mineral is brittle; he turned about, placed the specimen under a flat stone, and proceeded to test it by striking it with another. The substance did not crack or flake off; it simply bent under the blow.
>
> Marshall returned to the mill, his usually crusty face beaming, and cried:
>
> "Boys, by God, I believe I have found a gold mine!"
>
> He showed his nuggets as proof. His worker were not impressed. They continued about their tasks. On the morning of the fifth day, after having in the course of his inspection of the tailrace and of the shallow side of the river selected several more yellow nuggets, enough to make three ounces tied in his kerchief, Marshall started out on horseback to cover the fifty mountain miles down to the fort, sleeping that night under an oak tree. He was ostensibly searching for a wagonload of supplies which Sutter had promised.
>
> At nine o'clock on the morning of January 28, 1848, sopping wet from a cloudburst which had enveloped him the last eight miles, Marshall arrived at the fort. He asked Sutter where they could talk with privacy. Puzzled, Sutter took him to his bedroom–sitting room in the main building, locking the door. Marshall asked for two bowls of water, a stick of redwood and some twine and sheet copper to make a scales. Sutter told him that he had scales in the apothecary shop and went for them himself, failing to lock the bedroom door when he returned. A clerk walked in with some papers just as Marshall was about to dump the yellow nuggets onto the table. Marshall cried in consternation:
>
> "There! Didn't I tell you we had listeners!"
>
> Sutter quieted his overwrought partner, then gazed down at the yellow nuggets which James Marshall poured out of his kerchief onto the table. Sutter examined the specimens, pulled down a volume of his *Encyclopedia Americana,* studied it for a time, tested the nuggets with *aquafortis,** which had no effect on them, balanced them on the scales with a like amount of silver, then dipped the scales into water, the yellow nuggets quickly outweighing the silver.
>
> Sutter turned his now wide and flashing eyes up into the face of the wildly excited Marshall.
>
> "It's gold," Sutter said. "At least 23-carat gold!"

Captain Folsom sent a flake of gold described as Marshall's first piece to Washington in 1848. This piece is now preserved in the Smithsonian Institution.

With this event, a few bits of stream sediment changed the face of California and the history of the entire world. The gold rush was slow at first, because it took months for word of Marshall's discovery to travel. The first to arrive in the summer of 1848 were Hawaiians, followed by miners from Mexico, Peru, and Chile by the end of the year. By the end of 1848, California's non–Native American population was 20,000 and included 5000 miners. Waves of Americans from the eastern states, South America, and Europe followed. At the end of 1849, the population had grown to 100,000; by 1852, 224,000 non–Native Americans lived in California, and 100,000 were miners. At that time, 24 percent of California's population was foreign-born. Most Americans knew little about mining, but among the immigrants were experienced miners with origins in Mexico, Peru, Chile, Spain, and Cornwall in England. Between 1851 and 1855, the United States accounted for 45 percent of the world's gold output.

*Nitric acid; "strong water" in Latin.

Source: Excerpted from Stone, I. 1956. *Men to Match My Mountains.* Garden City, N.Y.: Doubleday & Co., Inc., pp. 105-106; with quoted materials from Marshall's biographer, George F. Parsons, and Sutter's biographer, James P. Zollinger.

Fig. 8-26 **A,** Gold-bearing quartz vein in metamorphic rocks of the western Sierra Nevada. **B,** An early gold miner. (Source: **A,** Photo by author; **B,** California Department of Conservation, Division of Mines and Geology.)

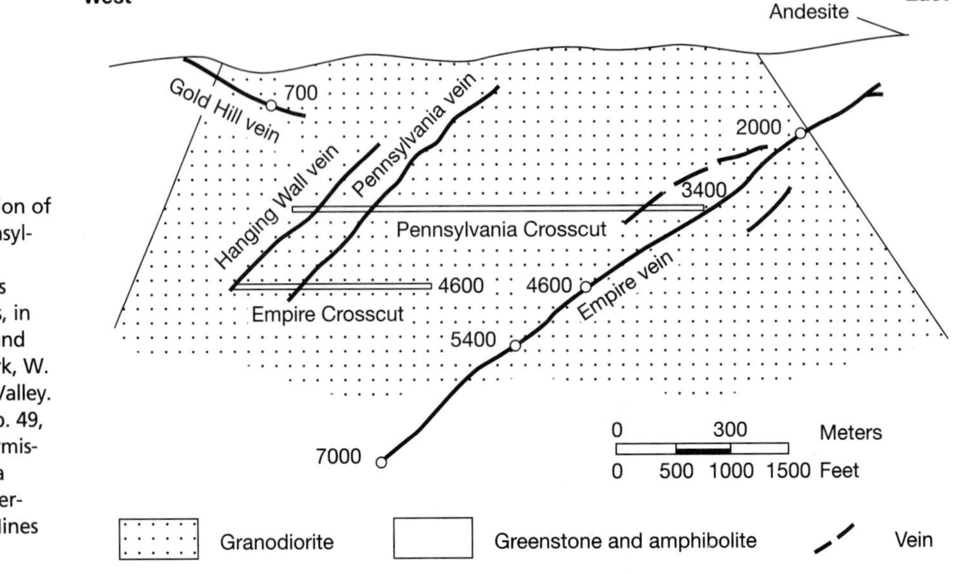

Fig. 8-27 Cross section of the Empire and Pennsylvania mines, Nevada County. The numbers show the mine levels, in feet, below the ground surface. (Source: Clark, W. Gold Mines of Grass Valley. *California Geology*, p. 49, 1984. (Used with permission of the California Department of Conservation, Division of Mines and Geology.)

lion in gold (Fig 8-27). Visitors can see many of its original above-ground workings at Empire Mine State Park.

Although the initial finds of gold in California were nuggets in stream gravels, it did not take Sierra Nevada miners long to learn that gold-bearing quartz veins were the source of the stream sediments and that the veins were most common in the metamorphic rocks of the western foothills (Fig. 8-28). Geologists later noted the association between lode deposits and oceanic volcanic terranes that have been intruded by younger plutonic rocks. Today, geologists believe that the Sierran gold was originally deposited in the rocks of the accreted oceanic terranes. The migration of gold-bearing fluids into the rocks probably occurred near submarine volcanic vents and resulted in the formation of mineralized zones similar to those seen at modern submarine spreading centers. Subsequently, the heat generated during the intrusion of the Sierra Nevada batholith mobilized the gold and caused it to be concentrated in the veins where it is found today. The gold-bearing veins cut across Mesozoic metamorphic rocks, as well as across some of the plutonic rocks, so the gold mineralization must postdate those rocks. By dating rocks which predate and postdate the gold-bearing veins, geologists estimate that the gold veins were emplaced between 115 and 120 million years ago.

Placer Gold

The gold discovered by James Marshall were nuggets of pure gold that were part of the alluvium of the American River. Like the other rocks and minerals in the streambed, the gold did not form there. It was eroded from sources higher in the mountains, transported by the stream, and deposited with the other sediments. Because it is durable and chemically stable, gold resists chemical weathering.

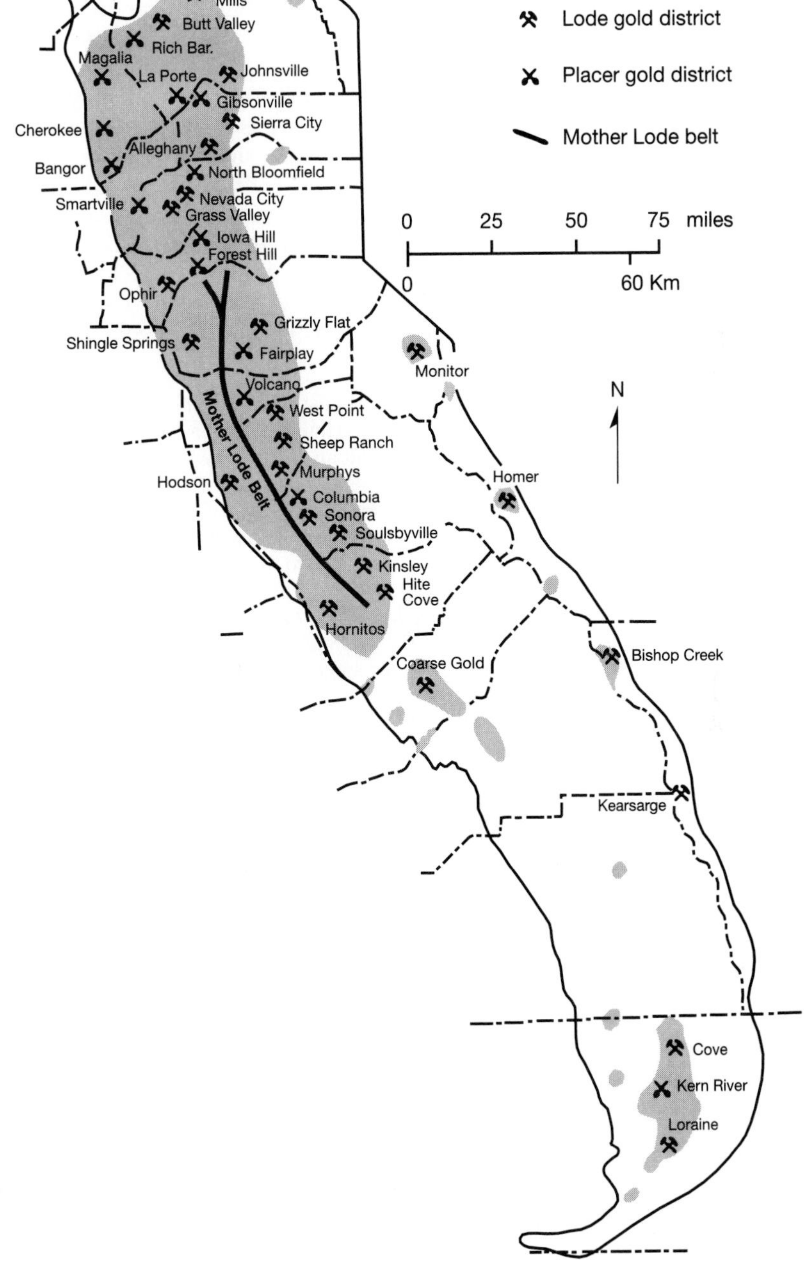

Fig. 8-28 Major gold deposits of the Sierra Nevada foothills. (Source: 1966. *Geology of Northern California.* Bulletin 190, p. 211. Used with permission of the California Department of Conservatin, Division of Mines and Geology.)

Although it is easily worn into fine particles because it is soft, gold persists in sedimentary environments like stream channels and beaches, where less resistant minerals do not survive. Sedimentary ore deposits formed in this way are known as ***placer deposits,*** from the Spanish word for sand bank. Other examples of valuable placer deposits include diamond placers in South America and beach placers of titanium oxide mined in various localities. Placer deposits of gold or other valuable ores may survive in sediments and sedimentary rocks long after the stream or beach where they accumulated has disappeared from the landscape.

Placer deposits accounted for more than 40 percent of California's total gold output. These include recent deposits found in and near the streams flowing from the Sierran crest to the Sacramento and San Joaquin Valleys. Sedimentary deposits of Tertiary rivers have also yielded rich placer gold in the northwestern Sierra. The ancient placers are the deposits of rivers that flowed westward from a previous Sierra Nevada range during Tertiary time (about 25 million years ago). These ancient deposits, called the "Tertiary gravels," were first studied by Waldemar Lindgren, who reconstructed the paths of the ancient rivers (Fig. 8-29). This description of placer mining is from R.W. Paul's 1947 book, *California Gold: The Beginning of Mining in the Far West*:*

> For countless centuries the streams had been dropping flakes and chunks of the precious mineral onto sandbars, into crevices in the banks, and into "potholes" in the beds of the rivers. Every stone along a river's course had provided an obstacle behind which fragments might lodge.
>
> Water had been the primary agent in storing up this accumulation of wealth, and water was to be the miner's chief weapon in his attack upon it. Sometimes the miner might be fortunate enough to find crevices richly piled with comparatively pure gold dust that could be scraped out with a knife and spoon, but in most cases he had to dig down into promising-looking ground with a pickaxe and shovel. After he had thrown up mounds of "dirt," he found himself facing the problem of separating the yellow grains from the gravel and sand with which they were intermixed.

During the early Gold Rush years, miners devised a number of methods for sorting gold flakes from the river sediments. All methods relied on the simple fact that gold is much denser than the other components of river alluvium—mainly silicate minerals. Because of its greater density, gold settles out more quickly from suspension in water; this is one reason for its concentration in placers. The simplest way to sort gold from other sediments was by ***panning,*** a method still popular with hobbyists today. California miners used either a flat-bottomed pan made of iron or tin or a wooden bowl called a batea. The pan was filled with sediment and water, and the larger gravel was picked out. Then the miner lowered the pan into stream or pool at a slight tilt and swirled the sediment.

Lighter materials were carried away from the heavier grains or "pay dirt," which often contained heavy minerals such as magnetite and pyrite ("fool's gold"), as well as the gold flakes or nuggets. If the miner saw a "flash in the pan," he was hopeful, but at times the sediment didn't "pan out."

Panning is a simple and inexpensive method of placer mining, but it is a low-volume operation. Miners next developed methods of sorting larger volumes of sediment. The ***rocker*** or ***cradle,*** in use as early as 1848, was a long box mounted on rockers

---

*From Rodman, P.W. 1947. *California Gold: The Beginning of Mining in the Far West.* Lincoln, Nebr.: University of Nebraska, p. 50.

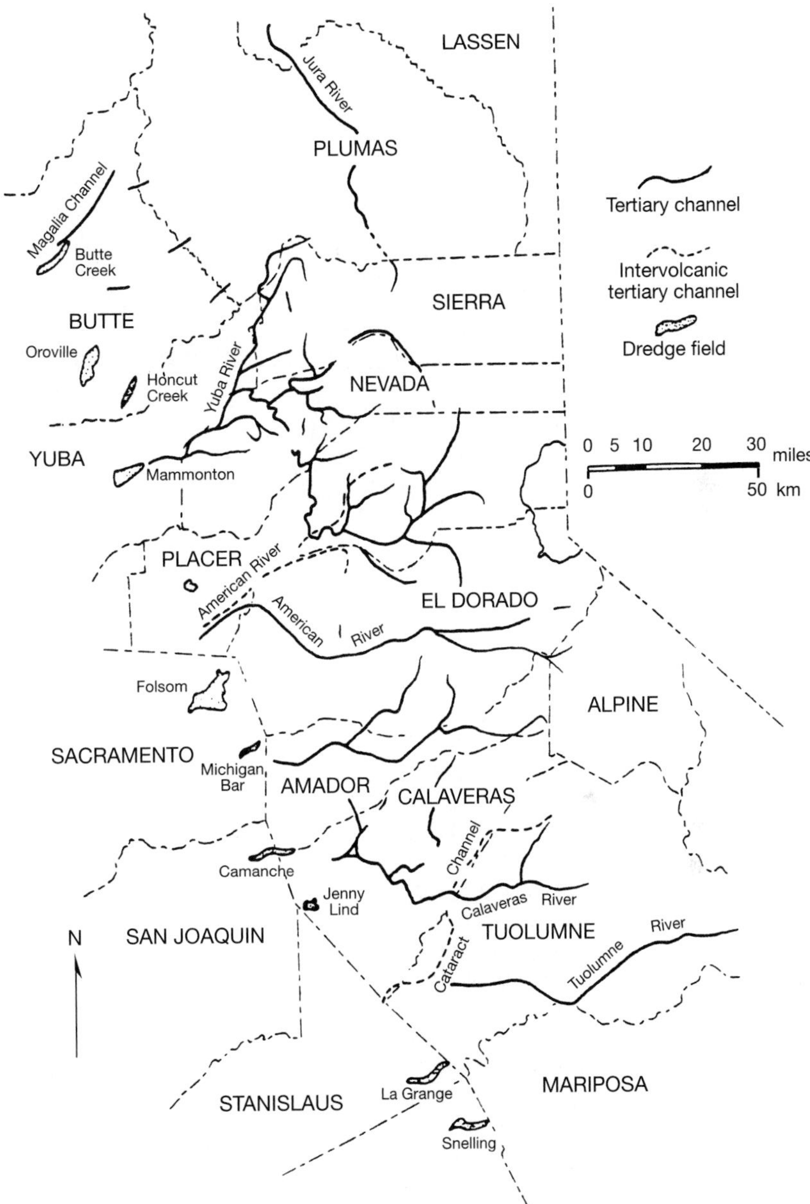

Fig. 8-29 Map showing the ancient rivers that deposited the Tertiary gravels. (Source: *Gold Districts of California.* Bulletin 193, p. 18, 1970. After Lindgren, 1911, and Jenkins, 1935. Used with permission of the California Department of Conservation, Division of Mines and Geology.)

like a baby's cradle. Positioned on a slope with its downhill end open, the cradle was fed at the top end by a hopper with a sieve on its base. Water was continuously poured into the hopper to create a current, and the gold tended to drop behind in the rocker box, where it was trapped by cleats. The operation of the cradle enabled three or more men to work together to wash a larger volume of sediment. Many of California's richest and most accessible placer deposits were mined using the pan and the cradle.

Fig. 8-30 An early drawing of California placer miners. (Source: Paul, R. 1947. *California Gold.* Lincoln, Nebr.: University of Nebraska Press.)

In late 1849 miners began to move vast quantities of sediment, diverting rivers to reach the gold that lay within the streambeds. Large mining associations formed to handle the cost of these operations, and new methods for moving vast tonnages of sediments came about. A 12-foot-long trough called a ***long tom*** operated in the same fashion as a cradle. Miners improved the design by adding a chain of riffle boxes at the lower end of the long tom to catch more of the gold with each washing. The end result of these modifications was the ***sluice box,*** an open wooden trough with a series a riffle boxes, fitted with cleats to catch the heavy sediment. By 1850, groups of miners were constructing ditches and flumes to move enough water to run their sluices (Fig. 8-30). These water diversions had lasting effects on California's water laws (Chapter 10).

One of the problems with gold washing is that the process does not completely separate the gold from other heavy minerals. Miners from South America purified their washings by first drying the mixture and then tossing them into the air or blowing on them. The process winnowed the heavier gold from the other grains. However, this was an inefficient if nonpolluting process, and it was soon replaced by the use of mercury. When gold-rich sediment is mixed with mercury, the mercury binds the gold into an amalgam. Subsequent heating of the mercury vaporizes it, leaving behind pure gold. This technique, which had been practiced by medieval alchemists, was used by late 1849 and is still used today. The demand for mercury fueled the development of mercury mines in the Coast Ranges (Chapter 12).

By the early 1850s, much of the easily mined placer gold had been extracted from the riverbeds of the western foothills. The entire beds of most stream channels had been overturned, some more than once, and the rivers themselves had been diverted and dammed to get to placer gold. Even today, piles of sediment persist throughout the foothills, looking like the diggings of countless large gophers.

Miners next turned their attention to older placer deposits, those formed by streams that flowed westward from an older Sierra Nevada mountain range. Cornish

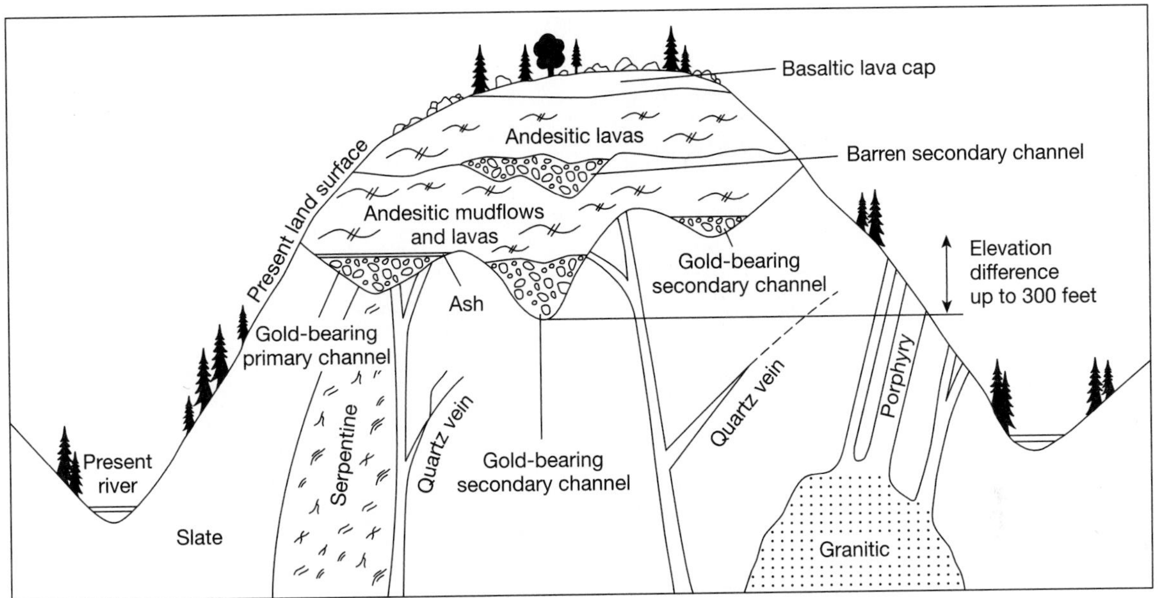

Fig. 8-31 Location of Tertiary placer deposits. (Source: Chakarun, J. *California Geology,* 1988.)

miners had been the first to look for older buried gravel deposits, which have long been referred to as the Tertiary gravels by geologists. Because of recent uplift of the Sierra Nevada, Tertiary gravels are now located above the modern river channels (Fig. 8-31). Consequently, it was necessary for the miners to find a way of washing these gravels to get to the gold that was concentrated near the bottom layers of sediment. In some areas, miners tunneled under the Table Mountain volcanic flows to reach the ancient gravels beneath them. One example of successful tunneling was the Omega Mine near Jamestown.

In 1853, ingenious miners came up with the idea of attaching a nozzle to a hose and forcefully washing the sediments down from the hillsides. This marked the birth of ***hydraulic mining,*** a practice responsible for changing the landscape from the Sierran foothills to San Francisco Bay. Instead of being able to wash 1 cubic yard of sediment in a day, or 4 cubic yards with a sluice, a miner could wash 50 to 100 cubic yards per day using a hydraulic system together with a sluice. Soon so much sediment had washed from the hillsides into the streams of the Sacramento Valley and into the San Francisco–San Joaquin delta that navigation and agriculture were affected. The Sawyer Decision of 1884 prohibited the dumping of mining-related sediment into the Sacramento and San Joaquin rivers and their tributaries. Sediment from hydraulic mining in the late 1800s can still be identified in the eastern part of San Francisco Bay more than a century later.

During the twentieth century, sediments have been extracted from riverbeds by ***dredging.*** Floating suction dredges are able to remove sediment for gold washing while the rivers are flowing, thus eliminating the need for diversion. Dredging operations were active in some Sacramento River tributaries until the early 1960s. Throughout the western foothills, great piles of sediment known as ***dredge fields*** can still be found (Fig. 8-32). Visitors can see an excellent example of hydraulic mine fields at Malakoff Diggins State Historic Park in North Bloomfield (Fig. 8-33).

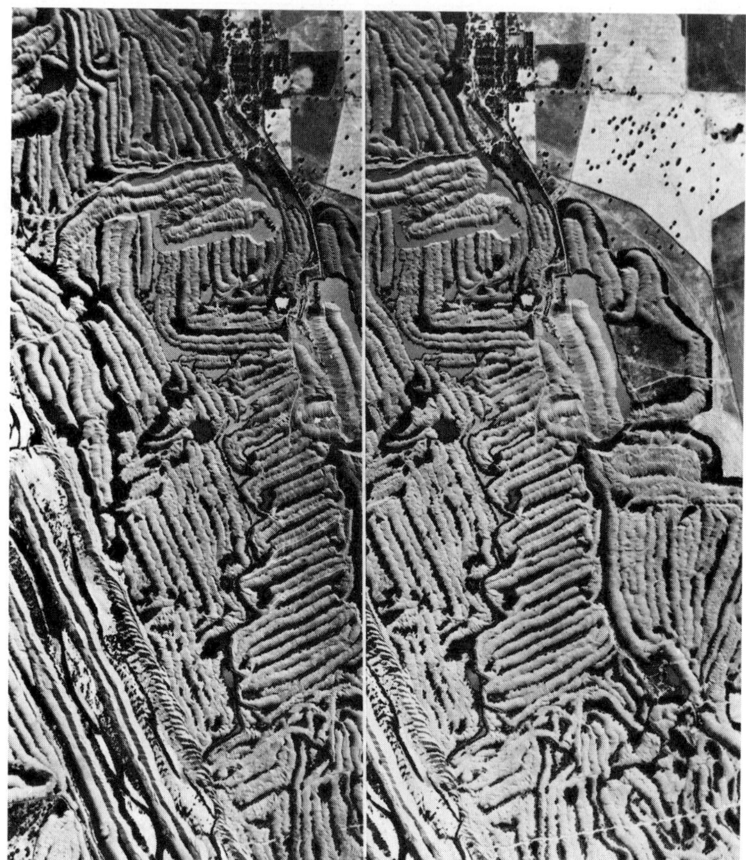

Fig. 8-32 A stereogram pair of dredge tailings along the lower Yuba River near Marysville in 1932. To view this figure three-dimensionally, use a pocket stereoscope. (Source: U.S. Department of Agriculture.)

Fig. 8-33 The hydraulic pits at Malakoff Diggins State Park, North Bloomfield, Sierra County, more than a century after hydraulic mining ceased. (Photo by author.)

Mining the Lode Deposits

California's lode deposits are much less accessible than most of the placer deposits, because they are exposed only where quartz veins come to the surface. Underground mining presented many more challenges than placer mining, so the lode deposits were not extensively mined until 1851, after the easy pickings offered by the rivers had dwindled. Early miners dug shafts and tunnels to follow the quartz veins, developing new technology for underground extraction as they went (Fig. 8-34). The problem of separating the gold from the quartz was handled by the ***stamp mill,*** a device which pulverized the quartz with a heavy iron pounder. After crushing, the gold was sorted using mercury and a variety of sluices in a manner similar to that used in placer mining. As placer mining declined, lode mining continued as new and deeper veins were discovered. Because of the high cost of underground mining, companies were formed and stock sold to finance the cost of operations.

Underground mining continued in the Sierran foothills until the 1940s. The Jackson-Plymouth mining district in western Amador County opened its first lode mines in 1850 and continued to produce gold until 1942. One of those mines was the Argonaut, a mile-deep mine with miles of tunnels. Two of the hazards of underground mining—poor ventilation and wooden timbers—combined with deadly results in 1922, when a disastrous fire at the Argonaut killed 47 miners. The fire began at the 3350-foot level of the mine, trapping an entire shift of men who were working at levels beneath the fire.

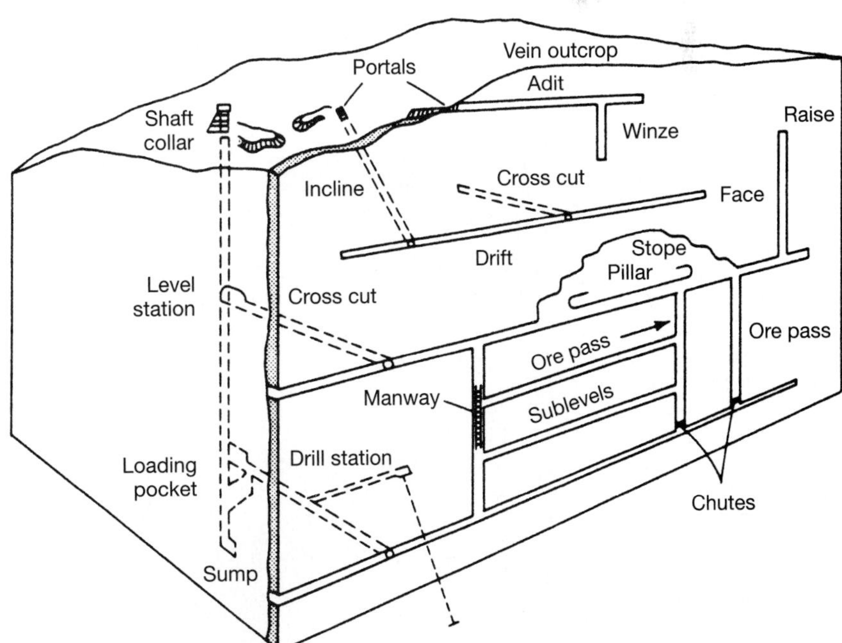

Fig. 8-34 Typical features of an underground mine. (Source: Peters, W.C. 1978. *Exploration and Mining Geology.* New York: John Wiley & Sons.)

Gold Production From 1848 to the Present

Two famous gold finds in California were the 195-pound mass found at Carson Hill in 1854, and a 54-pound, pure gold nugget found at Magalia in 1854, which was melted down shortly after its discovery. Unfortunately, masses of this size are exceptional. Most Sierran gold has been mined from deposits containing between 1/4 and 1/2 ounce of gold per ton of ***ore.*** An ore is any accumulation of a particular element, including metals like gold, silver, or mercury, that has been sufficiently concentrated to be profitably mined. It is easy to see from this definition that economics, accessibility, and technology define ores as much as geologic circumstances. What was discarded as waste rock during one period may be remined profitably in later years, if the price of the commodity rises or if new mining methods develop. California's gold production has fluctuated with changes in all of these conditions (Fig. 8-35). Mining ventures and investment in precious metals like gold are usually financially risky, partly because of the complexities of market conditions, and partly because of the geologic uncertainties associated with ore deposits.

In the 1980s, a rise in the price of gold, combined with a new method of extracting it, produced a new wave of mining at abandoned sites in the Sierra Nevada. Today, mining companies are using a chemical process called ***heap leaching*** to extract gold from deposits that have not been profitable in the past. Heap leaching, which uses cyanide to dissolve gold from the quartz veins, increased the gold recovery by 100 times over other methods of mining lode deposits. It is currently being used at several mines in California and Nevada.

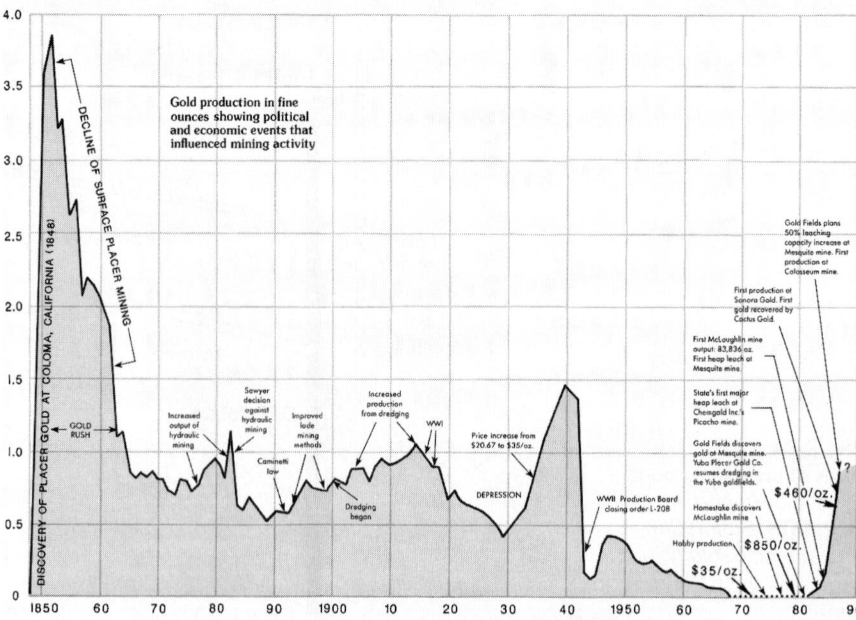

Fig. 8-35 Diagram showing California's gold production between 1848 and 1988. (Source: Fuller, W. California gold production 1848-1988. *California Geology* 41:180-181, 1988. Used with permission of the California Department of Conservation, Division of Mines and Geology.)

Fig. 8-36 Outcrops of the Mehrton Formation west of Jamestown along Highway 108. (Source: Sarna-Wojcicki, A. U.S. Geological Survey.)

CENOZOIC VOLCANIC ROCKS OF THE SIERRA NEVADA

During middle Cenozoic time, between about 28 and 22 million years ago, huge ash flows erupted from the Great Basin, from what is now Nevada. These volcanic flows covered Mesozoic rocks in parts of the Sierran Province, during a period when no major mountains existed in the area that is now the Sierra Nevada. The ash flows must have flowed downhill from Nevada to deposit the Valley Springs Formation, found today in the northwestern Sierran foothills. Younger andesitic volcanoes erupted about 9 to 11 million years ago from the high Sierra Nevada to deposit the distinctive Mehrton Formation in the western Foothills west of Jamestown, the Table Mountain flows (discussed on p. 145), and widespread volcanic flows in other areas (Fig. 8-36).

GLACIERS

Yosemite Valley, with its sheer granite walls, domes, and waterfalls, is probably the most famous scenery in the United States (Fig. 8-37 and Color Plate 15). Without the action of glaciers, there would be no Yosemite Valley as we know it. Glaciers are responsible for much of the landscape of the High Sierra, and although most of the ice has melted during the past 10,000 years, it had left a record of the Sierra Nevada's past climate for geologists to study.

A glacier is a body of ice that moves over a land surface. Glaciers form by the accumulation of snow over many years. As older snow is buried by new snowfall, it undergoes a transformation not unlike metamorphism. Snowflakes are compacted, their edges are rounded off, and air is driven from the snow. The ice recrystallizes, forming larger, more compact crystals. During its transformation to glacier ice, snow increases its den-

Fig. 8-37 Yosemite Valley and Half Dome, viewed from Glacier Point, along Highway 120. (Source: Sarna-Wojcicki, A.)

sity about 9 times and its strength about 50 times. Ice fields form by the continual accumulation of snow and ice. When the ice is sufficiently thick, it begins to flow under its own weight. As it flows, it carries rock and sediment with it—on the ice, within it, and along the sides and bottom of the glacier. The bits of rock are the agents that actually accomplish the grinding and scraping of they override. Although harder than snow, glacier ice is still much softer than most of the minerals found in the Sierra Nevada.

For a glacier to form and to grow, winter snowfall must exceed summer melting over a number of years. A glacier is thus a balanced system that responds to fluctuations in both temperature and precipitation. Favorable conditions for the formation and growth of a glacier are maximized when winter snowfall is high, and summer temperatures are low. If more snow melts in summer than accumulates in winter, then the glacier will shrink. This is often referred to as the ***retreat*** of the glacier, although, of course, the ice does not actually move back up the valley. A mountain glacier is also balanced between the ***accumulation area,*** at the highest elevations, and the ***ablation area,*** at the lowest points, because temperatures are lower at higher elevations (Fig. 8-38).

Sierran Glaciers

In the Sierra Nevada, three factors influenced the distribution of Pleistocene glaciers. Because the range is lower in the north than in the southern and central Sierra, glaciers were smaller there. On the other hand, the southern Sierra is warmer and drier than the northern latitudes, so glaciers were restricted in the southern part of the range. Because of the rain shadow effect, snowfall was far greater on the western slopes, causing much longer valley glaciers to form there. The largest ice fields in the Sierra Nevada during Pleistocene glaciation were in the vicinity of the upper Tuolumne, Merced, and San Joaquin rivers (Fig. 8-39). Valley glaciers flowed down from the Sierran crest for as much as 100 kilometers on the western slope. Using the evidence left behind by the most recent (Tioga) glaciers, geol-

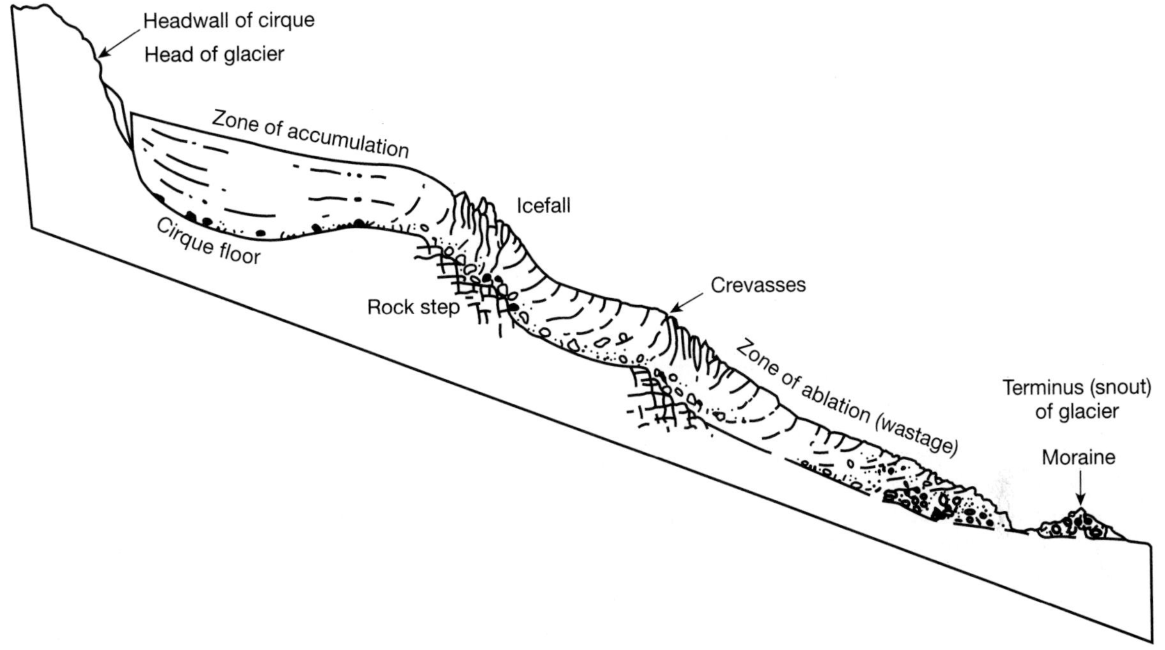

Fig. 8-38 Diagram of a valley glacier. (Source: Hill, M. Glaciers: A picture story. *California Geology* 27:28, 1974. Used with permission of the California Department of Conservation, Division of Mines and Geology.)

ogists have reconstructed the glaciers that once filled the Yosemite area (Fig. 8-40). In the box on p. 186, John Muir describes the scope of Sierran glaciers during the Pleistocene.

Today more than 100 glaciers remain in the Sierra Nevada, most of them on the sun-sheltered northern sides of peaks. The largest, Palisade glacier, covers less than 1 square kilometer. Compared with more than 500 km^2 covered by the largest glaciers about 20,000 years ago, today's glaciers are extremely small (Fig. 8-41). During this century, the Sierran glaciers have shrunk considerably, as indicated by comparison of early photographs with those taken more recently.

Glacial Erosion. In the High Sierra, most of the evidence of past glacial activity is in the form of erosional features unique to glaciers. As the moving glaciers dragged rock fragments over vast areas of bedrock, they created scratches called glacial ***striations,*** grooves, crescent-shaped gouge marks, and a very shiny ***glacial polish.*** Granitic surfaces along the Tioga Pass Road are polished as if they had been subjected to a giant lapidary wheel (Fig. 8-42). On a larger scale, granitic mounds called ***roche moutonees*** were streamlined underneath the ice, leaving asymmetrical domes that point in the direction of ice flow (Fig. 8-43).

Another important erosional feature of alpine glaciers are ***cirques,*** bowl-shaped depressions that the ice has carved at the head of glacial valleys (see Fig. 8-2). It is the elevations of the cirques that give geologists an estimate of the ancient position of the snow line during glacial periods. During the most recent glacial period in the Sierra Nevada, the snow line was about 750 meters lower than today. Today, many

Text follows on p. 188.

THE EARTH'S CLIMATIC CYCLES

Geologists have long been aware that the Earth's climate has experienced fluctuations far grander than the variability we see by examining the weather records of the past few centuries. Since the early 1800s, naturalists have recognized that great masses of ice had once extended far south from the Arctic Circle to cover much of northern Europe and North America. This has occurred several times during the past 2 million years, a geologic time period recognized as the great Ice Age, or the Pleistocene. Geologists also recognized that during the Pleistocene, mountain ranges in temperate latitudes, including the Sierra Nevada, the Rocky Mountains, and the Alps of Europe, were covered by ***alpine glaciers*** (mountain glaciers).

The Late Cenozoic period, about the last 10 million years, has been a time in which glaciers have periodically built up throughout the world. One of the requirements for the occurrence of worldwide glaciation is to have continents near the Earth's poles, where temperatures are cold enough for glaciers to form. Another requirement is the uplifting of young mountain chains to elevations sufficient to sustain alpine glaciers. Because of plate tectonics, the Late Cenozoic period has been favorable for glacial conditions. The Sierra Nevada has been uplifted high enough to sustain alpine glaciers for a least the past 3 million years.

Astronomic factors are thought to trigger the episodic global cooling that results in glacial periods. Variations in the tilt of the Earth on its axis, the precession in its elliptical orbit, and the solar sunspot cycle affect the amount of solar radiation that strikes Earth. About every 100,000 years, these variables combine to minimize the amount of solar radiation, and cooling occurs. If the continents and mountain ranges are in the correct positions, then glaciers can form and grow.

In the western United States, large alpine glaciers occupied the higher valleys of the higher mountain ranges during periods of global cooling. Large valley glaciers were present in the Rocky Mountains, the Cascade Range, and some of the higher Basin and Range Province. In California, geologists have found evidence of Pleistocene glaciation in much of the Sierra Nevada, in the higher parts of the Klamath Mountains, on the flanks of Mt. Shasta and Lassen Peak, and even in the high parts of the northern Coast Ranges, including high peaks of the Yolla Bolly Wilderness west of Redding.

During glacial periods, more of the Earth's water is held in the great ice sheets on land than during warm periods. As a result, worldwide sea level is lower during times of glaciation. Conversely, when the Earth warms and ice sheets melt, water is added back to the oceans, causing a rise in sea level. As we will see in Chapter 12, climatically caused changes in sea level had dramatic effects on San Francisco Bay and other coastal environments.

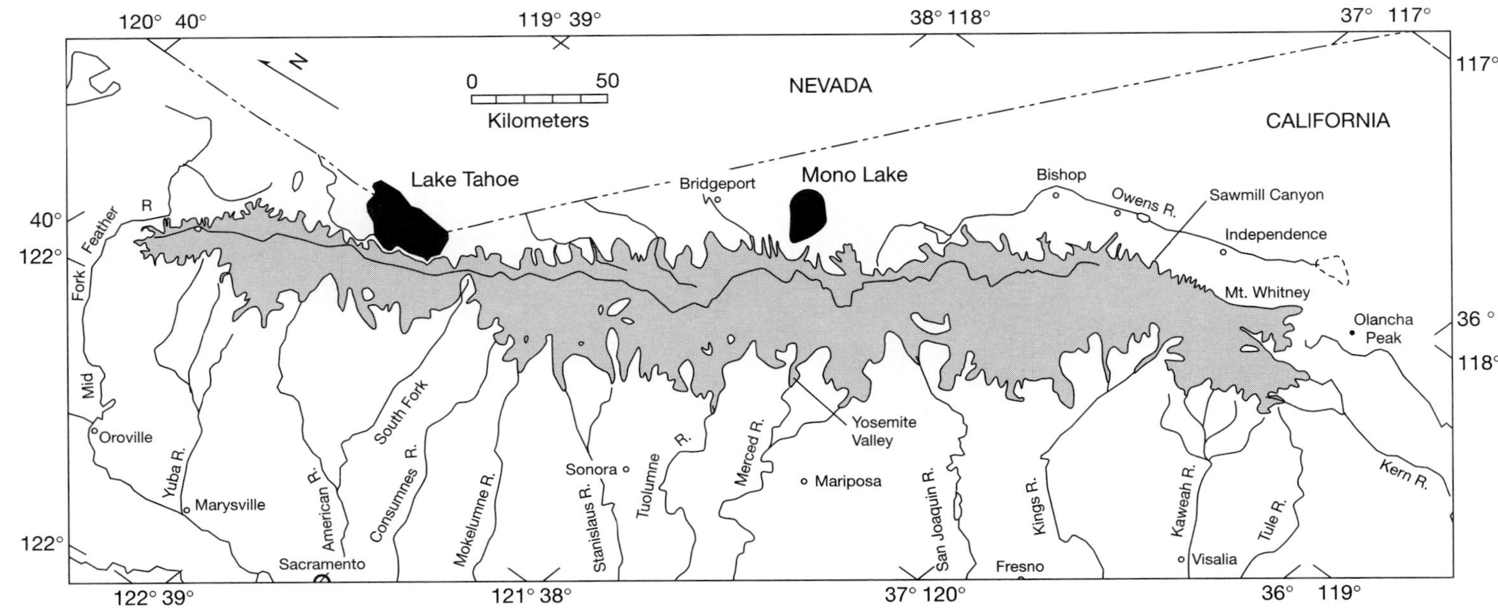

Fig. 8-39 This map shows the approximate extent of ice in the Sierra Nevada 20,000 years ago during the Tioga glaciation. (Source: Baily, E.H., editor. 1966. *Geology of Northern California*. Bulletin 190, p. 158. Used with permission of the California Department of Conservation, Division of Mines and Geology.)

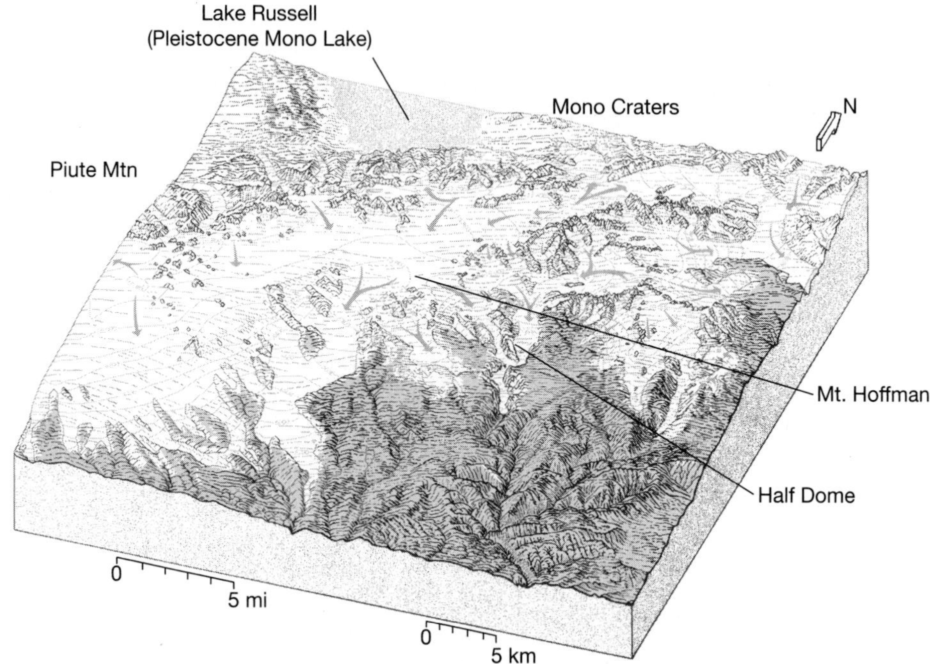

Fig. 8-40 Oblique view across the Sierra Nevada, showing the glaciers in the Yosemite area 20,000 years ago. (Source: Alpha, T.R. U.S. Geological Survey.)

THE WORK OF GLACIERS

John Muir, one of California's most famous naturalists, spent many years exploring the Sierra Nevada during the late 1800s and early 1900s. Muir was one of the first to observe the glaciers of the Sierra Nevada, and he firmly believed that glacial ice had been the main creator of Yosemite:

> In the beginning of the long glacial winter, the lofty Sierra seems to have consisted of one vast undulated wave, in which a thousand separate mountains, with their domes and spires, their innumerable canyons and lake basins, lay concealed. In the development of these, the Master Builder chose for a tool, not the earthquake nor lightning to rend asunder, not the stormy torrent nor eroding rain, but the tender snow-flowers, noiselessly falling through unnumbered seasons, the offspring of the sun and sea.

From Gunsky, F.R., ed. *South of Yosemite: Selected Writings of John Muir.* Garden City, N.Y.: Natural History Press, p. 269.

Fig. 8-41 Dana glacier in 1975. Note the crevasses in the glacier ice. (Source: Clark, M. U.S. Geological Survey.)

Fig. 8-42 Glacial striations and polish on granodiorite at Olmstead Point, south of Highway 120 in Yosemite National Park.

Fig. 8-43 Pothole Dome, a roche moutonée in Tuolumne Meadows, Yosemite National Park. (Source: Huber, N.K. U.S. Geological Survey.)

of the cirques contain small lakes called ***tarns,*** which are a favorite destination of the Sierra Nevada's backpackers.

Valleys eroded by glaciers have a characteristic U shape (see Fig. 8-37)—steep-sided and broad-bottomed—that contrasts with the normal V shape of a stream-eroded valley. The carving of glacial valleys also created another erosional feature—waterfalls. The great waterfalls of Yosemite are spectacular examples of ***hanging valleys.*** During the time when ice filled the Merced River canyon, tributary streams flow into the glacier at the level of the ice. When the ice later melted, the Merced River flowed near its much lower current elevation, and the tributary streams were left "hanging" hundreds of meters above the river (Fig. 8-44).

Some of the Sierra Nevada's spectacular landforms are only partly created by glacial erosion. The structure of the granitic rock itself—the Half Dome granodiorite—is partly responsible for the domes of Yosemite, including Half Dome. This particular plutonic rock is unusually massive, containing few fractures or joints. As the rocks were uplifted, they were unburdened of the huge weight of rock that once covered them. The release of the pressure, plus the release of the glacial ice, causes the rock to expand toward the free surface (Fig. 8-45). The rock expands upward or toward canyon walls, and the outside layers then flake off as uncovering proceeds. This process is called ***exfoliation,*** and the process is responsible for dome formation in the Sierra Nevada. In areas in which the rocks already have more typical, linear joint systems, the "release stress" is relieved along those fractures, and exfoliation domes do not develop.

Glacial Deposits. In the balanced system of alpine glaciation, it is logical to assume that the rock removed by glacial erosion must be deposited at some lower-elevation point. Glacial deposits are recognized throughout the Sierra Nevada, most of them at the ablation areas of the former glaciers. However, ice-

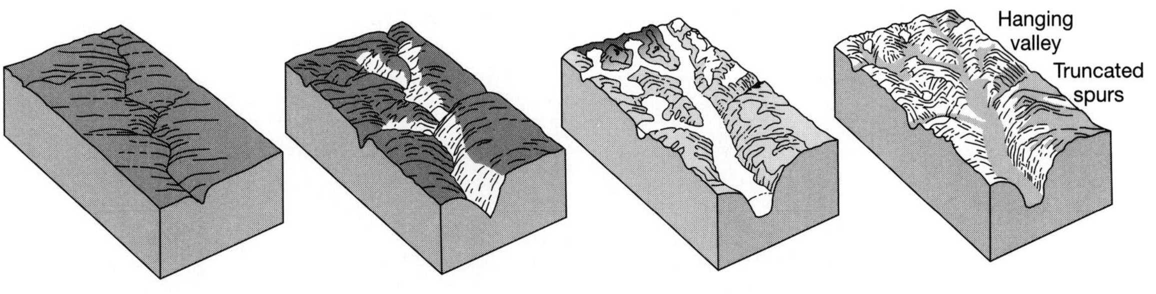

Fig. 8-44 The evolution of U-shaped and hanging valleys. (Source: Huber, N.K. 1987. *The Geologic Story of Yosemite National Park.* U.S. Geological Survey, Bulletin 1595.)

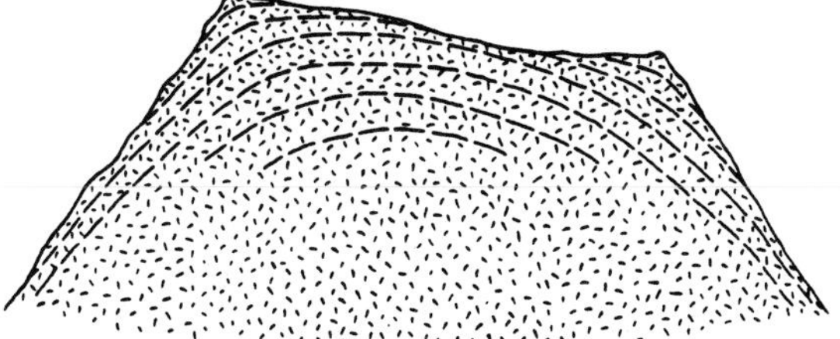

Fig. 8-45 The development of exfoliation on massive granitic surfaces. (Source: Huber, N.K. 1987. *The Geologic Story of Yosemite National Park.* U.S. Geological Survey, Bulletin 1595.)

transported boulders known as ***erratics*** have been dumped on glacially polished surfaces in the highest areas (Fig. 8-46). Glacial erratics are identified by their positions in places where no stream, avalanche, or other process could have deposited them. Rocks matching the composition of the erratic boulders are usually found higher in the ranges, where the boulders were originally picked up by the glaciers.

Along the sides and at the foot of a valley glacier, ridges of sediment known as ***moraines*** pile up. At the ends of glaciated valleys, moraines form characteristic loops (Fig. 8-47); these may enclose a lake or a meadow if streams have not cut a notch through them (see Fig. 8-41). Lake Tahoe is partly dammed by glacial moraines, although faults are largely responsible for the lake's existence. Moraines are composed of the material pushed along by the glacier and left behind after the ice melted. This poorly sorted material may contain huge boulders and is called glacial ***till.*** Even when the glacial moraines have been eroded, older deposits of till can be recognized along the slopes of the Sierra Nevada by their characteristically poor sorting and large size of the largest boulders.

By identifying and mapping moraines and till, geologists are able to reconstruct the extent of glaciers that have long since melted. If the deposits can be dated, then

Fig. 8-46 Glacial erratic in Yosemite Valley. (Source: Sarna-Wojcicki, A. U.S. Geological Survey.)

Fig. 8-47 Aerial view of glacial moraines along the eastern side of the Sierra Nevada. Grant Lake appears at the left edge of the photo. (Source: Beck, F. University of Colorado.)

the timing of glacial periods can also be established. Unfortunately, Sierran glacial deposits are very difficult to date because they are not completely preserved, and because they contain very little datable organic material. Most of the glacial deposits on the western slope of the Sierra Nevada have been eroded by the powerful rivers there. Only on the dry eastern side, where stream erosion is minimal because of the rain shadow effect, is a partial record of glaciation preserved in the form of moraines.

The History of Sierran Glaciation

Geologists are able to recognize two sets of glacial moraines in the canyons of the eastern Sierra Nevada, as well as older glacial deposits left by two or more earlier periods of glaciation. The most recent moraines are, of course, the least eroded and easiest to recognize. They were deposited during the Sierran glacial period named the ***Tioga glaciation*** by geologist Eliot Blackwelder. Tioga glaciers filled the valleys of the Sierra Nevada, between about 20,000 and 100,000 years ago, during the most recent episode of worldwide glaciation.

Before the Tioga glaciation, more extensive glaciers had occupied the Sierra Nevada. Moraines of these glaciers, called the ***Tahoe glaciers,*** are found further down the valleys than the Tioga moraines, and they form much larger ridges (Fig. 8-48). In fact, it is only because the Tahoe glacial period came before the less extensive Tioga glaciation that any evidence of the Tioga period has been preserved. The Tahoe glaciers would have completely destroyed any moraines that lay in their path. Geologists have disagreed vigorously about the age of the Tahoe glacial period, but many believe that it correlates with the worldwide glacial period that occurred between about 130,000 and 200,000 years ago.

Evidence for glacial periods before the Tahoe glaciation is found at several Sierran localities. The evidence consists mainly of older glacial tills which no longer

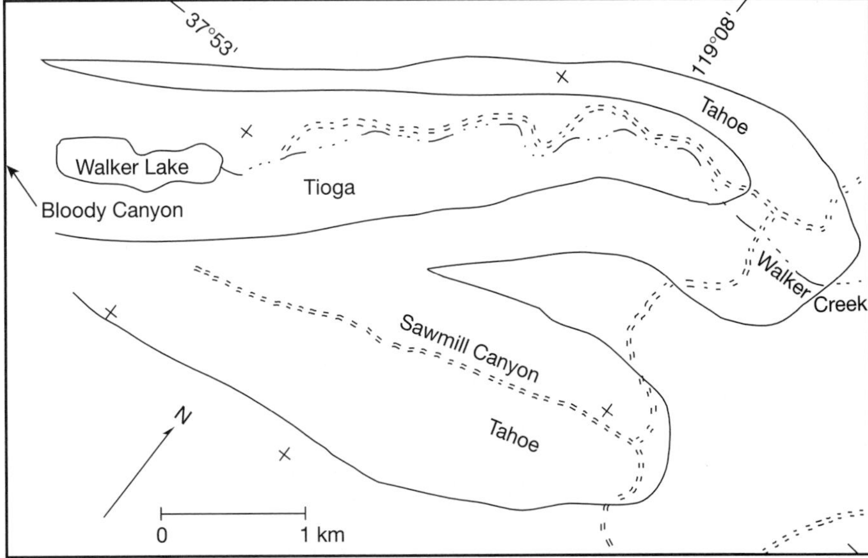

Fig. 8-48 Sketch of glacial moraines in Sawmill and Bloody Canyons on the eastern side of the Sierra Nevada. These moraines can be seen in Fig. 8-47. (Source: Burke, R.M., and Bierkeland, P. 1979. *Friends of the Pleistocene Field Trip Guidebook.*)

Fig. 8-49 Gravels of the Sherwin till (bottom of outcrop) were buried by the Bishop Ash when it erupted about 760,000 years ago. Photo taken on U.S. Highway 395 near Sherwin summit. (Photo by author.)

Table 8-1 SIERRA NEVADA GLACIATION

Time	Glacial/Interglacial Period	Event
10,000 to present	Holocene	Present period of global warming
18,000-22,000	Tioga	Peak of most recent glacial period
105,000-125,000	Unnamed in Sierra	Prior interglacial period worldwide
130,000-160,000	Tahoe (?)*	Age of older moraines in Sierran valleys
790,000	Sherwin	Till underlying Bishop Ash (see Chapter 5)
2.2-2.4 million		Beginning of worldwide Pleistocene glaciation

*Some investigators believe that the Tahoe moraines were formed after the 150,000- to 125,000-year warm period. The age of the Tahoe glaciation is still in dispute.

preserve any moraine-like topography. One locality where older Sierran till has been preserved is world-renowned among geologists. Along State Highway 395 near Sherwin Summit, glacial till is buried by thick deposits of the Bishop Ash, erupted 758,000 years ago from the Long Valley Caldera (Fig. 8-49; see also Chapter 5). The glacial period recorded by the older till is appropriately known at the Sherwin glaciation (Table 8-1). Older glacial deposits can also be sent further north along Highway 395 near Bridgeport. Geologic evidence indicates that once the Sierra Nevada had been uplifted to a high enough elevation, it was subjected to periodic episodes of glaciation during the past few million years.

Rivers below the alpine glaciers of the Sierra Nevada were also dramatically affected by glacial episodes. Fed by enormous volumes of summer meltwater and glacial sediment, these rivers carried huge amounts of ***glacial outwash.*** Some of these sediments were deposited in the Central Valley, where they can be seen today as alluvial deposits of at least four different ages (see Chapter 11). In the Basin and Range Province, large lakes formed during the cooler glacial periods (see Chapter 6). Next to plate tectonics, recent climatic changes may be the most important natural factor that controls California's landscape.

FURTHER READINGS

BATEMAN, P.C. 1992. *Plutonism in the Central Part of the Sierra Nevada Batholith, California.* U.S. Geological Survey, Professional Paper 1483, 186 pp.

CLARK, W.B. 1970. *Gold Districts of California.* California Division of Mines and Geology, Bulletin 193, 186 pp.

FARQUAR, F. 1965. *The History of the Sierra Nevada.* Berkeley, Calif.: University of California Press, 262 pp.

GUNSKY, F.R., ed. 1968. *South of Yosemite: Selected Writings of John Muir.* Garden City, N.Y.: Natural History Press, 269 pp.

HILL, M. Glaciers—a picture story. *California Geology* 2:23-44, 1974.

HILL, M. 1975. *Geology of the Sierra Nevada.* Berkeley: University of California Press, California Natural History Guide 37, 232 pp.

HUBER, N.K., BATEMAN, P.C., AND WAHRHAFTIG, C. 1989. *Geologic Map of Yosemite National Park and Vicinity, California.* U.S. Geological Survey, Miscellaneous Investigations Map I-1874, scale 1:125,000.

HUBER, N.K. 1987. *The Geologic Story of Yosemite National Park.* U.S. Geological Survey, Bulletin 1595, 64 pp.

PAUL, R.W. 1947. *California Gold: The Beginning of Mining in the Far West.* Lincoln, Nebr.: University of Nebraska Press, 379 pp.

CHAPTER

9

The Klamath Mountains

Accreted Terranes and a View of the Mantle

THE LANDSCAPE

When viewed on a relief map (Endpaper 4, inside the back cover) or from the air, the Klamath Mountains Province appears quite different from other areas of California. Several mountain ranges make up the Klamath Mountains: the Trinity Alps, the Marble Mountains (Fig. 9-1), the Salmon Mountains, and the Siskiyou Mountains are the major ones. These mountain chains are lower than the Sierra Nevada, and not as parallel as those of the Coast Ranges, although most run generally north-south. The mountain summits of the Klamath Province all have about the same elevations, between 1500 and 2100 meters, with the highest peaks standing above that level. The consistency in the elevation of the ridges, even though they are separated by deep canyons that wind between them, gives the impression of a carved-up plateau.

Early geologists working in northwestern California suggested that the plateau-like surface of the Klamath Province was once a much lower plain. As the ranges were uplifted during the past few million years, the rivers carved impressive gorges through this formerly low-lying surface. Today the Klamath River and its largest tributary, the Trinity River, flow through the province in canyons cut more than 1000 meters into the hard metamorphic rocks of the province. The high surfaces of the province remain as evidence of a time when the ranges were much lower.

ACCRETED TERRANES OF THE KLAMATH MOUNTAINS

A look at the California geologic map (Endpaper 1, inside the front cover) gives one the impression that the geology of the Klamath Mountains Province is similar to that of the northwestern Sierra Nevada, and in general, this impression is an accurate one. Like the metamorphic belts of the northern Sierra, the Klamath Mountains are a composite of accreted oceanic terranes. In fact, the concept of accreted terranes was first developed to explain the geology of the Klamath Province. After careful study of the rocks and fossils of the Klamath Mountains and northern Sierra Nevada, geologists have concluded that some of the rock units in each province are in fact fragments of the same oceanic plates.

Fig. 9-1 Marble Mountains Wilderness Area. (Source: Harwood, D. U.S. Geological Survey.)

As mapping and other studies allowed geologists to sort out the complexities of the Klamath Mountains, they have divided the rocks of the province into an increasing number of distinct terranes. At present, five major terranes are recognized, and several of these are subdivided into two or more units (Fig. 9-2). The terranes are generally younger from east to west, reflecting the succession of plate collisions that added each oceanic slice to the North American continent. Each terrane is bordered by major faults that represent the lines or sutures where plate fragments are joined. By dating the movement of the faults and the metamorphism that accompanied periods of deformation (Chapter 8), geologists have confirmed that terranes were accreted in a westward succession. Like the rocks of the western metamorphic belt in the Sierra Nevada, the rocks of the Klamath Mountains were intruded by Mesozoic plutons. Dating these plutons, which intrude across terrane boundaries, allows geologists to date tectonic events in the Klamath Mountains by applying the principles of cross-cutting relations (see Chapter 3).

All except the easternmost terranes of the Klamath Mountains are thought to be the remnants of far-traveled oceanic plates. All contain metachert, oceanic metavolcanic rocks (Fig. 9-3), and fragments from the mantle beneath the plates (discussed later in this chapter). The youngest accreted terrane, known as the western Jurassic belt, includes the Galice Formation, a long belt of metasedimentary rocks formed approximately 150 million years ago. To the west, these rocks are separated from rocks of the Coast Ranges by a major fault, the South Fork Mountain thrust. Traveling eastward, and back in time, one next encounters the Western Paleozoic and Triassic belt, a very complex unit which has been subdivided into at least three separate terranes: the Rattlesnake Creek–Marble Mountains, Hayfork, and the North Fork–Salmon River terranes. All represent fragments of oceanic plates, and all contain both oceanic metavolcanic rocks and metachert. The Stuart Fork terrane, a still-older belt of rocks metamorphosed at

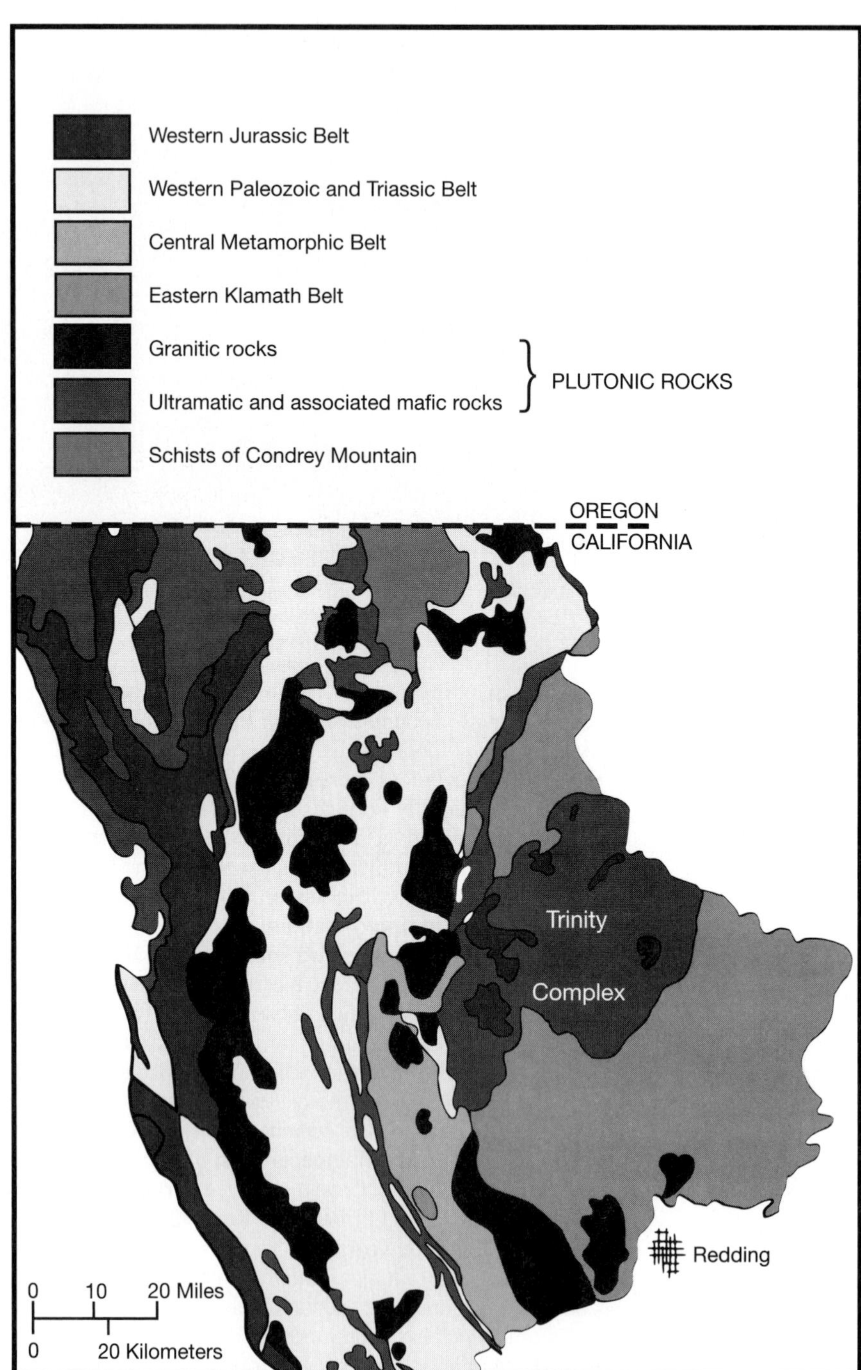

Fig. 9-2 Generalized terrane map of the Klamath Mountains province. (Source: Vennum, W. *California Geology* 47(2), 1993. Used with permission of the California Department of Conservation, Division of Mines and Geology.)

Fig. 9-3 Metavolcanic rocks of the Klamath Mountains at Condrey Mountain, showing well-preserved pillows. Note matchbook for scale. (Source: Hotz, P. U.S. Geological Survey.)

high pressures and relatively low temperatures, has been correlated with one of the subduction complexes of the Sierra Nevada. East of the Stuart Fork terrane, the central metamorphic belt contains rocks very similar to the Feather River terrane, another subduction zone complex in the Sierra Nevada. Based on the results of radiometric dating, geologists have determined that the metamorphic minerals in the central metamorphic belt formed during deformation in Devonian time. The original rocks of this terrane were oceanic sedimentary and basaltic volcanic rocks.

Finally, along the eastern edge of the province is the eastern Klamath Belt, which includes the Eastern Klamath terrane, the Trinity ultramafic belt, and the Yreka terrane. The oldest rocks in the Klamath Mountains Province are those of the Eastern Klamath terrane. These rocks, which range from Cambrian to Triassic in age, are thought to match the Northern Sierra terrane discussed in the previous chapter. Both terranes consist of metamorphosed andesitic volcanic rocks erupted from an oceanic arc, and both contain metasedimentary rocks that originated on a carbonate platform. The similarities between the Paleozoic formations in the Northern Sierra terrane and the Eastern Klamath terrane, especially the ages and geometric relations of individual rock units, have led geologists to conclude that both originated as parts of the same offshore volcanic arc.

THE McCLOUD FAUNA

Carbonate rocks of Permian age are found adjacent to and overlying the fractured metavolcanic rocks of the Eastern Klamath and Northern Sierra terranes. The McCloud Limestone, named for the town of McCloud in Siskiyou County, is an important Permian formation within the eastern Klamath terrane. The limestone, which can be viewed in Shasta Caverns in Shasta Lake Recreation Area, contains abundant corals, brachiopods, and fusulinids (Fig. 9-4). Geologists interpret its original environment of deposition as carbonate reefs fringing the eastern Klamath volcanic arc. These ancient reefs may have appeared similar to coral reefs found today near Indonesia (Fig. 9-5).

Paleontologists now recognize that the Permian-aged fossils preserved in the McCloud limestone are a unique assemblage of marine organisms. They have named this assemblage the McCloud fauna, because its distinctive corals and fusulinids are best seen in the rocks of the eastern Klamath Mountains near McCloud. The same distinctive assemblage of fossils is found in a belt of Permian rocks stretching from the Northern Sierra terrane in California north to British Columbia (Fig. 9-6). The fossils of the McCloud fauna are unlike those found in other accreted terranes of western North America.

Fig. 9-4 Large fossil corals of the Permian McCloud fauna. The sample is from Quinn River Crossing, Nevada. The larger coral measures about 4 centimeters across. (Source: Stevens, C. San José State University.)

Fig. 9-5 Modern corals from a fringing reef.

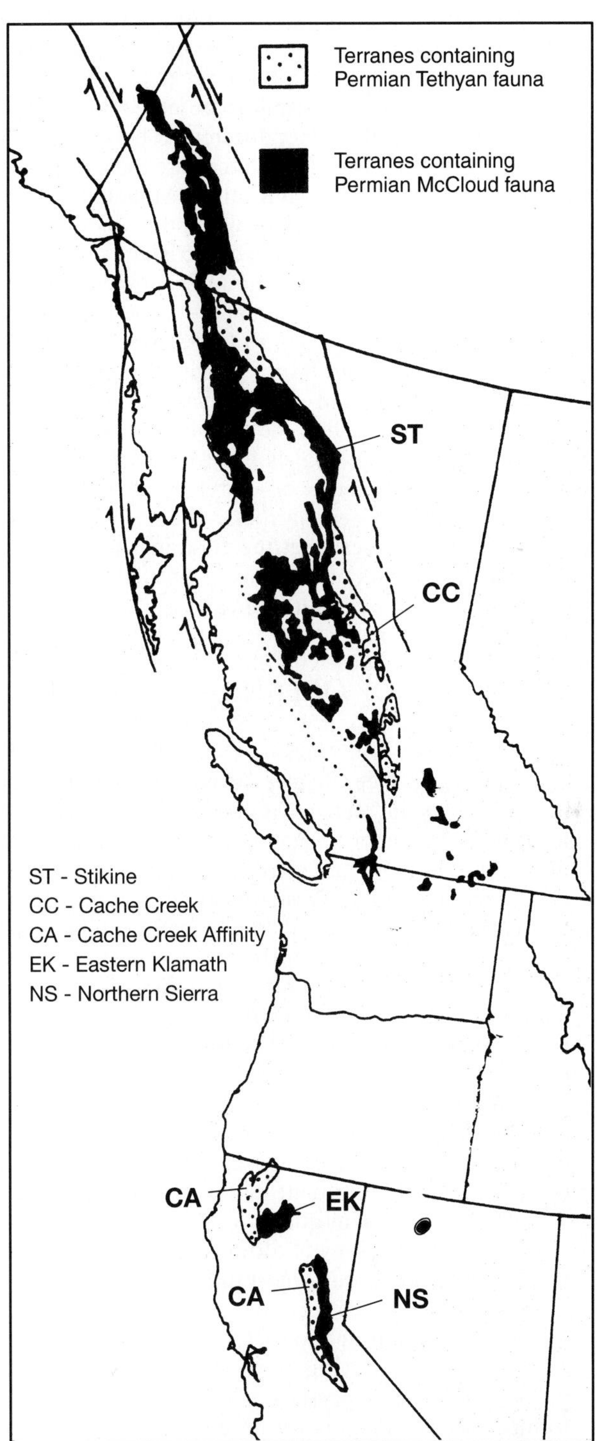

Fig. 9-6 Map showing the distribution of terranes containing fossils of the Permian McCloud fauna and terranes with Permian Tethyan fauna. (Source: Miller, E. 1992. Late Paleozoic paleographic and tectonic evolution of the western U.S. Cordillera. Geological Society of America. *Decade Of North American Geology.* Vol. G-3, p. 83.)

The carbonate rocks in the central Klamath Mountains contain a completely different community of marine organisms. The assemblage of corals, fusulinids, and other species found in those rocks are representative of the Tethyan fauna, a group of Permian organisms represented in rocks found in Japan, China, and southeast Asia. The distinct differences between the McCloud and Tethyan faunal assemblages, even though the rocks are the same age, supports the hypothesis that some of the terranes of the Klamath Mountains and the Sierra Nevada originated at great distances from Paleozoic North America. However, geologists are uncertain about the original position of the island arcs of the northern Sierra and Eastern Klamath Mountains. The community of reef organisms found in the McCloud fauna is also distinctly different from the community that inhabited the shallow ocean immediately adjacent to North America, as represented by Permian fossils found in the Basin and Range province a few hundred kilometers to the east (see Chapter 7). This would suggest that the Permian island arc was distant enough from North America that the populations were isolated from each other. On the other hand, many geologists believe that the volcanic arc was relatively close to North America because of correlations with the tectonic events recorded by Paleozoic North American rocks. Studies of modern coral populations and their ability to mix suggest that the Eastern Klamath terrane must have been several thousand kilometers offshore from North America during Permian time to prevent the populations from intermixing.

OPHIOLITE SUITES: REMNANTS OF OCEANIC BASEMENT

The sedimentary and volcanic rocks of the Eastern Klamath terrane rest on older ultramafic rocks (see Chapter 2) known as the Trinity peridotite, the largest exposed body of ultramafic rock in North America. Further west, belts of ultramafic rock also form the basement of several of the younger terranes of the Klamath Mountains (see Fig. 9-2). It is these rare, dark-colored ultramafic rocks that are responsible for the dark alpine peaks like Preston Peak (elevation 2700 meters) that are so unusual to those accustomed to light-colored granitic alpine scenery.

Geologists now recognize a characteristic assemblage of unusual rocks as the components of an ***ophiolite*** or ***ophiolite suite.*** At most localities, the rocks in an ophiolite suite are highly disrupted by faulting, folding, and metamorphism, but geologists have been able to reconstruct their original geometry by detailed mapping and other studies. At the base of a reconstructed, or idealized, ophiolite are rocks from the upper mantle beneath the oceanic crust. These are ultramafic plutonic rocks rich in iron and magnesium and low in silica (Chapter 2). Ultramafic rocks include ***dunite,*** composed entirely of the mineral olivine and ***peridotite,*** composed mostly of olivine with some plagioclase feldspar and pyroxene (Fig. 9-7). (The fact that gem-quality olivine is known as peridot gives some commonality to the terminology of ultramafic rocks.) After they have been displaced from the mantle into the crust, most ultramafic rocks are altered by hot (***hydrothermal***) fluids to serpentinite.

Above the ultramafic rocks in an ophiolite suite is a layer of mafic plutonic rock, mostly gabbro (Fig. 9-8). The lower section within the gabbro typically displays a layering of crystals. The layering is caused by the sinking of the first-formed, dense minerals to the bottom of the magma chamber, followed by the accumulation of minerals, which crystallized later (Fig. 9-9). The gabbro layer is

Fig. 9-7 Outcrop of ultramafic rocks along Highway 70 in the northern Sierra Nevada. (Photo by author.)

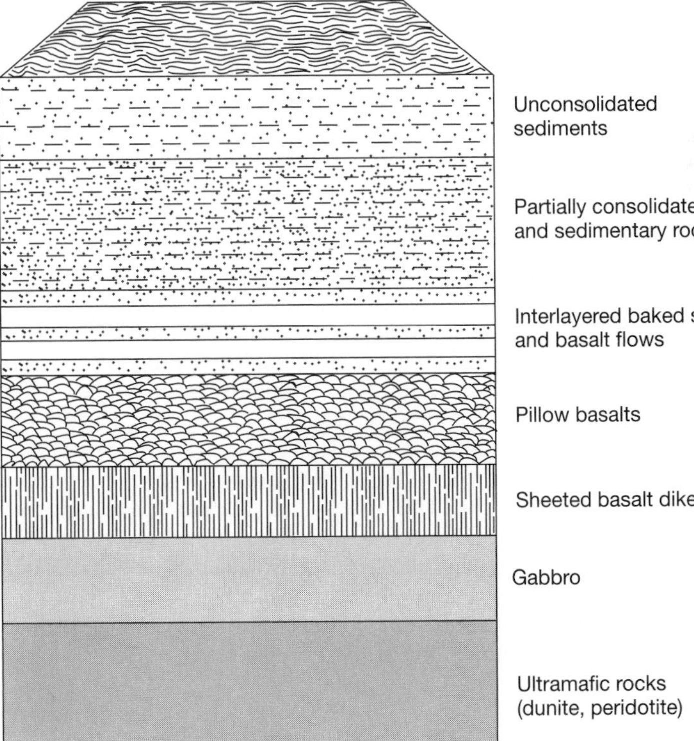

Fig. 9-8 Rock assemblage through the ocean floor that constitutes an ophiolite suite. (Source: Coch, N., and Ludman, A. 1991. *Physical Geology.* New York: Macmillan, p. 434.)

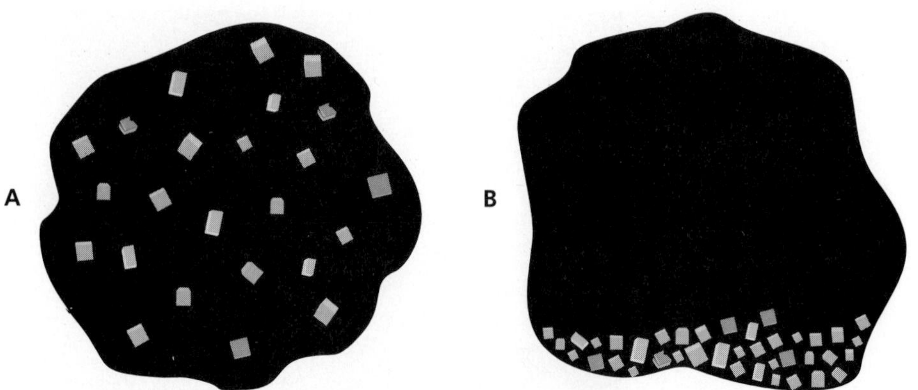

Fig. 9-9 Accumulation of early-formed crystals at the bottom of a magma chamber. The crystals have higher densities than the surrounding magma and **A,** sink to the floor of the magma chamber. **B,** The early-formed crystales accumulate and are segregated to form an ultramafic rock. (Source: Coch, N., and Ludman, A. 1991. *Physical Geology.* New York: Macmillan, p. 434.)

in turn overlain by a complex of vertical ***sheeted dikes,*** mafic plutonic rocks that crystallized at shallower depths beneath the ocean floor. On the top of this igneous pile is a section of mafic volcanic rocks, typically pillow basalt, erupted on the ocean floor. Geologists believe that the entire assemblage represents a cross section through the ocean crust into the upper mantle beneath it. Sedimentary rocks formed in the deep ocean, including chert and shale, are usually found at the top of such an ophiolite suite.

Direct observation and sampling of the ocean floor has enabled researchers to verify that oceanic crust is composed of basalt, both pillow lavas and sheet flows. Deep drilling has penetrated into the upper zone of sheeted basalt dikes, but no direct sampling has reached below about 1 kilometer. Indirect studies using seismic techniques confirm the basic three-layer model of ultramafic upper mantle, gabbro, and basaltic volcanic rocks. However, knowledge about the nature of oceanic crust and mantle relies heavily on studies of ophiolites that we interpret as on-land samples of oceanic crust and upper mantle. In part because they are commonly associated with volcanic rocks produced by explosive eruptions rather than those produced at mid-ocean ridges, California's ophiolites are thought to have formed beneath island arcs or basins fringing the arcs.

On the California geologic map (Endpaper 1), the largest ophiolite suites can be identified by the purple slivers and pods representing areas underlain by ultramafic rocks. The Trinity and Josephine ophiolites are the largest bodies in the Klamath Mountains province. The Feather River peridotite, the Smartville Complex, and the Kings-Kaweah ophiolite are examples of major bodies in the Sierran metamorphic terranes. The Coast Range ophiolite extends along the western margin of the Great Valley province (see Chapter 11). Ophiolitic rocks are also found within the Franciscan Assemblage in the Coast Ranges (see Chapter 12). Considering California's long history as a convergent plate margin, it is not surprising that many fragments of oceanic basement are present in the accreted terranes of these three provinces (Fig. 9-10).

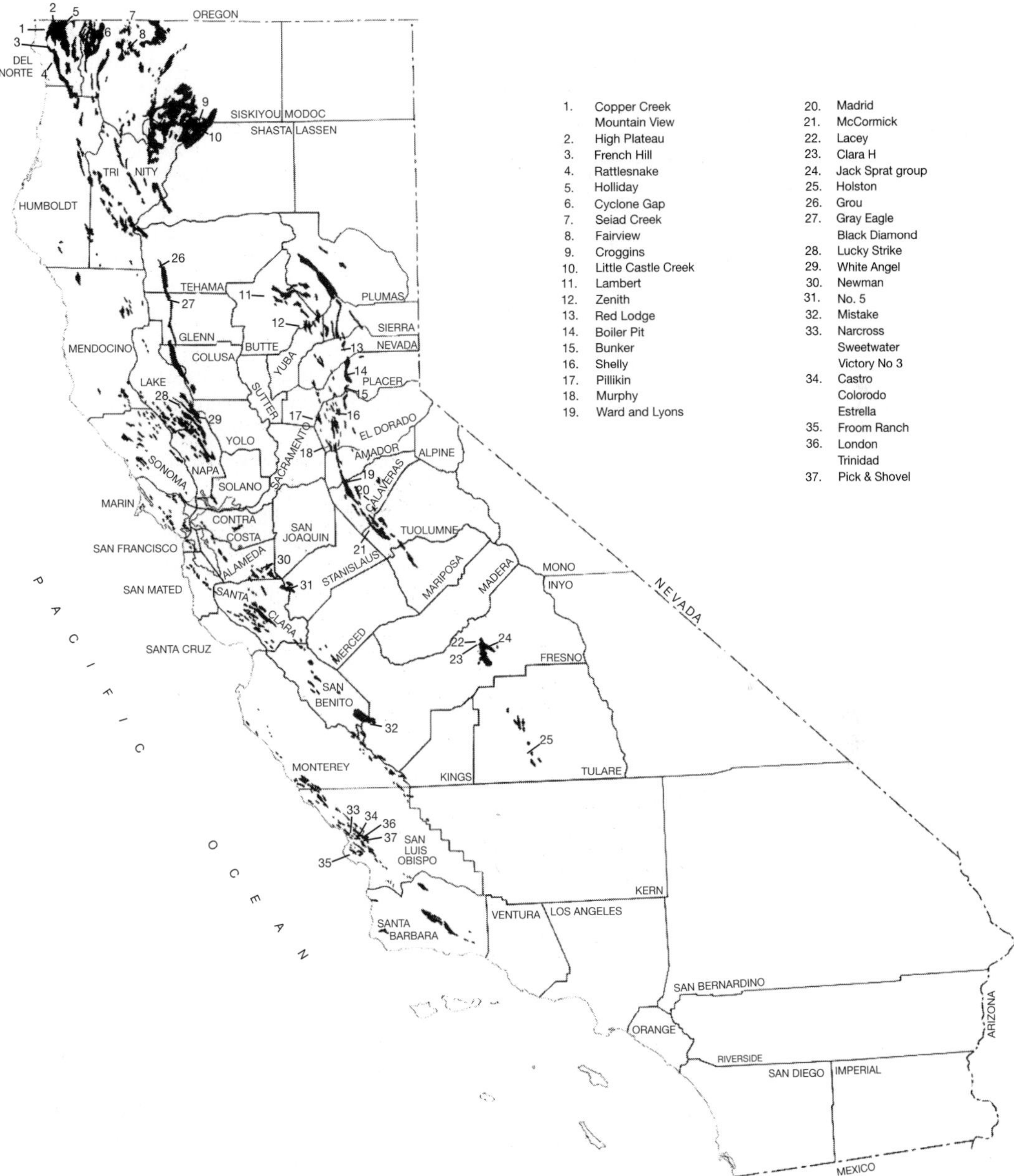

Fig. 9-10 The distribution of ultramafic rocks and chromite mines (shown by numbers) in California. (Source: Wright, L.A., editor. 1957. Bulletin 176, p. 124. Used with permission of the California Department of Conservation, Division of Mines and Geology.)

RICHES FROM THE BOTTOM OF A MAGMA CHAMBER

Found within many of California's ultramafic rocks are rich pods and slivers of ***chromite,*** an oxide mineral containing chromium and iron. Chromium is used as a steel alloy and for plating metal surfaces. Chromite has been mined from the Josephine and Trinity ophiolites, as well as from smaller ultramafic bodies in the Klamath Mountains, the western foothills of the Sierra Nevada, and the Coast Ranges. The most intense chromite mining in California was done during World War II, when supplies from overseas were cut off. California has been the second largest chromium supplier in the United States, although no commercial mining has occurred since the mid-1960s.

Geologists believe that chromite concentrations developed during the original crystallization of the ultramafic plutonic rocks. As magma crystallization began, the early crystals sank to the floor of the magma chamber because they were denser than the surrounding magma. Settled at the floor of the magma chamber, the early-formed minerals were isolated from further reaction with the magma (see Fig. 9-10). Early-formed silicate crystals, those with the highest temperature of crystallization, include the mafic minerals olivine and pyroxene. Geologists believe that many ultramafic plutonic rocks crystallize by this process of ***magma differentiation.*** Minerals that crystallize later may also settle out from the magma, ultimately producing a layered plutonic rock.

Chromium is one of the Earth's elements that is concentrated in the mantle and thus found associated with mantle rocks. Many of the ultramafic plutonic rocks in ophiolites have been altered to serpentinite, but the chromite remains mostly unaltered. Some of the richest concentrations yielded as much as 20,000 tons of chromite from a single pod deposit. Nickel, another valuable element that is rare in crustal rocks, is also associated with the ultramafic rocks of the Klamath Mountains Province but has not been mined at this time. The ultramafic rocks are also the probable source of reported platinum found in beach deposits near Crescent City.

Like the Sierra Nevada, the Klamath Mountains were a major destination of gold seekers during the 1850s. Early miners recognized the geologic similarities between the two provinces and were quick to discover both lode and placer deposits. Today visitors can see the remains of gold-mining operations near Weaverville, and dredge operations are still active in a few areas.

FURTHER READINGS

HACKER, B.R., DONATO, M.M., BARNES, C.G., MCWILLIAMS, M.O., AND ERNST, W.G. Timescales of orogeny: Jurassic construction of the Klamath Mountains. *Tectonics* 14(3):677-703, 1995.

HARWOOD, D.S., AND MILLER, M.M. 1990. *Paleozoic and Early Mesozoic Paleogeographic Relations: Sierra Nevada, Klamath Mountains, and Related Terranes.* Geological Society of America, Special Paper 255, 422 pp.

CHAPTER

10

Water in California

Water issues have dominated California's politics, history, and economy since the state's early years. Nowhere else is so much water stored, diverted, and transported in order to overcome four of nature's water-related limitations to human activities. The limitations are these: (1) most of the state is a semidesert that is naturally unsuitable for agriculture or major cities; (2) precipitation is seasonal, falling almost entirely between October and April; (3) years of drought and years of flood are part of the normal, natural pattern; and (4) *the water in California is not where the people are.*

Californians used more than 11 trillion gallons of water in 1985, yet too few people know where their water comes from or where their wastewater goes. As concern for the availability and safety of water grows, it becomes increasingly important to understand the fundamental principles of ***hydrology,*** the study of water, and the basic elements of California's hydrologic systems.

PRECIPITATION

Much of California has a ***Mediterranean climate,*** with warm, dry summers and mild, wet winters. The Pacific Ocean moderates the temperatures along the coast, but in the Sierra Nevada, winters are cold, and winter precipitation falls as snow that is stored until the spring snowmelt. The average annual precipitation of California is about 58 centimeters, with about two thirds falling in the northern third of the state (Fig. 10-1). Annual precipitation ranges from more than 254 centimeters in the northwestern Coast Ranges to less than 25 centimeters in the southern San Joaquin Valley and most of southeastern California. Most of the state is therefore semidesert or desert, and only great human effort has made it possible for 31 million people to inhabit California.

As discussed in Chapter 6, California's mountain ranges create a rain shadow effect for the slopes and valleys east of them (see Fig. 6-2). The effect of the rain shadow can be seen in the distribution of precipitation within the state, particularly on the dry eastern side of the Sierra Nevada. The winter storms that bring most of California's water originate in the northern Pacific Ocean and travel eastward. Storms from in and around the Gulf of California may also bring precipitation to southern California in late summer, and local thunderstorms provide an additional

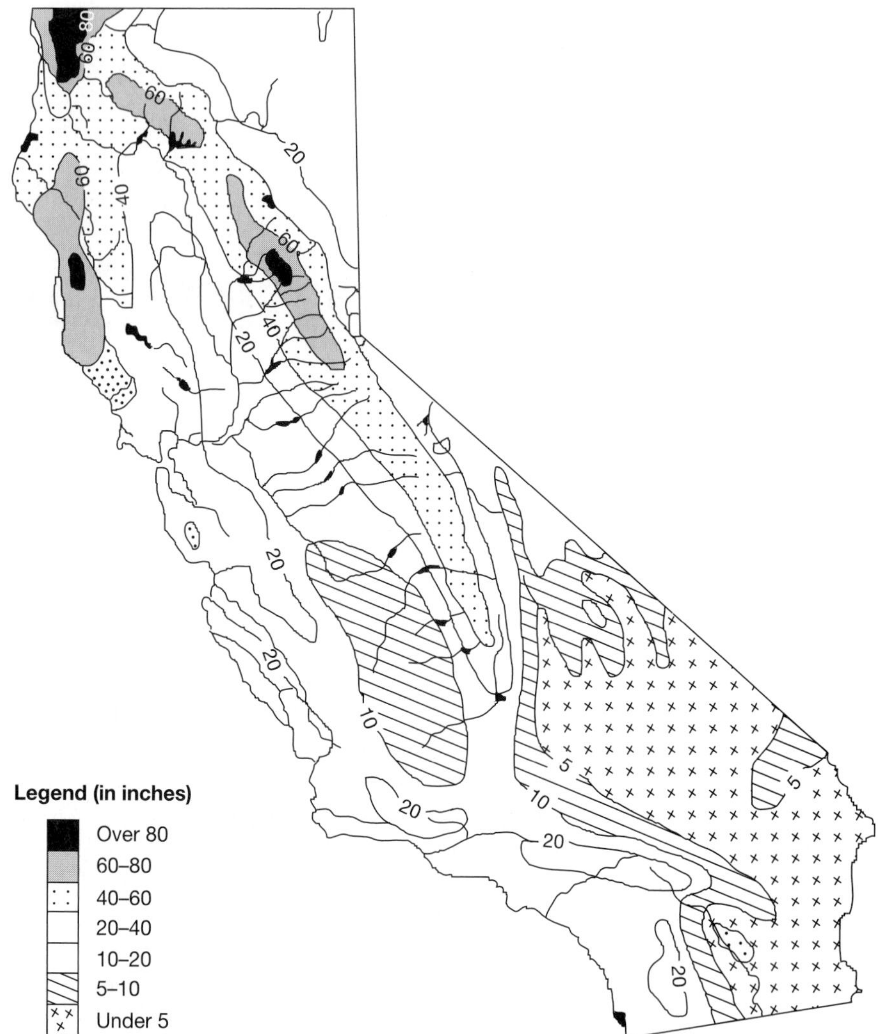

Fig. 10-1 Average yearly precipitation in California in inches. To convert the units on the map to centimeters, multiply values by 2.54. (Source: California Department of Water Resources, 1987, *California Water: Looking to the Future,* Bulletin 160-87.)

moisture source in the mountains during summer. However, in most of California, little or no precipitation falls between May and October. The lack of reliable rainfall during the main growing season creates one of California's major reasons for water storage and diversion.

Another important and often overlooked aspect of California's climate is the normal cyclic occurrence of both flood and drought. Prolonged, severe droughts are part of this recurrent, variable pattern; during 1928 to 1934, 1976 to 1977, and 1987 to 1992, insufficient water supplies have stressed the state's ability to meet the demand for water. By contrast, damaging floods produced by extremely high rainfall occur somewhere in the state at least once in every decade (Fig. 10-2).

A

B

Fig. 10-2 **A,** Effects of the 1964 flood on the town of Orick, Humboldt County. **B,** Flooding in a residential neighborhood of downtown San José in March, 1995, caused when the Guadalupe River overflowed its banks. (Sources: **A,** †Janda, R. U.S. Geological Survey; **B,** Santa Clara Valley Water District.)

THE HYDROLOGIC CYCLE AND DRAINAGE BASINS

Of all the precipitation that falls over California, only about 35 percent becomes the ***runoff*** to streams and lakes. Because of California's relatively high temperatures, ***evaporation*** from the land surface takes up much of the precipitation, and ***transpiration*** by the vegetation uses another substantial portion (Fig. 10-3). Unless the ground surface is completely impervious to water, a portion of the precipitation enters the soil by ***infiltration.*** Only the water which does not evaporate, is not used

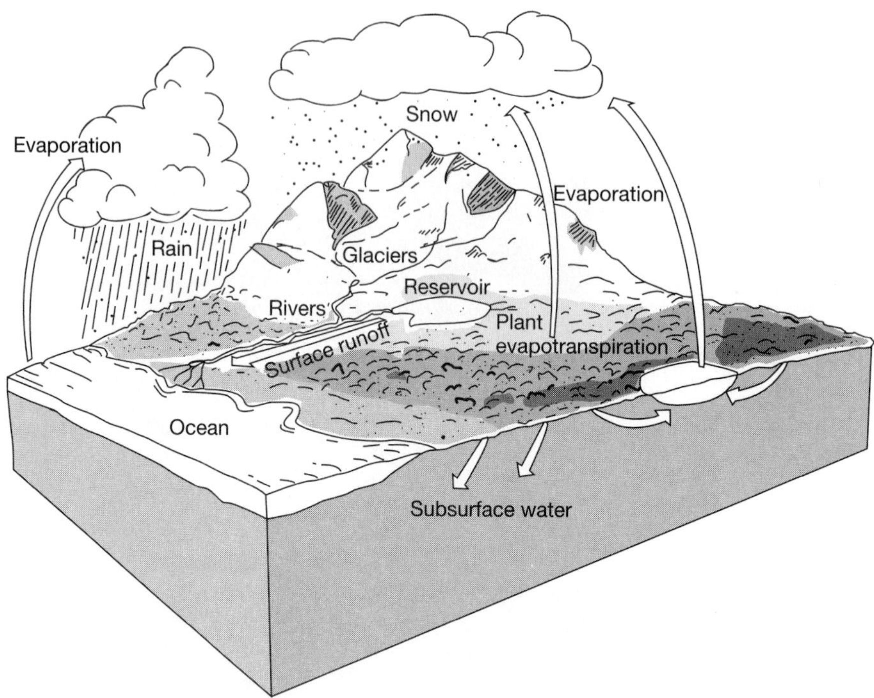

Fig. 10-3 The hydrologic cycle. (Source: Coch, N., and Ludman, A. 1991. *Physical Geology.* New York: Macmillan Publishing Co.)

by the plants, and does not infiltrate into the subsurface becomes runoff. In California, the mean annual runoff is only about 20 centimeters, and many of the hot, dry deserts may experience years without any runoff.

Over thousands of years, the typical amount of runoff that occurs in different parts of the California landscape is obviously extremely variable. However, in all landscapes, running water is the most effective agent of erosion, even in desert areas that experience runoff only once or twice in a decade. Runoff from the land surface flows downhill under the influence of gravity, and water collects to carve ***stream channels.*** Larger channels naturally are required to carry larger amounts of water that collect as small channels join.

It is not surprising that the network of stream channels in a given area would be adjusted to accommodate the typical runoff conditions. The natural stream network that develops is a system of ***drainage basins*** or ***watersheds*** (Fig. 10-4). Each stream channel, regardless of size, collects runoff from an area defined as its drainage basin. All of the precipitation that falls within the boundaries of that basin and becomes runoff will eventually pass through the mouth of the channel. All runoff outside of the boundaries flows into the channels of other drainage basins. Because the mouth of the channel is the lowest point in the drainage basin, it is often referred to as the ***base level*** for that stream. The boundary of a drainage basin is termed its ***divide,*** and high areas near the divide are often called the ***headwaters*** of a stream. Drainage basins may have areas much less than a few acres, or they may encompass huge portions of continents (Table 10-1). The drainage basin of a large river such as the

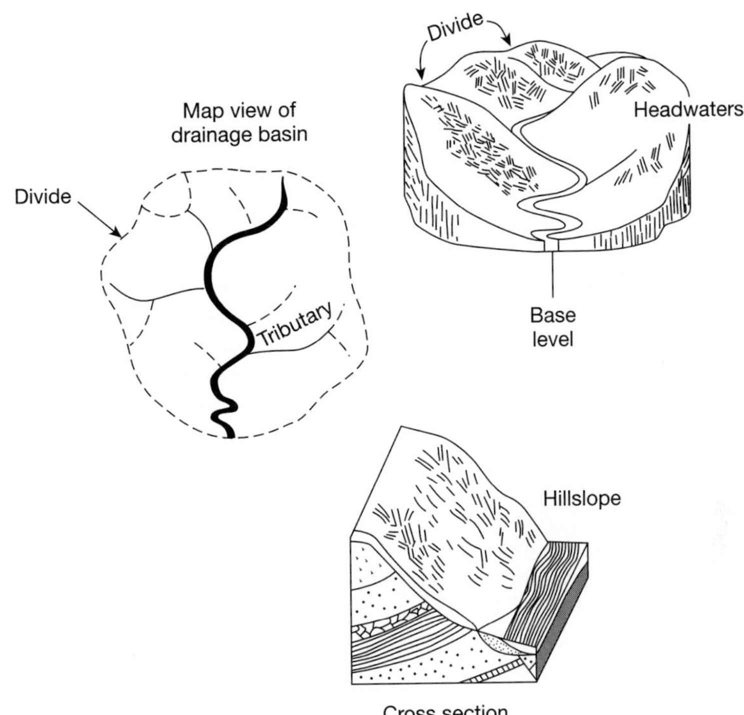

Fig. 10-4 Elements of a drainage basin. (Source: Modified from Ryan, D. 1976-77. *The Co-Evolution Quarterly,* Sausalito, Calif.)

Table 10-1 DATA FOR SELECTED RIVERS IN CALIFORNIA AND ELSEWHERE

River	Drainage Area, mi² (km²)	Average Discharge, cfs (m³/sec)	Average Annual Runoff (in thousands of acre-feet)
Klamath	12,000 (31,080)	17,785 (504)	12,900
Eel	3,113 (8063)	7410 (210)	5,379
Pit	4,711 (12,200)	3880 (110)	2,819
Sacramento	8,900 (23,050)	24,670 (699)	17,870
Russian	793 (2,054)	2,365 (67)	1,712
San Joaquin	1,676 (4,341)	4,375 (124)	3,179
Kings	1545 (4,001)	2,225 (63)	1,624
Colorado	242,000 (626,780)	17,750 (503)	12,860
Mojave	2,121 (5493)	6 (0.17)	4.3
Los Angeles	827 (2,142)	215 (6)	156
Mississippi	1,250,965 (3,240,000)	635,200 (18,000)	460,360
Amazon	2,374,500 (6,150,000)	7,100,000 (200,000)	5,115,000

cfs = cubic feet per second.
Sources: California Water Atlas, U.S. Geological Survey Water-Resources Data for California.

Sacramento River contains many smaller ***tributary*** drainage basins, such as the American and Yuba Rivers, and these tributaries collect water from even smaller tributaries (see Endpaper 3). The ultimate base level for most of California's streams is sea level, because most runoff eventually flows to the Pacific Ocean. However, in eastern California, streams commonly flow into closed basins, some of which contain lakes (see Chapter 6). In these cases, the lake or the dry floor of the basin is the base level for the stream.

The geologic work accomplished by streams is to carry materials from high areas of the landscape to low-lying areas, thus eroding the mountains and depositing sediments in deltas, lakes, dry basins, or along the continental shelf. The material carried by a stream is its ***load,*** which consists of the ***dissolved load*** of molecules and ions in solution and the ***mechanical load*** of solid particles. The mechanical load consists of the ***suspended sediment*** carried in the flowing water and the ***bed load*** of larger particles that roll, skip, and bounce along the bed of the stream. During floods, streams in California may carry very high sediment loads; for example, during the December 1964 flood, the Eel River carried as much suspended sediment in one day as it had during the previous 23 years.

A typical large stream in California undergoes radical changes as it flows from the mountains to the ocean or into a large river like the Sacramento or San Joaquin. In the mountainous headwaters, channels are steep, narrow, and irregular, and the steeply falling stream has sufficient energy to transport boulders and cobbles (Fig. 10-5). Rapids and riffles with pools separating them are common. As the stream col-

Figure 10-5 Large boulders and rapids are typical of a steep mountain stream; this is the Truckee River. (Photo by author.)

lects runoff from tributaries, its channel becomes larger, and as it flows to more gently sloping terrain, its channel becomes less steep (Fig. 10-6). In the Great Valley, a typical stream transports sand and finer sediment and flows in a broad and more gently sloping channel.

A Balanced System

The network of stream channels on the landscape can be thought of as a natural system of canals built over many thousands of years to be exactly designed to carry the natural runoff that is supplied. Within a given watershed, the channels of tributaries are adjusted to each other and to the main stream. A drainage system reflects the prevailing conditions of climate, vegetation, topography, and geologic materials that determine runoff. Changes in runoff, base level, or topography are accommodated by changes in the stream system, so that a condition of ***dynamic equilibrium*** exists in the drainage basin. The streams in the drainage basin have just the gradient needed to carry their sediment loads, without becoming either more or less erosive as they flow downstream.

The dynamic equilibrium of a drainage basin may be disturbed by a variety of events, both natural and human-induced. Some examples of natural causes of disturbance are (1) climate changes, which may change the runoff available in the system, the sediment load, or both; (2) rapid uplift or down-dropping, resulting in base-level changes, which cause the stream to cut down or fill with sediment (Fig. 10-7); and (3) major landslides, which may overwhelm the system with sediment. Examples of human disturbances are urbanization, deforestation, irrigation, hydraulic mining, and water diversions and dams (see Fig. 8-32).

Fig. 10-6 The lower channel and floodplain of Redwood Creek, Humboldt County. The town of Orick is built on the floodplain on the right side of the photograph. Following the 1964 flood (see Fig. 10-2, A), levees were constructed along the banks of lower Redwood Creek. (Photo by author.)

Fig. 10-7 River terraces along Colorada Baranca near Ventura. Tectonic uplift and base-level changes caused this stream to fill with alluvium and subsequently cut a new channel. (Source: Sarna-Wojcicki, A. U.S. Geological Survey.)

STREAMFLOW IN CALIFORNIA

Given the enormous demand and sometimes unreliable supplies of water in California, it is important for managers to have exact measurements of the volume of runoff in the state's rivers. Long-term records also provide an important assessment of the potential for drought and flood. Streamflow measurements are made regularly on dozens of streams by the U.S. Geological Survey, the California Department of Water Resources, irrigation companies, municipal water districts, flood control agencies, and others.

The volume of water flowing in a stream at any given time is termed the stream's ***discharge.*** The discharge is the volume of water flowing past a certain point in a given unit of time. The elements of a river's discharge are the average velocity of flow, the width of the channel, and the water depth. Discharge is usually expressed as cubic feet per second (cfs) or cubic meters per second (m^3/s) (Fig. 10-8).

To measure discharge, hydrologists measure the width and depth of the flowing water and use a velocity meter to record velocity. At many stations, discharge is estimated from the water depth, which is continuously recorded by a ***stage recorder.*** At flood warning stations, alarms are automatically triggered when the flow exceeds a certain height. Records of total discharge, average annual discharge, peak flood discharge, and record low flows are available for many of California's streams. Table 10-1 and Fig. 10-9 illustrate the average flows for selected streams.

Because water managers in California need to know the available volume of water at a given moment in time, they commonly use the unit of ***acre-feet*** to express

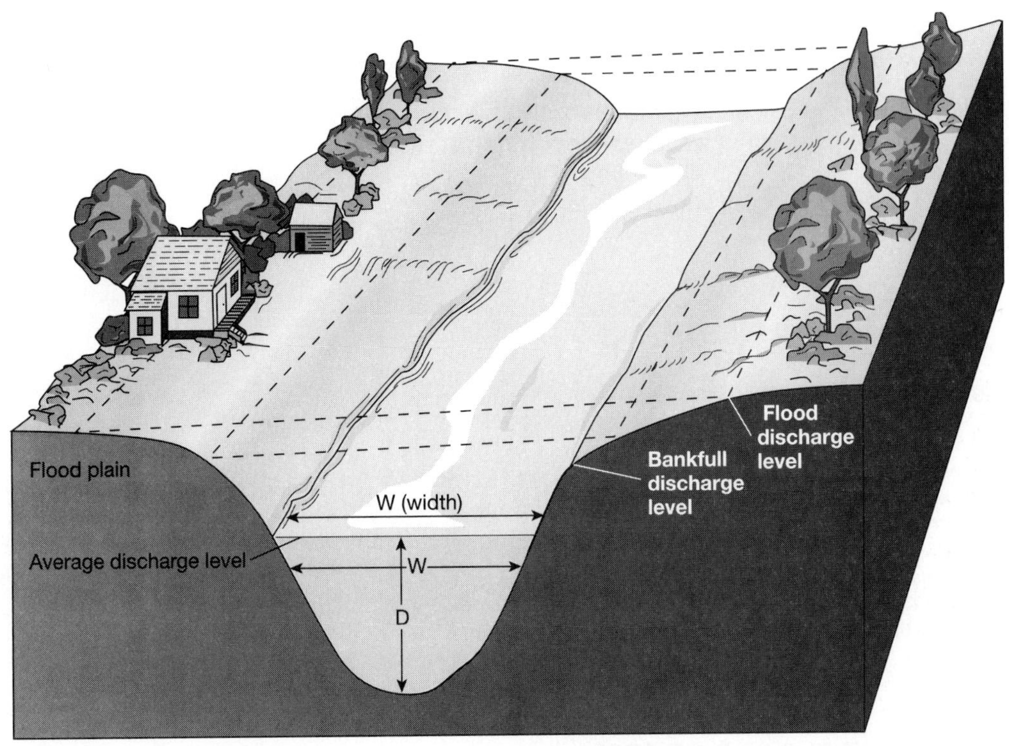

Fig. 10-8 Cross section of a stream channel showing important stream elements and the parameters used to calculate streamflow:

$$\text{Discharge} = \text{Average width} \times \text{Average depth} \times \text{Average velocity}$$
$$= \text{Cross-sectional area} \times \text{Velocity}$$
$$\frac{ft^3}{sec} = ft^2 \times \frac{ft}{sec}$$

(Source: Modified from Coch, N., and Ludman, A. 1991. Physical Geology. New York: Macmillan Publishing Co.)

annual runoff or volumes of water stored in a reservoir. One acre-foot is the amount of water that covers an acre of land to a depth of one foot, and it is equivalent to about 43,560 cubic feet of water, or about 325,851 gallons.

Water Units
1 cfs = 677,376 gallons per day
43,560 cfs = 1 acre-foot per second
3.07 acre-feet = 1 million gallons
34.2 million acre-feet (used in California in 1985) = 11,144,104,000,000 gallons

Because they have been carved by the streams, stream channels are sufficiently wide and deep to contain their average flows. Geologists have long debated about which "typical" flow determines channel size, but most agree that it is a discharge somewhat greater than the average yearly high flow. However, it is very clear that California's stream channels cannot always contain the infre-

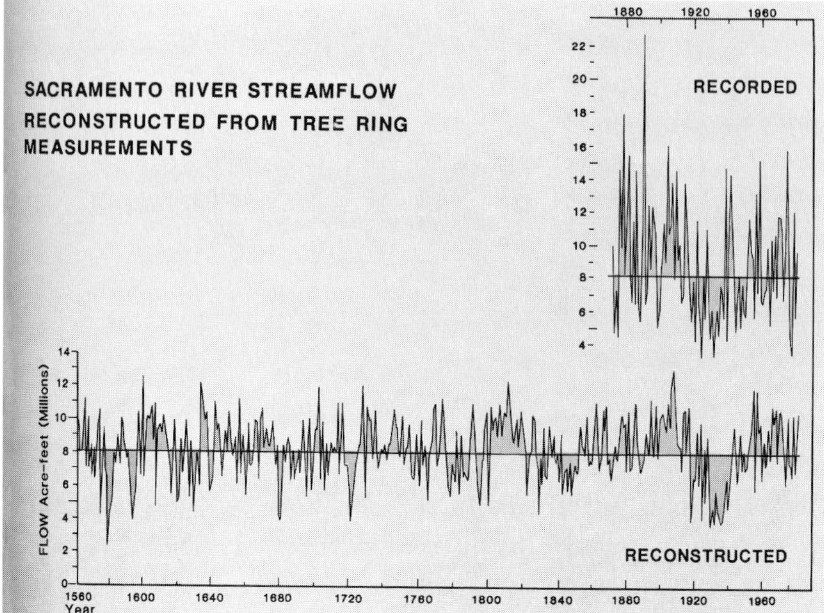

Fig. 10-9 The long-term streamflow record for the Sacramento River, reconstructed from tree-ring records. (Source: California Department of Water Resources. 1987. *California Water: Looking to the Future,* Bulletin 160-87.)

THE GREAT FLOODS OF 1862

In 1860, naturalist William H. Brewer began his assignment as a member of the State Geological Survey of California under the direction of State Geologist Josiah Whitney. Because his wife and only child had just died, Brewer welcomed the opportunity to leave the East for California. During his 4 years with the Survey, he meticulously recorded his observations of nineteenth century California, including details on everything from life at the mines and missions to earthquakes and fossil localities. His priceless collection of diaries, letters, and field notebooks are an important resource for historians, geologists, and naturalists. From *Up and Down California in 1860-1864, The Journal of William H. Brewer,* we get a vivid picture of the floods of January 1862:

January 19, 1862

The great central valley of the state is under water—the Sacramento and San Joaquin valleys—a region 250 to 300 miles long and an average of at least 20 miles wide, a district of 5000 or 6000 square miles, or probably 3 to 3 1/2 millions of acres! Although much of it is not cultivated, yet a part of it is the garden of the state. Thousands of farms are entirely under water—cattle starving and drowning.

January 31, 1862

In this city [San Francisco], 37 inches of water has fallen, and at Sonora, in Tuolumne, 102 inches, or 8 1/2 feet, at the last dates. These last floods have extended over this whole coast. At Los Angeles it rained incessantly for 28 days; immense damage was done—one whole village destroyed. It is supposed that over one fourth of all the taxable property of the state has been destroyed. The legislature has left the capital and has come here, that city being under water....

All the roads in the middle of the state are impassable, so all mails are cut off. We have had no "Overland" for some weeks, so I can report no new arrivals. The telegraph also does not work clear through, but news has been coming for the last 2 days. In the Sacramento Valley for some distance the tops of the poles are under water!

Source: Farquar, F., ed. 1966. *Up and Down California in 1860-1864, the Journal of William H. Brewer.* Berkeley: University of California Press, 583 pp.

TABLE 10-2 MAJOR CALIFORNIA FLOODS

Date	Area Affected	Lives Lost	Damage and Cost
December 1861-January 1862	Entire state	Unknown	Much of inhabited parts of state submerged for many weeks
December 1937	Northern two thirds of state	Unknown	$15 million
March 1938	Coast south of San Luis Obuspo, Mojave Desert	87	$79 million
December 1955	Northern two thirds of state	67	$166 million
December 1964	Northern half of state	24	Greatest known flood in northern California's recorded history; $239 million
January-February 1969	Southern, central coast, Mojave Desert	60	$400 million
January-February 1980	Central, southern coast	18	$350 million
January 1982	San Francisco	31	$75 million
February 1986	Northern half of state	14	$379 million
January and March 1995	Northern (January) central (March)	27	$3 billion; 42 (January), 57 (March) of 60 counties declared disaster areas; 10,000 residences damaged or destroyed
December 1996-January 1997	Sierra Nevada, Great Valley	8	$1.6 billion; 43 counties declared disaster areas; 16,000 residences damaged or destroyed

Sources: *California Floods and Droughts.* 1991. U.S. Geological Survey Water Supply Paper 2375; and National Climatic Data Center, 1997, *www.ncdc.noaa.gov.*

quent but normal peak discharges. When a stream overflows the banks of its channels, flooding occurs. Flooding is a natural part of the hydrologic cycle; over time, it can result in the development of a ***floodplain*** adjacent to the channel. As the water spreads out and loses velocity, it naturally drops its suspended sediment load, which is left as deposits of fine-grained sediment above and adjacent to the channel (see Figs. 10-2 and 10-6). By recognizing and determining the ages of floodplain sediments, geologists can identify areas likely to be flooded every 10 years, 50 years, 100 years, or less frequently.

Floodplains are highly desirable agricultural areas because they are relatively flat surfaces and because they periodically receive fine sediment with nutrient-rich organic and mineral material (see Chapter 11). Their level topography also makes them prime localities for urbanization in a state where flat land is relatively scarce. In an effort to minimize flood damage, current regulations in some counties prohibit development on floodplains that are inundated every 25 years or more frequently. However, larger floods with expected recurrence times of 50, 100, or even several hundred years (see the box on p. 214) are a predictable, damaging occurrence along low-lying areas in the state (Table 10-2).

The lives of Californians continue to be affected by floods, sometimes with catastrophic results. The normally dry canyons of southern California's mountains can be transformed into rivers of death, as evidenced by the narrative in the box on p. 216.

DEBRIS FLOW AND A FAMILY'S MIRACULOUS SURVIVAL IN THE SAN GABRIEL MOUNTAINS

On a February night some years ago, the Genofiles were awakened by a crash of thunder—lightning striking the mountain front. Ordinarily, in their quiet neighborhood, only the creek beside them was likely to make much sound, dropping steeply out of Shields Canyon on its way to the Los Angeles River. The creek, like every component of all the river systems across the city from mountains to ocean, had not been left to nature. Its banks were concrete. Its bed was concrete. When boulders were running there, they sounded like a rolling freight. On a night like this, the boulders should have been running. The creek should have been a torrent. Its unnatural sound was unnaturally absent. There was, and had been, a lot of rain.

The Genofiles had two teenage children, whose rooms were on the uphill side of the one-story house. The window in Scott's room looked straight up Pine Cone Road, a cul-de-sac, which, with hundreds like it, defined the northern limit of the city, the confrontation of the urban and the wild. Los Angeles is overmatched on one side by the Pacific Ocean and on the other by very high mountains. With respect to these principal boundaries, Los Angeles is done sprawling. The San Gabriels, in their state of tectonic youth, are rising as rapidly as any range on earth. Their loose inimical slopes flout the tolerance of the angle of repose. Rising straight up out of the megalopolis, they stand ten thousand feet above the nearby sea, and they are not kidding with this city. Shedding, spalling, self-destructing, they are disentegrating at a rate that is also among the fastest in the world. The phalanxed communities of Los Angeles have pushed themselves hard against these mountains, an aggression that requires a deep defense budget to contend with the results. Kimberlee Genofile called to her mother, who joined her in Scott's room as they looked up the street. From its high turnaround, Pine Cone Road plunges downhill like a ski run, bending left and then right and then left and then right in steep christiania turns for half a mile above a three-hundred-foot straightaway that aims directly at the Genofiles' house. Not far below the turnaround, Shields Creek passes under the street, and there a kink in its concrete profile had been plugged by a 6-foot boulder. Hence the silence of the creek. The water was now spreading over the street. It descended in heavy sheets. As the young Genofiles and their mother glimpsed it in the all but total darkness, the scene was suddenly illuminated by a blue electrical flash. In the blue light they saw a massive blackness, moving. It was not a landslide, not a mudslide, not a rock avalanche; nor by any means was it the front of a conventional flood. In Jackie's words, "It was just one big black hill coming toward us."

In geology, it would be known as a ***debris flow.*** Debris flows amass in stream valleys and more or less resemble fresh concrete. They consist of water mixed with a good deal of solid material, most of which is above sand size. Some of it is Chevrolet size. Boulders bigger than cars ride long distances in debris flows. Boulders grouped like fish eggs pour downhill in debris flows. The dark material coming toward the Genofiles was not only full of boulders—it was so full of automobiles that it was like bread dough mixed with raisins. On its way down Pine Cone Road, it plucked up cars from driveways and the street. When it crashed into the Genofiles' house, the shattering of safety glass made terrific explosive sounds. A door burst open. Mud and boulders poured into the hall. We're going to go, Jackie thought. Oh, my God, what a hell of a way for the four of us to die together.

The parents' bedroom was on the far side of the house. Bob Genofile was in there kicking through white satin draperies at the panelled glass, smashing it to provide an outlet for water, when the three others ran in to join him. The walls of the house neither moved or shook. As a general contractor, Bob had built dams, department stores, hospitals, six schools, seven churches, and this house. It was made of concrete block with steel reinforcement, sixteen inches on center. His wife had said it was stronger than any dam in California. His crew had called it "the fort." In those days, twenty years before, the Genofiles' acre was close by the edge of the mountain brush, but as developer had come along since then and knocked down thousands of trees and put Pine Cone Road up the slope. Now Bob Genofile was thinking, I hope the roof holds. I hope the roof is strong enough to hold. Debris was flowing over it. He told Scott to shut the bedroom door. No sooner was the door closed than it was battered down and fell into the room. Mud, rock, water poured in. It pushed everybody against the far wall. "Jump on the bed," Bob said. The bed began to rise. Kneeling on it, they could soon press their palms against the ceiling. The bed also moved toward the glass wall. The two teenagers got off to try to control the motion and were pinned between the bed's brass railing and the wall. Boulders went up against the railing, pressed it into their legs, and held them fast. Bob dived into the muck to try to move the boulders, but he failed. The debris flow, entering through windows as well as doors, continued to rise. Escape was still possible for the parents but not for the children. The parents looked at each other and did not stir. Each reached for and held one of the children. Their mother felt suddenly resigned, sure that her son and daughter would die and she and her husband would quickly follow. The house became buried to the eaves. Boulders sat on the roof. Thirteen automobiles were packed around the building, including five in the pool. A din of rocks kept banging against them. The stuck horn of a buried car was blaring. The family in the darkness in their fixed tableau watched one another by the light of a directional signal, endlessly blinking. The house had filled up in 6 minutes, and the mud stopped rising near the children's chins.

Source: Excerpted from McPhee, J. 1989. Los Angeles against the mountains. In *The Control of Nature.* New York: Farrar Straus & Giroux.

WATER CONTROL AND TRANSFER

Most of California's 31 million inhabitants live in cities and suburbs that are far too arid to support them. The state's enormous agricultural industry, particularly in the San Joaquin, Imperial, and other southern valleys, grows crops that require far more water than what is supplied naturally. In addition, most of California's industries require large volumes of water for manufacturing, cleaning, and processing. It has been only because of efforts to bring water to the people of California that the state can boast the greatest population and largest economy in the United States, and water development continues to be a major force that shapes the state's economy, politics, and history. Today California relies on about 1300 dams and reservoirs with a total storage capacity of about 43 million acre-feet, as well as on six major aqueduct systems and countless smaller structures to accomplish the task of water supply (Fig. 10-10). Unfortunately, basic lessons of the hydrologic cycle, stream equilibrium, and the variability of precipitation in California have often been overlooked during the development of California and the other states of the arid West, with the result that water crises are an almost yearly event.

Dams and reservoirs have been constructed for different purposes, although many may serve more than one purpose simultaneously. In the Mississippi River basin, many dams have been built for flood control. Some of California's first water projects, consisting mainly of ***levees*** and bypass channels, were built in the late 1800s on the American and Sacramento Rivers to protect Sacramento and surrounding areas. Dams built for flood control are designed to reduce peak flood discharges, retaining some of the flood runoff in the reservoir upstream of the dam. Levees are also commonly constructed along the banks of rivers, raising the height of the channel walls to contain high flows. In southern California, at the mouth of steep canyons flowing from the San Gabriel, San Bernardino, and other ranges, ***debris basins*** have been constructed to contain the huge volumes of sediment washed from canyons during floods.

A second purpose for constructing dams is for water storage and irrigation. In California, where peak runoff periods are directly out of phase with the main crop-growing season, it is essential to save runoff for the summer months. Runoff generated by the spring snowmelt can be caught in the reservoirs and later released. Many of California's streams have minimum flow increased or maintained by releases from reservoirs. Dams are also constructed to generate hydroelectric power. The energy released by the falling of the water through the dam's gates is converted to turbine-generated electricity. Finally, reservoirs may store water that is intended for transfer out of the drainage basin to another stream or to one of the aqueduct systems. Many of California's reservoirs also provide recreation as an additional benefit.

It is not surprising that the construction of dams has an impact on stream equilibrium. Because water flow ceases in reservoirs, streams drop their sediment loads there, with the predictable result that the reservoirs are filled by ***deposition,*** or "siltation." Without dredging of the sediment or enlargement of the dam, many reservoirs would be filled within 20 to 30 years of construction. Downstream effects of dams are more variable. Reduction of flood flows may lead to downstream reaches becoming choked with sediment. On the other hand, removal of the sediment load behind the dam may cause the stream to become more erosive below the dam as it attempts to restore its load. Removal of part of a stream's discharge to another drainage basin has obvious effects on both streams as they attempt to adjust to the changes.

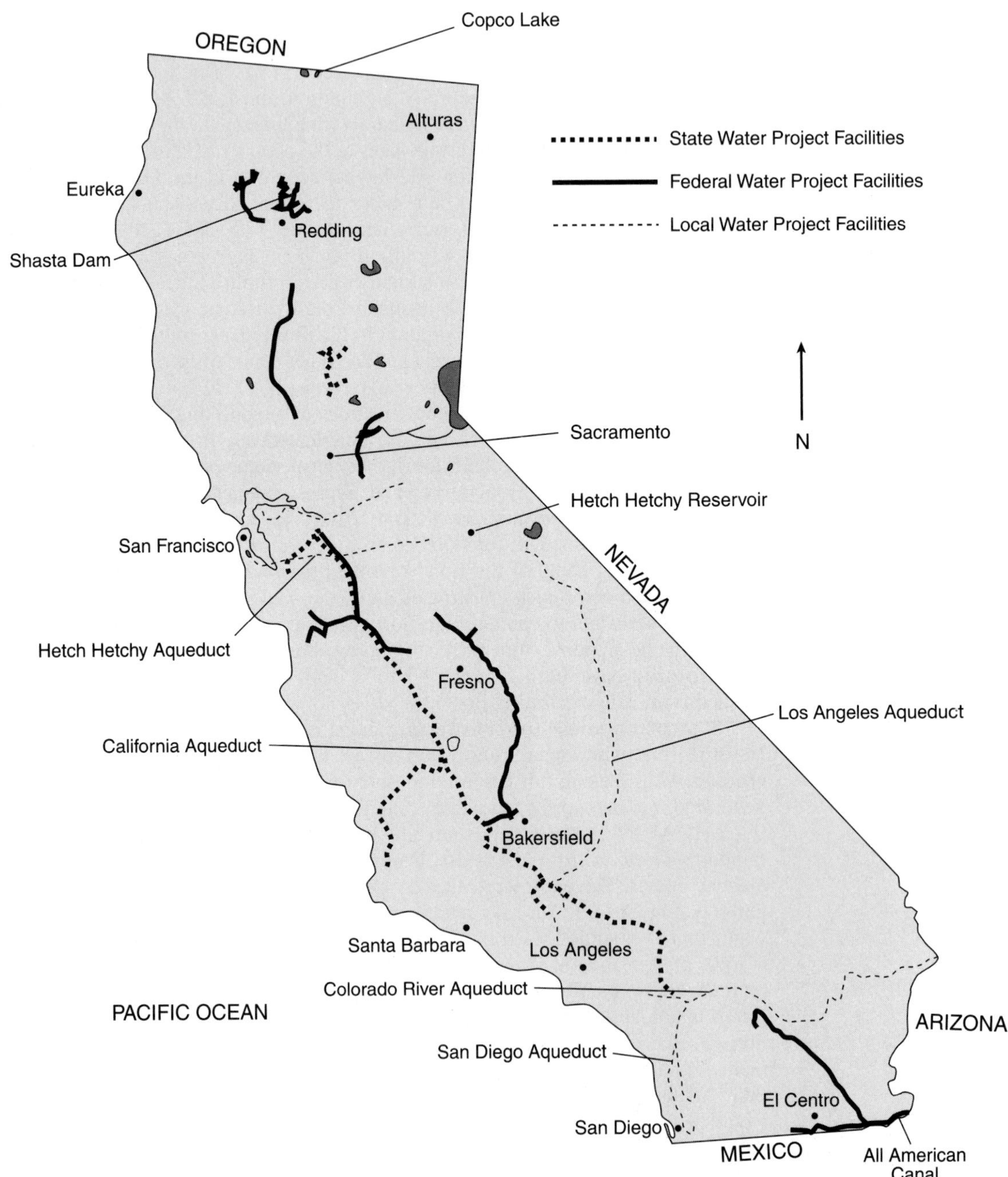

Fig. 10-10 California's manmade water network. (Source: California Department of Water Resources.)

CALIFORNIA'S AQUEDUCTS

The network of aqueducts that brings water to farms and people would not be possible without the system of water laws governing California. Water rights in California are established under two fundamentally different systems. ***Riparian rights,*** which prevailed in California under Mexican rule, stipulate that those owning property adjacent to a stream or other body of water own the right to use the water in that stream. In the arid West, the lack of flowing streams made riparian law unworkable, particularly when the Gold Rush brought a great demand to divert water from streams to wash the gold-bearing gravels from the hillslopes. The miners developed a system of ***appropriation rights,*** which allow water to be appropriated and used in a beneficial manner at any site. Under this system, the holder of the first appropriation right in a stream system is entitled to use all of the water that they have historically used before the second appropriator receives any. Those holding the rights receive their water allotment in the order of priority. In times of drought, those with lower priority may receive no water. Both types of rights are recognized today in California, and a complex body of laws regulates water use in the state.

What has made California's water diversion system possible is that appropriation rights may be sold or transferred without losing their priority. This possibility, together with the legality of moving water away from the streams, has enabled cities, agricultural districts, water companies, and government agencies to move surface water to the farms and cities needing water. Recent legal decisions have required that minimum flows be maintained in some streams in order to preserve important habitats and populations, but historically the use of California's water by fish, birds, and other wildlife did not have legal protection. The Wild and Scenic Rivers Act of 1972 also currently prohibits additional construction of major dams and diversions on the Smith, Klamath, Trinity, and Eel Rivers in the North Coast region.

The first of the aqueducts built in California delivered its first water from the Owens Valley to the city of Los Angeles, 388 kilometers away (see Fig. 10-10). The Los Angeles Aqueduct, opened in 1913, supplies about 80 percent of Los Angeles' water supply through a system of conduits, tunnels, and natural river channels. Because of the elevation of the Owens Valley, about 1220 meters above sea level, the water flows largely under the influence of gravity (see the box below).

THE HISTORY OF THE LOS ANGELES AQUEDUCT

The early pueblo of Los Angeles relied on the Los Angeles River for its water needs. At most times a very small stream nourished in part by groundwater from the San Fernando Valley, the Los Angeles River experienced occasional huge floods that overtopped the river's banks. As the city grew, the river's flow dropped because of pumping from groundwater wells, and city engineers began to look for additional water supplies. Fred Eaton and his successor William Mulholland, early superintendents of the Los Angeles City Water Company (later the Los Angeles Department of Water and Power), turned their sights eastward in a search for abundant and cheap water for their growing city. By 1900, the population of Los Angeles had reached 100,000, and within 4 years, it doubled. The search for water grew more intense as Mulholland launched a scare-tactics campaign in the local papers to gather support.

Continued.

HISTORY OF THE LOS ANGELES AQUEDUCT—cont'd

More than 400 kilometers away, along the eastern front of the Sierra Nevada, settlers had begun in the 1860s to irrigate the Owens Valley with water from the Owens River. By the late 1800s, the flourishing farmlands supplied the silver mines of Tonopah, Nevada to the east. Although the Owens River lies in the rain shadow of the Sierra Nevada, it collects abundant runoff from tributaries that flow eastward from the snowy High Sierra. The Owens River flowed into Owens Lake, a remnant of a formerly more extensive chain of lakes along the Sierra Nevada (see Fig. 6-17). Owens Lake lies at an elevation of about 1200 meters above sea level, a fact that tempted Eaton to consider tapping the water of the Owens Valley for Los Angeles. The elevation difference meant that water could be delivered by gravity, without any expensive pumping. At first, Mulholland was skeptical about the project, claiming that evaporation losses from the desert reservoirs and canals would be enormous. However, in 1905, the first exploratory visit to Owens Valley began one of the most colorful episodes in the West's water wars.

The U.S. government had been involved in the irrigation of the American West in a major way since passage of the Reclamation Act. In the Owens Valley, the Reclamation Service had been developing the Owens Valley Reclamation Project, which would aid the farmers in irrigating the valley. Convinced of the necessity of acquiring the Owens Valley water, Eaton and Mulholland began buying up land with senior priority water rights, often paying far more than the assessed value of the land. In an arrangement of questionable morality, the Los Angeles Power and Water Company hired the Reclamation Service's senior engineer as a consultant to investigate acquisition of Owens Valley water for the city of Los Angeles. By the time that Owens Valley farmers realized that Los Angeles was conducting an all-out grab, the Los Angeles City Water Company controlled many of the water rights. In 1907, after extensive lobbying by Mulholland, the Federal Reclamation Project was killed, and Owens Valley water rights were effectively placed in the hands of Los Angeles.

After 6 years of construction, the first water was delivered to Los Angeles in 1913 by the Los Angeles Aqueduct. The water-delivery agreements stipulated that the water could only be used by the City of Los Angeles, and that surplus water would remain in the Owens Valley. At the time of construction, the San Fernando Valley lay outside the city and was therefore ineligible to receive Owens Valley water. The actual water delivered by the aqueduct was far in excess of what was needed by the city, but the extent of the surplus was kept quiet. Mulholland had placed the end of the aqueduct in the San Fernando Valley, and, together with a group of wealthy investors, he had bought substantial land in the seemingly arid wasteland. The arrival of Owens Valley water made the land worth millions overnight, and the annexation of the San Fernando Valley to the City of Los Angeles completed the land and water grab.

The Owens Valley began to revert to its natural desert state, and the farms and orchards dried up. Irrigation ditches operated by small companies continued to tap water from the Owens River, but the City of Los Angeles fought for every drop.

Things came to a head during the drought years of the 1920s, and frustrated ranchers blew up parts of the aqueduct system on several occasions. On November 16, 1924, a group of ranchers closed the Alabama Gates that send water from the lower Owens River into the aqueduct. This action temporarily reestablished the natural flow from the river into Owens Lake. The sabotages were unsuccessful, and by the mid-1930s, the City of Los Angeles owned 95 percent of the farmland in the Owens Valley.

In the 1970s the Los Angeles–eastern California water wars heated up again, when Los Angeles' diversions of water farther to the north caused the level of Mono Lake to drop to new low levels. Biologists feared that the drop in the lake levels would destroy the safety of the breeding grounds of the Mono gull. The gull builds its nests on two islands in Mono Lake, inaccessible to predators unless the lake level becomes low enough for land bridges to form. A heated legal battle ensued, accompanied by a blast of environmental publicity, with the result that Los Angeles agreed to maintain sufficient flow into Mono Lake to protect the lake's habitats.

THE COLORADO RIVER AND THE SALTON SEA

A second major source of water in southern California is the Colorado River, which supplies water to 12 million people in seven states and parts of Mexico. Completed in 1940, the Colorado River Aqueduct carries water 400 kilometers to Los Angeles and, since 1947, via the San Diego Aqueducts to San Diego. Hoover Dam, completed in 1935, also supplies a large portion of southern California's electricity. Downstream on the Colorado River, the All-American Canal and the Coachella Canal irrigate the Yuma, Palo Verde, Imperial, and Coachella Valleys. California's total allotment from the Colorado River is currently 4.4 million acre-feet per year, of which 87 percent goes to irrigate the desert valleys, and 11 percent to the Metropolitan Water District, which serves 11 million people in San Diego and in communities adjoining Los Angeles.

Salton Sea Formation

Beginning in the late 1800s, the Imperial Valley had been irrigated by a system of canals taking water from the Colorado River. In 1901, a cut was opened along the river to bypass the intakes to the irrigation canals, which were choked with sediment. Unfortunately, five floods hit the area in 1905, causing the entire Colorado River to be diverted into the Alamo and New Rivers. Within weeks, the water formed the Salton Sea in the low-lying basin of the Salton Sink. In 1907, the river was rediverted back into its channel, but the Salton Sea remains as a part of the Imperial Valley's landscape (Fig. 10-11). Despite enormous losses by evaporation, the Sea is maintained by water diverted from the Colorado River.

Fig. 10-11 An infrared satellite image taken by Landsat-I showing the Salton Sea and vicinity. Note the contrast between irrigated fields south of the Salton Sea and sand dunes east of the fields.

OTHER AQUEDUCTS

In 1923, water dammed behind O'Shaughnessy Dam flooded Hetch Hetchy Valley of the Tuolumne River, signaling the end of a long and bitter battle for its preservation. John Muir had stated, "Dam Hetch Hetchy? As well dam for water tanks the people's cathedrals and churches, for no holier temple has ever been consecrated to the heart of man." But San Francisco was thirsty for reliable water supplies, and in 1934, the first Sierra Nevada runoff flowed through the Hetch Hetchy system to San Francisco. Today the system has been expanded to supply San Francisco and 50 additional communities and water districts. Communities of the East Bay Municipal Utilities District receive water from the Mokelumne Aqueduct, which also brings Sierran water to the Bay area.

As groundwater supplies became scarce in the farmlands of the San Joaquin Valley, farmers began to examine the possibility of importing water from the north. Although the state approved construction of the Central Valley Project (CVP), which would bring water to the valley from a storage reservoir in the upper Sacramento watershed, California could not fund the project, and the Federal Bureau of Reclamation took it over. Shasta Dam, in the upper Sacramento Valley, was completed in 1951, and Friant Dam on the San Joaquin River in 1954. The Delta-Mendota Canal, the Contra Costa Canal, Friant-Kern Canal, and the Madera Canal transported the stored water, and the San Luis Reservoir and canal were added to the system in 1967. About 83 percent of the water supplied by the project is used for irrigation. Under the Federal Reclamation Act of 1903, the water supplied by the CVP can only be used by farmers holding 160 acres of land or less, but violations of the 160 acre rule by individuals and large corporations have been numerous. In 1992, a landmark bill, the Central Valley Project Improvement Act, was passed by Congress. The bill mandates more water for wetlands and for maintenance of instream flows, allows water to be transported outside the project service area, and charges the farmers more for their water. The bill also provides funds for restoration and enhancement of fish and wildlife habitat. Implementation of the bill may bring major changes to agriculture in the San Joaquin Valley.

California's most recent water project, the State Water Project, resulted in construction of the 444-mile-long California Aqueduct (Fig. 10-12). This system sup-

Fig. 10-12 The California Aqueduct near Tracy, California. (Source: Sarna-Wojcicki, A. U.S. Geological Survey.)

plies water to the Napa, Livermore, and Santa Clara Valleys of northern California and to Kern County farmers in the San Joaquin Valley (see Fig. 10-10). The system also delivers water to Southern California even though water must be pumped uphill over the Tehachapi Mountains and across the San Andreas Fault.

GROUNDWATER

Precipitation that infiltrates into the ground may be stored in the spaces within sediments and rocks beneath the earth's surface. This stored water is ***groundwater,*** and it is an important part of California's water supply. At present, slightly less than half of the water used in California is groundwater, and in the drought years of 1987-1991, groundwater use rose to about 60 percent of California's total supply. Although no one knows the exact number of water wells in California, estimates are between 1 and 2 million.

As water seeps into the ground, some of it is held by the soil near the surface in the ***unsaturated zone*** (Fig. 10-13). This zone may be dry during periods when evaporation and transpiration are greater than infiltration. Beneath the unsaturated zone, groundwater fills all available spaces in the rock or sediment in the ***zone of saturation.*** The boundary between the unsaturated and saturated zones is the ***water table.*** The water table marks the depth at which one encounters saturated materials while digging a hole or drilling a well. The water table can be at the earth's surface in river valleys or at springs, or it may be 100 meters or more below the surface beneath the floor of a desert valley. The position of the water table changes seasonally, rising as the groundwater is ***recharged*** with infiltrated precipitation and falling during dry periods. During prolonged droughts in California, drops in the water table of tens of feet have been recorded.

The capacity of rocks and sediments as sources of groundwater is extremely variable. The ***porosity*** of a rock or sediment is a measure of the percentage of space in the material; porosity may be as low as 1 to 3 percent in granitic rocks or greater

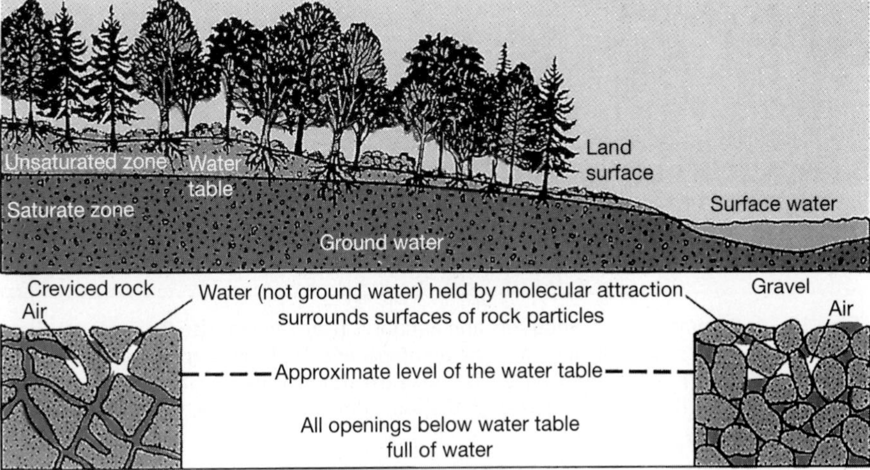

Fig. 10-13 Diagram illustrating the occurrence of groundwater. (Source: Waller, R. 1994. Groundwater and the Rural Homeowner. U.S. Geological Survey, General Interest publication 1994-380-615.)

that 50 percent in some sediments. In order for groundwater (or petroleum or other fluids) to flow through rock or sediment, the pores must also be connected. ***Permeability*** of rocks or sediments is the ability of the material to transmit fluids. Clay and shale, which both have high porosity, are generally impermeable because the water is tightly held to the clay particles in very small pores. In contrast, fluids are easily transmitted in gravel, sand, and sandstone. Even crystalline rocks such as limestone or granite may be permeable if fractures, weathering, or bedding create sufficient permeability (see Fig. 10-13). Permeable materials are called ***aquifers,*** whereas those which retard groundwater movement are called ***aquitards*** or ***aquicludes*** (Fig. 10-14).

Groundwater flows very slowly from areas of high water pressure to low-pressure areas. Usually the direction of flow is very similar to the slope of the land surface, so that groundwater generally flows from higher elevations to low-

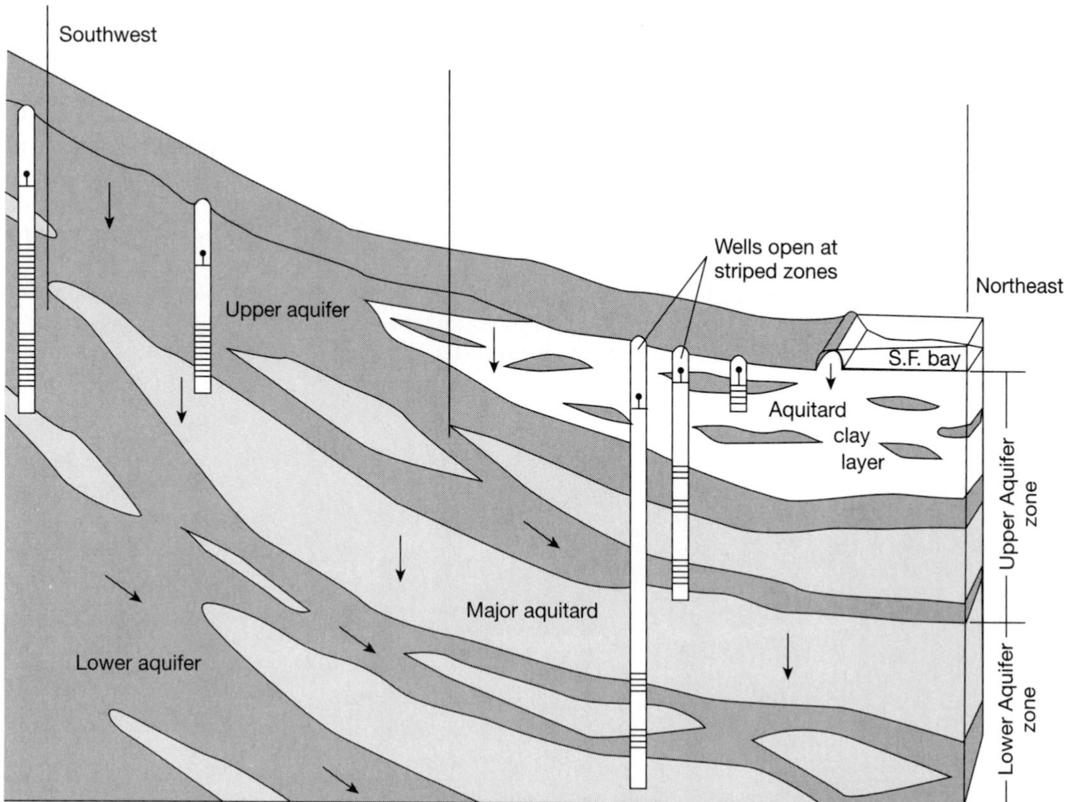

Fig. 10-14 Diagram showing aquifers and aquitards near San Francisco Bay in Santa Clara County. The aquifer layers are sandy alluvium, and the aquitards are clay layers deposited when the Bay was slightly more extensive than at present (see Chapter 12). Like the topography, the layers slope gently toward the Bay, and the groundwater flows in the aquifers in the same direction. (Source: Modified from Environmental Protection Agency diagram.)

lying areas. Groundwater may return to the land surface where it is discharged into streams or flows at the ground surface at springs, which form where the water table intersects the land surface. In California, the largest volume of groundwater is stored in unconsolidated, relatively young, sandy sediments. Commonly, these are alluvial deposits that underlie the valleys and basins of California. The aquifers used today for groundwater are relatively shallow, with water being pumped from depths of several meters to a few tens of meters. Some aquifers are ***confined*** between layers of less permeable sediment or rock. In this situation, the water may be stored under pressure. Groundwater is extracted by drilling a well into the saturated zone. In many areas of California, the groundwater was naturally confined under sufficient pressure that groundwater would flow to the surface without pumping, thus creating an ***artesian well*** (Fig. 10-15). With drops in the water table, artesian conditions disappear, and mechanical pumps are required to lift the water to the surface. Of course, with continued drops in the water table, wells may dry up.

It is important to recognize that the groundwater is an integral part of the hydrologic cycle. The level of the water table is maintained by recharge, and it will drop if water is pumped from aquifers at a rate exceeding the recharge. Changes to another part of the hydrologic cycle often have major impacts on the groundwater system. For example, if a significant part of a drainage basin is urbanized, the amount of recharge of the groundwater may be significantly reduced because asphalt and concrete do not allow infiltration to occur (Fig. 10-16). Because water that formerly seeped into the ground may now run off, streamflows may increase.

Fig. 10-15 A flowing artesian well in the northern Santa Clara Valley near Milpitas. Under normal conditions, groundwater is pumped into the pipe to the right of the well. Because of heavy rains during 1995-1997, the water table has risen to the point that artesian conditions allow water to flow under its own pressure from below the pipe. (Source: Calhoun, J.B. Santa Clara Valley Water District.)

Fig. 10-16 Percolation ponds along Los Gatos Creek, San Jose. Surplus runoff is released into Los Gatos Creek from Lexington Reservoir upstream, and then allowed to infiltrate into the groundwater system from these ponds. The groundwater is later pumped from the groundwater system as needed for municipal use. (Source: Santa Clara Valley Water District.)

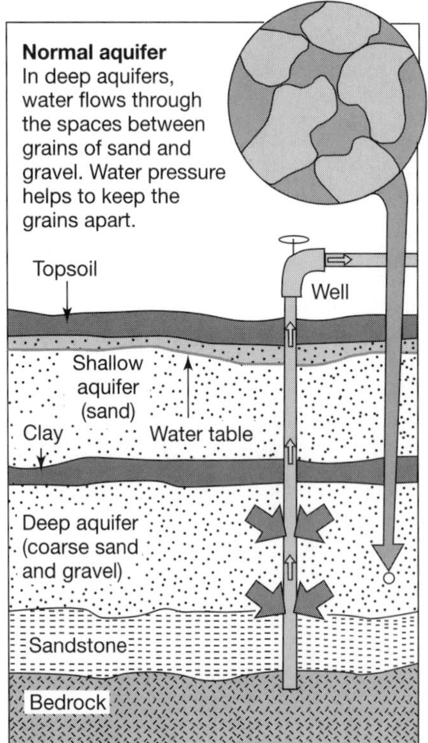

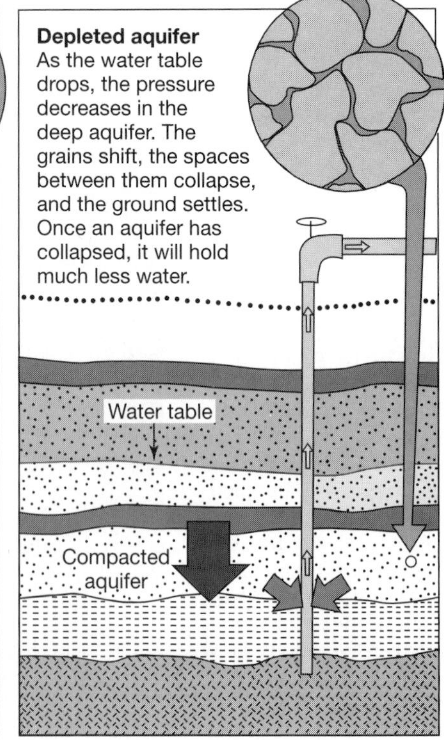

Fig. 10-17 The effects of compaction on an aquifer. (Source: *San José Mercury News,* April 7, 1991.)

GROUNDWATER PUMPING, SUBSIDENCE, AND SALTWATER INTRUSION

In most large aquifers in California, water is stored in the pore spaces between grains of sand or gravel, and the aquifers are commonly confined by clay-rich layers. In the deeper aquifers, the pressure of the water helps to keep the grains separated. When pumping from an aquifer exceeds the rate of recharge, the water table drops and the water pressure decreases. The pore spaces are collapsed, the grains shift together, and compaction occurs (Fig. 10-17). Because pore spaces in the aquifer are lost, the capacity of the aquifer to hold water is reduced and will not be recovered, even with reduced pumping and a series of wet years.

The land surface above a compacted aquifer may also collapse, especially if the area is underlain by young, unlithified sediments. The process of land-surface sinking is known as ***subsidence.*** Evidence of subsidence includes ground cracking and damage to roadways, aqueducts, and structures. Subsidence caused by excessive groundwater pumping is a common occurrence in areas of California where groundwater is pumped for agricultural and municipal wells. Described below are a few examples:

- In the China Lake basin in the Mojave Desert, a series of long cracks developed during 1990, damaging the runways at the China Lake Naval Air station. At first, geologists were uncertain about the origin of the cracks and included faulting in their list of possible causes. Investigations of the geology and groundwater use in the basin revealed that the cracks are the expression of subsidence in the alluvial aquifer being pumped to supply the base and surrounding community.
- The San Joaquin Valley and the Delta area have a long history of subsidence resulting from groundwater pumping. On the western edge of the San Joaquin Valley near Mendota, an area of several square miles under the California Aqueduct and Interstate 5 subsided by 25 centimeters during 1991. Some areas of the San Joaquin Valley have sunk as much as 9 meters since the 1920s. In the Delta, farmlands now lie well below sea level because of subsidence, caused by the disintegration of organic layers within the sediments as well as by agricultural development. Only a system of dikes and levees stands between these farmlands and disastrous flooding.
- In the central third of the Santa Clara Valley, almost 2.5 meters of subsidence occurred between 1934 and 1967 (Fig. 10-18). As a result, areas near sea level around the margins of San Francisco Bay are more frequently flooded by storms and high tides. Some parts of the city of Santa Clara have subsided below sea level, and only levees along the shore of San Francisco Bay prevent flooding by bay water.

In aquifers along the coast, fresh groundwater normally flows to the sea to a point where it sits in contact with salt water. Because it is lighter than seawater, which contains more dissolved load, the fresh groundwater overlies the saltwater (Fig. 10-19). Pumping fresh water at a rate greater than the recharge rate causes the water table to drop and the pressure in the aquifer to drop. Salty groundwater is then drawn inland and may replace the fresh water in wells near the coast. This process is known as ***saltwater intrusion,*** and it occurs in coastal aquifers along with subsidence. Well water in parts of the Santa Clara Valley, the Salinas Valley, and the Long Beach areas has been made unusable by saltwater intrusion. In parts of the Los Angeles Basin, groundwater levels fell to 30 meters below sea level in the 1950s. Although pumping fresh water back into the aquifers has halted the spread of saltwater intrusion in some areas, it has not managed to push back the intruding seawater.

In most large groundwater basins in California, pumping rates are beginning to be restricted to minimize the problems of ***overdraft,*** or overpumping, of the aquifers. With this management, water managers hope to attain a ***safe yield*** of groundwater extraction to minimize the problems of water table lowering, subsidence, and saltwater intrusion.

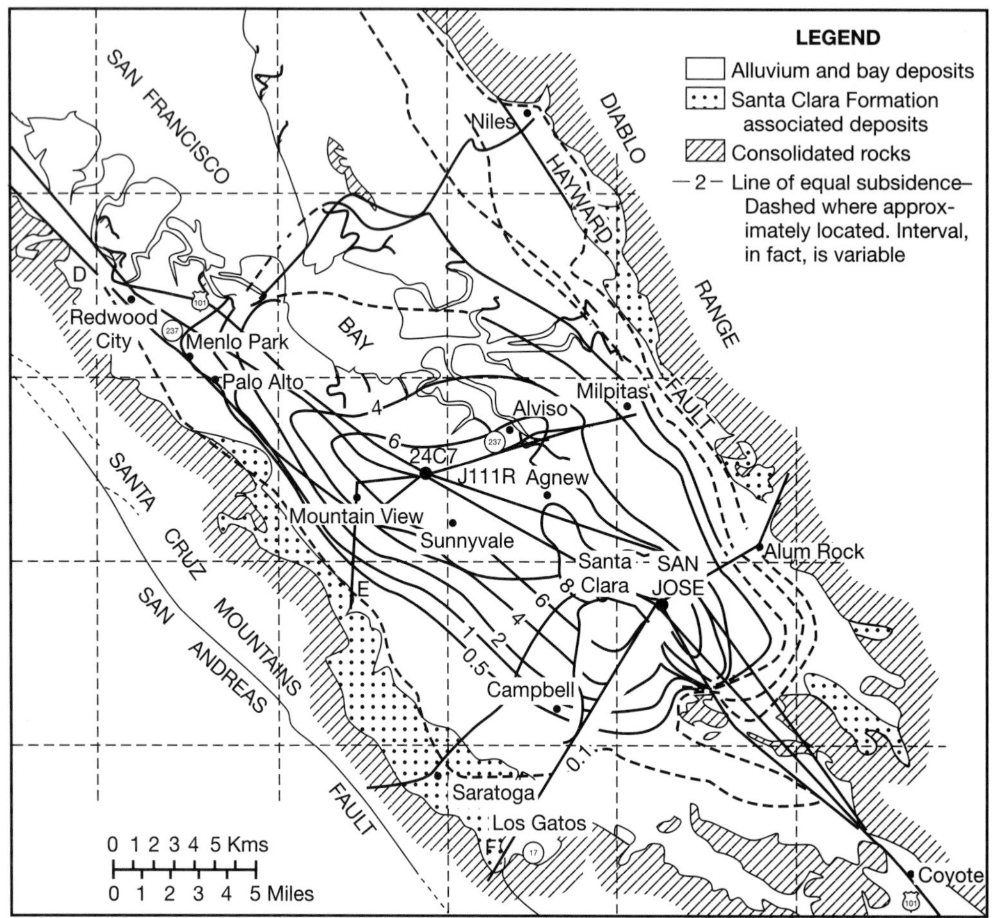

Fig. 10-18 Subsidence in the Santa Clara Valley between 1934 and 1967. Subsidence values are given in feet. (Source: Poland, J. U.S. Geological Survey, Professional Paper 497-F.)

WATER QUALITY

Both surface water and groundwater may naturally contain a variety of dissolved and solid materials. Dissolved substances may be carried in rainfall, or they may be derived from the chemical weathering of rocks within the drainage basin. The vegetation and animals of the drainage basin contribute a variety of organic compounds. Solid particles may be part of a stream's sediment load, but are not normally present in slow-moving groundwater. The solid particles in an aquifer generally act as a filter that removes some of the substances from the water.

The materials contained in water give it a taste and appearance unique to that water source. Many dissolved substances, such as iron, are essential for life. However, water pollution occurs when undesirable or toxic substances are present in the water, or when abnormally high levels of naturally occurring substances are introduced. Water may be naturally unsuitable for drinking because of the presence

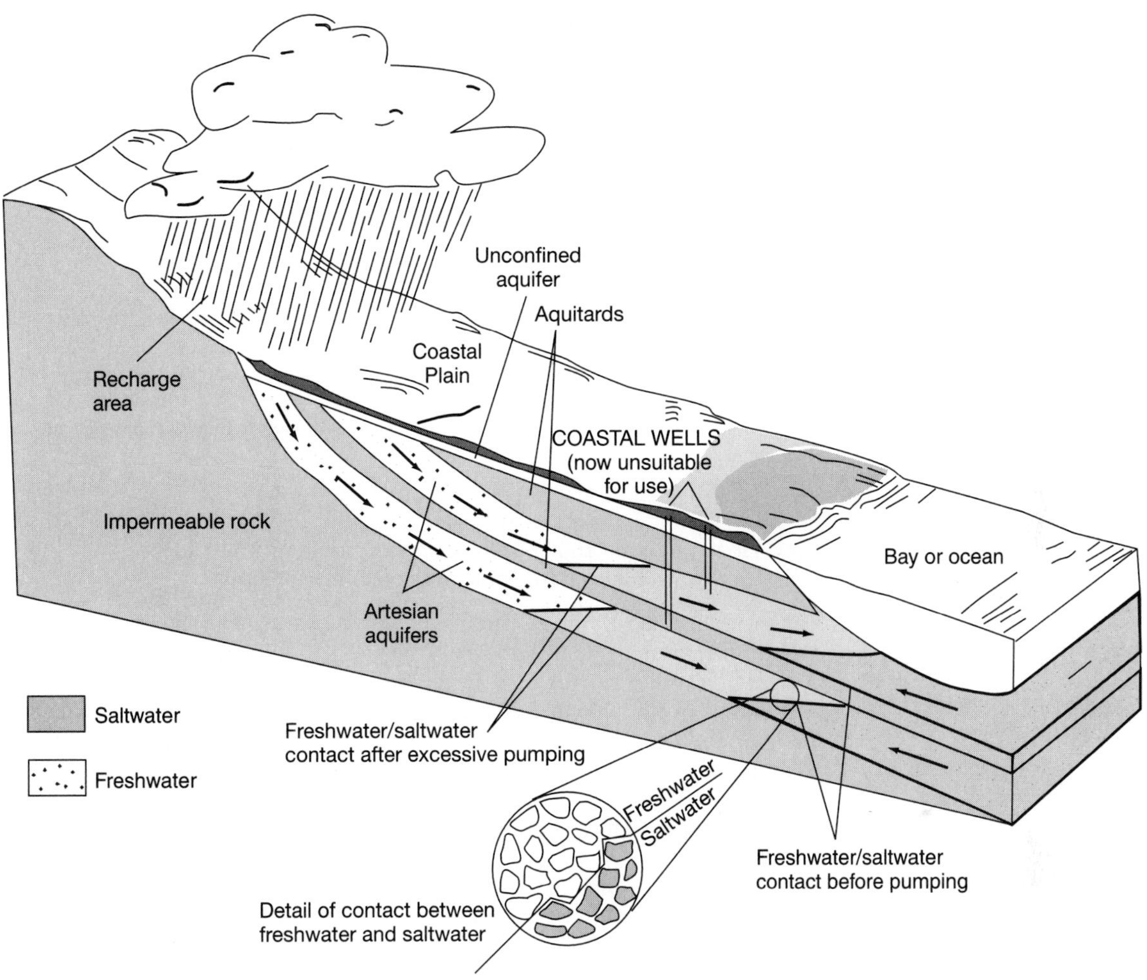

Fig. 10-19 Saltwater intrusion in coastal aquifers. (Source: Modified from Coch, N., and Ludman, A. 1991. *Physical Geology.* New York: Macmillan.)

of volcanic gases, dissolved metals, or high amounts of salts, or it may become contaminated by human activities. Because of public concern about the safety of drinking water, the U.S. Environmental Protection Agency (EPA) has established safe levels of contaminants for public drinking water, known as ***maximum contaminant levels (MCLs).*** These standards specify the maximum concentration allowed for particular substances, and the list is continually being revised as new contaminants are identified and as studies of the risk of exposures are conducted (Table 10-3).

In the United States we have come to take for granted the availability of unlimited, safe drinking water. Potentially deadly problems of water contamination by sewage and animal waste are rare in California, and the State's water is generally abundant, clean, and cheap. Municipal water is usually treated and filtered to ensure

Table 10-3 SELECTED WATER CONSTITUENTS AND DRINKING WATER STANDARDS

Constituent	Federal MCL	Constituent	Federal MCL
Total dissolved solids (TDS)	500 ppm[1]	Coliform, per 100 ml	1-4
Asbestos	7.1 ppb[2]	1,1Trichlorethane (TCA)[4]	200 ppb
Cadmium	10 ppb (5)[3]	Trichloroethylene (TCE)[4]	5 ppb
Lead	50 ppb (5)[3]	Trihalomethanes (THM)[5]	100 ppb
Mercury	2 ppb	Enarin[6]	0.2 ppb
Selenium	10 ppb	Heptachlor[6]	0.4 ppb (.01)[3]
Benzene	5 ppb (1)[3]	Nitrate (ion)[7]	45
PCBs	0.5 ppb		

[1]Parts per million; [2]parts per billion; [3]number in parentheses is the lower California standard; [4]solvents; [5]byproduct of chlorination; [6]pesticide; [7]fertilizer, animal waste.
Source: Purin, G., and Fono, E. 1989. *Consumer's Guide to California Drinking Water.* Sacramento: Golden Empire Health Planning Center, 81 pp.

its safety. However, concern about the safety of drinking water supplies has grown in recent years as problems have come to the public's attention:

- Agriculture, industry, and communities are all sources of water contamination in California. Runoff from agricultural lands in California's valleys introduces harmful levels of nitrates from fertilizers, as well as pesticides and high concentrations of potentially toxic elements flushed out of soils by excessive irrigation.
- Thousands of leaking underground storage tanks are sources of groundwater pollution by gasoline, solvents, and other hazardous chemicals. In 1984, the California Water Resources Control Board identified 165,000 leaking underground storage tanks, about 80% containing gasoline or other petroleum compounds.
- Landfills, both municipal and industrial, are sources of a host of contaminants ranging from animal wastes to used paint, batteries, and highly toxic solvents. In all, roughly 2 million tons of hazardous waste was produced by factories, businesses, and government operations in California in 1989. In 1983, California had 1800 surface storage impoundments for industrial wastes.

Current regulations require proper disposal or recycling of hazardous compounds, but enforcement has not yet been perfected. Current regulations also require liners for landfills and leakproof containers for underground tanks, but a legacy of thousands of sites remains to be cleaned up in California.

Groundwater contamination has turned up in wells in all parts of California. In 1986, the Department of Health Services tested 538 wells and found that 18 percent showed some level of contamination. Cleanup of the sites with large concentrations of highly toxic chemicals is presently being managed by the EPA's Superfund program. These sites include the notorious Stringfellow Acid Pits in Riverside County, where 35 million gallons of chemical wastes were dumped into 22 ponds during the 1950s and 1960s. In southern California, other sites are contaminated by oil refinery waste. Santa Clara County has the dubious distinction of having the country's largest number of Superfund Sites in any county (28 in 1992). Many of those sites involve leaking storage tanks for solvents used to manufacture silicon chips for the computer industry (Fig. 10-20). In the Sacramento Valley, Superfund sites include the aban-

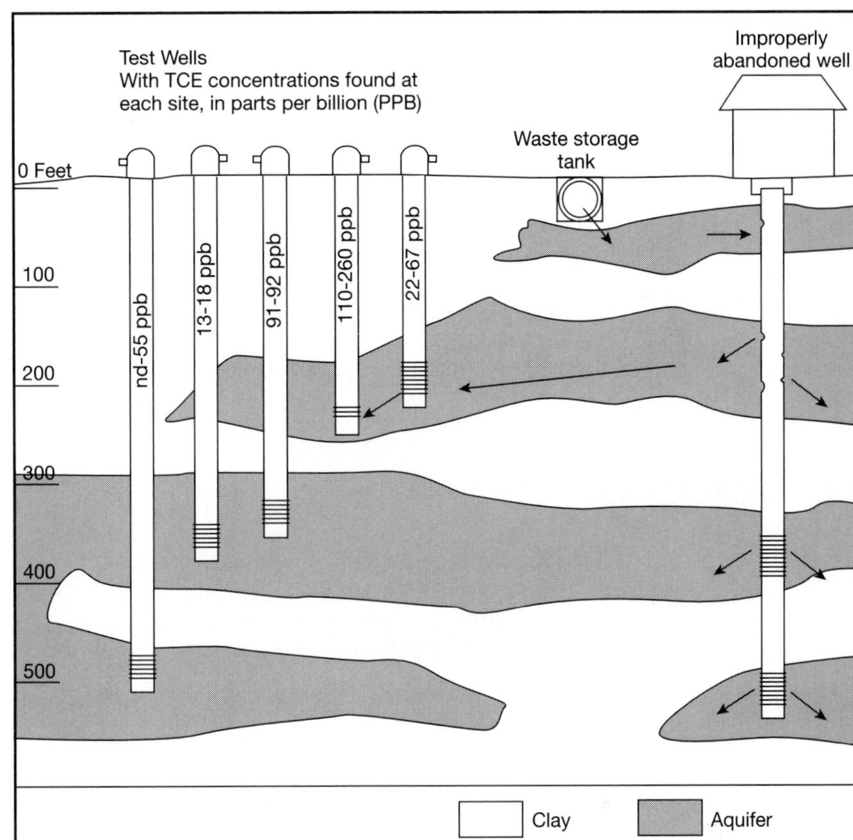

Fig. 10-20 Groundwater contamination by trichloroethylene from a waste storage tank in Santa Clara County. The arrows show possible the possible pathways of contaminants (Source: *San José Mercury News.*)

doned Iron Mountain copper mine near Redding, an area near Oroville contaminated by wood-treatment chemicals, and a plume of solvents near Sacramento.

While groundwater cleanup is progressing at numerous sites in California, it will take decades to complete the task. Once a contaminant site is identified, the extent of contamination must be determined by careful study of the aquifers. The direction and rate of groundwater flow must be precisely known, a difficult task that requires drilling and sampling to obtain detailed knowledge of the geologic picture in three dimensions. It is also important to estimate the rate of spreading of the contaminant plume to protect the groundwater in the vicinity of the spill. Cleanup may be accomplished by pumping out contaminated water and treating it, a costly process that may take several years of continuous pumping.

WATER FOR THE FUTURE

In a year with average runoff, California's current water supplies are adequate to meet the needs of the population. However, the drought of 1987-1992 stretched the ability of the system to supply water for much of the state and raised the issue of additional water for the future. As it has been during California's past, the issue of water development is a controversial one. Many feel that increased conservation, limited expansion of industry, and limited population growth are the steps necessary to preserve the ecosystems of California's rivers and wetlands. Others believe that continued development of water projects is a fundamental aspect of guaranteeing California's economic growth. As water supplies become scarce, communities are investigating the possibility of desalinization, conservation programs, and water purchases from other water districts. Many believe that agriculture will have to relinquish some of the 83 percent of California's water supply that it currently uses to facilitate the growth of cities and industry. The cost of the original Hetch Hetchy system was $100 million and that of the Los Angeles Aqueduct approximately $200 million. As of 1988, the State Water Project had cost $3.7 billion. Even if other obstacles to further water development could be overcome, the cost of such projects is formidable.

FURTHER READINGS

CALIFORNIA DEPARTMENT OF WATER RESOURCES. 1987. *California Water: Looking to the Future,* Bulletin 160-87, 122 pp.

KARHL, WL., ed. 1978-1979. *The California Water Atlas.* Governor's Office of Planning and Research, in cooperation with the California Department of Water Resources, 118 pp.

KAHRL, W.L. 1982. *Water and Power: The Conflict Over Los Angeles' Water Supply in the Owens Valley.* Berkeley: University of California Press, 583 pp.

MOUNT, J. 1994. *California Rivers.* Berkeley: University of California Press, 359 pp.

PURIN, G., AND FONO, E. 1989. *Consumer's Guide to California Drinking Water.* Sacramento: Golden Empire Health Planning Center, 81 pp.

REISNER, M. 1986. *Cadillac Desert: The American West and its Disappearing Water.* New York: Viking-Penguin, Inc., 582 pp.

CHAPTER

11

The Great Valley

Sediments and Soils

Viewed from above, California's Great Valley is a long, flat-bottomed hollow, smoothed out between the rugged mountains of the Coast Ranges and the Sierra Nevada (see the cover image and Endpaper 4). From Red Bluff to Bakersfield it is a relatively featureless plain, broken most noticeably by the Sutter Buttes near Marysville. The Great Valley, also known as the Central Valley, is about 650 kilometers long and averages about 80 kilometers in width. Its 15 million acres of farmlands constitute the richest agricultural region in the history of the world. In 1986 alone, the $14 billion agricultural harvest exceeded the value of all the gold mined from California.

The Sacramento Valley forms the northern third of the Great Valley, and the San Joaquin Valley makes up the southern two thirds of the province. The south-flowing Sacramento River and the north-flowing San Joaquin River come together in the Delta area where they flow into San Francisco Bay. Both valleys are watered by the large rivers flowing west from the Sierra Nevada, with smaller tributaries flowing eastward into the Sacramento and San Joaquin Rivers from the Coast Ranges (Endpaper 3 and Fig. 11-1). South of Fresno is a region of low-lying bottomlands where rivers drain onto the floors of ***intermittent*** lakes, the largest of which is Tulare Lake. Gently rolling plains and hills as high as 200 meters rise from the valley floor along the borders with the surrounding mountainous provinces.

Before cultivation and the introduction of cattle, the Great Valley was a vast area of native grasslands, riparian forests, and extensive tule marshes. It was home to one of the largest Native American populations in North America, estimated by historians at more than 100,000. Because of the mild climate and ample food and water, the valley also supported a wide variety of wildlife. In 1852, James H. Carson, a writer for the *San Joaquin Republican* in Stockton, in *Tulare Plains* described the

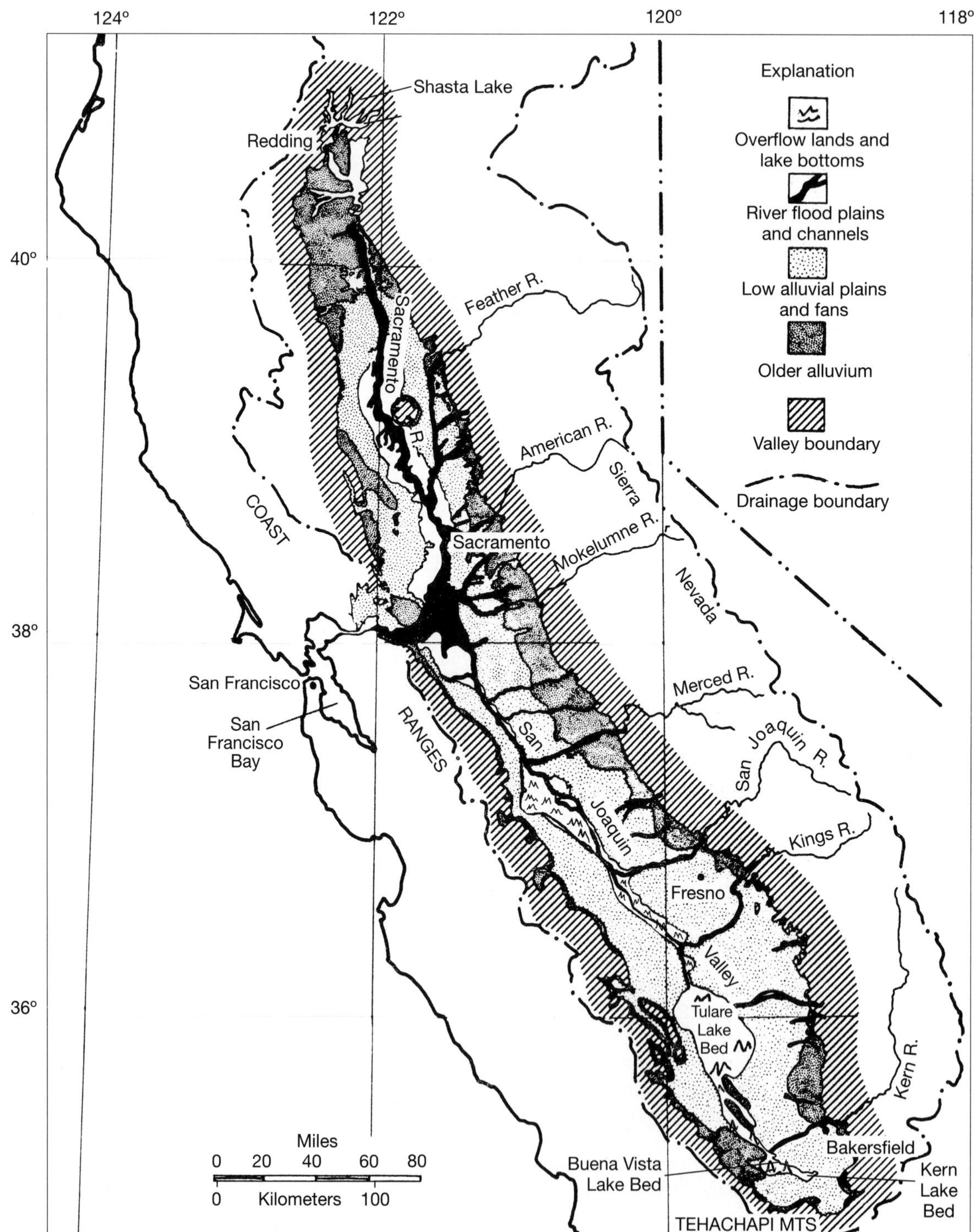

Fig. 11-1 Map of the geomorphic features of the Great Valley. (Source: Bailey, E.H., editor. Geopmorphic units after Davis and others (1959) and Olmstead and Davis (1961). California Department of Conservation, Division of Mines and Geology, Bulletin 190, p. 240.)

wildlife of the Tulares (San Joaquin Valley). A century and a half later, it is difficult to recognize even the traces of this former world*:

> Every beast and bird of the chase and hunt are to be found in abundance on the Tulares. Horses, cattle, elk, antelope, black tail and red deer, grizzly and brown bear, black and grey wolves, cayotes, ocelets, California lions, wild cats, beaver, otter, minks, weasels, ferrets, hare, rabbits, grey and red foxes, grey and ground squirrels, kangaroo rats, badgers, skunks, musk rats, hedge hogs, and many species of small animals not here mentioned; swan, geese, brant, and over twenty different description of ducks also cover the plains and waters in countless myriads, from the first of October until the first of April, besides millions of grocus (sand hill crane,) plover, snipe, and quail. The rivers are filled with fish of the largest and most delicious varieties, and the sportsman and epicurean can find on the Tulares every thing their hearts can desire.

FLOODPLAIN SEDIMENTS

Most of the surface of the Great Valley is covered by recent and Pleistocene alluvium (Endpaper 1). Sediments eroded from the Sierra Nevada, and to a lesser extent the Coast Ranges, are deposited on the floodplains and bottomlands as the mountain streams greatly decrease their velocity in the flat-bottomed valley. The broad, flat floodplains of the Central Valley are fertile agricultural land, periodically replenished with fine-grained alluvium by spring floods, and these lands are one of California's greatest natural resources.

As discussed in Chapter 8, periodic glaciation of the Sierra Nevada had drastic affects on the landscape. The impacts of global climate change were also strongly felt by Great Valley rivers. Huge quantities of sediment were piled up in moraines at the edges of alpine glaciers during times of glaciation. Sierra Nevada rivers, made even more powerful by the abundant glacial meltwater and the considerably wetter climate during these periods, carried much of this sediment downstream to the Sierran foothills and into the Central Valley. Geologists term this episodically deposited alluvium ***glacial outwash.*** At least four major pulses of deposition took place as sediment washed out from the Sierra Nevada during the glacial episodes of the past 2 million years.

In the central, lower parts of the Great Valley, the younger alluvium buried older deposits, which geologists can identify in cores obtained from beneath the Valley floor (Fig. 11-2). Today the floodplains of the Sacramento and San Joaquin Rivers and their tributaries receive sediment during times when the rivers overflow their channels, further burying the older deposits. However, along the Sierran foothills, the older alluvium can be seen at the surface. Because the Sierra Nevada continues to be uplifted relative to the Great Valley, rivers are constantly cutting downward (Fig. 11-3). As they cut or ***incise*** their channels, the rivers leave behind remnants of the older floodplain surfaces as ***stream terraces.*** Stream terraces can be recognized by their flat surfaces (Fig. 11-4; see also Fig. 10-7) and by the presence of alluvium beneath the surfaces. Along the edges of the main river valleys, several terrace levels are preserved at different elevations above the modern rivers. These terrace levels record a time sequence of river down-cutting. With increased elevation above the modern rivers, one finds progressively older valleys and deposits.

---

*Source: Browning, P., ed. 1991. *Bright Gem of the Western Seas: California 1846-1852*, p. 90. Great West Books, P.O. Box 1028, Lafayette, California.

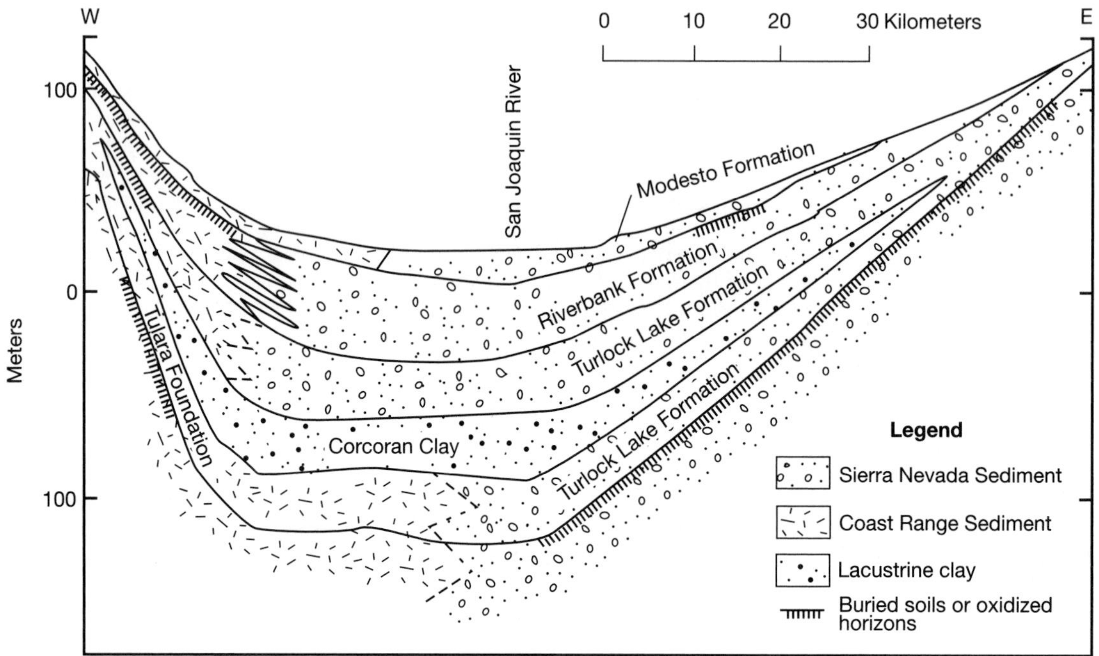

Fig. 11-2 Diagram of buried alluvium identified in core samples from the northern San Joaquin Valley. The Corcoran Clay layer is discussed in Chapter 12. (Source: Modified from Bartow, J.A. 1991. U.S. Geological Survey, Professional Paper 1501.)

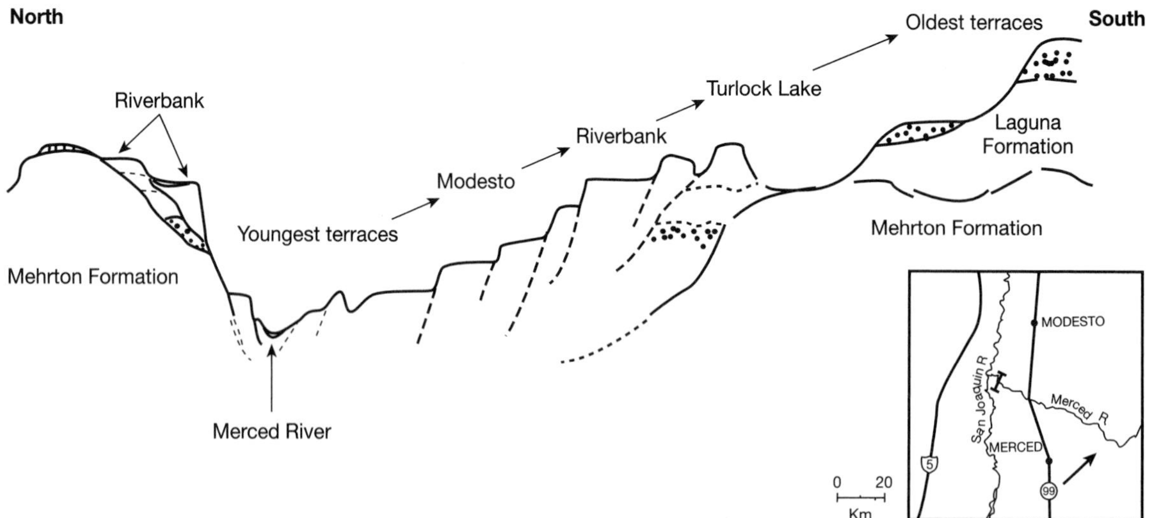

Fig. 11-3 A schematic cross section of river terraces above the Merced River near Snelling, Merced County, in the area shown on the inset map. As the Sierra nevada has been uplifted and tilted, the Merced River has cut progressively downward to lower levels. The terraces and their deposits mark the former, now elevated, positions of the river floodplain and channel. The highest alluvium, the Turlock Lake formation, is the oldest. (Source: Modified from Marchand, D. and Allwardt, A. 1981. U.S. Geological Survey, Bulletin 1470.)

Fig. 11-4 A relatively flat younger terrace of the Merced River, near Dry Creek. In the background, an older terrace shows the rolling hills typical of older surfaces. (Source: †Marchand, D. U.S. Geological Survey.)

GREAT VALLEY SOILS

The soils of California's Great Valley are a direct result of the geological processes discussed above. Fertile soils are one of California's greatest resources, and they are particularly abundant on young alluvial surfaces. Soil forms as the result of both geologic and biologic processes that alter sediment and rock at the Earth's surface. In the Great Valley, most soils are formed on floodplain deposits. Following deposition of sediment on a floodplain, vegetation becomes established on the surface. As plant material decays as a result of the activity of microbes living in the soil, organic matter is added to the upper soil in the form of ***humus.*** The addition of humus to the original deposit increases the fertility of the soil, changes the texture of the soil, and increases the potential for mineral weathering because of the formation of organic acids. The accumulation of organic matter in the soil is recognized by the brown, gray, or black colors in the upper part of the ***soil profile.*** The layer or ***soil horizon*** characterized by organic matter accumulation is known as the ***A horizon***. A horizons of soils can be thought of as the zones of maximum biologic activity.

At the same time that biologic processes are altering the sediment, chemical, and physical weathering processes act on the minerals. As water percolates downward, it becomes acidified by the organic acids in the A horizon. Beneath the A horizon, chemical weathering breaks down silicate minerals. Some easily dissolved elements may be moved completely through the profile if sufficient water moves through the deposit, while other less soluble elements may simply be redeposited lower in the profile when the percolating water evaporates. Weathering of common silicate minerals like feldspar and mica produces clay minerals, which cause the soil to become stickier and better aggregated. Downward-percolating water also carries fine solid particles of clay, iron oxide, and salts lower in the profile. The soil beneath the dark A horizon becomes increasingly red as iron oxide (rust) builds up. The soil horizon characterized by the accumulation of one or more weathering products is

Fig. 11-5 A closeup view of a soil profile formed on the Riverbank Formation. Note the quarter in the upper right-hand area. The blocky aggregated zone immediately beneath the quarter is the B horizon, where sufficient clay has accumulated to form soil peds. At the base of the photo, accumulated silica has whitened the soil color. (Source: †Marchand, D. U.S. Geological Survey.)

known as the ***B horizon.*** B horizons can be thought of as the zones with the greatest evidence of weathering activity (Fig. 11-5).

Because the processes that form B horizons occur more slowly than the buildup of organic matter, young soils display A horizons but have not yet developed B horizons (Fig. 11-6). In the Great Valley, soils on floodplain sediments deposited during the past several thousand years typically show A horizons overlying unchanged floodplain sediments. Soils on older floodplain sediments, found on higher terraces dating to the last glacial period or before, display progressively redder B horizons with progressively more accumulated clay (see Figs. 11-5 and 11-7). In some parts of the Valley, these older soils also have developed ***hardpans,*** hard layers formed by the accumulation of silica, iron, or calcium carbonate. The hardpans pose difficult problems for irrigation engineers in some areas. In order to plant orchards in some areas along the margin of the San Joaquin Valley, farmers have dynamited through the hardpans.

The oldest soils found in the Great Valley can be seen on ancient fans and terraces high above the modern rivers along the eastern side of the valley. Here one can see several meters of highly weathered, bright red soil that probably has taken more than a million years to form.

Soil profiles develop on other types of deposits, as well as on rock surfaces, in a similar manner to that described above. The type of soil that forms on a surface, the depth of the soil profile, and the rate at which it forms all depend on several soil-forming factors. These include the climate of the region, the organisms present (especially plants and microbial organisms), the topography of the surface, the nature of the rock or deposit, called the ***parent material*** by soil scientists, and the time that has elapsed since the surface formed. The great fertility of California's alluvial soils in the Central Valley and other valleys throughout the state, is the result of an ideal combination of all of these factors.

Fig. 11-6 A relatively undeveloped soil profile. (Source: Sarna-Wojcicki, A. U.S. Geological Survey.)

Fig. 11-7 A relatively well developed soil profile on granitic rocks can be seen on the left side of this outcrop on Highway 120, western Sierra foothills. (Source: Sarna-Wojcicki, A. U.S. Geological Survey.)

Continued intensive agriculture and irrigation have created several water-quality problems in the Great Valley and other agricultural areas of California. Partly as a result of fertilizers and irrigation runoff, high levels of nitrates, pesticides, and herbicides in streams pose a water quality problem in some areas. However, not all water-quality problems are caused by the application of chemicals to farmlands.

In arid regions like the San Joaquin Valley, less water percolates through the soil profile every year than in more humid regions. Over time, relatively soluble minerals—calcium carbonate, sulfates, and other ***salts,*** or soluble minerals—accumulate in the lower part of the soil profile at the depth where percolating water evaporates. The buildup occurs because rainfall is not sufficient to flush these dissolved materials through the soil. In the western San Joaquin Valley, the bedrock source for the alluvial parent materials is the marine sediments of the Coast Ranges. As a result, the natural salinity of the soils is high, and soils in this area are particularly high in selenium-bearing minerals.

Irrigation of arid soils causes salts to be dissolved and carried to the lower horizons at a higher rate than would occur under natural conditions. In many areas, including the western San Joaquin Valley, problems occur because of the presence of clay layers in the alluvial parent materials. The clay layers form a barrier to percolating water, causing salt-rich irrigation water to build up near the surface. It is estimated that approximately 10 percent of the soils in the Great Valley have been damaged by the buildup of salts (Fig. 11-8). Salinization may inhibit plant growth, damage leaves, or even kill salt-sensitive crops.

One solution to the problem of salinization is to install subsurface drains beneath fields to carry off salt-laden irrigation water (Fig. 11-9). Drains were used by farmers to remedy salt buildup at least 7000 years ago in ancient Mesopotamia. In the western San Joaquin Valley, a 135-kilometer-long concrete canal known as the San Luis Drain was built in the early 1970s to carry agricultural drainage water from the fields north to Kesterson Reservoir near Los Banos. In the early and mid-1980s, wildlife biologists began finding alarming numbers of dead and deformed chicks of water birds near the reservoir and nearby ponds. Studies revealed that water in the reservoir and ponds contained selenium concentrations as high as 400 parts per billion (ppb), far above the drinking water standard of 10 ppb. Remediation of the Kesterson site,

SOILS, FOOD, AND THE ROCK CYCLE

Agricultural scientists have demonstrated that sixteen chemical elements are essential for plant growth. Three of these ***nutrients,*** carbon, oxygen, and hydrogen, are obtained directly from the atmosphere by the leafy parts of plants. Nitrogen, sulfur, and phosphorous are made available to plant roots by the decay of organic matter. For the remainder of the essential nutrients, plants depend directly on mineral weathering. These elements are released from minerals into solution and then absorbed from the soil by roots. Many of the most essential elements—potassium, magnesium, calcium—are common constituents of silicate minerals such as feldspar and mica, as well as of other common minerals (see Table 2-1). The remaining essential micronutrients—iron, chlorine, manganese, zinc, boron, copper, and molybdenum—are also supplied by minerals.

Animals, including humans, that rely on plants for some or all of their food supply also require essential chemical nutrients for survival and growth. Small amounts of elements not required by plants, such as ***selenium*** and iodine, are necessary for the growth of many animal species. Studies have shown that soils deficient in these elements can lead to disease and death. On the other hand, large concentrations of some elements, including sodium, boron, selenium, and copper, may be toxic to animal species. Both deficiencies and high concentrations of certain elements in soils can be a natural result of the weathering of certain types of rocks and minerals.

From what we learned in Chapter 2 about the rock cycle, it is clear that the supply of nutrients to soils by organic decay, and especially by mineral weathering, is a slow process. In addition to direct solution from minerals, nutrients can be replenished by other geologic processes, including addition from wind-blown dust and deposition of fine-grained flood sediments. The natural plant community in a given area is well balanced with its nutrient supply. However, when humans cultivate the land, they often replace the natural plants with crops that require a greater nutrient supply. As soil nutrients are "mined" more quickly than they are naturally replaced, soils become depleted in some nutrients, thus requiring farmers to use fertilizers, and thereby increasing the cost of farming. Most crops grown in the Central Valley today require the addition of nitrogen, phosphorous, and other nutrients.

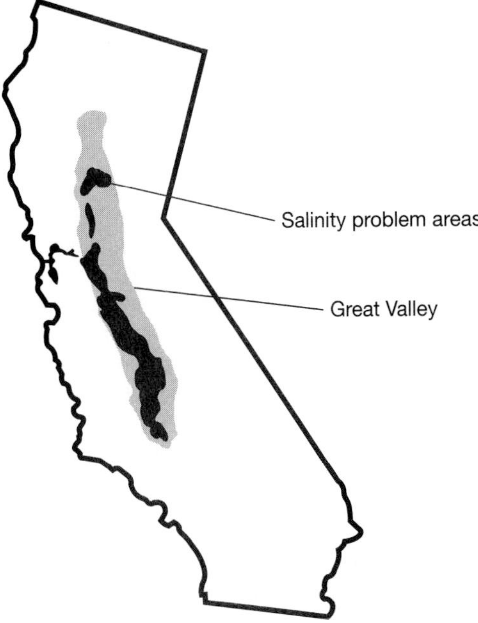

Fig. 11-8 Map showing the extent of salinization in the Great Valley. (Source: Johnson, S. and others. 1993. *The Great Central Valley.* Berkeley: University of California Press.)

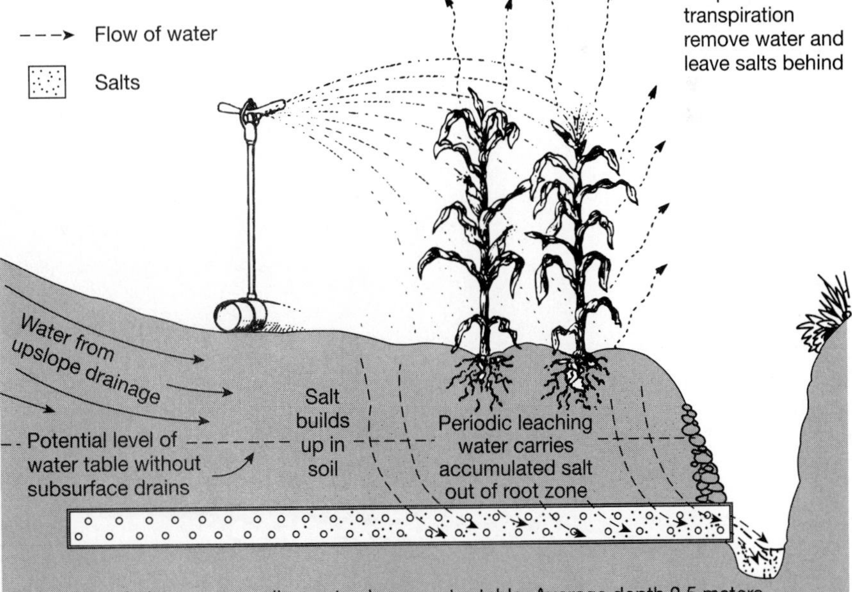

Fig. 11-9 The design of subsurface drains constructed to flush salts from soils in the western San Joaquin Valley. (Source: U.S. Department of Interior, Bureau of Reclamation. 1984. *Drainage and Salt Disposal.* Sacramento: Central Valley Project, Information Bulletin 1.)

including filling of the selenium-contaminated ponds, is in progress. The long-term solution to the problem of salinization of soils in the San Joaquin Valley had not been found, and it is of great importance to the economy of California.

SUTTER BUTTES

Rising dramatically from the flat floor of the Sacramento Valley, Sutter Buttes are the major topographic feature of the otherwise nearly flat Great Valley (Fig. 11-10). They were originally named the Marysville Buttes for the nearest city, about 12 kilometers to the southeast. Sutter Buttes are about 17 kilometers in diameter, and the highest peak is about 3300 meters above the valley floor.

The geologists who first studied Sutter Buttes in detail, Professors Williams and Curtis of the University of California at Berkeley, drew an analogy between the Buttes and a medieval castle. At the center of Sutter Buttes is a group of towering domes called the *Castellated Core* because of their resemblance to a castle (Fig. 11-11). Surrounding the core is a low circular area underlain by older, more easily eroded sedimentary rocks: this is appropriately called the *Moat.* The smooth slopes leading up to the Moat on all sides are formed by pyroclastic flows and alluvial deposits—the *Rampart* leading to the "castle."

Igneous activity at Sutter Buttes began during late Cenozoic time with the intrusion of silicic magma into the existing late Mesozoic and Tertiary sedimentary rocks. The first shallow intrusions did not reach the surface, but they caused the sedimentary rocks to be uplifted and tilted. Volcanic eruptions of intermediate and silicic lava followed, beginning about 1.6 million years ago. The first volcanic eruptions formed rhyolite domes along a circular system of faults and fractures that underlie the Moat (Fig. 11-12). The final volcanic activity formed the rugged core of Sutter Buttes, a

Fig. 11-10 Sutter Buttes, viewed toward the east. (Photo by author.)

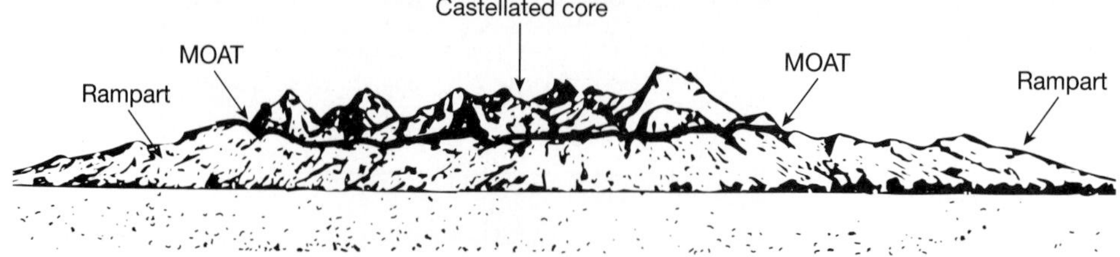

Fig. 11-11 A cross-sectional sketch of Sutter Buttes, looking toward the east from near Williams. (Source: Williams, H. 1929. *Geology of the Marysville Buttes, California.* Berkeley: University of California Publications, Department of Geological Sciences Bulletin, Vol. 18, pp. 103-220.)

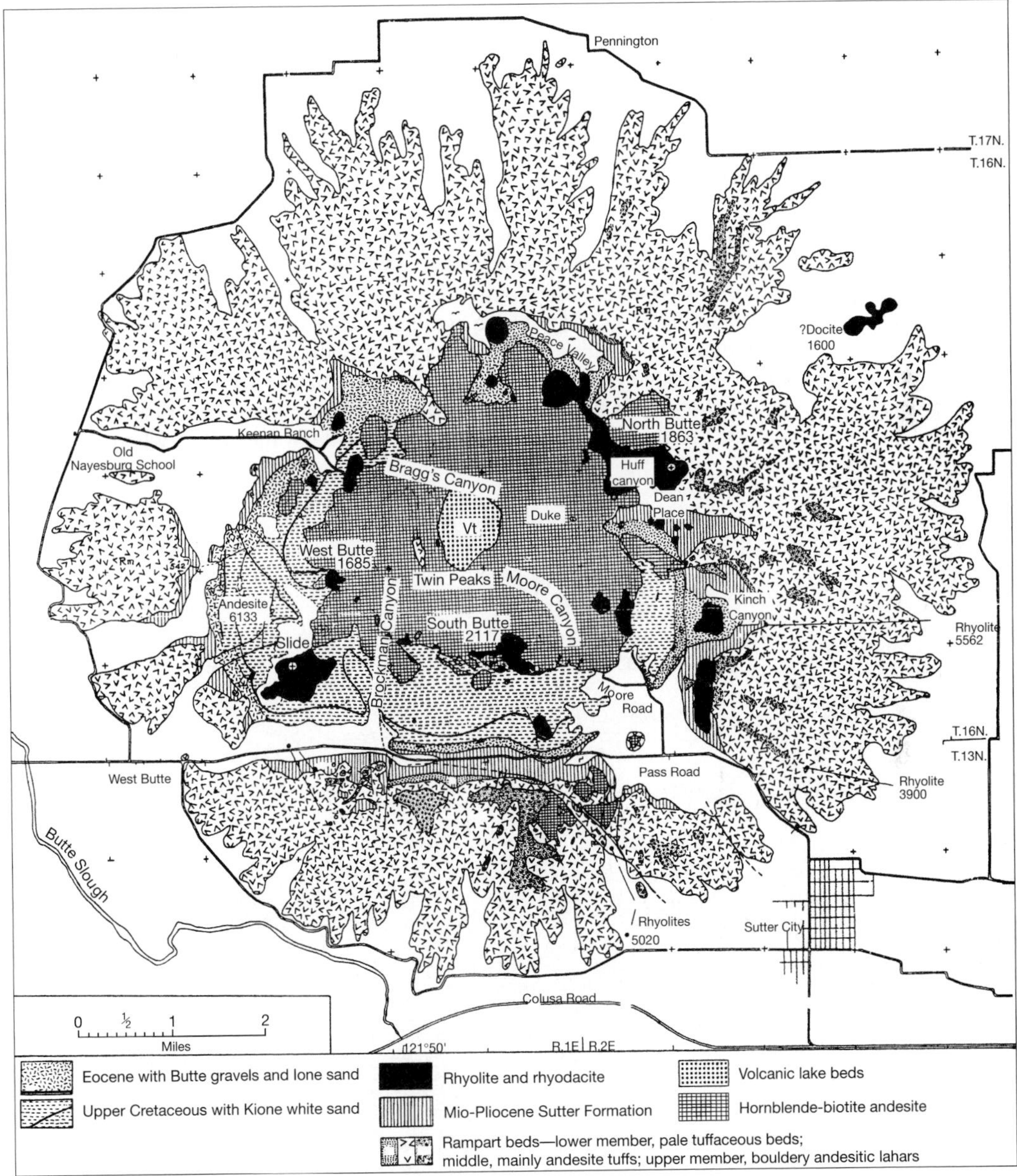

Fig. 11-12 Geologic map of Sutter Buttes. (Source: Williams, H., and Curtis, G.H. 1977. *The Sutter Buttes of California.* Berkeley: University of California Press, Publications in Geological Sciences.)

group of andesite domes. At the center of the Core is a thick section of sediments that filled a deep crater lake during the eruptions of the pyroclastic material. All volcanic activity apparently ceased at Sutter Buttes by about 1.3 million years ago.

The eruptions that produced the domes also generated pyroclastic materials that were deposited around the flanks of Sutter Buttes. As would be expected, the older pyroclastic materials found on the Rampart are more silicic, coinciding with the formation of the rhyolite domes. Younger pyroclastic materials are andesitic.

THE GREAT VALLEY SEQUENCE

Beneath the young Cenozoic alluvium of the Great Valley lies an incredible thickness of older sediments and sedimentary rocks. The geology beneath the surface can be pieced together from the many wells that have been drilled to explore for oil and natural gas. Geologists are also able to examine the older rocks directly where they are exposed along the margins of the valley and around the margins of Sutter Buttes. The three-dimensional picture that emerges shows an elongate, asymmetrical ***sedimentary basin*** with approximately the same borders as the modern Great Valley. The basin is deeper on the western side of the valley, where it is filled by as

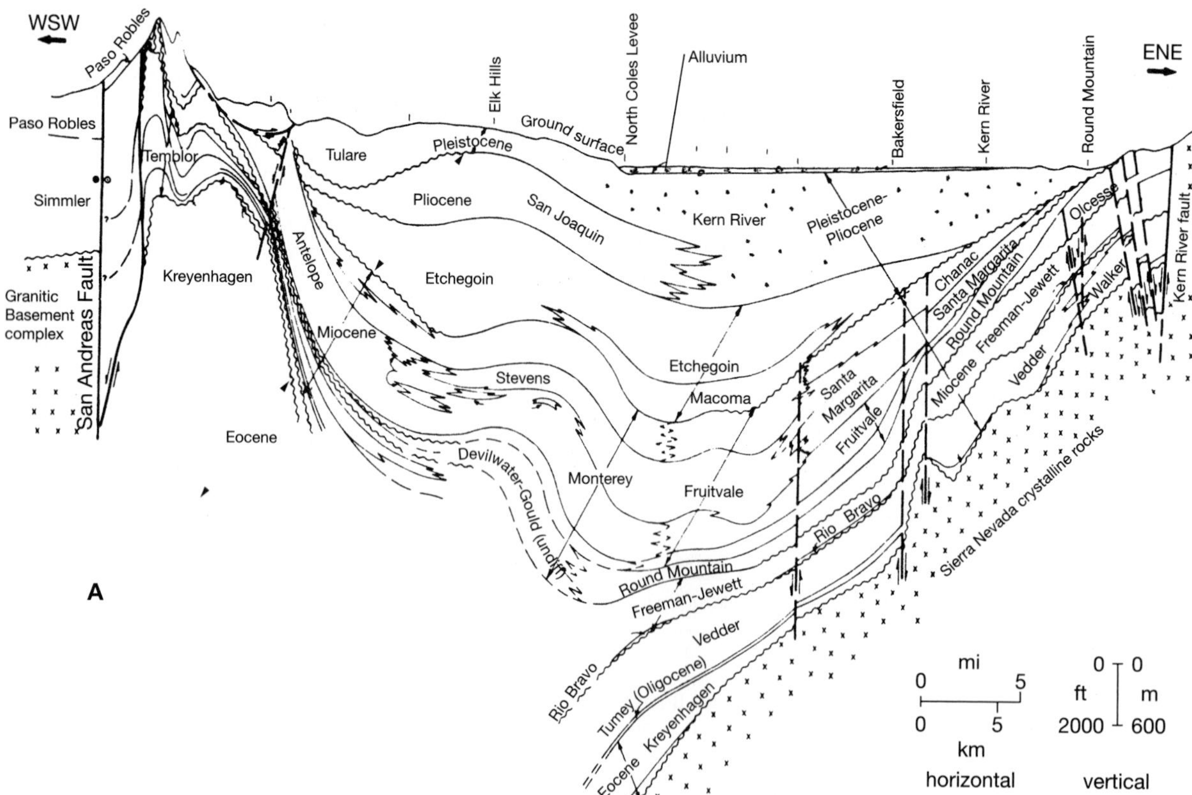

Fig. 11-13 A, A simplified cross section across the San Joaquin Valley near Bakersfield, showing the formations found in the upper part of the sedimentary basin.

much as 20,000 meters of sediment (Fig. 11-13). Along the western margin of the basin, the sedimentary formations have been steeply upturned so that many of the older units are exposed (Fig. 11-14). The rocks on the eastern flank of the valley dip inward more gently toward the axis of the basin.

The Mesozoic (Jurassic and Cretaceous) and earliest Cenozoic rocks of the Great Valley, known as the Great Valley Sequence, consist of sandstone, shale, and conglomerate. Marine fossils and characteristic sedimentary features indicate that they accumulated in an ocean basin that lay west of the Mesozoic North American margin. On the eastern side of the valley, the rocks are characteristic of deposits formed deltaic and shallow ocean waters. In contrast, those found on the western side of the valley were apparently deposited deeper on the ocean floor by turbidity currents (which are discussed in more detail in Chapter 12). Using the nature of the sedimentary rocks to interpret the environments where they were originally deposited, geologists conclude that the shoreline of the ancient ocean basin lay along the western edge of the Sierra Nevada (Fig. 11-15).

The sedimentary basin where the Great Valley Sequence accumulated was a ***forearc basin*** sandwiched between the Sierran arc (Chapter 8) and the Mesozoic subduction zone. Today similar modern basins can be found along the margins of the Pacific Plate between volcanic arcs and ocean trenches; one example is the Sunda-Java forearc basin of Indonesia. Marine sediments of the Great Valley Sequence continued to be deposited in the forearc basin in early Cenozoic time, after the magmatic activity in the Sierra Nevada had ceased.

By examination of the minerals present in the Great Valley Sequence, geologists can detect the sources of the sedimentary formations. Sandstone in the older, lower

B

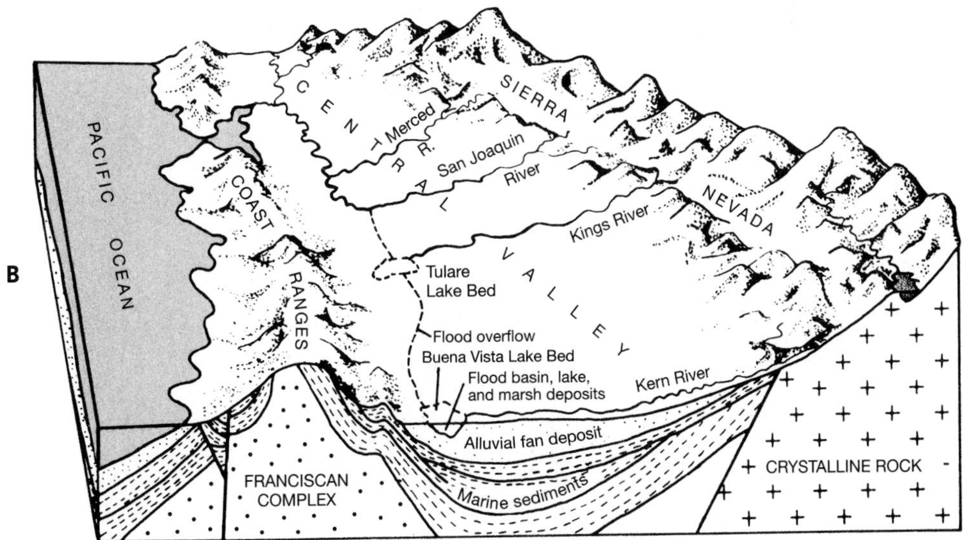

Fig. 11-13, cont'd **B,** A perspective drawing of the same region. (Source: **A,** Modified from Christiansen, R., and Yeats, R. *Decade of North American Geology,* Vol. G-3, p. 331. **B,** Modified from Page, R. 1986. U.S. Geological Survey, Professional Paper 1401-C.)

Fig. 11-14 Steeply dipping beds of the Great Valley Sequence along the western edge of the Sacramento Valley along Cache Creek at Monticello Dam. Note dog at far right for scale.

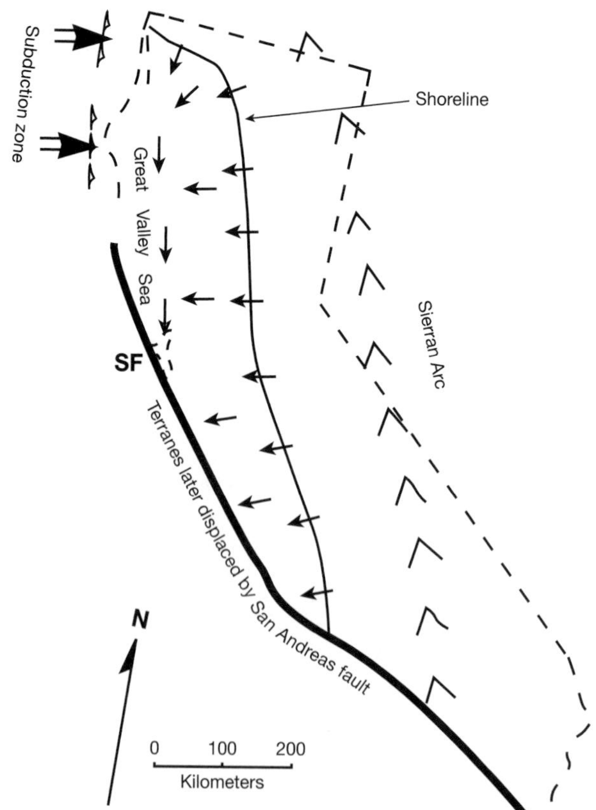

Fig. 11-15 Map showing the approximate position of the Great Valley sea during Late Cretaceous time. The arrows show the direction of sediment transport, as indicated by sedimentary features in the rocks. (Source: Modified from Ingersoll, R. 1978. Paleogeography and paleotectonics of the Late Mesozoic forearc basin of northern and central California. In *Mesozoic Paleogrography of the Western U.S.: Pacific Section.* Los Angeles: Society of Paleontologists and Mineralogists.)

formations is rich in volcanic rock fragments eroded from the volcanoes of the Jurassic and early Cretaceous Sierran arc. The volcanic fragments are proof that the chain of volcanoes, long since eroded from the Sierra Nevada, existed during Jurassic and Cretaceous time. Younger sandstones of the Great Valley Sequence are rich in feldspar and quartz eroded form the granitic pluton. In the northern Great Valley, the Klamath Mountains were also a source of sediment.

CENOZOIC SEDIMENTS: THE END OF THE GREAT VALLEY SEA

During Jurassic and most of Cretaceous time, the Great Valley sea persisted because the basin continued to subside as sediments filled it in. However, by the end of Mesozoic time, the northern part of the Great Valley basin began to fill up as uplift and deposition out-paced the subsidence of the basin. Sediments typical of increasingly shallow marine environments reflect this change. By about 24 million years ago, the sediments deposited in the Sacrament Valley were mostly of terrestrial origin. At that time, a basaltic lava erupted from the northern Sierra Nevada and flowed west and then south for at least 260 kilometers to the vicinity of Orland Buttes. Today, a volcanic flow called the Lovejoy Basalt can be seen near Oroville, where it forms Table Mountain, similar to the Table Mountain discussed in Chapter 8. The lava buried alluvium across today's Sacramento Valley, providing evidence that by about 24 million years ago, the basin had emerged from the sea.

Younger pyroclastic flows and ash erupted from the Cascades in Pliocene time are widespread in the Sacramento Valley, particularly at the northern end. About 3.5 to 4 million years ago, volcanic debris flows and pyroclastic lavas flowed into the northeastern side of the Sacramento Valley from sources in the Cascades; these deposits make up the Tuscan Formation (Fig. 11-16). During the same time, on the western side of the Sacramento Valley, alluvial deposits eroded form the Coast Ranges were deposited as the Tehama Formation. Both units contain distinctive, widespread ash layers erupted from the vicinity of the present-day Lassen Peak. One of these is the 3.4-million-year-old Nomlaki Tuff (see Chapter 5). Streams continue to deposit alluvium along floodplains of the Sacramento to this day.

In the south, the deep marine basin persisted much longer, as indicated by the continued deposition of marine shale and sandstone during early and middle Cenozoic time. In the longest-lived part of the Great Valley Sea, the San Joaquin basin, marine sediments from the Sierra Nevada and the newly formed Coast Ranges accumulated until late Pliocene time. Three million years ago, the San Joaquin basin was a large embayment in the coastline (Fig. 11-17). Today, the San Joaquin basin still persists as a lowland, and it is not mere coincidence that this is the site of the Great Valley's lake basins (see Fig. 11-1).

About 2 million years ago, the San Joaquin Valley emerged above sea level, and terrestrial sediments began to cover the marine sediments that had already been deposited. Between about 1 million and 600,000 years ago, after the San Joaquin Valley was closed off from the Pacific Ocean, much of the Great Valley was periodically occupied by a large lake. Geologists infer the existence of this lake from an extensive layer of clay, about 10 to 15 meters thick, but as much as 50 meters thick in some places, found at about the same depth beneath the San Joaquin and southern Sacramento Valleys. The clay contains Pleistocene fossils typical of shallow lakes. Known as the Corcoran clay, it is an important piece of evidence for the history of the Sacramento-San Joaquin river system, a topic discussed in more detail in Chapter 12.

Fig. 11-16 Pyroclastic flows of the Tuscan Formation. (Source: Harwood, D. U.S. Geological Survey.)

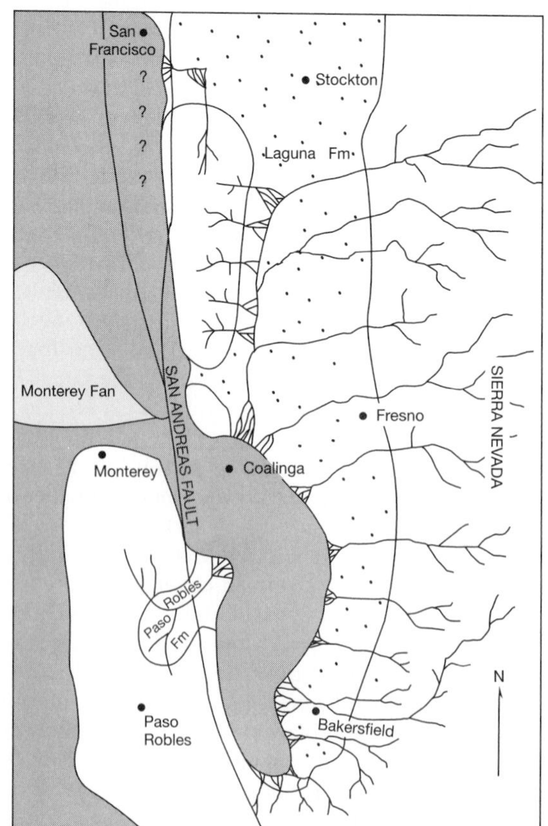

Fig. 11-17 The San Joaquin–Sacramento Delta and the marine embayment in the San Joaquin Valley about 5 million years ago. (Source: Modified from Bartow, J.A. 1991. U.S. Geological Survey, Professional Paper 1501.)

BENEATH THE GREAT VALLEY SEQUENCE

In the eastern half of the Central Valley, where the sediments are thinner, the sediments of the Great Valley sequence were deposited on the igneous and metamorphic rocks of the western Sierra Nevada. At Sutter Buttes, the Sierran rocks are covered by about 2100 meters of sedimentary rocks of the Great Valley Sequence and younger rocks. Further north, along the northern end of the Sacramento Valley, the Great Valley rocks were deposited on the rocks of the Klamath Mountains. Most geologists interpret the relations between the rocks of these areas as further evidence that the Great Valley forearc basin lay west of the North American continent during the time that the Great Valley sequence was deposited.

Along the western margin of the Great Valley, where the rocks have been tilted steeply and exposed, geologists are able to observe the rocks that lay beneath the Great Valley basin when sediments first began to accumulate. The best exposure of the base of the Great Valley sequence is along the banks of the South Fork of Elder Creek, a small tributary to the Sacramento River, about 10 kilometers northwest of the town of Paskenta. Because the rocks are steeply tilted, the different units and the contacts between them can be seen by walking along the creek. At this locality, mudstone of the Great Valley sequence was deposited on a sedimentary ***breccia*** containing angular clasts of mafic and ultramafic rocks. The contact is an uncomformity, indicating that a period of erosion separated the formation of the breccia and the initial deposition of mud in the Great Valley basin (see Chapter 3). The rubble of ophiolite debris in turn overlies layered ultramafic rocks of the Coast Range ophiolite (see Chapter 9). Near Mt. Diablo in the Coast Ranges east of San Francisco Bay, the sheeted dike section of the Coast Range ophiolite is about 2.5 kilometers thick and is overlain by 1.5 kilometers of pillow basalt.

Igneous rocks which intrude the Coast Range ophiolite in the western Sacramento Valley crystallized about 155 million years ago, and the oldest rocks of the Great Valley sequence contain *Buchia* fossils that are about 152 million years old (see Fig. 3-8). The Coast Range ophiolite is therefore bracketed between these two ages, and it is about the same as the Josephine ophiolite in the Klamath province.

Another excellent exposure of the Coast Range ophiolite, termed the Point Sal ophiolite remnant, can be found along the California coast in northwestern Santa Barbara County (Fig. 11-18). The ophiolite at Point Sal contains many of the rocks typical of an ophiolite suite (see Chapter 9), including submarine basalt flows, sheeted dikes, gabbro, and serpentinized ultramafic rocks. The Point Sal ophiolite remnant has been displaced from the main Coast Range ophiolite by Cenozoic faulting (discussed in Chapter 16).

At the western edge of the Coast Range ophiolite, where it borders the rocks of the Coast Ranges, is a major fault system known as the Coast Range fault (Fig. 11-19). This Mesozoic fault system separates the rocks of the Great Valley Sequence and its basement ophiolite from those of the subduction complex known as the Franciscan Assemblage (see Chapter 12). Many geologists believe that the Coast Range fault marks the line along which oceanic rocks were thrust underneath the western edge of the forearc basin (Fig. 11-20). However, the Coast Range fault shows evidence of multiple types and ages of movement, and its complete history is far from understood.

Fig. 11-18 Basalt dikes cutting gabbro of the Point Sal ophiolite near Santa Barbara. (Source: †Bailey, E. U.S. Geological Survey.)

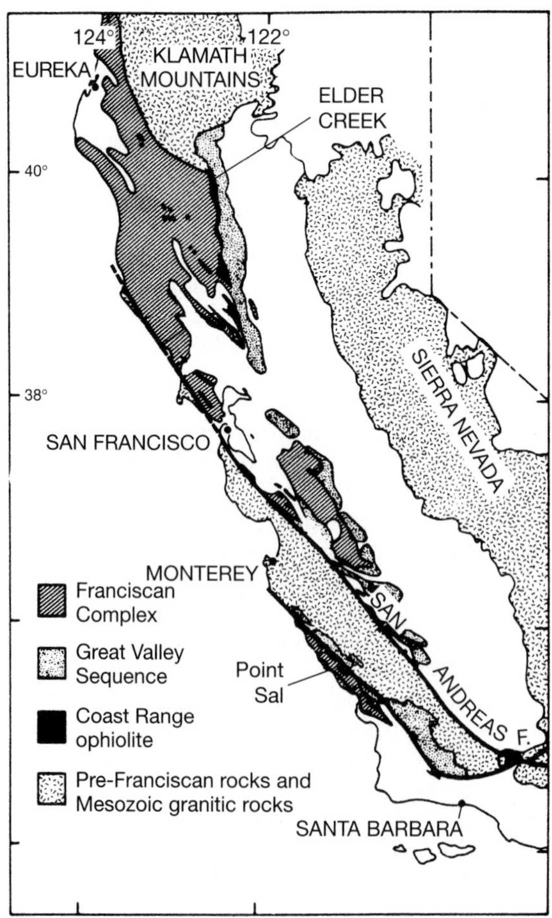

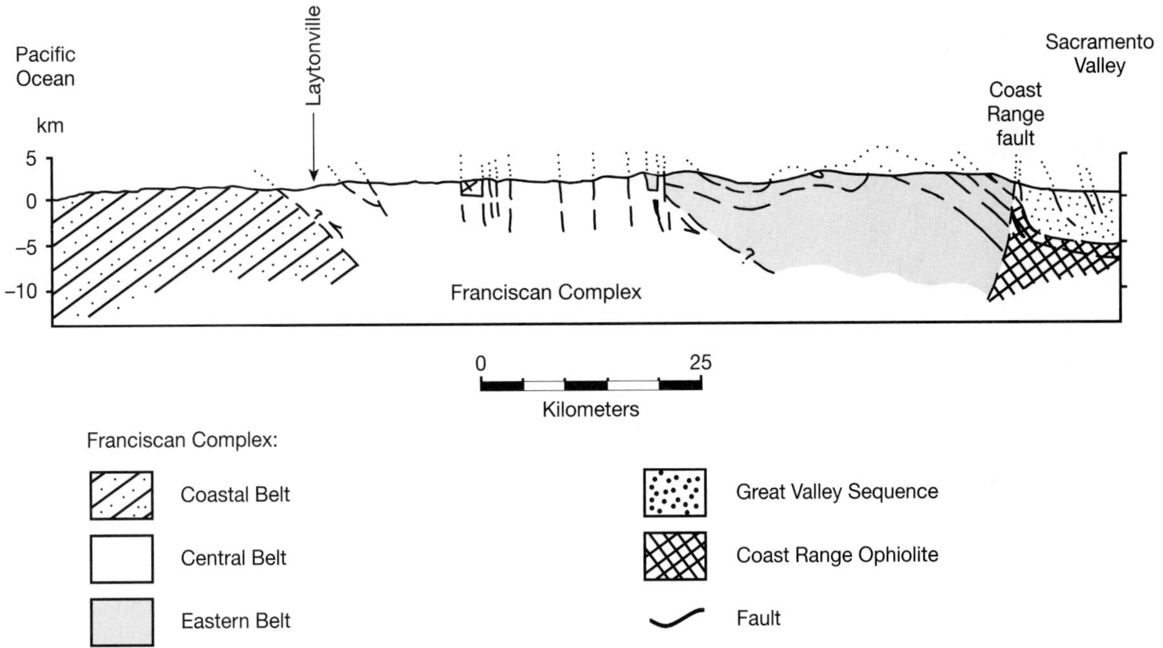

Fig. 11-20 A cross section across the northern Coast Ranges and the western Sacramento Valley, showing the Coast Range fault near Elder Creek. (Source: Modified from Cowan., D., and Bruhn, R. 1992. Late Jurassic to Early Late Cretaceous Geology of the U.S. Cordillera. In Geological Society of America, *Decade of North American Geology,* Vol. G-3, p. 186.)

FURTHER READINGS

ANDERSON, W. 1983. *The Sutter Buttes: A Naturalist's View.* Chico, Calif.: The Natural Selection Publishers, 326 pp.

BLAKE, M.C., JR., JAYKO, A.S., JONES, D.L., AND ROGERS, B.W. 1987. Unconformity between Coast Range ophiolite and part of the lower Great Valley sequence, South Fork of Elder Creek, Tehama County, California. In *Geological Society of America Centennial Field Guide.* Vol. 1, pp. 279-282.

DICKINSON, W.R. 1981. Plate tectonics and the continental margin of California. In Ernst, W.G., ed. *The Geotectonic Development of California.* Upper Saddle River, N.J.: Prentice Hall, Inc., pp. 1-28.

HAUSBACK, B.P., AND NILSEN, T.H. 1989. Sutter Buttes Field Trip Guide. Washington, D.C.: American Geophysical Union, pp. 101-111.

JOHNSON, S., HASLAM, G., AND DAWSON, R. 1993. *The Great Central Valley: California's Heartland.* Berkeley: University of California Press, 253 pp.

PAGE, R.W. 1986. *Geology of the Fresh Ground-water Basin of the Central Valley, California, With Texture Maps and Sections.* U.S. Geological Survey, Professional Paper 1401-C.

Salinity in California (Special Issue). 1984. *California Agriculture* 38(10).

SINGER, M.J., AND MUNNS, D.N. 1987. *Soils: An Introduction.* New York: Macmillan, 492 pp.

CHAPTER

12

The Coast Ranges

Mountains of Complexity

California's Coast Ranges do not reach the heights of the Sierra Nevada, but in places they rise steeply from the Pacific Ocean to elevations of 1000 meters within 2000 meters of the coast. To visitors from the plains of the Midwest or the rolling mountains of the eastern United States, these are rugged and impressive mountains. Some of the highest peaks are Big Pine Mountain (2083 meters) in Santa Barbara County, Cone Peak (1572 meters) near Monterey, Mt. Tamalpais (784 meters) in Marin County, and Solomon Peak (2312 meters) in Trinity County. The Coast Range Province extends about 1000 kilometers from the Transverse Ranges to the Oregon Border and about 130 kilometers from the Pacific Ocean to the Great Valley in the south and the Klamath Mountains in the north (see the map on p. 54).

The northwest-oriented grain of the landscape (Endpaper 4) is a striking characteristic of the Coast Range Province. The Coast Ranges are a series of long, northwest-trending ranges separated by parallel river valleys. Prominent ranges are the Mendocino Range in the north, the Diablo Range east of San Francisco Bay, the Santa Cruz Mountains west of the Bay, and the Santa Lucia Range and Gabilan Range, which flank the Salinas Valley. California's large coastal rivers flow in the valleys between the ranges; these include the Eel, Russian, and Salinas Rivers.

Like many of California's landscape features, the orientation of rivers and ranges is controlled by tectonics. The boundaries between ranges and valleys are generally defined by faults that separate more resistant rocks from weaker ones. In some cases, the faults are active members of the San Andreas system; these faults are discussed in Chapter 14. In other areas, the faults are older and mark the junction between two accreted blocks of rock. A landscape where the orientations of different rock types and faults control the drainage pattern is called a ***structurally controlled topography*** (see Color Plate 18).

PLATE TECTONICS AND COAST RANGE GEOLOGY

Beginning about 140 million years ago, in the mid-Jurassic period, a new plate boundary formed along the western edge of the North American plate. The accretion of the oceanic terranes in the Sierra Nevada and the Klamath Mountains had been accomplished (see Chapters 8 and 9), and the subduction zone had moved west of the modern Great Valley. From this time until about 28 million years ago, oceanic crust was accreted to North America along a subduction zone that ran the length of western North America (Fig. 12-1, *A*). Fragments of the oceanic plates that collided with the North American plate along this subduction zone are preserved-today: these are the Mesozoic and early Cenozoic rocks of the Coast Ranges. The exact movement of the ancient plates is not completely understood, but at least some of the terranes traveled north from near the equator. One of these distant oceanic plates has been named the Kula Plate. The slower, eastward-moving ancient plate has been named the Farallon Plate.

As the Farallon Plate was consumed along the subduction zone, the Pacific Plate came into direct contact with the North American Plate (Fig. 12-1, *B*). This contact began about 28 million years ago, but it did not occur at the same time along the entire length of western North America. When the meeting did occur, the plate boundary along the western edge of the North American Plate changed from a subduction zone to a transform boundary. As more of the Farallon Plate came into contact with North America, the transform boundary grew (Fig. 12-1, *C*). Today the transform boundary runs from Cape Mendocino to the Gulf of California—the San Andreas fault (Fig. 12-1, *D*). The consistent northwest orientations of the faults, the different rock units, and the topography demonstrate that the orientation of the plate boundary remained fairly constant from Mesozoic time until the present.

As we saw in Chapter 1, the type of active plate boundary controls the geologic processes that take place along it. The change from subduction and accretion to transform motion is recorded in the rocks of the Coast Ranges. In fact, the brief plate tectonics history given in the preceding paragraphs was put together only after decades of studying the rocks and the faults in the Coast Ranges.

Even subtle changes in the movement of the plates are reflected in the geology of the Coast Ranges. During the past 28 million years, the Pacific Plate has shifted its course slightly, and at times its rate of movement has also changed. One of these shifts produced the mountains of the Coast Ranges about 3 to 4 million years ago. At that time, the Pacific Plate began to move obliquely past the North American Plate, rather than sliding past it. The change caused some convergence between the plates, and compression squeezed up the mountains. Like the Sierra Nevada, the Coast Range mountains are geologically young; in fact, they are still being uplifted today. During the 1989 Loma Prieta earthquake, for example, the Santa Cruz Mountains west of the San Andreas fault grew approximately 1.2 meters. As we will see in the following pages, the geology of the Coast Ranges reveals much about the changing plate boundary along western North America.

The Franciscan Assemblage: A Geologic Puzzle

Geologic maps of California show that the Franciscan Assemblage of Mesozoic age forms the heart of the Coast Range Province. This unit forms two northwest-trending belts, one extending almost the length of the province along the eastern edge of the

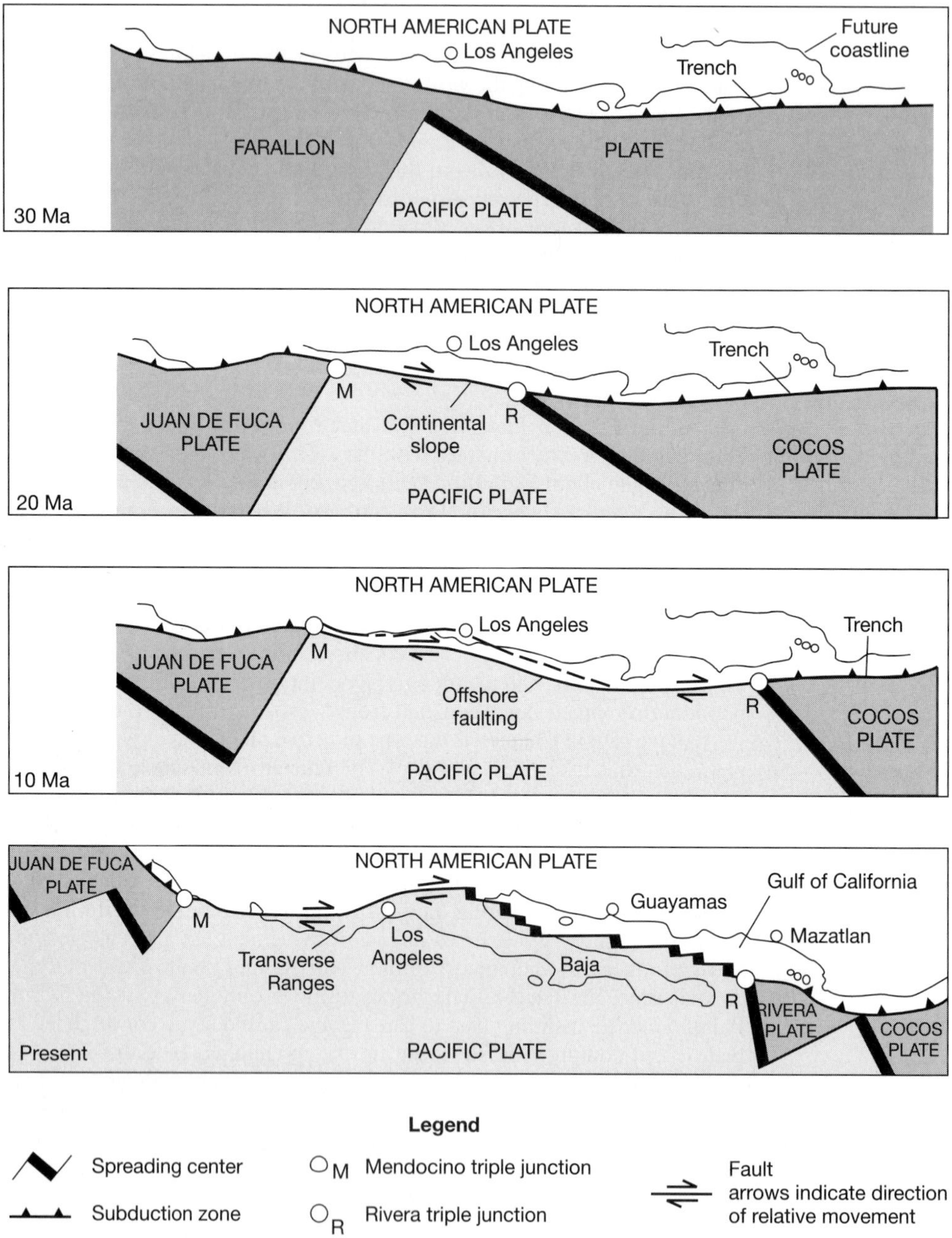

Legend

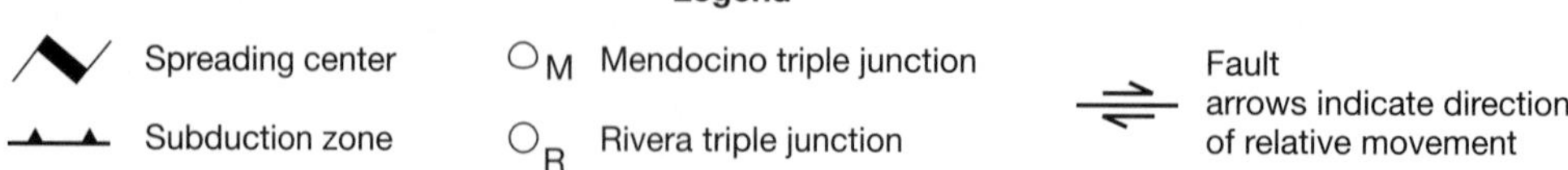

Fig. 12-1 A, Plate tectonics setting of California at about 28 million years before present. The Farallon Plate was being subducted beneath North America along the length of California. As the Pacific Plate came into direct contact with North America, the San Andreas transform boundary was created. **B-D,** As more of the Pacific Plate came into contact with North America, the San Andreas transform lengthened. (Source: Irwin, W.P. 1990. Geology and plate-tectonic development. In Wallace, R., ed. U.S. Geological Survey, Professional Paper 1515.)

HOW OLD IS A MOUNTAIN? THE STORY OF MT. DIABLO

Mt. Diablo is a 1173-meter-high peak in Contra Costa County where a variety of rocks can be found (Fig. 12-2). At the center of the mountain is Mesozoic serpentine, flanked by a variety of younger Mesozoic and Cenozoic sedimentary rocks. Because the landscape cuts across all of these rocks, the mountain itself must be younger than the rocks. Once again, the principles of superposition and cross-cutting discussed in Chapter 3 can be applied.

But how much younger than the rocks is the mountain? South of Mt. Diablo is a bed of volcanic rock that represents an ancient ash flow, similar to those forming on the flanks of Cascade volcanoes today. By analyzing its chemistry (Chapter 5), geologists have traced it to similar rocks erupted from the Sonoma Volcanic Field to the north (Fig. 12-3). The ash flow, called the Lawlor Tuff, is about 5 million years old. Applying the principles of gravity, it is clear that the ash could not have flowed over Mt. Diablo, nor could it have flowed across the Carquinez Straits of San Francisco Bay. The logical conclusion is that neither obstacle was there when the volcano erupted. The uplift of Mt. Diablo and the folding and faulting of the Lawlor Tuff and the other rocks must have occurred less than 5 million years ago. On the sides of Mt. Diablo, the alluvial layers containing the Lawlor Tuff are tilted away from the mountain. The layers were flat when deposited, so their tilting must have resulted from the growth of Mt. Diablo, again confirming that the mountain rose after these young layers were deposited.

Fig. 12-2 Aerial view of Mt. Diablo from north of the Carquinez Straits, looking south. (Source: Sarna-Wojcicki, A. U.S. Geological Survey.)

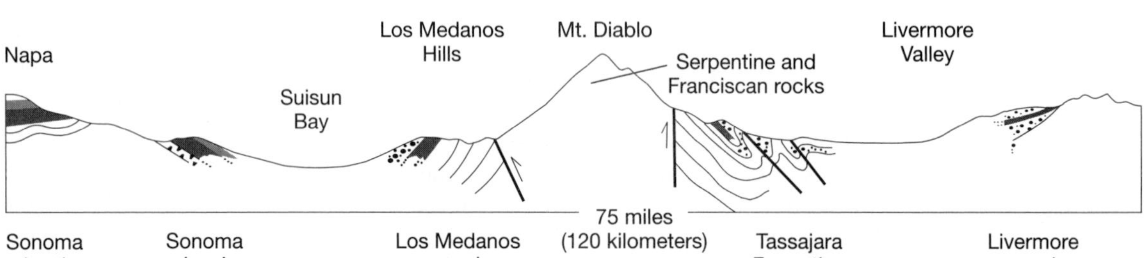

Fig. 12-3 Cross section from north to south across Mt. Diablo, showing the 4-million-year-old Lawlor Tuff (indicated here as the black layer) as it occurs beneath the modern landscape. When the ash flow erupted, it must have flowed downhill from its source in the Sonoma volcanic field. At that time, neither Suisun Bay nor Mt. Diablo was present. (Source: Modified from Sarna-Wojcicki, A. 1972. *Friends of the Pleistocene Field Trip Guidebook.*)

San Andreas fault. The second belt of Franciscan rocks is found along the western edge of the Province and appears on land only in the southern Santa Lucia Range (Fig. 12-4; see also Endpaper 1).

The bewilderment of early geologists who encountered the rocks of the California Coast Ranges was probably not much different than that of a student who completes a general course in geology and then attempts to explain the rocks seen in a typical road cut through the Coast Ranges. By the early 1900s, the "textbook" principles of geology (outlined in Chapters 2 and 3) had developed to explain the origins of many different rock types. However, the rules of classical geology were not sufficient for geologists to explain the origin of the rocks they saw in the Coast Ranges. Without the theory of plate tectonics, these rocks are a baffling mixture of different rocks types, jumbled together with enormous complexity. The solution of one geologist, Andrew Lawson, who first mapped these rocks in 1895 and prepared a geologic map of San Francisco in 1914, was to try to make unity out of seeming chaos and name the rocks the Franciscan Formation.

Until the early 1970s, however, the origin of Franciscan rocks remained a mystery. It is interesting to view the Franciscan problem by examining the different rock types and reviewing their origin. From a perspective enlightened by our knowledge of plate tectonics, it now is clear that the Franciscan rocks represent pieces of oceanic crust that have been accreted to North America by subduction and collision. Although plate tectonics explains much about the Franciscan, geologists working today in the Coast Ranges are still developing new ideas.

Most abundant in the Franciscan Assemblage are sandstone and shale (or mudstone). In addition to quartz and feldspar, the sandstone contains fragments of shale, chert, and other rocks, giving it a gray, brown, or "dirty" color. The color gives rise to its name, ***greywacke sandstone.*** The shale or mudstone is also typically dark gray and is commonly interlayered with sandstone beds (see the box on p. 258). Geologists now know that the sequences of sandstone and shale accumulated as marine sediments in relatively deep water by a process that was completely unknown during Lawson's time.

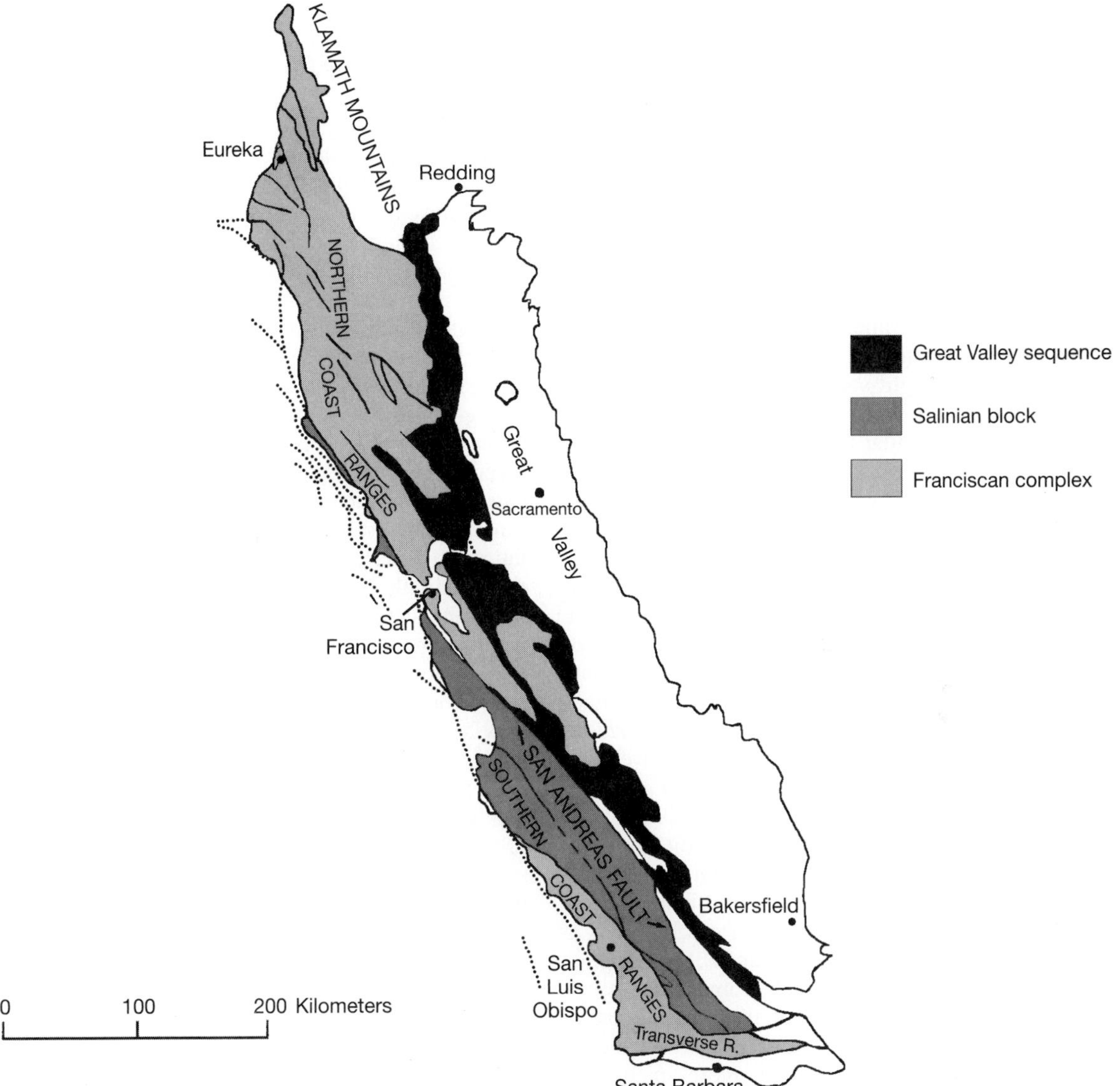

Fig. 12-4 The distribution of Franciscan rocks, the Great Valley sequence, and rocks of the Salinian block. (Source: Irwin, W.P. 1990. Geology and plate-tectonic development. In Wallace, R., ed. U.S. Geological Survey, Professional Paper 1515.)

MONTEREY CANYON AND FAN: A MODERN ANALOG FOR FRANCISCAN SANDSTONE AND SHALE

Just a kilometer off the coast near Monterey is the head of Monterey Canyon, a winding gorge that reaches depths of 3650 meters. The canyon, deeper in places than Arizona's Grand Canyon, is well known to divers, fishermen, and oceanographers as a haven for diverse and abundant marine life. Monterey Canyon extends from the continental shelf onto the floor of the deep ocean, and during the geologically recent past it has carried enough sediment to form the giant Monterey submarine fan at its lower end. The fan is a submarine delta similar in shape and origin to a large river delta (Fig. 12-5).

The processes that carry sediment from the beach and shelf into the canyon and out onto the deep ocean floor differ from sediment transport in a river. Sediments are carried to the canyon head from nearby rivers by nearshore ocean currents flowing on the shallow continental shelf. The sediments accumulate on the gentle slopes near the outer edge of the continental shelf until they become unstable. Then a mass of water-laden sediment breaks loose in a ***turbidity current,*** a dense, fast-moving mixture that travels into deeper water like a submarine landslide (Fig. 12-6). Turbidity currents can reach velocities of more than 90 kilometers per hour and can carry sediment as far as 1000 kilometers down the continental slope. Along the shelf near Monterey Canyon, turbidity currents may be triggered by earthquakes on nearby faults as well as by the gradual buildup of unstable sediment.

As a turbidity current travels from the submarine canyon onto the flat surface of the ocean floor, its velocity decreases, and particles settle out onto the sea floor. Heavier large particles settle more quickly and form the base of the turbidity current's deposit. The finest particles settle last, perhaps days after the first particles. The deposit that results from a single turbidity current is thus coarse-grained at the base, gradually becoming finer grained toward the top. This type of bedding, ***graded bedding,*** is diagnostic of turbidity currents and thus identifies a deposit as one formed in an offshore marine environment. Each subsequent turbidity current produces a similar deposit with gradually smaller particles toward the top, known as a ***fining-upward sequence*** (Fig. 12-7). Turbidity currents may rip up portions of the underlying mud layers, incorporating chunks of mud known as ***rip-up clasts.*** As the submarine channel shifts laterally over time, repeated turbidity currents build up a submarine fan with a shape similar to that of a delta seen at a river's mouth. The Monterey submarine fan contains more than 100 times the volume of sediment that would be required to fill Monterey Canyon, indicating that it has taken as long as a few million years for the fan to form and that the sediment was derived from a very large drainage area.

Over geologic time, a rhythmic sequence of alternating sand and mud, representing the deposits of hundreds or thousands of turbidity currents, may be lithified and preserved as sandstone and shale. These sequences, which are common to many formations in the Coast Range Province, are known as ***turbidites*** (Fig. 12-8). Their diagnostic characteristics, such as graded bedding and rip-up clasts, have been preserved to tell geologists that turbidity currents were active in the geologic past. The greywacke sandstone and shale today seen as part of the Franciscan Assemblage were deposited in environments similar to those found today in the Monterey submarine canyon and fan.

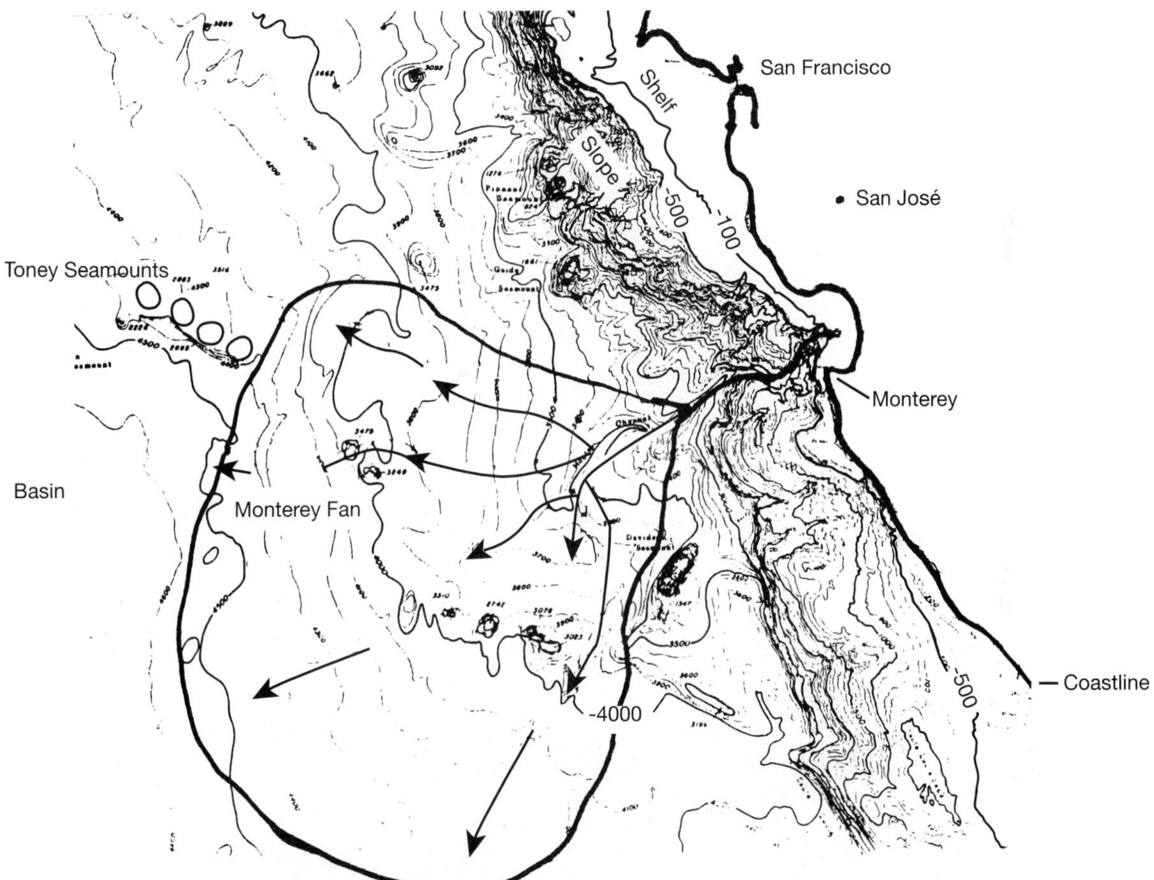

Fig. 12-5 A bathymetric map showing the outline of the Monterey submarine canyon and fan. The depths on contour lines are given in feet below sea level.

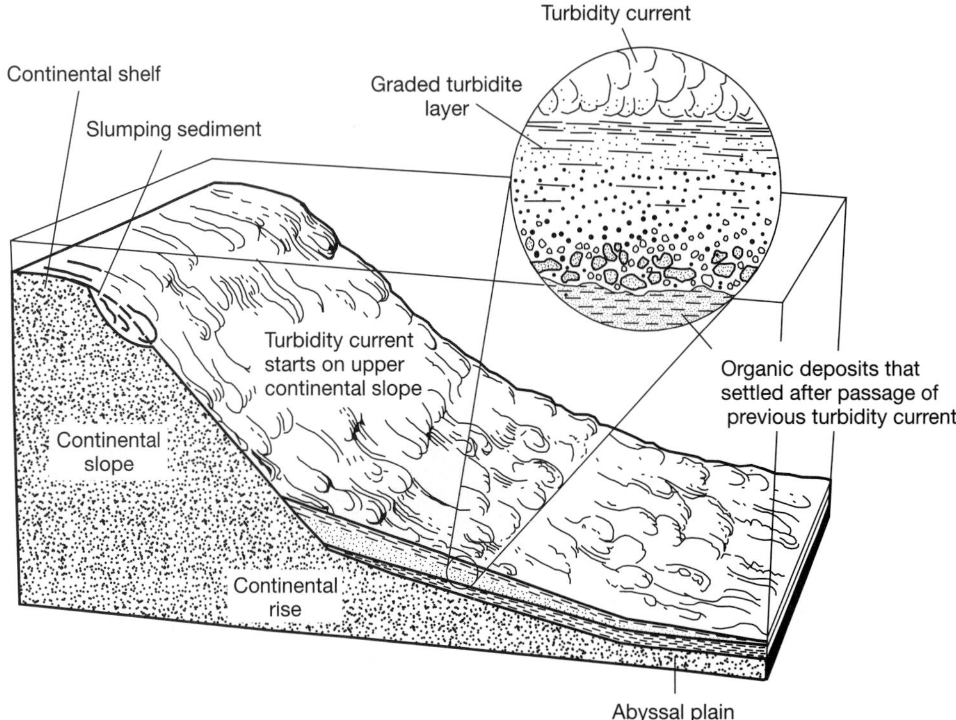

Fig. 12-6 Diagram showing turbidity currents at the edge of the continental shelf. (Source: Coch, N., and Ludman, A. 1991. *Physical Geology.* New York: Macmillan Publishing Co., Inc.)

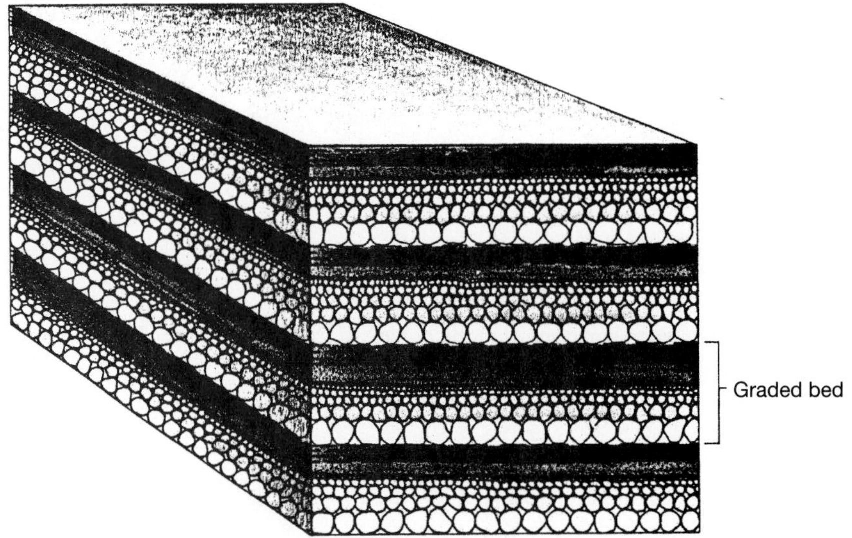

Fig. 12-7 Formation of graded beds by repeated turbidity currents. (Source; Coch, N., and Ludman, A. 1991. *Physical Geology.* New York: Macmillan.)

Fig. 12-8 Outcrop of nearly vertical turbidites, showing the characteristic rhythmic alternation of sandstone and shale beds. Cretaceous Pigeon Point Formation, San Mateo County coast. (Photo by author.)

A third type of sedimentary rock found in the Franciscan Assemblage is ***chert,*** a dense, fine-grained rock composed almost entirely of silica (SiO_2). Because chert is a very hard rock, it resists erosion and in places like the Marin Headlands on the northern side of the Golden Gate (Fig. 12-9 and Color Plate 19), chert holds up higher ridges and hills in the Franciscan landscape. Chert originates as siliceous ***ooze,*** a very fine-grained rain of radiolarian skeletons onto the deepest ocean floors. Although it makes up only about 1 percent of all Franciscan rocks, the chert has provided some of the most important clues about the origin of the Franciscan. The clues lie in the remains of radiolaria whose skeletons make up the chert (Fig. 12-10; see also Fig. 3-9). The types of radiolaria present in a chert sample provide evidence about the water depth, temperature, and chemistry of the ancient Mesozoic ocean that was their home. Two important and surprising conclusions have resulted from analyzing Franciscan cherts during the past two decades: (1) the Franciscan Assemblage consists of a number of separate blocks of oceanic plate; and (2) some of the blocks may have originally formed in oceanic water near the equator.

A last sedimentary rock type found in the Franciscan in very minor amounts is ***limestone,*** which occurs as isolated outcrops. It is most abundant on the San Francisco Peninsula, where it occurs together with volcanic rocks indicative of an ancient submarine volcanic plateau. Limestone is also found near Laytonville in Mendocino County, where it appears to represent a deep-sea ooze composed of foraminifera. Limestone in the Franciscan has been of great commercial interest because it is a relatively rare rock in much of California. Limestone is necessary for making cement and concrete, and many of the limestone blocks within the Franciscan are rock quarries. A good illustration of the value and scarcity of these

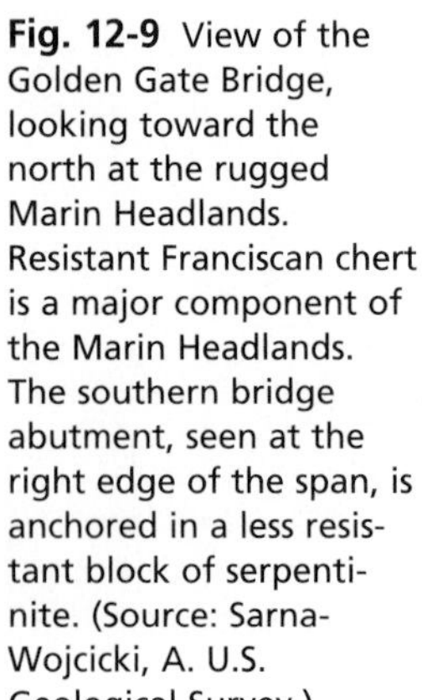

Fig. 12-9 View of the Golden Gate Bridge, looking toward the north at the rugged Marin Headlands. Resistant Franciscan chert is a major component of the Marin Headlands. The southern bridge abutment, seen at the right edge of the span, is anchored in a less resistant block of serpentinite. (Source: Sarna-Wojcicki, A. U.S. Geological Survey.)

Fig. 12-10 Resistant chert beds at Baker Beach, Marin Headlands. (Source: Sarna-Wojcicki, A. U.S. Geological Survey.)

Fig. 12-11 Pillow lavas in Franciscan greenstone, Marin County near Nicasio Reservoir. A rock hammer provides scale. (Photo by author.)

limestone blocks is the fact the Permanente Quarry, located on the San Francisco Peninsula, was first opened to supply the concrete to build Shasta Dam, 500 kilometers to the north. Like the chert, the limestone appears to have originally formed at latitudes far to south of present-day California.

Volcanic rocks make up about 10 percent of the total area where Franciscan rocks occur. Most of the volcanic rocks are basalt that has been slightly metamorphosed by interactions with seawater. These changes produced green metamorphic minerals, resulting in a characteristic Franciscan volcanic rock logically named ***greenstone.*** At some localities, the greenstone shows well-developed rounded forms called ***pillows*** (Fig. 12-11) (see Chapter 2). The origin of these features would have been a mystery to early geologists, but in recent years, geologists have explored active submarine volcanoes, like the mid-Atlantic ridge and the Hawaiian islands, using submersible vessels. They have observed pillow lavas being formed as hot lava enters the ocean and is rolled into blobs. Studies of the chemistry of modern lavas and comparison with Franciscan greenstones confirm that they formed in various submarine volcanic environments, including ancient midocean ridges, sea mounts, and oceanic islands. Like chert, greenstone is a relatively hard rock, and both have been extensively quarried for road base.

Metamorphic rocks of the Franciscan Assemblage are perhaps the most unusual member of the mixture. These are most abundant in the northeastern Coast Range, but can also be found on the Tiburon Peninsula in Marin County, near Cazadero in Sonoma County, and in the eastern Diablo Range. These rocks contain rare silicate minerals, including the blue mineral ***gloncophane,*** which gives a blue color to the schist in which it occurs. The rocks are known as ***blueschist,*** and early geologists were very puzzled by their origin. The rare metamorphic mineral ***jadeite*** is also found in Franciscan metamorphic rocks, and small amounts of gem-quality jadeite have been found on central California beaches.

During the 1970s, geologists conducted laboratory experiments to determine the pressure and temperature conditions that would produce the blueschist minerals. Rocks were heated and cooled under different pressures and temperatures to determine which groups of silicate minerals were stable under different conditions. These experiments indicated that conditions of high pressure but relatively low temperature were required for blueschist formation. Plate tectonics has given us an ideal environment for these unusual conditions: a subduction zone where an oceanic plate is being quickly forced underneath another plate.

A final and important component of the Franciscan mixture is serpentinite, a soft green rock composed mainly of the serpentine group of minerals. As discussed in Chapter 9, serpentinite originates in the Earth's mantle as a more brittle, ultramafic rock known as peridotite. Alteration of peridotite resulted in the formation of plastic, lower-density serpentinite. The weak serpentinite of the Coast Ranges is notorious for creating landslide problems, and in outcrops it commonly shows an extreme degree of shearing and flow (Color Plate 4). Serpentinite is also noted for the rare plant species that grow only on serpentinite because of its unusual chemistry. ***Eclogite*** is another rare metamorphic rock found associated with Franciscan rocks. Eclogite contains garnet and probably originates in the mantle.

If you review the various types of rocks described above, you can well imagine the confusion of early geologists when they attempted to explain how all of these diverse rocks came to be found together. Once the plate tectonics theory came to light, however, the task became less difficult. We now know that submarine volcanic rocks, deep-sea chert, marine limestone, and sediments deposited by turbidity currents could all be expected to form on the dynamic surface of an oceanic plate. By observation of modern subduction zones, where oceanic crust can be driven beneath a continent, geologists came to realize that the Franciscan rocks could represent fragments of such a collision. Pieces of the subducting oceanic crust, together with metamorphic rocks formed in the subduction zone and fragments from the upper mantle, could be scraped off against the margin of the continent and not be completely consumed in the subduction zone. As early as the 1970s, geologists recognized that the Franciscan Assemblage represented the remains of an ancient oceanic plate, which they named the Farallon Plate, accreted to North American during Mesozoic time just as pieces of the Sierra Nevada Province had been accreted earlier (Fig. 12-12).

Continuing detailed studies of the Franciscan have created an even more complex picture of its formation. It appears that Franciscan rocks represent not one, but several, accreted terranes. These separate Franciscan blocks represent the remnants of several oceanic plates that met their end against the edge of North America. So far, at least nine different blocks or terranes have been identified (Fig. 12-13). Studies of radiolaria contained in the chert, paleomagnetic analysis of the limestone, and detailed studies of the mineral content of the greywacke sandstone are some of the tools that geologists are using to complete this picture.

The area where Franciscan rocks are seen can be broadly divided into three belts (Fig. 12-14). In the Coastal Belt, a northwest-trending zone between Eureka and the Russian River, Franciscan rocks are mainly greywacke and shale

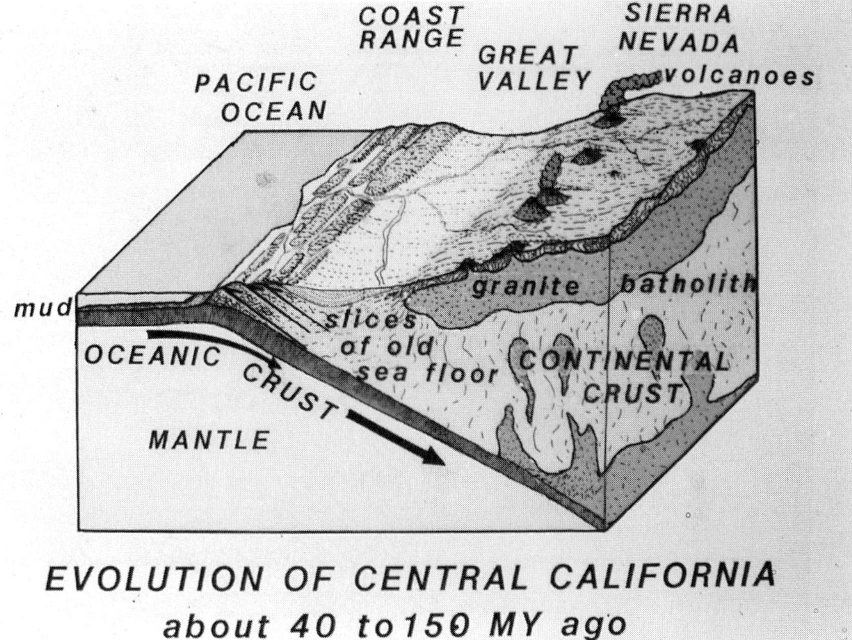

Fig. 12-12 A simplified cross section showing the subduction and accretion that formed the Franciscan complex in the Coast Ranges during Mesozoic time. (Source: Blake, M.C. U.S. Geological Survey.)

sequences that are not as disrupted as other parts of the Franciscan. The Central Belt, including the San Francisco Bay Area, is found east of the Coastal Belt, and this part of the Franciscan Assemblage is a jumble of rocks. Isolated small blocks of exotic greenstone, blueschist, eclogite, chert, or greywacke "float" in a matrix of highly sheared mudstone. Geologists term this mixture ***melange,*** adopting a cooking term used to describe a soup that contains floating chunks of goodies. In the Franciscan melange, the "soup base" or ***matrix*** is sheared mudstone, and the floating chunks are the more resistant blocks (Fig. 12-15). Geologists now realize that different terranes are present even within the melange (see the box on p. 268). For example, at the northern end of the Golden Gate Bridge, sequences of greenstone, chert, and sandstone represent deposition on a Mesozoic sea floor over a time span of 100 million years.

The Eastern Belt of the Franciscan extends east to the Coast Range fault, the major fault boundary with rocks of the Great Valley sequence (see Chapter 11). Here the rocks are typically blueschist and related metamorphic rocks. Faulted slices of these rocks are also found in the San Francisco Bay area on the Tiburon Peninsula and west of Los Banos. The Eastern Belt and the melange matrix of the central Belt are older than the blocks within the melange, and the Coastal Belt contains the youngest Franciscan rocks, with the entire Franciscan spanning a period of about 150 million years of accretion from middle Mesozoic (late Jurassic) to early Tertiary (mainly Eocene) time.

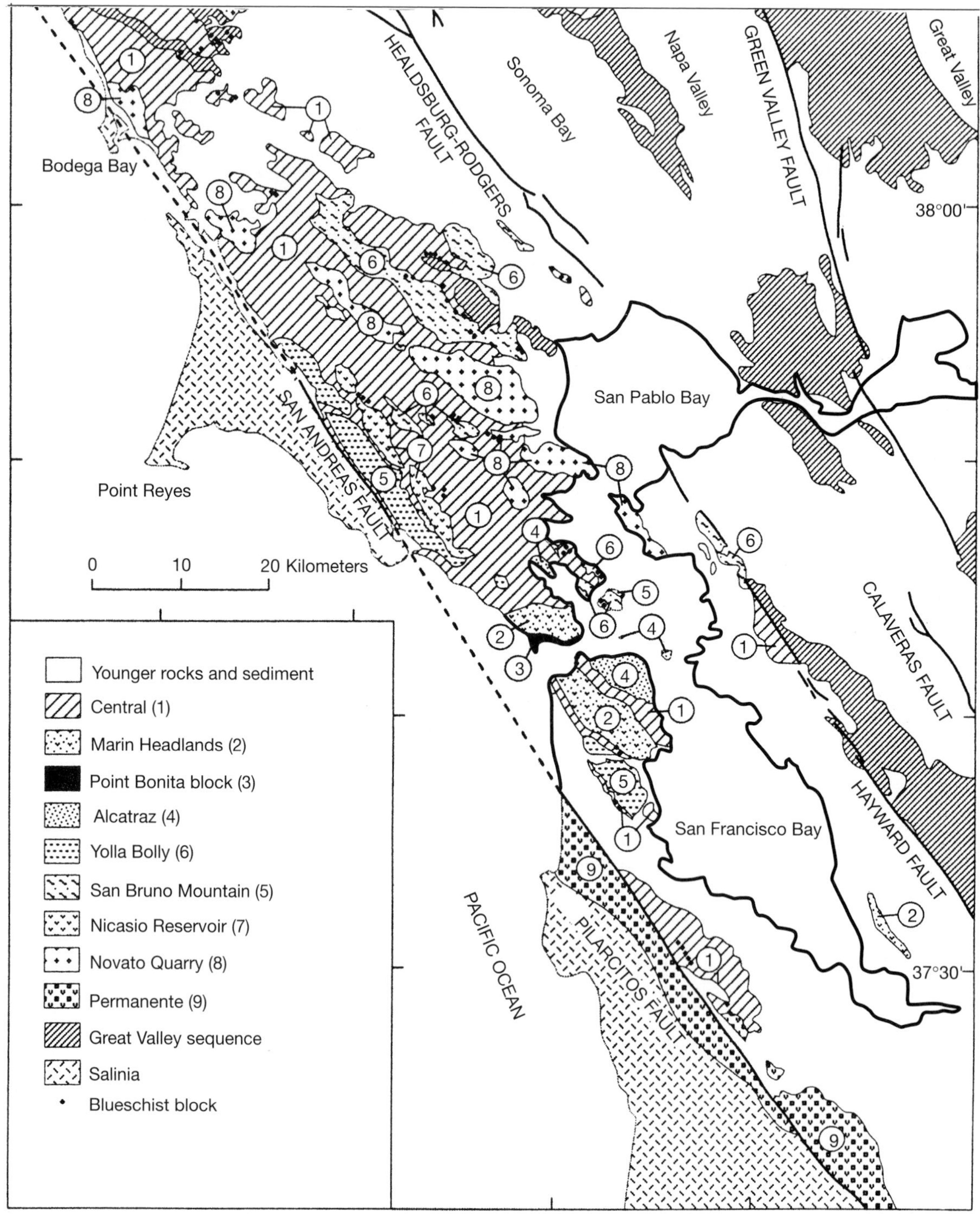

Fig. 12-13 Franciscan terranes in the San Francisco Bay area. (Source: Wahrhaftig, C., and Sloan, D. 1989. *Geology of San Francisco and Vicinity.* American Geophysical Union, Field Trip Guidebook T105.)

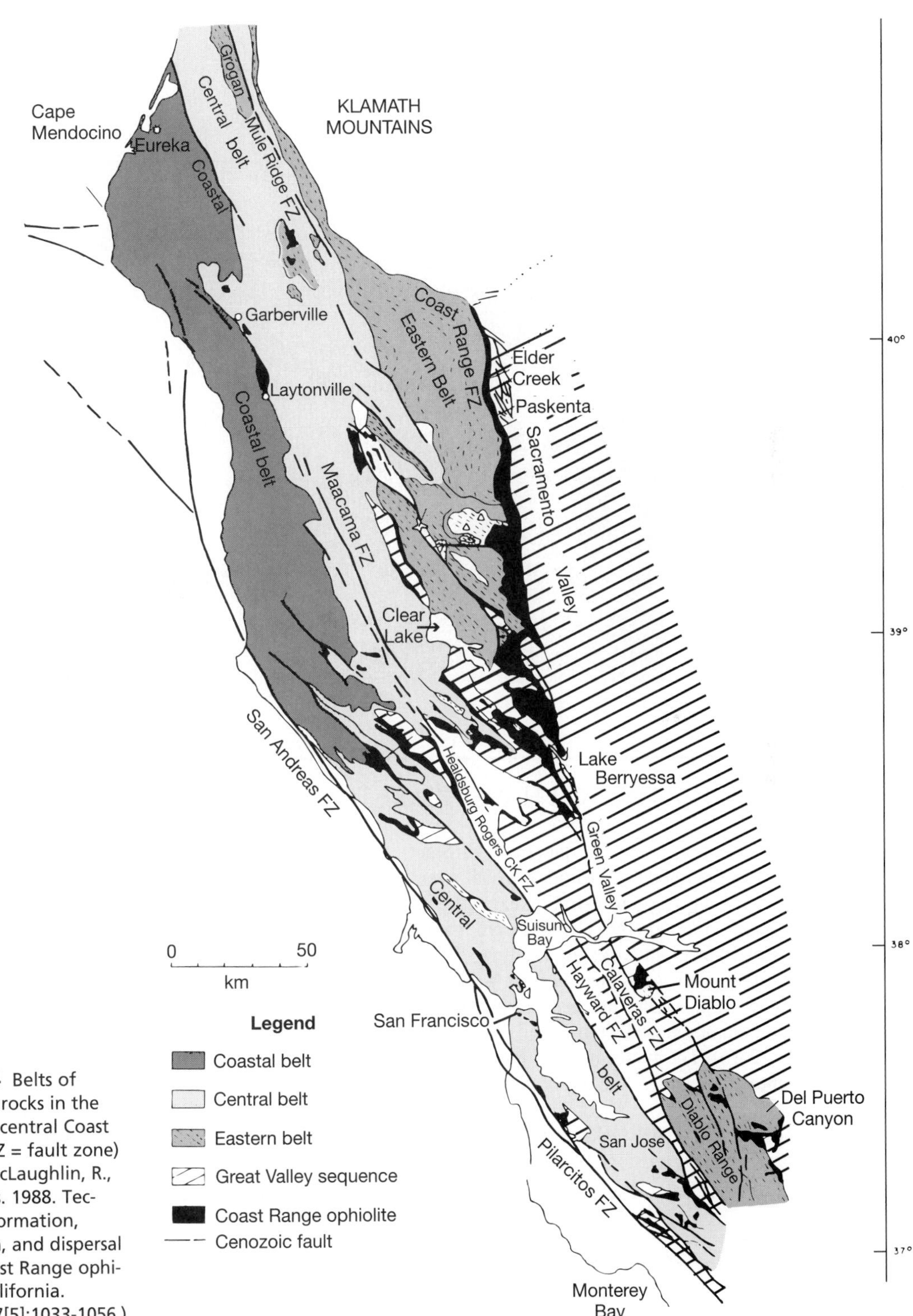

Fig. 12-14 Belts of Franciscan rocks in the north and central Coast Ranges. (FZ = fault zone) (Source: McLaughlin, R., and others. 1988. Tectonics of formation, translation, and dispersal of the Coast Range ophiolite of California. *Tectonics* 7[5]:1033-1056.)

Fig. 12-15 Typical Franciscan melange, Van Duzen River, Humboldt County. (Source: Blake, M.C. U.S. Geological Survey.)

WHY DOES SAN FRANCISCO HAVE HILLS? THE FRANCISCAN LANDSCAPE

The steep hills that pop up from the flat streets of San Francisco remain in the minds of visitors from all over the world. The city's landscape is a striking example of how geology shapes topography. The hills are more resistant blocks of Franciscan rock; for example, Telegraph Hill is resistant greywacke, Twin Peaks a sequence of radiolarian chert and basalt, and in the Bay, Angel Island contains metamorphosed Franciscan greywacke and chert (Fig. 12-16). The flatter parts of the city and some areas submerged under San Francisco Bay are underlain by weaker Franciscan rocks, typically sheared shale of the melange unit. The weaker Franciscan rocks have been worn down by erosion. Younger sediments, such as dune sand, alluvium, and Bay mud, then covered these lower parts of the landscape.

Throughout the Coast Ranges where Franciscan rocks occur, particularly the melange, the mountains have a characteristic appearance. Where the rocks are weak, large, deep landslides are common and the vegetation is dominated by grasslands; the landslides give the topography a lumpy, bumpy look. Resistant blocks commonly appear as isolated rocky knobs along ridges, or even poke up from the sides of hills (see Color Plate 20). Geologists familiar with the Franciscan topography commonly refer to it as a "melted ice cream" landscape (Fig. 12-17).

Fig. 12-16 Angel Island, San Francisco Bay, with the city of Tiburon in the foreground and the Berkeley Hills on the far horizon. (Photo by author.)

Fig. 12-17 The Geysers steam field, Sonoma County, illustrating a typical Coast Range landscape.

THE SALINIAN BLOCK

During the early years of work in the San Francisco area, geologists also recognized that Franciscan rocks are completely absent in an entire section of the Coast Ranges. This northwest-trending belt is bordered by the San Andreas fault on the east and the Sur-Nacimiento fault on the west. Geologists have long referred to it as the Salinian Block, named for the city of Salinas. Much of the northern Salinian Block lies offshore, but parts of it can be seen on the Point Reyes Peninsula and at Point Arena (see Fig. 12-4 and Endpaper 1). The older rocks in the Salinian Block are fundamentally different from the Franciscan Assemblage, except in one way: most of the older rocks in both blocks are of Mesozoic age.

The major component of the Salinian Block is granitic plutonic rocks (Fig. 12-18). These crystallized between about 110 and 78 million years ago, during late Mesozoic (Cretaceous) time. The mineral composition, chemistry, and age of the granitic rocks in the Salinian Block are very similar to those of the central Sierran Batholith. Like the Sierran Batholith, the Salinian Block also contains remnants of older rocks intruded by the magmas (see Chapter 8). Some of these are roof pendants within the granitic rocks. These are seen today as schist, gneiss, marble, and other high-grade metamorphic rocks formed from a variety of sedimentary rocks. They are abundant in the Gabilan and Santa Lucia Ranges near Big Sur. The metamorphic rocks are difficult to date, but geologists estimate that the original sediments were deposited in Paleozoic time. However, in the southeastern part of the Salinian Block, near Mt. Pinos, metamorphic rocks are of Precambrian age.

The similarities between rocks of the Salinian Block and the Sierra Nevada have convinced geologists that the granitic rocks of the Salinian Block are indeed pieces of the Sierran Batholith that have been separated and transported north by the San Andreas fault system (see Chapter 14). The granitic and metamorphic rocks represent part of the Mesozoic batholith intruded along the entire length of the North American continent as oceanic crust was being subducted to the west. The Franciscan rocks, pieces of Mesozoic oceanic crust, and the Salinian Block, pieces of continental crust, were brought together in the Coast Ranges by the San Andreas fault system long after they formed.

Fig. 12-18 Granitic rocks of the Salinian block form rounded outcrops along the northern Point Reyes peninsula, Marin County. (Photo by author.)

UNDER THE COAST RANGES: BASEMENT GEOLOGY

The older dense, crystalline rocks in any area are commonly referred to as the ***basement rocks*** because they form the foundation for the geologic package of rocks and sediment. In the Coast Ranges, two very different ***basement complexes*** have been brought together by faulting after they formed. The Franciscan basement consists largely of accreted ocean-floor and subduction zone rocks. The vastly different granitic-metamorphic basement of the Salinian block represents continental crust intruded by granitic plutons. It is important to remember that all of the basement rock in the Coast Ranges is relatively young, predominantly Jurassic and Cretaceous in age. Geologists and geophysicists are currently debating the relationships between the basement complexes of the Coast Ranges and the basement rocks underlying the Great Valley and the Sierra Nevada (see Chapter 11).

QUICKSILVER: MERCURY MINING IN THE COAST RANGES

In 1845, Captain Andres Castillero, a Mexican Army officer who had studied at the College of Mines in Mexico City, picked up some curious red rock from the yard of Mission Santa Clara about 5 kilometers northeast of the present city of San Jose. After questioning the mission's padre, Castillero found that the Ohlone Indians of the area used the red rock to paint both their bodies and the mission buildings. The people called the red dirt, which we know as mercury ore, *mohetka.* Although other explorers had discovered the curious red mineral as early as 1824, they had not been able to identify it. Urged on by the Mexican government's offer of the equivalent of $100,000 for discovery of valuable minerals, Castillero found some mission Ohlones to guide him to the source of the red rocks, about 25 kilometers southwest of the mission. After performing tests on the local rocks, he realized that they contained neither gold nor silver, but he did notice their similarity to rocks he had seen at La Mancha, Spain, the site of the world's largest mercury mines. By heating the red mineral and then condensing the escaping vapor, Castillero extracted small silver globs of native mercury. Pure mercury is liquid at room temperature, and the silver globs were identical to those seen in today's thermometers. Its common name, ***quicksilver,*** describes its appearance very well. Castillero filed a claim to the mining site in 1847, becoming the first person to file for a legal claim in California.

In 1847, the mines came under the direction of Alexander Forbes, who named them Nuevo (New) Almaden after the famous Almaden quicksilver mine in Spain. By 1848, New Almaden was producing 45 to 70 kilograms of mercury daily, which was stored in flasks holding 34 kilograms each. The mercury was used mainly to extract gold and silver from crushed ore. The ore was washed with liquid mercury, which combined with gold and silver. By vaporizing the mercury, miners could then extract the pure gold or silver. Obviously, the gold rush in the Sierra Nevada created a boom in the quicksilver industry. Mercury from the New Almaden mines was taken in wagons to Alviso at the southern end of San Francisco Bay, where it was ferried by boat to San Francisco and on to the gold fields or to Mexico or South America.

At the height of production, the New Almaden mines produced more than $2 million of quicksilver annually, and $15 million worth was mined between 1847 and 1864. In 1865, the New Almaden Mines were home to 2400 people from many countries. However, the demand for quicksilver fell as gold production dropped, and by 1912, the Quicksilver Mining Company had filed for bankruptcy. Brief flurries of demand for mercury to produce bombs during the World Wars resulted in activity at New Almaden, but in 1975 the mines closed. Today some of the historic buildings and a small museum preserve remnants of the New Almaden mines, but much of the hillside area has returned to chaparral (Fig. 12-19).

New Almaden is one of the three largest of the mining districts that produced mercury from the Coast Ranges. New Idria to the southeast was the largest pro-

Continued.

QUICKSILVER: MERCURY MINING IN THE COAST RANGES—cont'd

ducer of quicksilver, and the region around Clear Lake, known as the Mayacamas District, was a third major district. Mercury has been mined from the southernmost Coast Ranges near Santa Barbara north to Lake County. Most of the mercury is found as the red sulfide mineral cinnabar (chemical formula HgS_2; see Color Plate 21), although droplets of pure mercury are found in some samples. The cinnabar is most commonly found in a Franciscan host rock called *silica carbonate rock.* This unusual rock is associated with the serpentinite within the Franciscan and is thought to have formed by alteration of serpentinite by hot springs active along faults. The formation of the veins, masses, and disseminated bodies of cinnabar is thought to have occurred during mid-Tertiary (Miocene) and younger time. Applying the fundamental principles of cross-cutting relationships, the fact that significant mercury deposits occur in Miocene and younger rocks proves that the mineralization was younger. In fact, mercury mined in Lake County is found in Pleistocene sediments. All together, the mercury mines of the Coast Ranges account about 85 percent of the total U.S. mercury production.

Although it has a variety of uses, mercury has long been known to be highly toxic to humans and other organisms. Prolonged exposure to mercury vapors has a cumulative toxic effect on humans, one of the major effects being permanent brain damage. During the 1800s, when tall beaver-skin hats were the height of fashion for men, workers used mercury to tan the beaver hides, and the common occurrence of dementia caused by mercury poisoning gave rise to the expression "mad as a hatter." The levels of mercury present in the sediments of San Francisco Bay and in some of the streams of the Coast Ranges are naturally higher than the drinking water standards. A century of mercury mining increased the mercury levels, and many of the abandoned mines continue to pose a hazard.

The following excerpt from Brewer's *Up and Down California in 1860-1864* describes the recognition of the problem in New Almaden in 1861:

> An old furnace has been taken down, and the soil beneath for 25 feet down (no one knows how much deeper) is so saturated with the metallic quicksilver in the minutest state of division, that they are now digging it up and sluicing the dirt, and much quicksilver is obtained in that way. Thousands of pounds have already been taken out, and they are still at work.

Source: Farquar, F., ed. 1966. *Up and Down California in 1860-1864, the Journal of William H. Brewer.* Berkeley: University of California Press, 583 pp.

A

B

Fig. 12-19 **A,** The New Almaden mercury reduction works in the early 1880s. **B,** Abandoned works at New Almaden in 1991. (Source: **A,** New Almaden Quicksilver County Museum. **B,** Ly, R. San José State University.)

CENOZOIC SEDIMENTARY BASINS

Cenozoic sedimentary rocks can be found covering the basement rocks on both sides of the San Andreas fault. Most of the rocks were deposited in marine basins, and some of the basins remain offshore along the continental shelf today (Fig. 12-20). The formation of these basins, some of which received more than 6000 meters of sediment, was controlled by motion along the transform boundary. Some geologists believe that basins formed by extension during a period when the Pacific plate was moving in a more northwesterly direction relative to North America. As we will see in Chapter 16, the sedimentary rocks in California's late Cenozoic basins are important producers of petroleum.

Some of the older marine basins, which were being filled with sediment during Eocene time about 25 million years ago, were later separated by as much as 315 kilometers of right-lateral movement on the faults of the San Andreas system (see Chapter 14). The marine sedimentary rocks seen today in the central and northern Coast Ranges west of the San Andreas fault were adjacent to the San Joaquin Basin

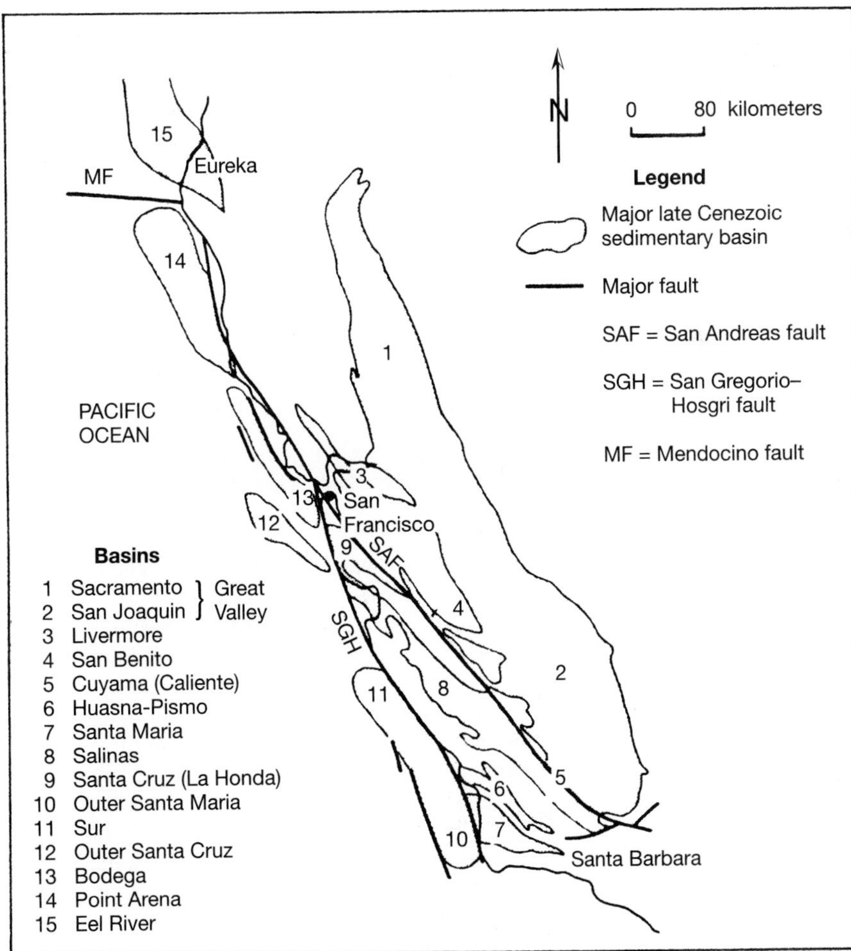

Fig. 12-20 Late Cenozoic sedimentary basins of the California Coast Ranges and adjacent offshore areas. (Source: Christiansen, R., and Yeats, R. 1992. In Geological Society of America. *Decade of North American Geology,* Vol. G-3.)

Fig. 12-21 Vertical chert beds in the Miocene Monterey Formation, Berkeley Hills. (Photo by author.)

Fig. 12-22 Dipping beds of the Rio Dell Formation south of the Eel River, Centerville Beach, Humboldt County. (Photo by author.)

when the sediments were deposited in Eocene time. Between about 22 and 11 million years ago, sediment rich in silica and organic matter, including sediments rich in diatoms, accumulated in deep marine basins. These sediments were subsequently uplifted and are seen today as the Monterey Formation, a distinctive and widespread sedimentary unit discussed in Chapter 16 (Fig. 12-21).

Because of recent faulting, compression, and uplift near the Mendocino Triple Junction, part of the Eel River Basin in northwestern California now lies on land. The sedimentary rocks which were deposited in the basin have been highly folded and faulted. Along the beach south of the mouth of the Eel River, the rocks have been tilted (Fig. 12-22). The tilting of the originally horizontal layers makes a walk along the beach from south to north a tour through the basin's history. One can "see" the Eel River basin fill with sediment as younger sedimentary formations represent increasingly shallow-water deposits (Fig. 12-23). The youngest unit is nonmarine, thus signaling the end of marine conditions in this part of the basin.

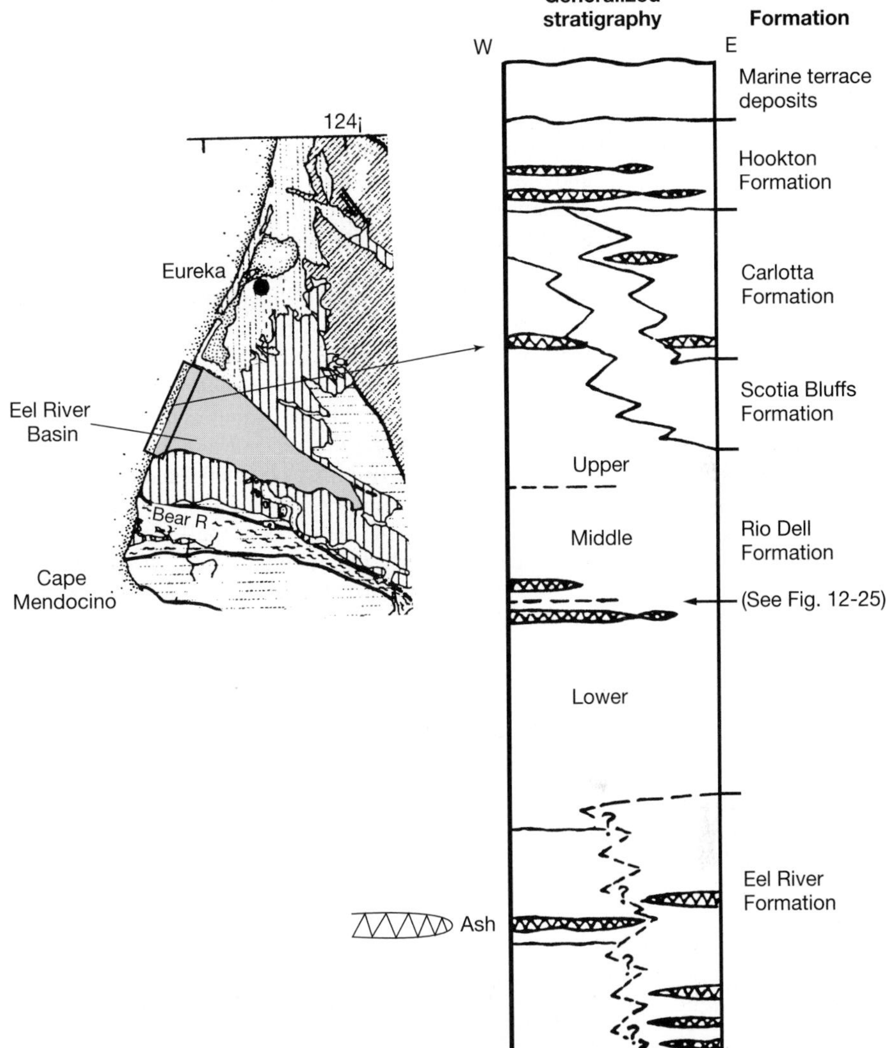

Fig. 12-23 A simplified geologic map and stratigraphic column showing formations of the Eel River basin, Humboldt County. The approximate positions of Fig. 12-22 and Color Plate 7 are shown in the cross section. (Source: McLaughlin, R., and others. 1982. *Friends of the Pleistocene Field Trip Guidebook.*)

Marine sedimentary rocks of Cenozoic age in other parts of the Coast Ranges reveal similar histories to that of the Eel River Basin. In the Berkeley Hills, for example, sedimentary units contain large amounts of debris shed from Sierra Nevada volcanoes. One of these formations is the Neroly sandstone, made up of sand-sized pieces of volcanic rock eroded from Miocene volcanoes that once covered the northern Sierra. In the Santa Cruz Mountains, submarine volcanic rocks, known as the Mindego Formation, are part of the basin-filling sequence. The San Joaquin Basin was a very deep marine basin between 35 and 15 million years ago, and then filled with material washed in from the rising Coast Range and Sierra Nevada. A shallow marine embayment persisted in the San Joaquin until 2.5 to 3 million years ago (Chapter 11).

CENOZOIC VOLCANIC ACTIVITY IN THE COAST RANGES

On September 8, 1769, the expedition commanded by Don Gaspar de Portola camped along a creek near a bay in the southern Coast Ranges. At the entrance to the bay stood a large, rounded hill of rocks, or a ***morro,*** derived from the Spanish word for "animal snout" or "rounded hill," which gives the name to both Morro Rock and the town of Morro Bay (Fig. 12-24). Morro Rock is actually the westernmost in a chain of 14 morros that form a northwest-trending line between San Luis Obispo and Morro Bay. The hills are the remains of plugs, domes, and volcanic necks of a chain of volcanoes; the rocks have been dated between 20 and 27 million years old. Volcanic rocks with similar chemistry and ages are also found in the Santa Cruz Mountains north of Santa Cruz and at Point Arena, where they are known as the Iverson Basalt.

At Pinnacles National Monument at the southern end of the Gabilan Range southeast of Hollister, visitors can see a spectacular array of volcanic towers and caves. The rocks at the Pinnacles are mostly pyroclastic flows and banded flows of rhyolite. Because erosion was concentrated along the original cooling cracks in the lava, the interiors were less eroded and left standing as towers (Fig. 12-25). As discussed in Chapter 14, the rocks at the Pinnacles, which are about 23 million years old, appear to match well with the Neenach volcanics 315 kilometers (195 miles) to the southeast near Lancaster, on the North American side of the San Andreas fault in the Mojave Desert.

Younger Cenozoic volcanic rocks are also found in the Coast Ranges (Fig. 12-26). East of the San Andreas Fault, rocks from a series of volcanic fields get systematically younger from south to north along the Coast Ranges. Southernmost of the group are the 9- to 15-million-year-old Quien Sabe volcanics in the central Diablo Range, followed by the 12 to 8 million year old basalt and rhyolite in the Berkeley Hills.

Volcanic rocks in the Sonoma volcanics are between 8 and 3 million years old, and at the northern end of the series are the Clear Lake volcanics, which are between

Fig. 12-24 Morro Bay and Morro Rock, San Luis Obispo County. (Photo by author.)

Fig. 12-25 Volcanic rocks at Pinnacles National Monument. (Photo by author.)

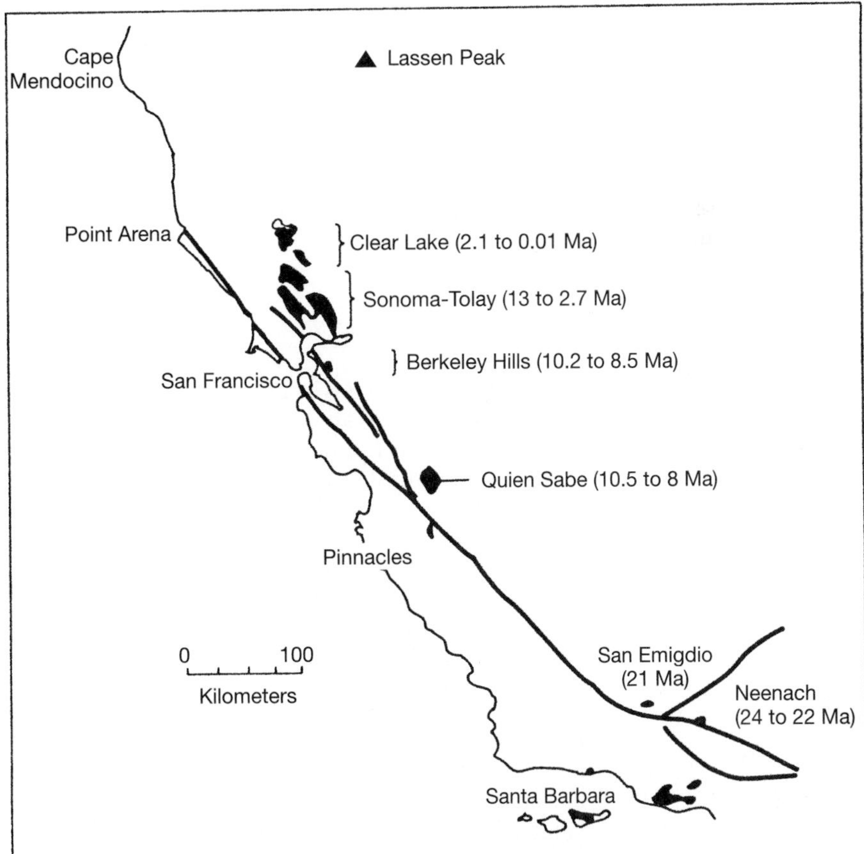

Fig. 12-26 Map showing locations and ages of Cenozoic volcanic rocks in the Coast Ranges. (Source: Wahrhaftig, C. and Sloan, D. 1989. *Geology of San Francisco and Vicinity.* American Geophysical Union, Field Trip Guidebook T105.)

GEYSERS GEOTHERMAL FIELD

Beneath the central part of the Clear Lake volcanic field, a large magma chamber appears to remain somewhat active. The area, which is south of Clear Lake along Geyser Creek, is known as "The Geysers" (see Fig. 12-17). Here magma close to the surface results in the production of large amounts of steam and water heated by the ***geothermal*** properties of magma. The steam is produced when groundwater percolates downward and is heated by magma. The heated groundwater travels toward the surface along faults and fractures in the overlying rock, which is mostly Franciscan greywacke. Numerous hot springs and boiling springs are active in the area, and in places the hot springs have produced hydrothermal mercury deposits within the volcanic field. At The Geysers area, relatively permeable greywacke is capped by impermeable rock that traps the geothermal energy in a steam reservoir located several thousand feet below the surface (see Fig. 4-1). Since 1960, the geothermal energy in The Geysers area has been tapped for electric power. Wells have been drilled into the stream, and the steam is converted by turbines into electricity.

2 million and 10,000 years old. In both the Sonoma and Clear Lake volcanic fields, early eruptions produced basalt, while later volcanism produces andesite and rhyolite. The Clear Lake area was an important source of obsidian for early Native Americans in the region (see Fig. 5-8).

Most geologists agree that the northward movement of volcanic activity over time is related to the northward movement of the Mendocino Triple Junction. However, the volcanic activity is not as clearly related to position of the subduction zone west of the North American Plate as are the presently active Cascade volcanoes. The chemistry of the volcanic rocks is more diverse than the subduction zone volcanoes, and the volcanic activity lags behind the passing of the triple junction. It is possible that a period of extension and volcanism followed the movement of the triple junction past each of these areas.

THE HISTORY OF SAN FRANCISCO BAY

Today San Francisco Bay is the outlet for the entire drainage of the Sacramento and San Joaquin Rivers. About 40 percent of California's runoff passes through the Golden Gate to the Pacific Ocean (Fig. 12-27; see also Chapter 10 and Endpaper 3), and most of this fresh water enters the Bay from the Sacramento–San Joaquin Delta. The southern portion of the Bay, a northwest-trending extension of the Santa Clara Valley, receives fresh water from streams that flow from the Santa Cruz Mountains and the Diablo Range. San Francisco Bay is California's largest ***estuary,*** which is a body of water, partly isolated from the open ocean, where both fresh water and marine water circulate. Cold, saline marine water flows into San Francisco Bay from the Pacific Ocean. Because salt water is denser than fresh water, it sinks beneath the fresh water flowing into the Bay from the Delta. The movement of the marine water is controlled by the tides, so that currents beneath the Golden Gate Bridge reverse twice daily. During major winter storms, large volumes of fresh water flow through the Sacramento–San Joaquin Delta and circulate into the South Bay. The floodwaters are important there because they flush contaminants from an area where circulation is normally poor.

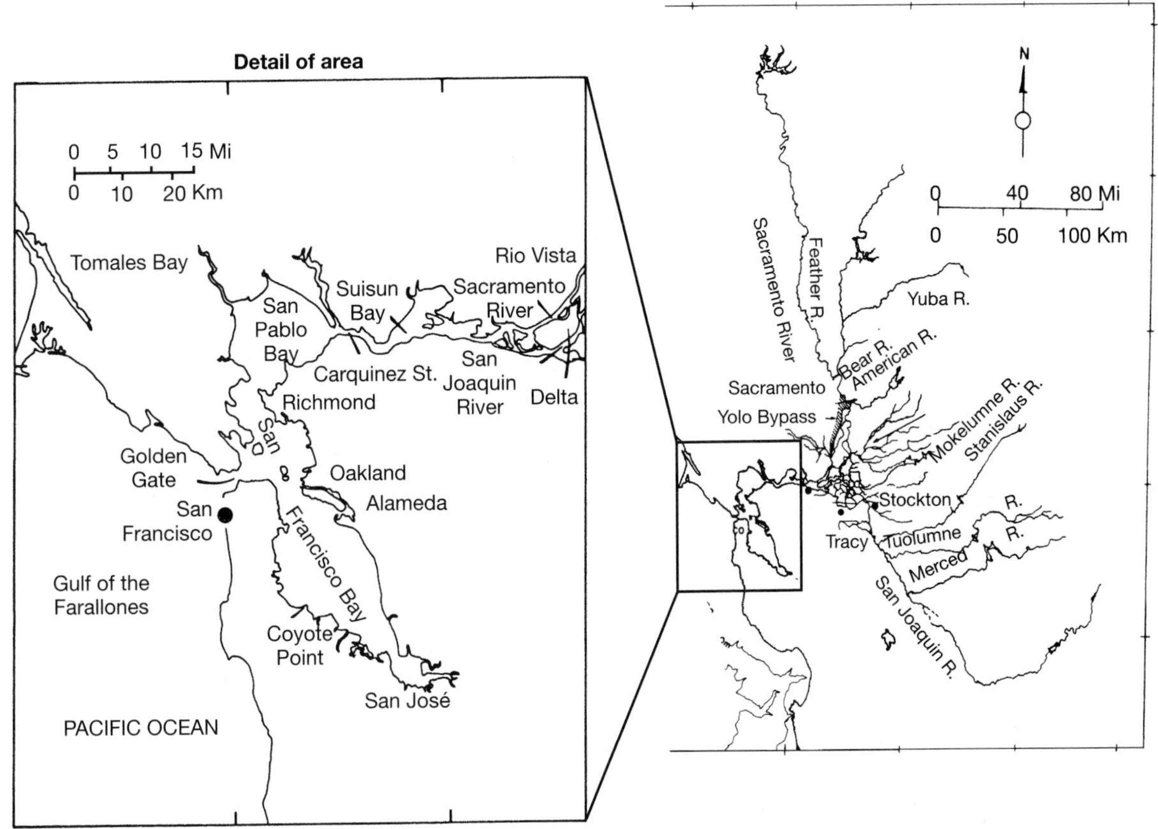

Fig. 12-27 San Francisco Bay (*left, exploded view*) and the tributaries to the Sacramento-San Joaquin Delta (*right*). (Source: Conomos, T.J. 1979. *San Francisco Bay: The Urbanized Estuary.* Pacific Division, American Association for the Advancement of Science.)

Because San Francisco Bay is a such a vital part of northern California's economy and its landscape, it is important to investigate its geologic past. Three forces have combined to shape San Francisco Bay and the surrounding landscape: tectonics, recent sea level changes, and human activities since about 1850.

TECTONICS, RIVER SYSTEMS, AND SAN FRANCISCO BAY

When the present Sierra Nevada was uplifted, approximately 5 million years ago, a system of rivers carried runoff from the highest parts of the range westward to the Great Valley, just as today's rivers do (see Chapter 10 and Endpaper 3). However, the exact configuration of this original river system is not known. Because the faults of the San Andreas system are so active, today's topography in the Coast Ranges is quite different from what the landscape would have looked like a few million years ago. It is likely that the Carquinez Straits and the Golden Gate did not exist at that time. In addition, before the recent tilting in the southern San Joaquin Valley, the rivers there would have flowed further south than at present.

During late Tertiary time, the streams of the Great Valley flowed south to a marine embayment in the San Joaquin Valley. The embayment, the remains of the Great Valley sea (see Chapter 11), was connected to the Pacific Ocean by a ***strait*** or narrow inlet. The strait crossed the San Andreas through a low area of the Coast Ranges, which had created a barrier by the time they were uplifted 3 million years ago. The rivers must have flowed through a low pass or gap in the range.

Geologists searching for this former position of the Sacramento–San Joaquin outlet have found clues in the sedimentary formations along the western margin of the San Joaquin Valley near Coalinga (Fig. 12-28). In the Kettleman Hills, sediments of Pliocene Age—the San Joaquin, Etchegoin, and Tulare Formations—contain minerals and rock fragments derived from sources in the southern Cascades, Sierra Nevada, and the northern Sacramento Valley. For example, in the San Joaquin Formation pumice and fine ash, transported by water, is concentrated in a thick bed. Chemical "fingerprinting" of the ash showed that it was the Nomlaki Tuff, produced by a very large eruption, 3.5 million years ago, from a source near today's Mt. Lassen. The sediments in the Kettleman Hills also show the sedimentary characteristics of delta deposits.

The sediments were deposited in shallow marine water that became shallower over time, so that the Tulare Formation is mostly marshy coastal plain deposits. Remembering that a marine embayment persisted in the San Joaquin Basin until Pliocene time (see Chapter 11), one can envision a river emptying into the embayment and forming a delta (see Fig. 11-17). The age of the deposits in the Kettleman Hills is fixed between about 5 to 2 million years before present by fossils and by volcanic ashes interbedded with the sediments.

One of the most fascinating ideas about this ancient river delta is that it was continuous with the Monterey submarine fan at the time that it was forming. The Monterey canyon and fan are very large features to have formed from the relatively small Salinas River (see the box on p. 258). Geologists have observed that moving the San Andreas Fault "backward" in time at the rate it presently moves, would place the fan and the deposits of the Kettleman Hills close to each other. Although this theory has not yet been proved, recent coring in the younger sediments of the Monterey Fan may reveal that it is indeed composed of Sierra Nevada materials.

Geologists speculate that it was movement on the San Andreas fault that caused the Kettleman Hills outlet to be blocked, forming Lake Clyde (named in honor of the

---

THE CORCORAN CLAY: THE SAN ANDREAS DEFEATS CALIFORNIA'S MIGHTY RIVERS

If the San Joaquin River once flowed to the Pacific in the vicinity of the Kettleman Hills, how and why did the drainage pattern change? The evidence is incomplete, but sediments in cores drilled throughout the San Joaquin and southern Sacramento Valleys provide some clues. A distinctive clay layer, named for the town of Corcoran, documents the presence of an ancient large lake in the San Joaquin and the southern Sacramento valleys. Fossils present in the clay indicate that the lake was not necessarily very deep, but it covered a very large area (see Fig. 12-30). The 758,000–year-old Bishop Ash is found within the older lake deposits, and an ash layer erupted from Yellowstone about 665,000 years ago is found near the top of the clay. At the time the Corcoran Clay was deposited, the Sierran streams must have had no direct outlet to the Pacific Ocean.

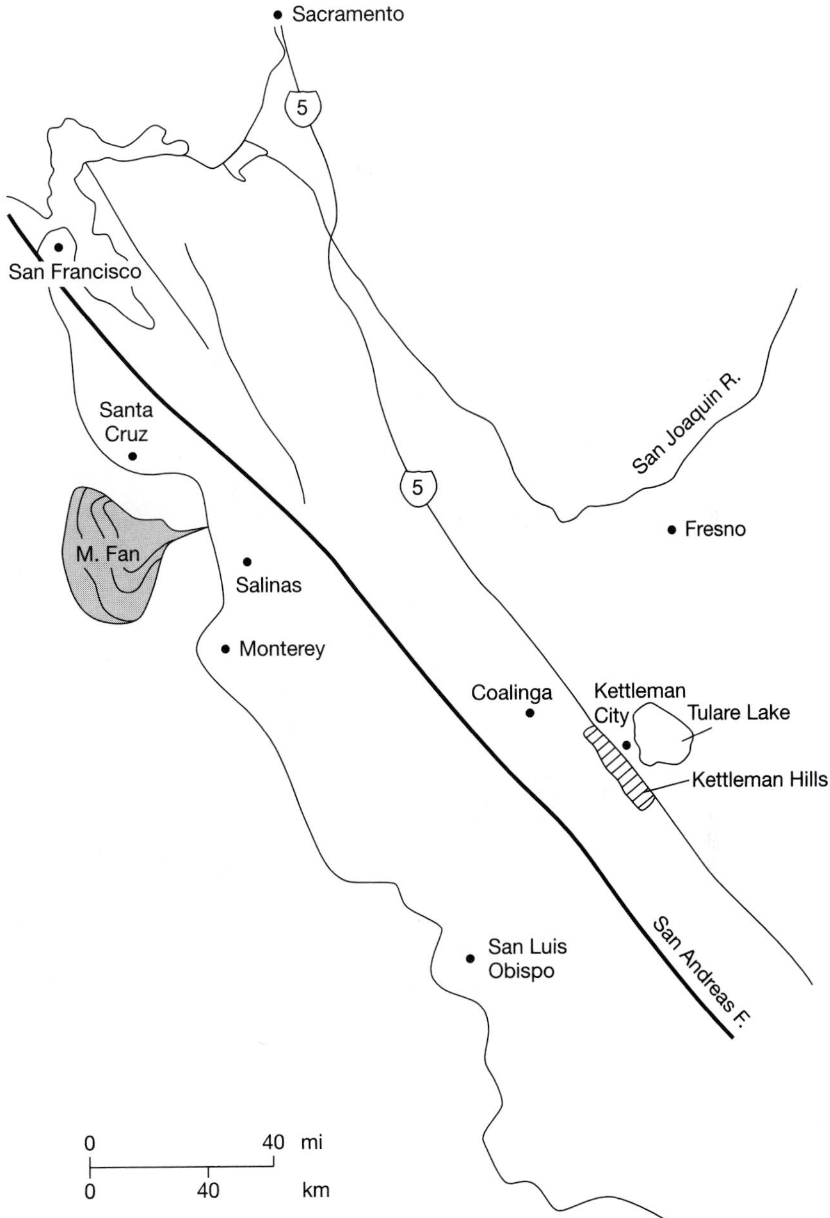

Fig. 12-28 Map showing the Kettleman Hills and Monterey Fan today.

late geologist Clyde Wahrhaftig). Right lateral movement and associated compression on the fault progressively distorted and constricted the strait connecting the embayment with the open ocean. The geography of the central Coast Ranges and Great Valley must have been vastly different than it is today.

Sediments found in the beach cliffs south of San Francisco provide evidence that the Sacramento–San Joaquin river system had established an outlet to the Pacific

Fig. 12-29 Closeup view of the Rockland Ash within sand and pebble layers of the Merced Formation. The penny provides scale. Beach cliffs near Daly City, San Mateo County. (Source: Sarna-Wojcicki, A. U.S. Geological Survey.)

through San Francisco Bay by about 400,000 years ago (see the box). Marine sediments found in the Merced Formation show an increase in the supply of sediment and a distinctive change from minerals and rocks that originated in the Coast Ranges to material that must have come from the Sierra Nevada. The Sierra Nevada–type sediments are found below the Rockland Ash (Fig. 12-29), which was erupted from the Mt. Lassen area 400,000 years ago, as discussed in Chapter 5. The Rockland Ash was also discovered in ***estuarine*** sediments from 120 meters beneath the modern Bay. The ash and the sediments were collected in cores drilled during construction of the bridges that span the Bay. The location and thickness of the sediments in the San Francisco region suggest that the original outlet was probably a few kilometers south of the Golden Gate. Geologists believe that Lake Clyde may have drained because it overflowed during an episode of glacial melting.

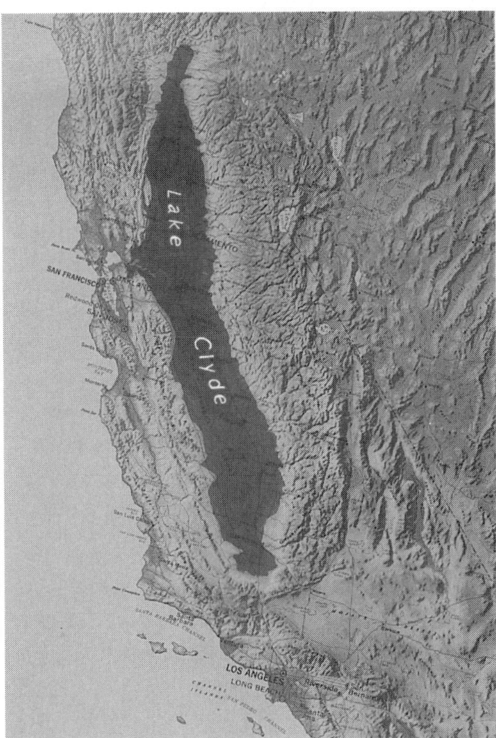

Fig. 12-30 The possible maximum extent of Lake Clyde 600,000 to 700,000 years ago. (Source: Sarna-Wojcicki, A. U.S. Geological Survey, unpublished.)

RECENT SEA LEVEL CHANGES

The global climate fluctuations that resulted in episodic glaciation of the Sierra Nevada during the last million years were reviewed in Chapter 8. It is not surprising to find that these worldwide fluctuations had a dramatic effect on San Francisco Bay. During glacial periods, more of the earth's water is stored in the ice sheets and alpine glaciers, and worldwide sea level is low. When the glaciers melt, more water is stored in the oceans, and sea level rises. Worldwide sea level was approximately 120 meters lower than at present during the last major glacial episode 18,000 to 22,000 years ago. The 120-meter lowering of sea level in the San Francisco area placed the shoreline just west of the Farallon Islands, about 32 kilometers west of today's beaches (Fig. 12-31). At that time, the combined Sacramento and San Joaquin Rivers, as well as the smaller streams flowing from the Santa Clara Valley, flowed past San Francisco, depositing their sediment load on what is now the continental shelf. During times of low sea level, rivers flowed through areas that are now San Francisco Bay.

Sediments deposited in the river valleys occupying San Francisco Bay during times of low sea level were river alluvium and, in some areas, dune sand. Alluvial deposits are typically sandy and similar in composition to the sediments carried by the streams that flow into the Bay today. Sand dunes are common in western San Francisco and Alameda. The sand that makes up the dunes was probably

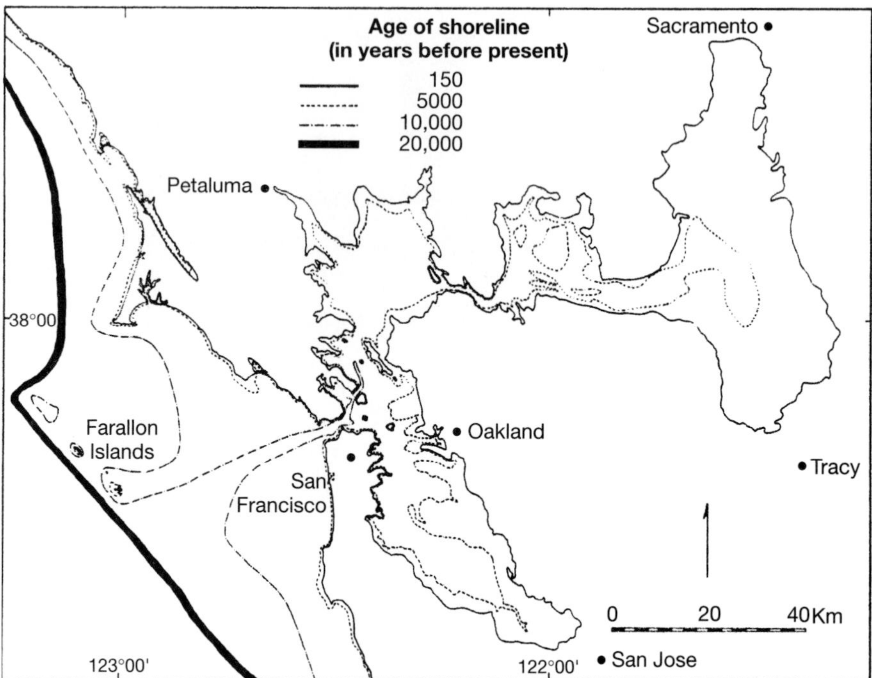

Fig. 12-31 The shoreline near San Francisco about 20,000 years ago, during the Tioga glaciation. (Source: Conomos, T.J. 1979. *San Francisco Bay: The Urbanized Estuary.* Pacific Division, American Association for the Advancement of Science.)

blown in from the area west of San Francisco during times when the continental shelf was exposed.

The present estuary that Bay Area residents take for granted formed less than 10,000 years ago, as the earth's climate warmed and the melting of glaciers caused worldwide sea level to rise. By 10,000 years ago, the rising marine water began to flood the Bay, and by about 4000 years ago, the Bay occupied its historic limit. The inland Delta had formed at the position of the higher sea level. Sediments deposited in the estuary are typically dark-colored clay and silt, and they contain foraminifera and other marine microfossils that could not survive in fresh water. The sediments are commonly called ***Bay mud.*** Sandier sediments deposited by channels in the estuary are commonly interfingered with the estuarine deposits.

After the Sacramento–San Joaquin outlet was established in its current location, San Francisco Bay came and went with climatically controlled sea level rises and falls. During glacial episodes, sea level was low, and river valleys occupied the Bay (Fig. 12-32). During warm periods, sea level rose and created an estuary. The oldest known estuary existed San Francisco Bay 400,000 years ago, based on the presence of the Rockland Ash in estuarine deposits identified beneath the Bay.

Drilling and coring beneath the Bay in preparation for constructing several of the bridges resulted in the collection of thousands of sediment samples. By careful analysis of the types of sediments in these samples, geologists were able to reconstruct the history of the Bay and confirm the alternating pattern of estuary and alluvial valleys (Fig. 12-33) (see the box).

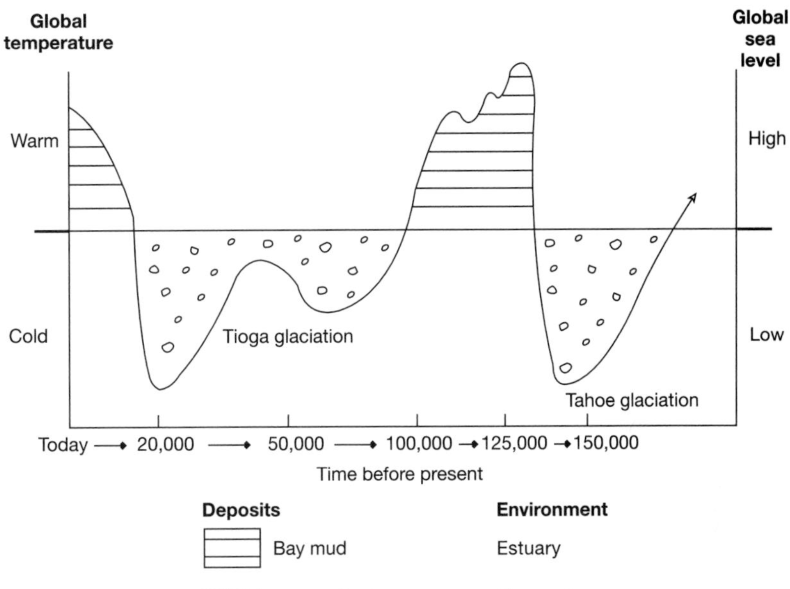

Fig. 12-32 Climatic controls on the sedimentation in San Francisco Bay.

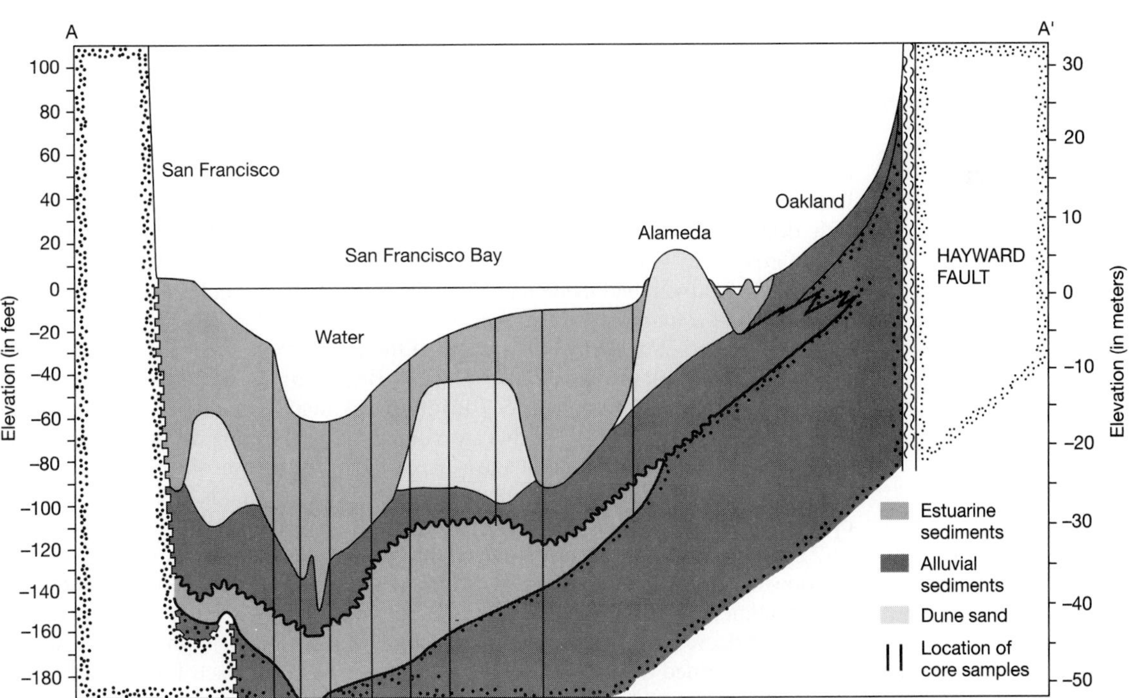

Fig. 12-33 Cross section showing the sediments beneath San Francisco Bay between San Francisco and Alameda. The vertical lines show the locations of cores. (Source: Atwater, B. 1977. U.S. Geological Survey, Professional Paper 1014.)

IMPLICATIONS FOR GROUNDWATER CONTAMINATION AND CLEANUP

Bay mud, sometimes containing shells, forms the upper layers of sediment along much of the margin of the southern part of San Francisco Bay. The density and stickiness of the clay-rich mud, a great frustration to many gardeners in the area, make the Bay mud an effective aquitard. In contrast, the alluvial deposits in the area, found in small channels and uphill from the areas where Bay mud is found, are permeable and much looser material because of their coarser grain size.

Beneath the surface of the Bay margins, alternating layers of alluvium and Bay mud record the sea-level changes discussed above. The alluvial layers, which slope from the hills toward the Bay, are excellent aquifers, and they met local demands for municipal and irrigation water until the 1950s. The aquifers are effectively confined by the intervening Bay mud, and artesian wells were abundant in parts of the Santa Clara Valley (Chapter 10). Unfortunately, wells in the upper aquifer and even some in the lower aquifer have become contaminated by leaking storage tanks in Santa Clara County.

A clear understanding of the geologic history of the Bay sediments is essential to understanding the flow of groundwater in the Santa Clara Valley and to cleaning up contamination of the aquifers. By recognizing that the confining Bay mud layer is only present in areas which were covered by the older estuaries, it is easy to see how contaminants could leak into the lower aquifer. Channels of permeable sand are present even in the modern estuary; channels within the older Bay mud also enable contaminants to spread. The resulting three-dimensional picture of aquitards and aquifers is very complex, and a thorough understanding of the recent geologic history of the Bay is required in order to ensure the future protection of groundwater resources.

HUMAN ACTIVITIES

Since about 1850, human activities have made enormous modifications to the San Francisco Bay region and to the estuary itself. The first drastic impact was the effect of hydraulic mining in the Sierra Foothills (see Chapter 8). In 1917, pioneering geologist Grove Karl Gilbert calculated that more than a billion cubic yards (0.9 billion m^3) of this sediment had made its way to the upper part of San Francisco Bay, where it filled in the estuary by several feet.

Various activities along the margins of the Bay have reduced the wetlands habitats in the Bay by approximately 85 percent. Beginning in the 1800s, farmers in the Delta constructed dikes and levees to drain the marshy wetlands for agriculture. These low-lying areas (Fig. 12-34) are vulnerable to flooding during major storms, and concern is mounting about the seismic stability of the old levees and dikes. Wetlands have also been diminished by salt evaporating ponds and by urban development on filled lands.

Human activities have also affected the Bay's water. Because of water diversions, the flow of fresh water into the Bay is only about 60 percent of its prediversion flow. The combined effect of drought and the diversions has been increased salinity in some areas and decreased circulation in the South Bay. Biologists and geologists first became concerned about the ecological health of San Francisco Bay as early as the turn of the century, when the threat to world-famous commercial oyster beds was first realized. In the 1960s, a grassroots effort succeeded in limiting construction activities in wetlands, and in recent years some wetland areas have been restored. Recent legislation requires maintenance of minimum flow to the delta, and the discharge of municipal and industrial waste is now strictly controlled (Fig. 12-35).

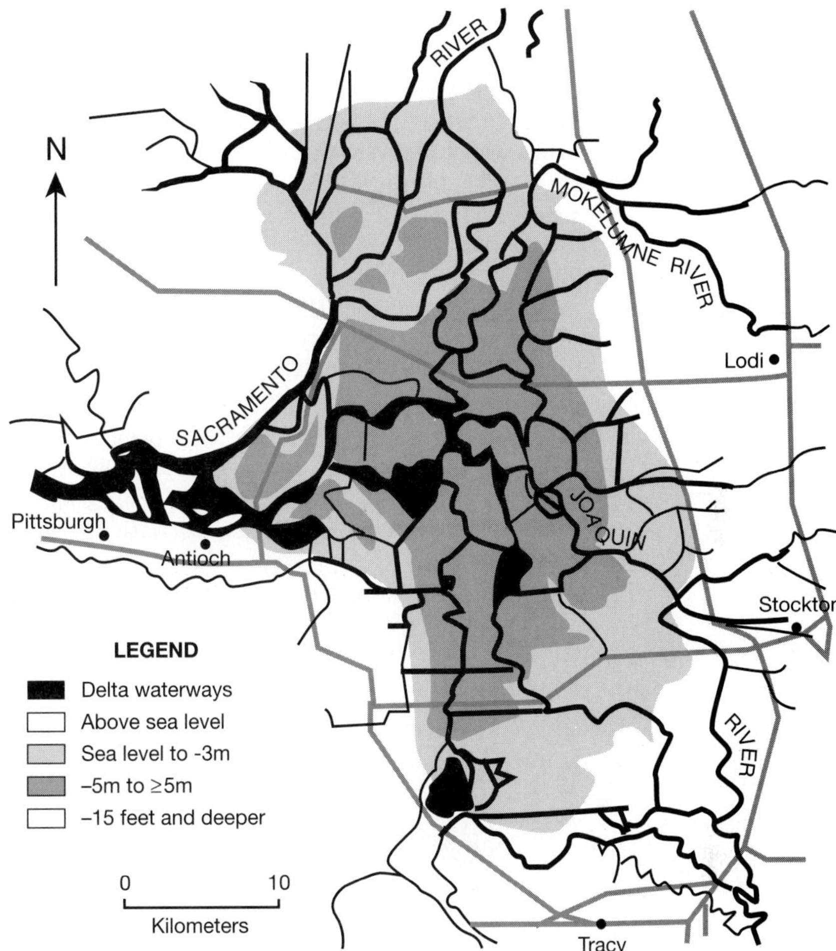

Fig. 12-34 Areas below sea level in the Sacramento–San Joaquin Delta. (Source: California Department of Water Resources. *California Water: Looking to the Future.*)

Fig. 12-35 Aerial view of San Francisco Bay near Foster City, San Mateo County. Wetlands were filled in to construct the city during the 1960s. The sloughs flowing through wetland areas into the Bay can still be seen. The western end of the San Mateo Bridge appears in the background. (Photo by author.)

FURTHER READINGS

AALTO, K.R., AND HARPER, G.D. 1989. *Geologic Evolution of the Northernmost Coast Ranges and Western Klamath Mountains, California.* Washington, D.C.: American Geophysical Union, Field Trip Guidebook T308, 82 pp.

BAILEY, E.H., IRWIN, W.P., AND JONES, D.L. 1964. *Franciscan and Related Rocks and Their Significance in the Geology of Western California.* California Division of Mines and Geology Bulletin, Vol. 183, 177 pp.

BURCHFIEL, B.C., LIPMAN, P.W., AND ZOBACK, M.L., eds. 1992. The Cordilleran orogen: Conterminous U.S. In *The Geology of North America,* Vol. G-3. Boulder, Colo.: Geological Society of America.

CONOMOS, T.J., ed. 1979. *San Francisco Bay: The Urbanized Estuary.* Pacific Division, American Association for the Advancement of Science, 493 pp.

SCHNEIDER, J., 1992. *Quicksilver: The Complete History of Santa Clara County's New Almaden Mine.* San José, Calif.: Zella Schneider, 178 pp.

WAHRHAFTIG, C. 1984. *A Streetcar to Subduction and Other Plate Tectonic Trips by Public Transport in San Francisco.* Washington, D.C.: American Geophysical Union, 76 pp.

WAHRHAFTIG, C., AND SLOAN, D. 1989. *Geology of San Francisco and Vicinity.* American Geophysical Union, Field Trip Guidebook T105, 69 pp.

WATERS, T. 1995. The other Grand Canyon. *Earth Magazine,* December 1995, pp. 46-51.

CHAPTER

13

Earthquakes, Faults, and Seismic Safety

It was 4:30 in the morning on Martin Luther King Day, January 17, 1994, when for the first time since 1933, an earthquake struck directly beneath an urbanized part of California. When the shaking was over, the Northridge earthquake had killed 33 people (the final toll was 60), injured more than 7000, left more than 20,000 homeless, and caused between $13 and $20 billion in damage (Fig. 13-1). The Santa Susana Mountains and the northern San Fernando Valley gained as much as 70 centimeters in elevation. The Northridge earthquake was followed by 3000 aftershocks during the next 3 weeks, and the personal and financial aftershocks continue today. But according to seismologists, the Northridge earthquake was unusual for one reason only: miraculously, it hit at 4:30 in the morning. The timing probably saved thousands of lives that would have been lost in collapsed parking structures, malls, and freeways had it struck during the day or evening.

Fig. 13-1 Collapsed parking garage, California State University Northridge campus. (Source: Rymer, M. U.S. Geological Survey.)

Most Californians will experience at least one significant earthquake during their lifetimes, and even a small tremor can be an unsettling experience. For most people, a general understanding of earthquakes alleviates their fears and aids in making more effective earthquake preparations. A constantly improving understanding of earthquakes has also made California's civic agencies among the world's best prepared for a large earthquake.

CAUSES OF EARTHQUAKES

Earthquakes occur when rocks in a section of the Earth's crust are stressed beyond their breaking point by tectonic forces. As discussed in Chapter 1, California's rocks are subjected to stress because of the continual movements of the Pacific and North American Plates. Because the rocks that make up the Earth's plates are brittle, they do not move continuously with the asthenosphere beneath them. As plate motion progresses, the blocks of rock above are strained, or pulled, accumulating ***strain energy.*** The strain energy builds up until the strength of the rocks is overcome, and the rocks break at their weakest points along fault planes. The rocks give way, then spring back into a new position (Fig. 13-2). Some of the stored strain energy is released in the form of frictional heat as the rocks grind against each other, and some is released as vibrations or ***seismic waves.*** This explanation for earthquakes was first developed by H.F. Reid, who studied the patterns of displacements along the San Andreas fault before and after the 1906 earthquake.

The behavior of rocks before and during an earthquake is frequently likened to the actions of a bow. As an archer pulls on the bowstring of a bow, strain energy is being stored. When the string is released, the arrow is propelled forward by the release of stored energy, and the bow and string spring back into position. The distance that the arrow flies is proportional to the amount of strain stored in the bow, just as the size of an earthquake reflects the amount of strain energy accumulated in the rocks.

An earthquake is initiated when fault rupture begins, and the point of original rupture is known as the earthquake's ***focus*** (Fig. 13-3). Although earthquakes do not originate right at the Earth's surface, 90 percent of all earthquakes originate at depths of less than 100 kilometers. Shallow-focus earthquakes have focal depths between a

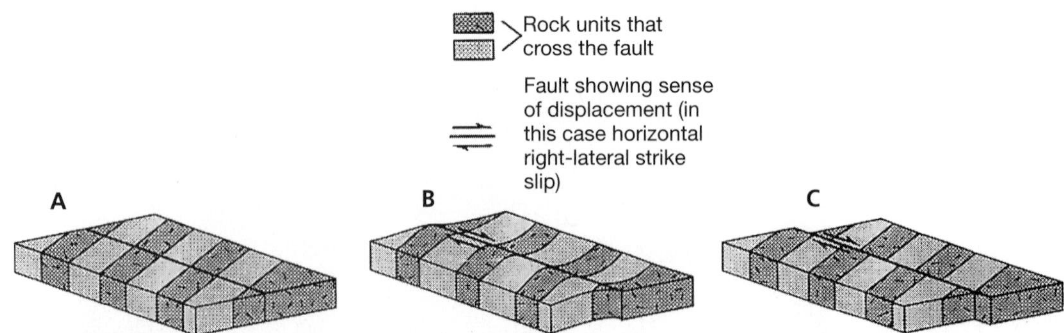

Fig. 13-2 Schematic of rock behavior during an earthquake. **A,** Starting position of no strain and no displacement. **B,** As the blocks are subjected to stress, the rocks accumulate strain energy and the layers are bent. **C,** Rupture occurs on the fault, and the rocks are displaced. (Source: Keller, E., and Pinter, N. 1996. *Active Tectonics.* Upper Saddle River, N.J.: Prentice Hall, Inc.)

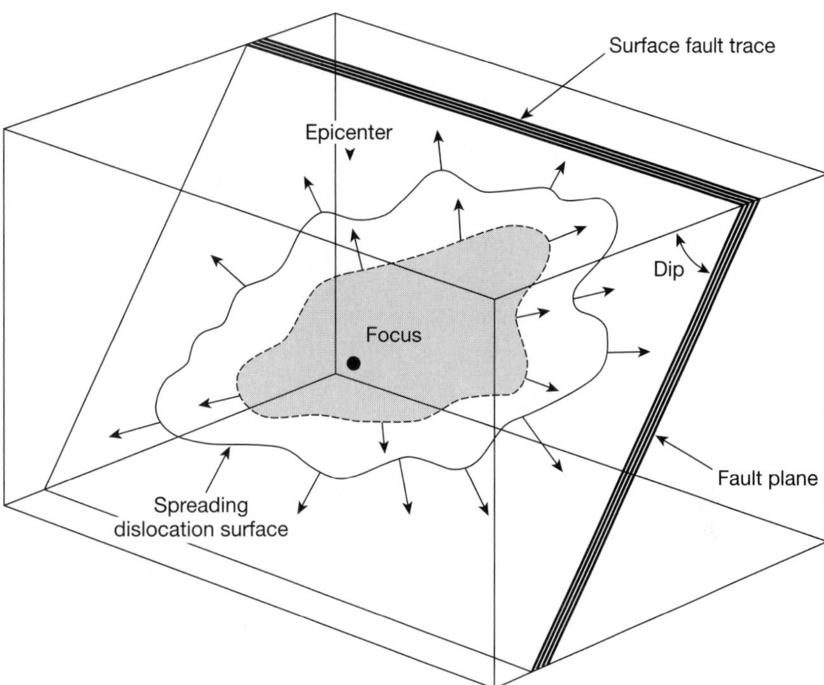

Fig. 13-3 Focus and epicenter of a schematic earthquake. (Source: Bolt, B. 1993. *Earthquakes,* 3rd ed. New York: W.H. Freeman.)

few kilometers and 70 kilometers. All of California's large historic earthquakes fall into this category, and most earthquakes along the San Andreas fault system originate in the upper 10 kilometers of the crust. The 1989 Loma Prieta earthquake, with a focal depth of 18 kilometers, was unusually deep for the central San Andreas fault. The 1994 Northridge earthquake originated at a depth of 19 kilometers.

Seismologists classify intermediate-focus earthquakes as those originating at depths between about 70 and 300 kilometers, and deep-focus earthquakes at depths between 300 and 700 kilometers. Below these depths, temperatures and pressures are so high that rocks are plastic rather than brittle. However, deep earthquakes are generated at subduction zones. Even along subduction zones, earthquakes with relatively shallow focal depths cause the greatest destruction. The 1964 Alaska earthquake, the strongest recorded in the United States, originated at a depth of about 30 kilometers.

EARTHQUAKES AND FAULTS

A casual look at almost any road cut, quarry, or sea cliff in California will reveal a ***fault***—an abrupt line or zone breaking through the rocks or sediments (Fig. 13-4). The rocks or sediments on either side of a fault do not match up, indicating that there has been movement along the fault of the materials on opposite sides. Geologists recognize faults by the evidence of displacement or offset across them. Sudden movement along a fault may take place during earthquakes, or the movement may be gradual.

Many faults seen at the Earth's surface have not moved in millions, or even hundreds of millions of years. These ***inactive faults,*** generated by forces that are no longer operating, are unlikely to be the source of earthquakes today. Others are con-

Fig. 13-4 A normal fault cuts through horizontal layers of sediment and volcanic ash in the Owens Valley near Bishop. (Photo by author.)

sidered to be ***active faults,*** that is, capable of generating earthquakes in today's tectonic environment. Active faults can be recognized in several ways:

- Earthquakes occur on them today
- Earthquakes have occurred on them in historic time
- They displace young features or materials

Using the principle of cross-cutting relations (see Chapter 3), it is clear that fault movement must be more recent than the materials it displaces (see Fig. 7-5). Repeated slippage weakens the rocks along a fault zone, making them the likely site of repeated earthquakes. For this reason, recognition of active faults is an important step in evaluating earthquake hazards in a region.

Faults can be meters or hundreds of kilometers long, and they may indicate centimeters or hundreds of kilometers of movement. During a single earthquake, only a part of a fault may slip; the slipped surface is known as the ***rupture surface*** for that earthquake. As would be expected, large-magnitude earthquakes result from ruptures along a great area of a fault. Many faults have ruptured to the Earth's surface (see Fig. 13-4), allowing geologists to examine their direction and age of movement. Other troubling active faults lie concealed beneath the surface; these ***blind faults*** pose particular seismic hazards to California's urban areas (see Chapter 16).

The boundary between the Pacific and North American Plates is far from a simple geometric line. In previous chapters, we have seen that California is made up of a complex assortment of rocks, many of which are separated by faults. It is not surprising, then, that California's active faults are also complex (Fig. 13-5 and Endpaper 2). Along the San Andreas system, many faults exhibit right-lateral motion. In some areas, this motion is accompanied by vertical movement as rocks are pushed together by compressional forces. In contrast, normal faults in the Basin and Range Province result from tensional forces (see Chapter 7). A more detailed view of the San Andreas fault system is presented in the following chapter.

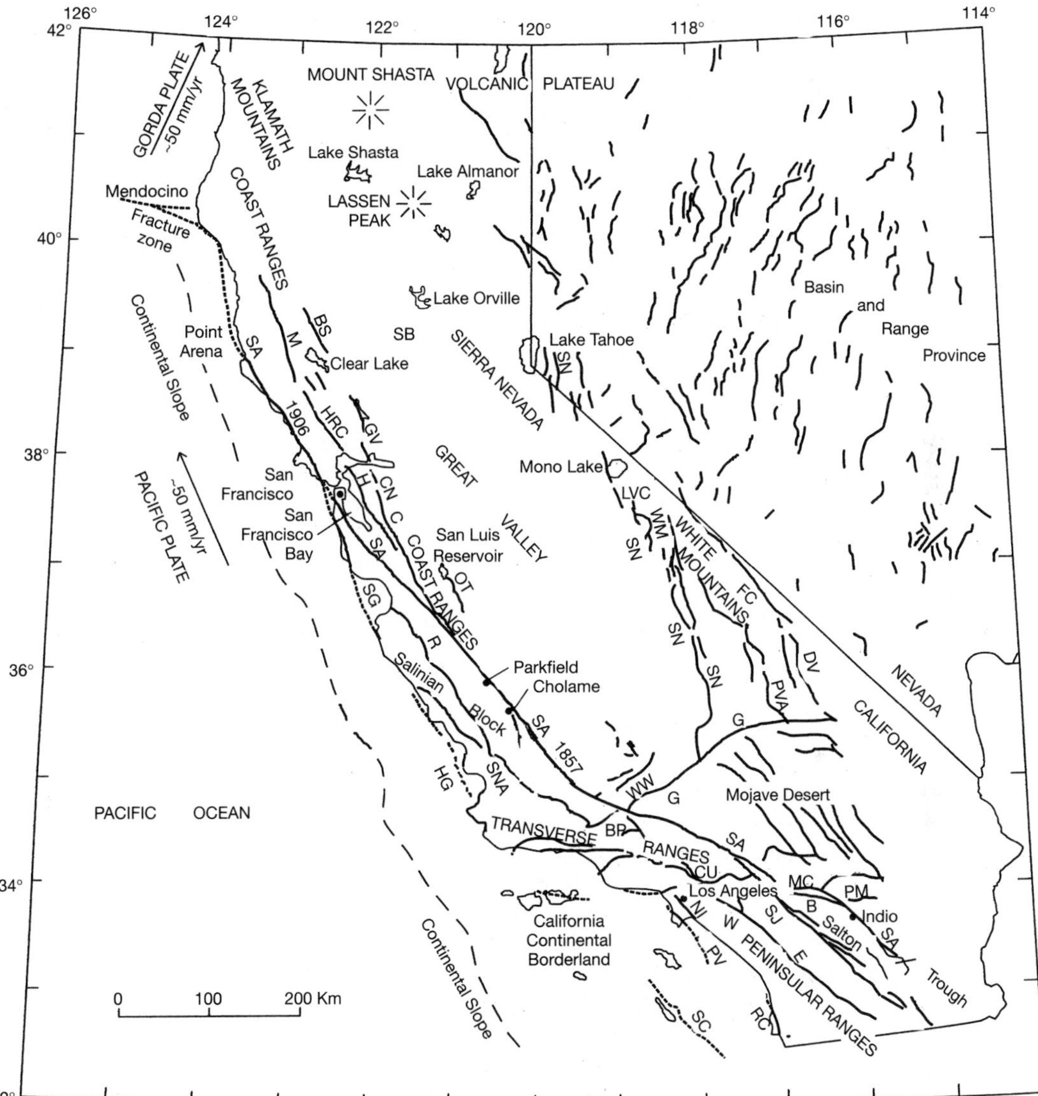

Fig. 13-5 A map of active faults in California (dotted where concealed). *B*, Banning; *BP*, Big Pine; *BS*, Bartlett Springs; *BZ*, Brawley seismic zone; *C*, Calaveras; *CN*, Concord; *CU*, Cucamonga; *DV*, Death Valley; *E*, Elsinore; *FC*, Furnace Creek; *G*, Garlock; *GV*, Green Valley; *H*, Hayward; *HG*, Hosgri; *HRC*, Healdsburg-Rodgers Creek; *IM*, Imperial; *LVC*, Long Valley Caldera; *M*, Maacama; *MC*, Mission Creek; *NI*, Newport-Inglewood; *OT*, Ortigalita; *PM*, Pinto Mountain; *PV*, Palos Verdes; *PVA*, Panamint Valley; *R*, Rinconada; *RC*, Rose Canyon; *SA*, San Andreas; *SC*, San Clemente Island; *SG*, San Gregorio; *SJ*, San Jacinto; *SN*, Sierra Nevada; *SNA*, Sur-Nacimiento; *W*, Whittier; *WM*, White Mountains; *WW*, White Wolf. Arrows and numbers indicate direction and amount of motion of Pacific and Gorda plates with respect to the North American plate to the east. (Source: Adapted from Wallace, R. 1990. Professional Paper 1515.)

EARTHQUAKE WAVES

During an earthquake, seismic waves travel outward from the point of original rupture. They arrive at the Earth's surface first at the ***epicenter,*** the point closest to the earthquake's focus. All earthquakes generate four types of seismic waves: two types of ***body waves*** that propagate through solid rock, and two types of ***surface waves*** that travel at or near the ground surface. ***P waves*** (or primary waves) travel the fastest of all earthquake waves. A P wave's motion is an alternating push and pull (compression and dilatation) like that of a sound wave (Fig. 13-6, *A*). When P waves emerge at the Earth's surface, some of them may be transmitted into the atmosphere as sound waves. These may sometimes be heard at the beginning of an earthquake as a rumbling or a sonic boom by animals or humans. The second, slower type of body wave is ***S waves*** (or secondary waves), which propagate through rock by a sideways motion at right angles to the direction of travel, in a fashion similar to the shaking of a jump rope from one holder to another (Fig. 13-6, *B*). When it arrives at the surface, the S wave shakes the ground surface with both vertical and side-to-side motion. S waves are responsible for much of the damaging ground shaking during earthquakes. The surface waves generated by earthquakes are called ***Love waves*** (Fig. 13-6, *C*) and ***Rayleigh waves*** (Fig. 13-6, *D*). Both are slower than the body waves, and the Rayleigh wave is the slower of the two. Love waves move in a manner similar to S waves, except that their motion is only side-to-side. Like S waves, the motion of Love waves can be very damaging to structures. Rayleigh waves move across the Earth's surface like ocean waves, producing both vertical and horizontal motion.

Because the S wave is slower than the P wave, it arrives at the Earth's surface later than the P. As the distance away from the earthquake's focus increases, the S waves lags farther and farther behind the P wave. If an observer is 100 kilometers from the earthquake's epicenter, the S wave may arrive 10 to 15 seconds after the P wave. Seismologists use the time difference between the first arrival of P waves and the first arrival of S waves to determine the distance between an earthquake's focus and a ***seismograph station*** (Fig. 13-7). If you know this distance for three or more stations, it is possible to locate the earthquake's point of origin. Because the speed of body waves is affected by the type of rock through which they travel, the exact location of an earthquake's epicenter may not be pinpointed until the data from tens of seismograph stations are processed, sometimes days after the quake.

An important characteristic of both types of body waves is that they are reflected or refracted (bent) when they encounter boundaries between different rock layers, including the boundary at the ground surface. The reflected waves amplify ground shaking at the surface. The patterns of ground shaking near the epicenter of an earthquake are very complex as waves travel outward from different areas of the rupture zone, reflect and refract at geologic boundaries, and are amplified by particular conditions like water-saturated soils.

Several federal, state, and local agencies have established networks of instruments designed to measure earthquakes. A network of seismograph stations records large earthquakes worldwide; data from these stations are transmitted to the National Earthquake Information Service in Golden, Colorado. The U.S. Geological Survey and Cal Tech in Pasadena are responsible for the seismograph network in Southern California, and the U.S. Geological Survey in Menlo Park and U.C. Berkeley are responsible for Northern California stations. Arrays of highly sensitive, portable ***seismometers*** are positioned in several localities where seismologists are studying

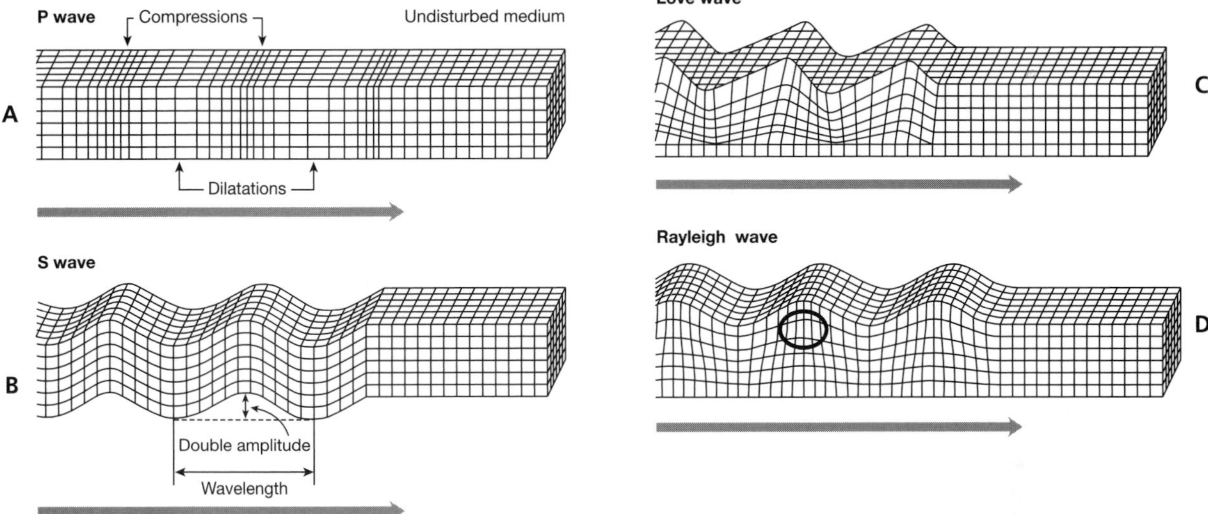

Fig. 13-6 Diagram illustrating the motions of four types of earthquake waves. (Source: Bolt, B. 1993. *Earthquakes,* 3rd ed. New York: W. H. Freeman.)

Fig. 13-7 Array of seismograph recorders at the U.S. Geological Survey, Menlo Park. The drums display the signals from seismographs at locations throughout northern California. (Photo by author.)

the behavior of specific faults or areas. Other sensors measure the frequencies of shaking in structures. Strong-motion seismographs are designed to record ***ground acceleration*** during an earthquake. Over a thousand strong-motion sensors are in place throughout the state.

EARTHQUAKE SIZE

Seismographs record earthquakes by capturing the relative motion between a suspended pendulum and the vibrating earth. Incoming seismic waves are recorded as electrical signals on tape or as lines on a paper drum. Seismographs can record hor-

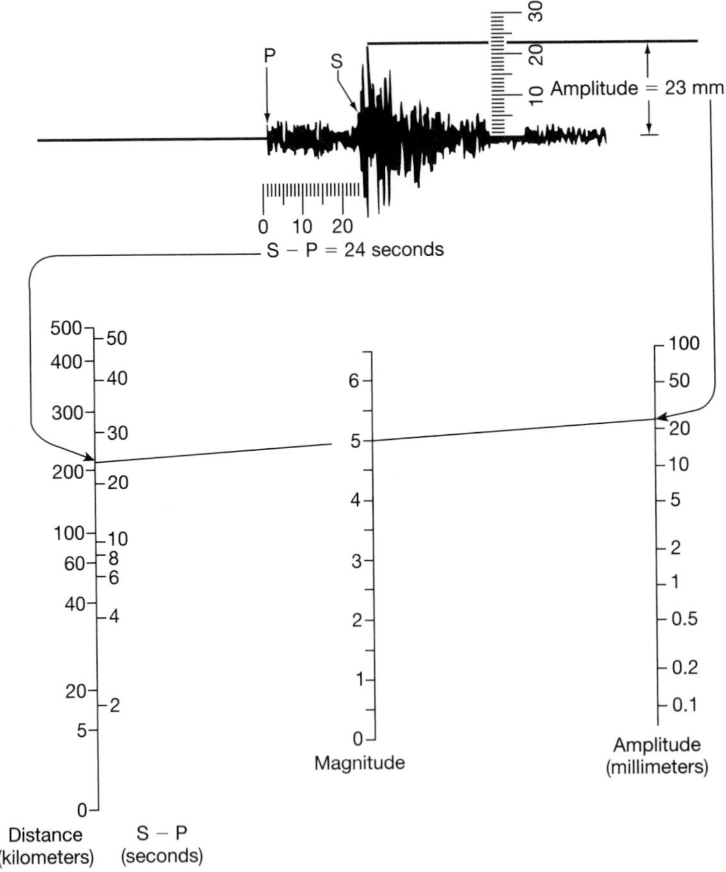

Fig. 13-8 An example of the calculation of Richter magnitude (R_L) from a seismogram. The procedure for calculating the magnitude is as follows:

1. Use the time difference between the arrival of the S and P waves (P–S=24 sec) to measure the distance to the focus.
2. Measure the height of the maximum wave motion on the seismogram (23 mm).
3. Place a straight edge between the correct points on the distance (left) and amplitude (right) scales to obtain the magnitude (M_L= 5.0).

(Source: Bolt, B. 1993. *Earthquakes,* 3rd ed. New York: W.H. Freeman.)

izontal or vertical motions, and they are capable of detecting small earthquakes that take place at great distances from the station.

Professor Charles Richter of the California Institute of Technology developed the first system for measuring earthquake size using the wiggly lines of seismograph records (Fig. 13-8). The ***Richter magnitude*** of an earthquake uses the amplitude of waves recorded by a seismograph as an indicator of the energy released during a quake. Because the Richter scale is related to strain energy release, the magnitude is a reflection of the earthquake itself. The Richter scale was developed specifically for southern California, based on the behavior of seismic waves in that region. Since its invention in 1935, the Richter scale has been used worldwide as a measure of earthquake size, but different methods or modifications of the Richter scale have been adapted for measuring deep earthquakes and earthquakes affecting different regions. In California, the Richter magnitude is often called the local magnitude (M_L).

Earthquake magnitudes are logarithmic, meaning that each whole-number increase on the scale is a 10-times increase in the amplitude of the seismic waves. Each whole-number increase in magnitude also signifies approximately a 30-fold increase in the amount of energy released (Fig. 13-9). "Great" earthquakes of

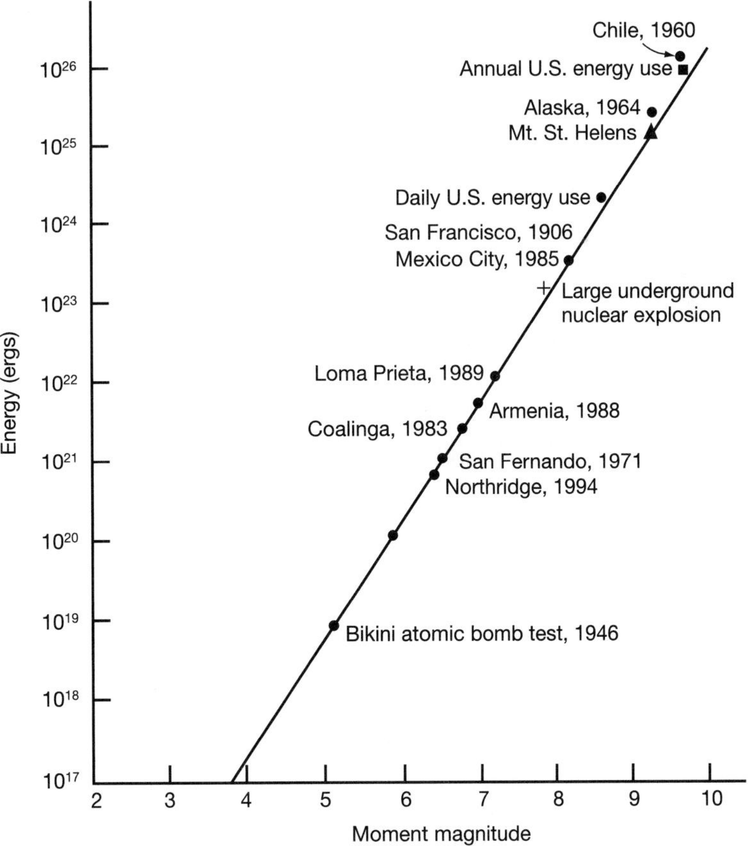

Fig. 13-9 Moment magnitudes and energy release of some moderate and large California earthquakes. (Source: Modified from Bolt, B. 1993. *Earthquakes,* 3rd ed. New York: W.H. Freeman.)

Table 13-1 COMPARISON OF EARTHQUAKE MAGNITUDES USING DIFFERENT SCALES

Earthquake Location and Date	Modified Magnitude (M_L)	Moment Magnitude (M_W)
Alaska, 1964	8.3	9.3
San Francisco, 1906	8.3	7.7
Loma Prieta, 1989	7.1	7.0
San Fernando, 1971	6.4	6.7
Northridge, 1994	6.4	6.7

magnitude 8.0 release *900 times* more energy than earthquakes with magnitudes of 6.0! A great earthquake strikes the Earth about once every 5 years, whereas 26 earthquakes with magnitudes between 5.7 and 7.0 struck California between1979 and 1989.

More recently, seismologists have used the ***moment magnitude*** (M_W) as a comparative measure of earthquake size (see Fig. 13-9 and Table 13-1). An earthquake can be thought of as a moment: a force couple created by equal but opposite forces acting across a fault, and separated by a distance that is proportional to the size of the earthquake. The moment magnitude takes into account both the area of the rupture surface and the amount of displacement during an earthquake along the fault. The moment magnitude provides a means of assessing the release of energy along the entire rupture zone of the fault, which may be hundreds of kilometers during a great earthquake. The moment magnitudes are consistent for earthquakes of all sizes, whereas the Richter magnitude values are less consistent for larger earthquakes. The largest earthquake that can be accurately recorded by M_L is about 7.2. As a result, earthquake magnitudes for both new and previous events are more likely to be reported as moment magnitudes. When comparing earthquakes, it is important that the same magnitude measurement be used.

EARTHQUAKE INTENSITY AND THE MERCALLI INTENSITY SCALE

If you ask California earthquake veterans to rank the earthquakes they have experienced, chances are good that they would use a scale based on earthquake ***intensity.*** The intensity of an earthquake is determined from the types and patterns of damage and the experiences of people. Although intensity is linked to the magnitude of an earthquake, other factors are also important. For example, the intensity of shaking, as perceived by an observer or as evaluated from structural damage, decreases with distance from an earthquake's epicenter. In general, earthquake intensity decreases, or ***attenuates,*** away from the epicenter.

Two additional factors unrelated to the earthquake are equally important in determining earthquake intensity. The nature of the soil or bedrock in an area has a great effect on the severity of ground shaking. Areas underlain by unconsolidated sediments may sustain greater damages than bedrock areas much closer to an earthquake's epicenter, particularly if the sediments are water-laden. A final factor that influences intensity is a human factor: the type and density of structures near the epi-

TABLE 13-2 MODIFIED MERCALLI INTENSITY SCALE (ABRIDGED)

Intensity	Effects
I	Not felt by people; detected by seismometers.
II	Felt only by a few persons at rest, or those who are on upper floors of buildings. Delicately suspended objects may swing.
III	Felt quite noticeably indoors, especially on upper floors of buildings; hanging objects swing. Standing cars may rock slightly; vibration like the passing of a light truck.
IV	During the day felt indoors by many, outdoors by a few. At night some awakened. Dishes, windows, doors rattle; walls make cracking sounds. Vibration like a heavy truck passing; standing cars rock noticeably; car alarms triggered.
V	Felt by nearly everyone; many awakened. Some dishes, windows, etc., broken; a few instances of cracked plaster; unstable objects overturned. Disturbance of trees, poles, pictures, and shutters sometimes noticed. Doors swing open or closed.
VI	Felt by all; many are frightened and run outdoors. Some heavy furniture moved; books fall off shelves; a few instances of fallen plaster or damaged chimneys. Windows, glassware, dishes broken.
VII	Difficult to stand. Damage negligible in buildings of good design and construction; slight to moderate in well-built ordinary structures; considerable in poorly built or badly designed structures. Some chimneys broken. Loose bricks and tiles fall. Noticed by persons driving cars.
VIII	Damage slight in specially designed structures; considerable in ordinary substantial buildings with partial collapse; great in poorly built structures. Houses moved off foundations if not bolted. Chimneys, water tanks, columns, monuments fall. Heavy furniture overturned. Sand and mud ejected in small amounts. Changes in well water. Difficulty steering cars.
IX	General panic. Damage considerable in specially designed structures; well-designed frame structures thrown out of plumb; some buildings collapse. Buildings shifted off foundations. Ground cracked. Underground pipes broken.
X	Some well-built wooden structures destroyed; most masonry and frame structures with foundations destroyed; ground badly cracked. Rails bent. Landslides from river banks and steep slopes. Shifted sand and mud. Water splashed (slopped) over banks.
XI	Few, if any, masonry structures remain standing. Bridges destroyed. Broad fissures in ground. Underground pipelines completely out of service. Earth slumps and land slips in soft ground. Rails bent greatly.
XII	Damage total. Waves seen on ground surfaces. Lines of sight and level distorted. Objects thrown upward into the air.

Source: Modified from U.S. Geological Survey, 1974. *Earthquake Information Bulletin* 6(5):28.

center. In many regions of the world, thousands of people are killed nearly every year when their poorly constructed residences are destroyed by moderate-magnitude earthquakes. In California, many similar earthquakes only rattle nerves and knock things off shelves.

The ***modified Mercalli scale*** is used to categorize earthquake intensity (Table 13-2). Following an earthquake, seismologists use questionnaires to obtain information from inhabitants. An intensity value is assigned to each response, and the values are plotted on a map. The values are contoured, producing a generalized view of the damage pattern (Fig. 13-10).

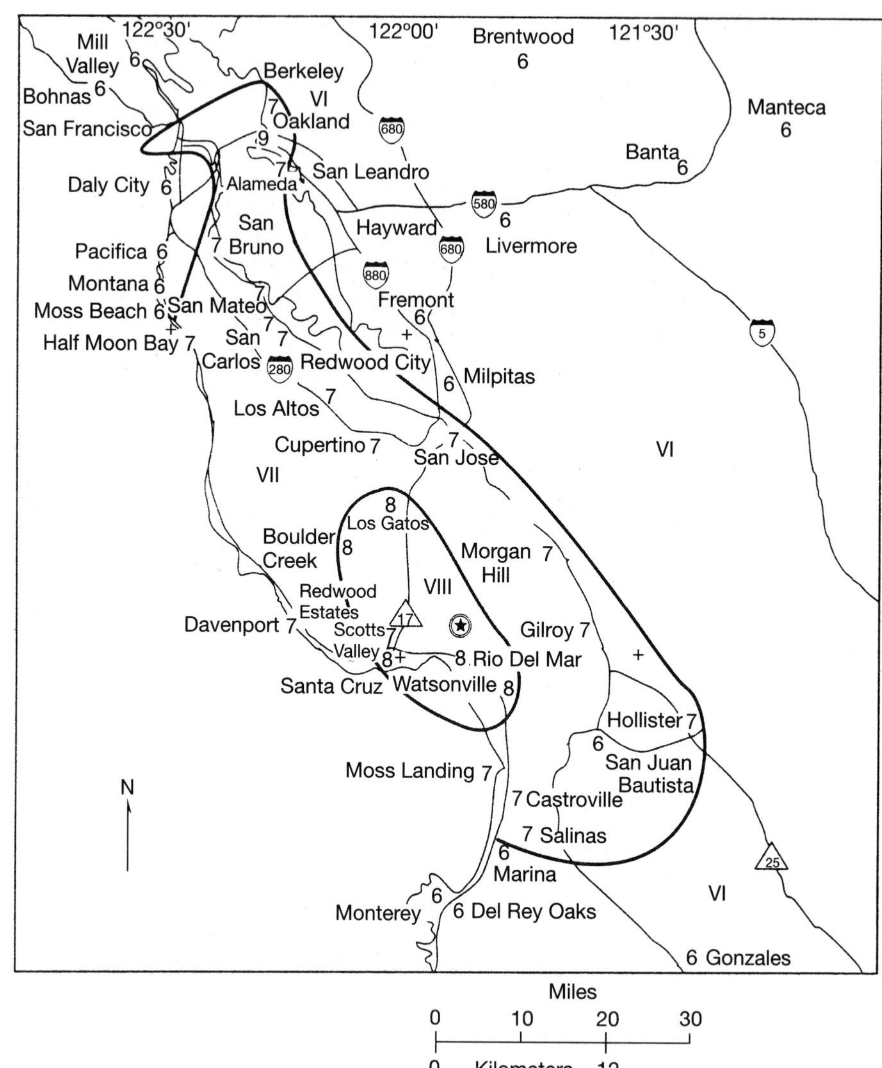

Fig. 13-10 Intensity map for the 1989 Loma Prieta earthquake. (Source: Plafker, G., and Galloway, J. 1989. *Lessons Learned From the Loma Prieta Earthquake.* U.S. Geological Survey.)

DESTRUCTIVE EVENTS DURING AN EARTHQUAKE

If a fault ruptures to the ground surface during an earthquake, then the ground surface is broken (see Fig. 7-5). The ground slips in the same direction as the fault, displacing everything that lies across the surface trace. In a major earthquake, this ***ground rupture*** may involve several meters of vertical or horizontal separation (Fig. 13-11). Buildings and other structures lying directly on the ruptured surface clearly could not withstand such motion if the displacement is large. In open areas, surface rupture may be seen as a crumpled line often called a "moletrack" because it resembles the burrowed passage of a mole (Fig. 13-12).

Fig. 13-11 Ground rupture on the Homestead Valley fault created by the 1992 earthquake, showing right lateral displacement. The dirt road is offset approximately 2.6 meters. (Source: Rymer, M. U.S. Geological Survey.)

Fig. 13-12 The "moletrack" created by ground rupture in the 1906 earthquake, Olema, Marin County, near the epicenter. (Source: Gilbert, G.K. U.S. Geological Survey.)

If differential vertical motion occurs, a linear cliff called a fault scarp results. As illustrated in Chapter 7, a fault scarp several meters high may be produced during a single, great earthquake. Surface rupture may also produce much more subtle zones of cracks, rumples, and horizontal displacements. Examples are discussed further in the following chapter. Ground rupture can cause great destruction to structures, but it is confined to the zones directly above the rupturing faults.

Ground shaking generally causes the greatest damage during an earthquake. Even a moderate earthquake shakes the ground many kilometers away from the earthquake epicenter. In areas where unconsolidated sediments are saturated with

Fig. 13-13 Fire resulting from ruptured gas lines along Balboa Boulevard, Granada Hills. (Source: Rymer, M. U.S. Geological Survey.)

water, ground shaking is more severe and lasts longer than the shaking in bedrock areas. Artificially filled land, created by filling in natural wetlands, may experience the most severe shaking of all. Ground shaking is responsible for collapsed and damaged buildings, elevated freeways, and other structures. Ground shaking may rupture gas lines pipelines and tanks to rupture, causing fires to start (Fig. 13-13). Furthermore, it may be difficult to put out fires if water mains are ruptured. About 80 to 90 percent of the destruction in San Francisco during the 1906 earthquake resulted from fire.

Liquefaction is a dangerous result of ground shaking in areas underlain by water-saturated fine sand. The wet sand becomes a slurry of sandy liquid as repetitive ground shaking causes the sediment to lose strength. A walk along the wet sand at the edge of a beach can reproduce the effect. Repeated tapping on the wet sand turns it to a jellylike mush. Pushing on the wet sand may also force water to the surface in adjacent areas. Both phenomena are possible during liquefaction. Water and sand may be forced to the ground surface, producing sand boils.

Liquefaction was widespread during the 1906 earthquake, which struck at the end of a wet winter, when soils and sediments were water-laden. During the 1989 Loma Prieta earthquake, liquefaction was common around the margins of San Francisco Bay and along the low-lying, sandy floodplains of coastal rivers (Fig. 13-14). The most notable liquefaction damage was in the Marina area of San Francisco, where apartment buildings are built in an area where sandy material had been used to fill in a lagoon. In an ironic example of history repeating itself, the liquefied fill brought with it to the surface some of the old rubble from 1906.

Fig. 13-14 Liquefaction of sandy sediments along the Pajaro River resulted in the collapse of the Highway 1 bridge. Note that the bridge pilings punched up through the pavement on the right side of the photo. (Source: Nakata, J. U.S. Geological Survey.)

Another potentially deadly side effect of ground shaking is the generation of earthquake—created ***landslides.*** Many of California's steep hillsides are underlain by relatively weak rocks. Ground shaking may push these areas over their threshold of stability, resulting in multiple landslides during a single earthquake. For example, the 1994 Northridge earthquake triggered widespread landslides in the steep Santa Susana Mountains to the north (Fig. 13-15).

Areas along California's coast are vulnerable to ***tsunamis.*** A tsunami is created when an earthquake ruptures the ocean floor, or as a result of huge submarine landslides or volcanic eruptions. Earthquakes with vertical displacement cause the water above the sea floor to be displaced. A series of long waves radiates outward from the point of rupture. The waves are long and low in the open ocean, but they can pile up along the shoreline to create a wall of water more than 30 meters high (Fig. 13-16). Because they may originate anywhere in the Pacific Ocean, tsunamis can arrive at the California coast without warning.

Following the 1964 Alaska earthquake, the harbor at Crescent City, California was destroyed by a 7-meter-high tsunami that killed 11 people. Today, the Pacific Tsunami Warning Center records seismographic data from around the Pacific, giving warnings to coastal areas whenever a Pacific earthquake with tsunami potential strikes. Along the San Andreas fault system, most of California's earthquakes involve primarily horizontal motion, so their tsunami potential is relatively low. Nevertheless, the 1906 San Francisco earthquake, whose epicenter was just southwest of San Francisco, generated a wave about 1 meter high in San Francisco Bay. In 1812, an earthquake centered in the Santa Barbara Channel apparently generated a tsunami that affected the Santa Barbara and Ventura areas.

A

B

Fig. 13-15 Landslides generated during the 1994 Northridge earthquake. **A,** Pacific Palisades, Santa Monica. **B,** Santa Susana Mountains. (Source: **A,** Jibson, R., and **B,** Harp, E. U.S. Geological Survey.)

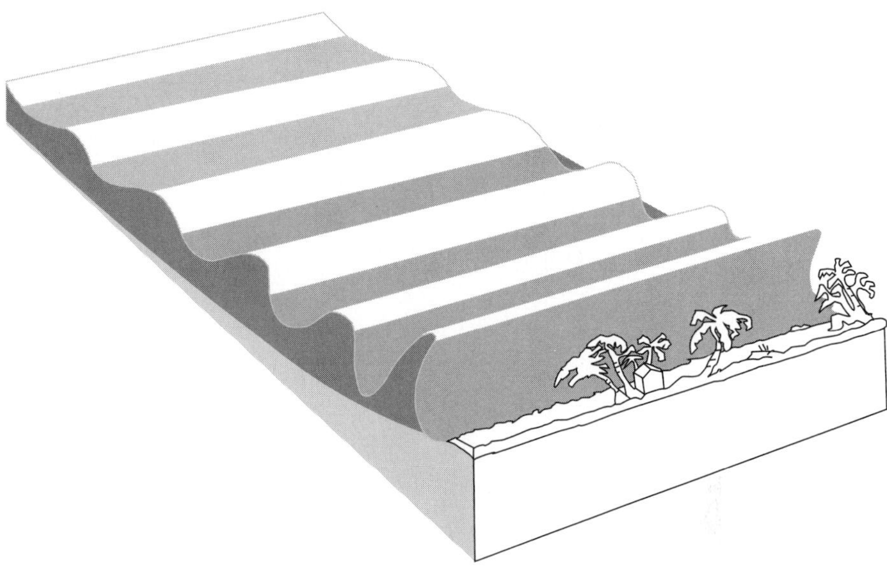

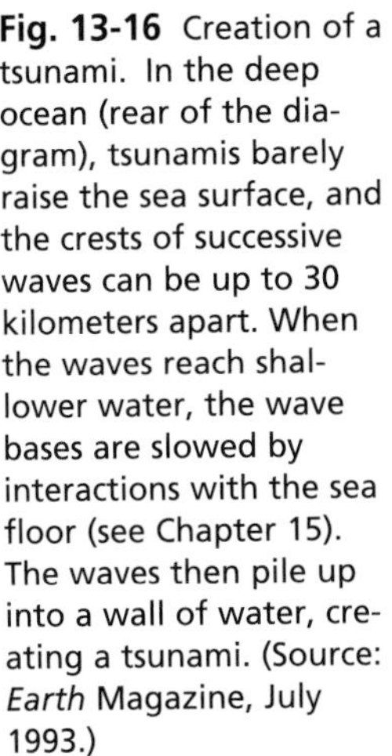

Fig. 13-16 Creation of a tsunami. In the deep ocean (rear of the diagram), tsunamis barely raise the sea surface, and the crests of successive waves can be up to 30 kilometers apart. When the waves reach shallower water, the wave bases are slowed by interactions with the sea floor (see Chapter 15). The waves then pile up into a wall of water, creating a tsunami. (Source: *Earth* Magazine, July 1993.)

EARTHQUAKE PREDICTION AND EVALUATION OF SEISMIC HAZARDS

During the past 25 years, researchers have begun a number of programs designed to increase our understanding of California's earthquakes, with the ultimate goals of hazard reduction and earthquake prediction. The most comprehensive of these programs is the National Earthquake Hazards Reduction Program (NEHRP), established in 1977 to reduce the risk to life and property from earthquakes in the United States. Although the scientific community is far from the goal of reliable earthquake prediction, significant progress has been made. The stated goals of the NEHRP are to:

Understand the earthquake source
Determine earthquake potential
Predict the effects of earthquakes
Disseminate research results

As discussed above, the first step in evaluating earthquake hazards is to identify active faults. The newly updated California fault map (Endpaper 2) is a detailed compilation of all of California's known and suspected active faults that can be mapped at the surface. The fault map and its accompanying explanation summarize the historic and geologic history of movement along each fault. More detailed maps, published by the California Division of Mines and Geology, are available to cities, counties, and development agencies (see below).

The best way of predicting the future behavior of California's faults is to better understand the patterns of their past behavior. The patterns cannot be seen by looking just at the historic earthquakes of the past century or two, because many of the most dangerous fault segments break less frequently. By remembering that strain energy accumulates between strain-releasing earthquakes (Fig. 13-17), we can easily conclude that the seismically quiet sections of active faults may pose the greatest danger.

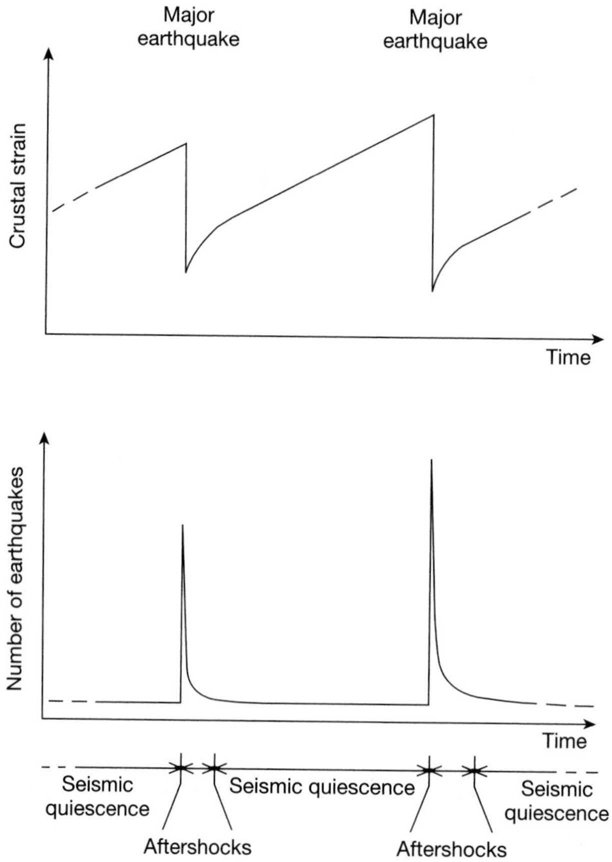

Fig. 13-17 The earthquake cycle. (Source: National Earthquake Hazards Reduction Program, U.S. Geological Survey.)

To study suspected active faults, geologists may excavate trenches across them (Fig. 13-18). By carefully mapping the sediments and soils in the excavated trench, geologists can measure the displacement of layers across the fault. They can also determine the youngest materials that are displaced by faulting, thus pinpointing the most recent ground-rupturing earthquake. In some trenches, geologists can find evidence of earthquakes that disturbed older sedimentary layers in the lower walls of the trench before the more recent layers were deposited (see Fig. 14-39). By dating the materials using radiocarbon dating or other methods, the timing of past earthquakes can then be determined with some accuracy. These detailed studies of ***paleoseismicity*** provide critical information about the characteristic frequency and magnitude of earthquakes along different segments of California's active faults.

Seismologists from several agencies maintain a network of seismograph stations and other instruments designed to monitor California's active faults. Highly sensitive seismometers (portable seismographs) monitor the microearthquakes along some fault segments to detect possible ***foreshocks*** of a large earthquake. To date, foreshocks have not proved to be reliable predictors of major quakes, although the

Fig. 13-18 Geologists have excavated and then peeled back one side of a trench to examine sediments offset by the San Andreas fault. (Source: Clahan, K.)

1975 Haicheng, China earthquake was predicted by foreshocks that began 4 days before the 7.3 event.

Scientists are also examining other possible earthquake precursors, including changes in electrical resistivity, in the amount of dissolved radon gas in groundwater, in water-well levels, and bulging of the ground surface. Some of these phenomena may result from changes in the rocks in the area of imminent rupture, particularly an influx of water prior to the quake. Other possible precursors include radio-wave signals and anomalous animal behavior.

In 1989, the U.S. Geological survey compiled a map showing the conditional probability of a major earthquake along various segments of the San Andreas fault system (Fig. 13-19). The map shows the probabilities during a 30-year period from 1988 to 2018, based on the known patterns of historic and prehistoric ruptures along various faults. The map had indicated a high probability of rupture on the segment of the San Andreas fault that later produced the Loma Prieta earthquake.

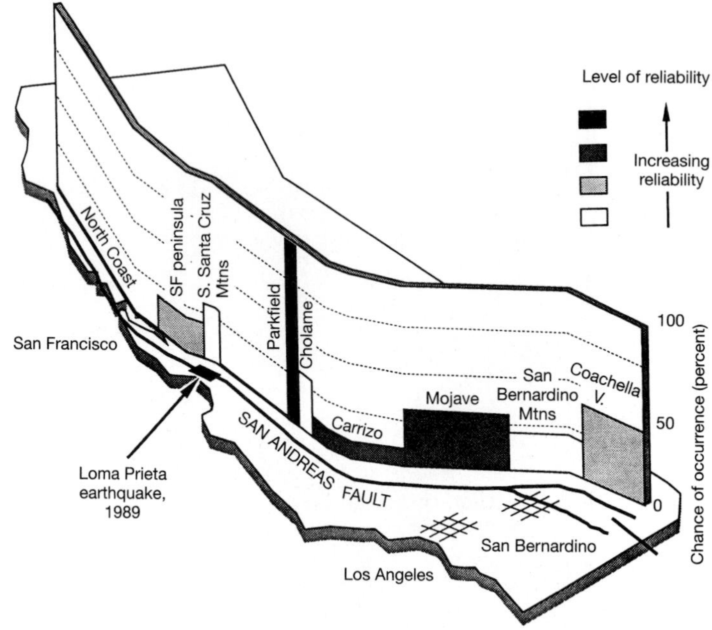

Fig. 13-19 Thirty-year probabilities for the San Andreas fault system for the period 1988-2018. (Source: U.S. Geological Survey, National Earthquake Hazards Reduction Program.)

PREDICTING GROUND SHAKING

Several characteristics of motion determine how much damage will result from ground shaking during an earthquake. These include the ***amplitude*** of ground motion, the duration of shaking, the frequency of the vibrations; and the peak acceleration of the ground. In general, large-amplitude waves cause greater damage to structures than low-amplitude waves. Duration of shaking is a critical factor, because as shaking continues, many structures lose their ability to withstand the motion. Engineers estimated that if the 15 seconds of shaking during the Loma Prieta earthquake had lasted a few additional seconds, the Embarcadero Freeway in San Francisco would have entirely collapsed. During large earthquakes like the 1964 Alaska event, ground shaking may last for three or more minutes.

The frequency of earthquake vibrations is one of the most important factors controlling damage because some materials and structures respond to vibrations by amplifying the motions at particular frequencies. If the frequency of some of the earthquake vibrations matches the natural ***resonant frequency*** of the soil beneath a structure, or of the structure itself, then the stress on the structure is significantly increased. Unconsolidated sediments or artificial fill with a high water content may greatly amplify ground shaking. Amplification is increased further if the natural frequency of the building matches that of the materials beneath it. For example, during the 1985 Mexico City earthquake, the greatest damage occurred in the areas where an ancient lake had deposited thick layers of clay. The clay layers amplified the motion of waves with a 2-second period. However, not all structures within the ancient lake area suffered equally. Because waves with a 1- to 2-second period match the natural frequency of 10- to 14-story buildings, these buildings showed far greater damage than either shorter or taller structures.

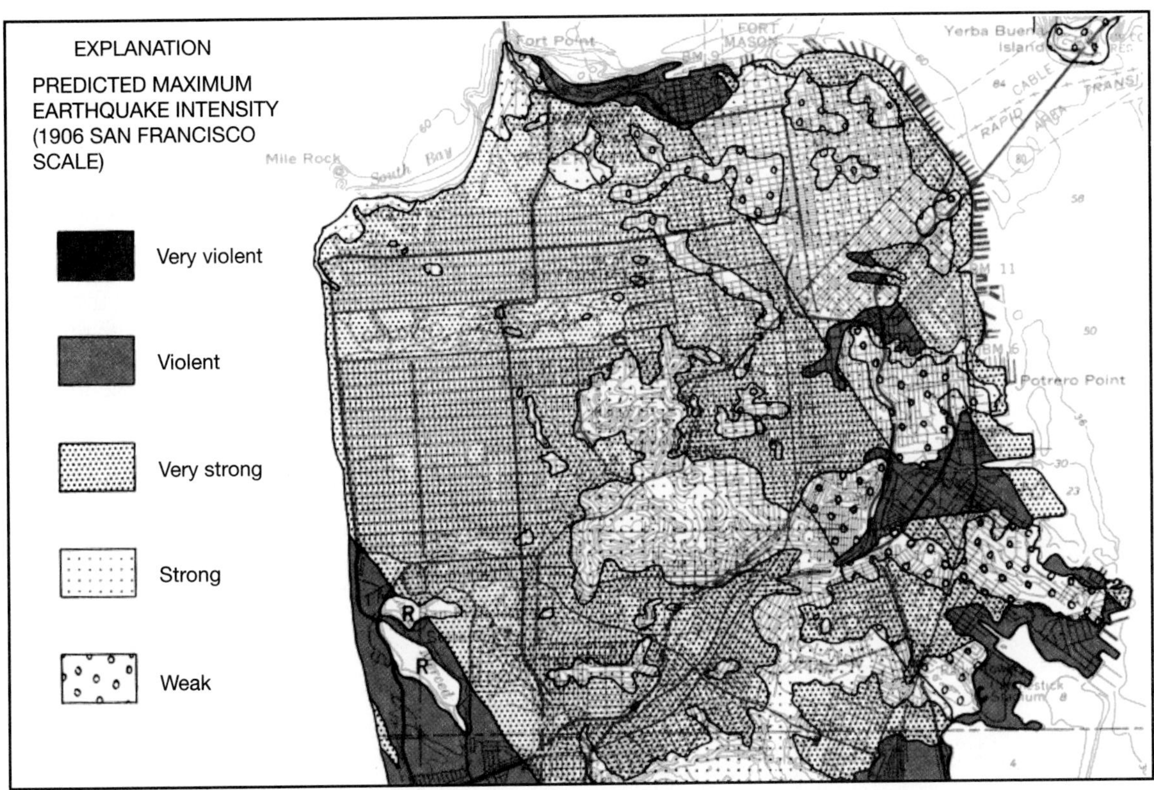

Fig. 13-20 Predicted severity of ground shaking in San Francisco. (Source: U.S. Geological Survey.)

The peak acceleration of the ground during an earthquake is another critical factor that influences the destructiveness of earthquakes. A structure exposed to vertical accelerations greater than 1*g*, or 100% of the force of Earth's gravity, would be airborne unless it was fastened to the ground. Until the 1971 San Fernando earthquake, seismologists had measured accelerations between 0.35 and 0.5*g* near the rupture zones of moderate-magnitude earthquakes. Until 1971, the peak acceleration of 0.5*g* was used by building engineers as the expected peak value. However, a peak horizontal acceleration of 1.25*g* was recorded during the 1971 San Fernando earthquake (discussed below). During the 1994 Northridge earthquake, surprisingly high horizontal accelerations of 1.8*g* and vertical accelerations of 1.2*g* were recorded at Tarzana, about 7 kilometers from the epicenter. The strong accelerations contributed to the extensive damage of modern structures during this earthquake. Understanding the correlation between geologic materials and the amplification of earthquake waves is an important step in designing earthquake–resistant structures. In San Francisco, a published map shows the predicted ground shaking for different parts of the city (Fig. 13-20). At the same time, continued instrumentation along California's faults is providing better estimates of expected ground motions, including peak acceleration, peak velocity, and duration of shaking. With this improving understanding of how the ground shakes during earthquakes, engineers are increasingly able to design earthquake-resistant structures.

BUILDING DESIGN

During an earthquake, structures respond to ground shaking in ways that reflect the overall building architecture and the characteristics of the materials used in construction. It is not possible to build structures that remain motionless during an earthquake, but it is possible to build structures that can withstand strong ground shaking. One of the more important factors in designing an earthquake-resistant structure is structural integrity. If all elements of a building are tied firmly together, then the structure will shake with a unified motion. If different elements of a building—for example, the foundation and the walls, the floors and supporting columns, two wings at right angles, or two ends of a bridge or freeway overpass—vibrate with different motion, then the building may shake apart.

Certain design elements can make a building particularly susceptible to earthquake damage. Examples are multistory houses with an open garage at the street

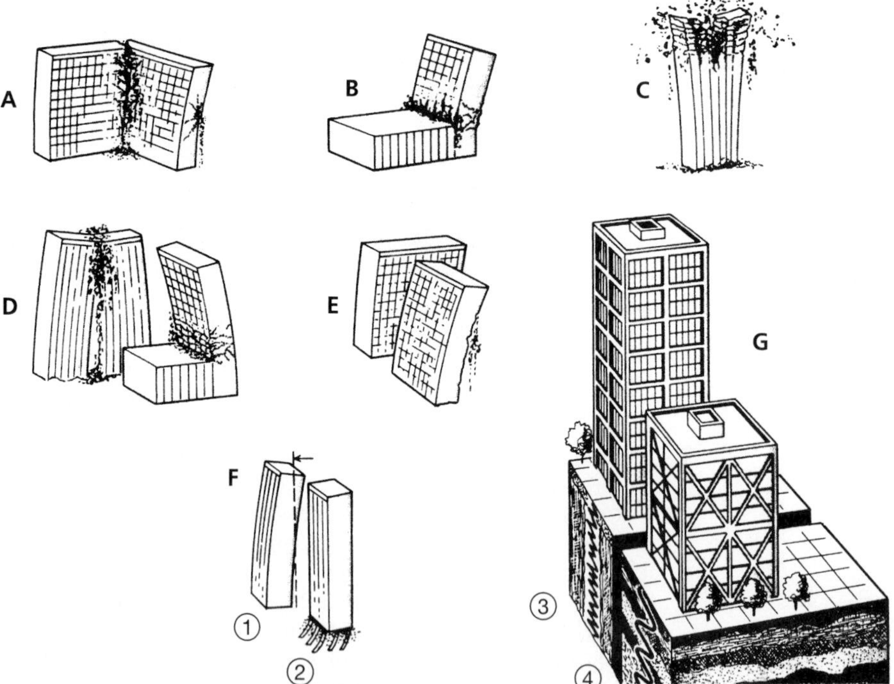

Fig. 13-21 Effects of ground shaking on high-rise buildings of various types. **A,** A building with wings at right angles experiences damage at the connection points because the wings respond differently; **B,** building elements with varying heights have different resonance frequencies; **C,** shaking is amplified at the top of a tall building; **D,** adjacent buildings with varying heights and shaking frequencies experience pounding; **E,** swaying is enhanced because of alignment relative to the direction of incoming waves; **F,** flexible high-rise buildings with different architecture: design *1* remains elastic, wherease the soft ground floor in *2* has no shear resistance; **G,** a pair of buildings on different soils: *3* on rock, which transmits higher-frequency waves, and *4* on softer soil layers, which are subject to wave resonance and require special bracing. (Source: Bolt, B. 1993. *Earthquakes,* 3rd ed. New York: W.H. Freeman, p. 251.)

level, split-level houses with large open areas on the lower walls, and houses constructed on stilts. In urban areas, strong ground shaking may damage high-rise buildings for several reasons, as illustrated in Fig. 13-21.

Wood-frame buildings are relatively resistant to earthquake damage because wood is relatively light and flexible. However, even these buildings can be destroyed or badly damaged if they have inadequate load-bearing walls, improper shear-wall bracing, or weak foundations. Many wood-frame buildings suffer needless damage during earthquakes because they are not securely anchored to the building foundation. Many pre-1940 buildings were not originally connected to their foundations. Fortunately, it is relatively inexpensive to attach wood-frame buildings to their foundations using anchor bolts.

The most dangerous structures in California are unreinforced brick buildings built before 1940. Masonry is heavy, and the strength of brick walls depends on the mortar holding the individual bricks together. For the same reason, brick chimneys and fireplaces, slate or tile roofs, decorative stonework hanging from upper stories, and masonry veneers can pose deadly hazards during an earthquake. Modern masonry is strengthened with reinforcing rods, and other brickwork built within the past 20 years is generally well anchored.

SEISMIC SAFETY AND GOVERNMENT REGULATIONS

The Long Beach earthquake, centered about 6 kilometers offshore of Newport Beach on the Newport-Inglewood fault zone, struck on March 10, 1933. Although it was a moderate event with a moment magnitude of 6.2, the earthquake killed 120 people, injured hundreds, and caused more than $40 million in damage. Water-saturated alluvial sediments underlying the Long Beach area were partially responsible for the destruction. But it was poor construction that accounted for the "tragic and scandalous failures of school structures" in the Long Beach area (Gordon B. Oakeshott, California Division of Mines and Geology, as quoted in *California Geology,* March 1973, p. 57). Luckily, the schools were almost vacant at 5:54 PM when the earthquake struck: officials estimated that the casualties would have been 10 times greater if schools had been in session. Following the Long Beach earthquake, the California legislature passed the ***Field Act.*** The Field Act requires state approval and inspection of both plans and construction of school buildings. This was California's first attempt to ensure seismic safety through government regulation. The Field Act has been amended twice since its passage. A 1967 addition required inspection of older school buildings, with replacement or strengthening mandated by 1975, but many financially strapped school districts have not yet come into full compliance. A second amendment, also adopted in 1967, requires all proposed school sites to be inspected to avoid siting within a fault zone or in a landslide area.

Early on the morning of February 9, 1971, a moderate earthquake (M 6.4) shook the densely populated San Fernando Valley of southern California. Although the shaking lasted only 15 seconds, the earthquake killed 58 people, destroyed two hospitals and a modern freeway overpass, and caused more than $500 million in damage. Had the shaking lasted 5 seconds longer, the tragedy would have been much worse. Overlooking the 80,000 residents of the San Fernando Valley stood the Van Norman Dam, built in 1915. During the earthquake, the upstream side of the dam surface broke away (Fig. 13-22). Partly because the reservoir was only slightly over half full, the dam held, and the residents below were spared from a disastrous flood.

If a single earthquake can be credited for California's current legislation regarding seismic safety, it would have to be the 1971 San Fernando earthquake.

Fig. 13-22 The Van Norman Dam, San Fernando Valley, following the 1971 earthquake. The geologists are standing immediately below cracks caused by the earthquake. (Source: U.S. Geological Survey.)

Because this earthquake and its effects were more accurately measured and thoroughly studied than any previous California earthquake, the San Fernando earthquake provided the basis for a number of legislative actions and code modifications designed to reduce earthquake hazards. The San Fernando earthquake and the 1983 Coalinga earthquake also helped to alert scientists to the threat posed by "blind" faults beneath the Los Angeles basin and other parts of California.

In 1927, the International Conference of Building Officials established the ***Uniform Building Code*** (UBC) in an attempt to bring standardization to all aspects of building construction. The UBC is updated and revised every 3 years, and since 1975, California cities and counties have been required by the California Health and Safety Code to adopt the most recent edition of the UBC. Specific sections of the UBC are designed to increase the resistance of structures to ground motion, for example with design requirements for resisting lateral forces. Building officials and municipal departments follow UBC guidelines when permitting and building new construction.

Extensive studies of relatively modern structures that were partly collapsed or seriously damaged during the 1971 earthquake revealed some troubling problems. Modern concrete-frame structures had inadequate reinforcing or insufficient shear bracing, and the walls and roofs of "tilt-up" concrete buildings were insufficiently tied together. Because modern buildings were damaged—despite being built to the standards in effect at the time—the UBC was extensively modified in 1973 and further strengthened in 1976. Future modifications are likely after studies of structural damage during the 1994 Northridge and 1995 Kobe, Japan earthquakes are completed. The surprising amount of damage to the welds in steel-frame buildings has been of particular concern.

The ***Alquist-Priolo Earthquake Fault Zoning Act*** (before 1994, called the Special Studies Zones Act) was passed into law in 1972. The Alquist-Priolo Act prohibits the construction of most human-occupied structures within 50 feet of an active fault. The faults zoned under the Alquist-Priolo Act are shown on official maps issued by the California Division of Mines and Geology (Fig. 13-23). These maps show the location of all faults with known surface displacement during the past

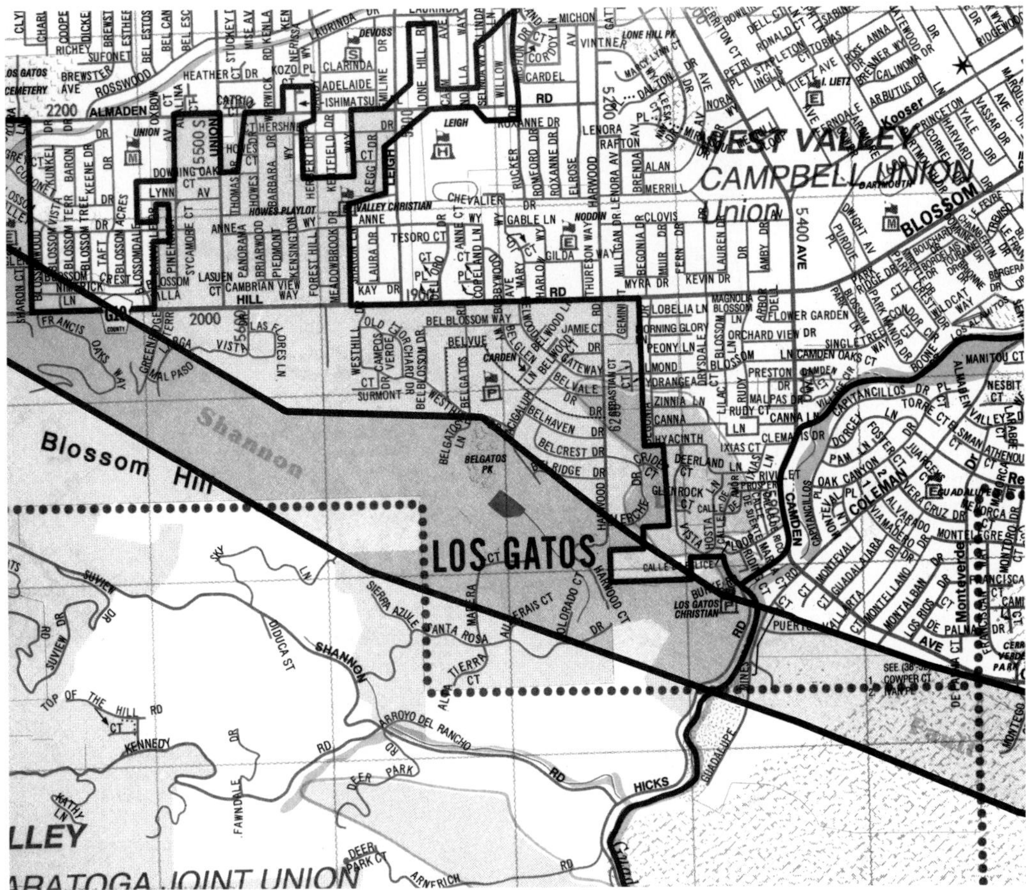

Fig. 13-23 Special study zone along the Shannon fault in Los Gatos, Santa Clara County, required by the Alquist-Priolo Act. (Source: Barclay Mapworks, Inc., Hayward, Calif., 1997.)

11,000 years, as well as faults considered to be potentially active with a high potential for ground rupture. Because the Alquist-Priolo Act prohibits most new construction across active faults, owners of pre-1972 buildings within the 50-foot setback zone find themselves unable to obtain permits for any new construction, including remodeling. Another important provision of the Alquist-Priolo Act is that property owners and their agents are required to disclose the existence of any active faults in the vicinity of a property being sold. Enforcement of the Alquist-Priolo Act and the granting of exemptions are the responsibility of local planning agencies.

The ***Hospital Safety Act,*** also passed following the San Fernando earthquake, called for the strengthening of construction standards for hospitals. The ***Dam Safety Act*** of 1973 required evaluations of the safety of existing dams in California. Part of the evaluation included preparation of maps showing the potential areas that would be flooded if a dam did fail. The Dam Safety Act also requires preparation of evacuation plans in the event of dam failure. The statewide dam inspections that followed passage of the Act resulted in a number of existing dams being strengthened for increased seismic safety. An amendment to the California Government Code, also passed fol-

Fig. 13-24 The Pacific Garden Mall, Santa Cruz, following the Loma Prieta earthquake. (Source: Nakata, J. U.S. Geological Survey.)

lowing the 1971 earthquake, requires every county and city to adopt a *Seismic Safety Element* as part of its general plan. Despite the development of seismic safety elements in the state's laws and building codes, older unreinforced brick or masonry (URM) buildings still pose one of our greatest hazards. For example, 46 URM buildings were demolished in the city of Santa Cruz during the 1989 Loma Prieta earthquake. This included historic Pacific Garden Mall, where several people died beneath the brick rubble (Fig. 13-24). The ***Unreinforced Masonry Building Law*** was signed into law in 1986. This law required California's local governments to make an inventory of unreinforced masonry (URM) buildings and to develop earthquake hazard mitigation programs for those buildings by 1990. As of 1992, 209 local agencies had complied with the law, but many communities had not completely developed URM programs, and only 53% of cities and counties had mandatory strengthening programs. Just in the seismically vulnerable metropolitan areas of the San Francisco Bay Area and the Los Angeles Basin, about 25,000 URM buildings with an average size of 10,000 square feet have been inventoried to date. In 1992, California established a uniform building code for earthquake hazard reduction in URM buildings with load-bearing walls. The URM Law could result in an estimated $4 billion effort to reduce earthquake hazards by the beginning of the next decade. In addition to the financial burdens of strengthening the buildings, many local governments would like to exempt both their historic buildings and their affordable housing from the code.

EARTHQUAKE PREPAREDNESS

Since 1769, 206 known earthquakes with magnitudes estimated or measured at 6 or greater have occurred in California, western Nevada, and the northern part of Baja, California, Mexico. Of these, 117 took place on the San Andreas fault system (see Chapter 14). Despite a greater public awareness of earthquake risk and increasingly earthquake–resistant building construction, it is not likely that California's vulnerability to earthquakes will decrease during the coming decades. As the population

COMMON EARTHQUAKE FEARS

One of the most frequently asked earthquake questions is "Is California going to fall into the ocean?" Another common fear is that an earthquake will cause the ground to open up and suddenly swallow entire buildings. Fortunately, neither of these popular myths is in California's earthquake scenarios. Although seaside cliffs and their resident structures may indeed tumble to the sea during an earthquake-induced landslide, these are localized catastrophes. As we will see in Chapters 15 and 16, the vertical element of motion along most of California's faults is slowly causing much of the coast to rise relative to sea level. Furthermore, even the largest expected earthquake will rupture only a portion of the San Andreas system, and most of that motion will be right lateral.

increases, urban areas and freeway corridors become increasingly congested, and this trend alone means that future moderate and large earthquakes will cause greater loss of life and property.

The U.S. Geological Survey, the California Office of Emergency Services, and local governments have prepared a series of "scenarios" that would follow a likely earthquake in the San Francisco Bay Area or the greater Los Angeles area (see Further Reading at the end of this chapter). In all cases, the scenarios predict large residential and commercial areas left without power or transportation access. Large numbers of people would be unable to obtain emergency services for 72 hours or more, and thousands could be stranded far from home by disabled transportation systems and freeways. It is important for all Californians to be prepared for earthquakes, not only at home, but also for earthquakes that may strike when they are away from home. Earthquake planning and preparedness may not eliminate all risks, but it can reduce the anxieties of people stranded far from their loved ones, pets, and households (see Chapter 19).

FURTHER READING

General Publications

Association of Bay Area Governments. 1993. *On Shaky Ground.* Oakland, Calif.
Maps showing seismicity and hazards in the San Francisco Bay Area.

Blair, M.I., and Spangle, W.E. 1979. *Seismic Safety and Land-Use Planning: Selected Examples from California.* U.S. Geological Survey, Professional Paper 941-B.

Boatwright, J., and others. 1995. *Ground Shaking in San Francisco.* U.S. Geological Survey.
Colored map and text illustrating predicted ground shaking in San Francisco.

Bolt, Bruce, 1993, Earthquakes: W.H. Freeman and Company, 3rd edition, 331 p.

Dvorak, J., and Peek, T. 1993. *Swept Away: The Deadly Power of Tsunamis.* Earth Magazine, July 1993, pp. 52-59.

Gore, R. Living with California's faults. *National Geographic* 187(4): 2-35, 1995.

GOVERNOR'S OFFICE OF EMERGENCY SERVICES. *Beat the Quake: California Earthquake Preparedness.* California Office of Emergency Services.

Brochure with important contacts to assess and assist in preparedness.

JENNINGS, C.W. 1994. *The Fault Activity Map of California and Adjacent Areas.* California Division of Mines and Geology, scale 1:750,000.

JENNINGS, C.W. 1995. New fault map of California and adjacent areas. *California Geology* 48(2):31-42.

An explanation of the California fault map cited above.

Surviving the Big One (video). Los Angeles, Calif.: KCET-TV.

This video, produced by Los Angeles television station KCET, is available from local libraries.

MONASTERSKY, R. Abandoning Richter: How a white lie finally caught up with seismologists. *Science News* 146:250-252, 1994.

U.S. GEOLOGICAL SURVEY. 1994. *The Next Big Earthquake in the Bay Area May Come Sooner Than You Think. Are You Prepared?* Menlo Park, Calif.

Available in English, Spanish, Chinese, Braille, and Recordings for the Blind.

YANEV, P.I. 1991. *Peace of Mind in Earthquake Country.* San Francisco: Chronicle Books, 218 pp.

ZIONY, J.I., ed. 1985. *Evaluating Earthquake Hazards in the Los Angeles Region—an Earth-Science Perspective.* U.S. Geological Survey, Professional Paper 1360, 505 pp.

Earthquake Planning Scenarios, Published by the California Division of Mines and Geology

DAVIS, J., AND OTHERS. 1982. *Earthquake Planning Scenario for a Magnitude 8.3 Earthquake on the San Andreas Fault in Southern California.* Special Publication 60, 128 pp.

DAVIS, J., AND OTHERS. 1982. *Earthquake Planning Scenario for a Magnitude 8.3 Earthquake on the San Andreas Fault in the San Francisco Bay Area.* Special Publication 61, 160 pp.

REICHLE, M., AND OTHERS. 1990. *Planning Scenario for a Major Earthquake in the San Diego–Tijuana Metropolitan Area.* Special Publication 100, 189 pp.

STEINBRUGGE, K., AND OTHERS. 1987. *Earthquake Planning Scenario for a Magnitude 7.5 Earthquake on the Hayward Fault, San Francisco Bay Area.* Special Publication 78, 242 pp.

TOPPOZADA, T., AND OTHERS. 1988. *Planning Scenario for a Major Earthquake on the Newport-Inglewood Fault Zone.* Special Publication 99, 200 pp.

TOPPOZADA, T., AND OTHERS. 1993. *Planning Scenario for a Major Earthquake on the San Jacinto Fault Zone, San Bernardino Area.* Special Publication 102, 208 pp.

TOPPOZADA, T., AND OTHERS. 1995. *Planning Scenario in Humboldt and Del Norte Counties, California, for a Great Earthquake in the Cascadia Subduction Zone.* Special Publication 115, 159 pp.

CHAPTER

14

The San Andreas Fault System

SETTING OF THE SAN ANDREAS SYSTEM

The San Andreas fault system marks the boundary between the Pacific and North American Plates. From the northern tip of the Gulf of California at its southeastern end, the San Andreas system follows a northwest trend for 1350 kilometers to its northwestern end at the Mendocino triple junction. At the heart of the system lies the San Andreas itself (Fig. 14-1), arguably the world's most famous fault. Running parallel to the San Andreas fault are several other active members of the system (Fig. 14-2 and Endpaper 2). These include the Rodgers Creek, Hayward, and Calaveras faults in the San Francisco Bay area, the Newport-Inglewood, Whittier, and Palos Verdes faults in the Los Angeles Basin, and the Elsinore, San Jacinto, and Imperial faults in southernmost California. Some geologists and seismologists also consider the Eastern Mojave Shear Zone in the Mojave Desert to be a member of the San Andreas system.

Fig. 14-1 Aerial view of the San Andreas fault, Carrizo Plain. (Source: Wallace, R. U.S. Geological Survey.)

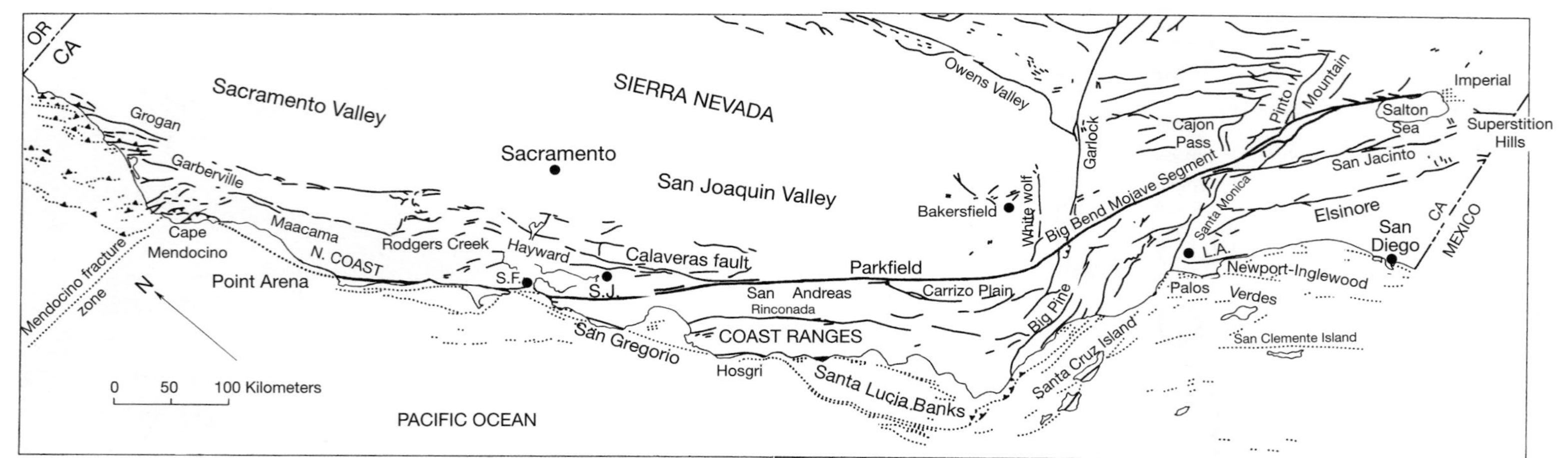

Fig. 14-2 Map showing segments of the San Andreas fault and subsidiary faults of the San Andreas fault system. (Source: Wallace, R. 1990. U.S. Geological Survey, Professional Paper 1515.)

As outlined in Chapter 1, the direction of movement of the two plates results in relative right-lateral motion between them, and most of the faults in the San Andreas system show this same movement characteristic. Along its entire length, movement takes place along one or more active faults. But here the similarities along the San Andreas fault system come to an end. In some areas, movement is confined to a single, narrow fault zone, whereas in others, multiple, complex ***fault strands*** mark the plate boundary. At its northern end, the San Andreas system is about 100 kilometers wide, whereas in southern California, the zone encompassing the active faults of the system is 300 kilometers across.

Sections of the San Andreas fault where faulting is confined to one or more linear zones of right-lateral movement are seen where the fault is parallel to the plate boundary (see Fig. 14-2). These relatively simple ***fault segments*** include the North Coast section and the segment near Parkfield. In contrast, sections of the fault system where the San Andreas fault system is not parallel to the plate boundary are segments of complexity. The most obvious of these segments is known as the Big Bend, but a less pronounced bend is found near San José. As shown in Fig. 14-3, left-stepping bends in the San Andreas fault cause convergent motion between the two sides of the fault rather than pure right-lateral slip. The convergence results in compression between the two sides of the fault, and these fault segments show folding, faulting, and uplift associated with compressional forces (see Chapter 1). Right-stepping bends in some of the faults of the San Andreas system produce local areas of extension, referred to as ***pull-apart basins.*** In these areas, the crust stretches, forming a basin where young sediments may accumulate (see Fig. 1-14).

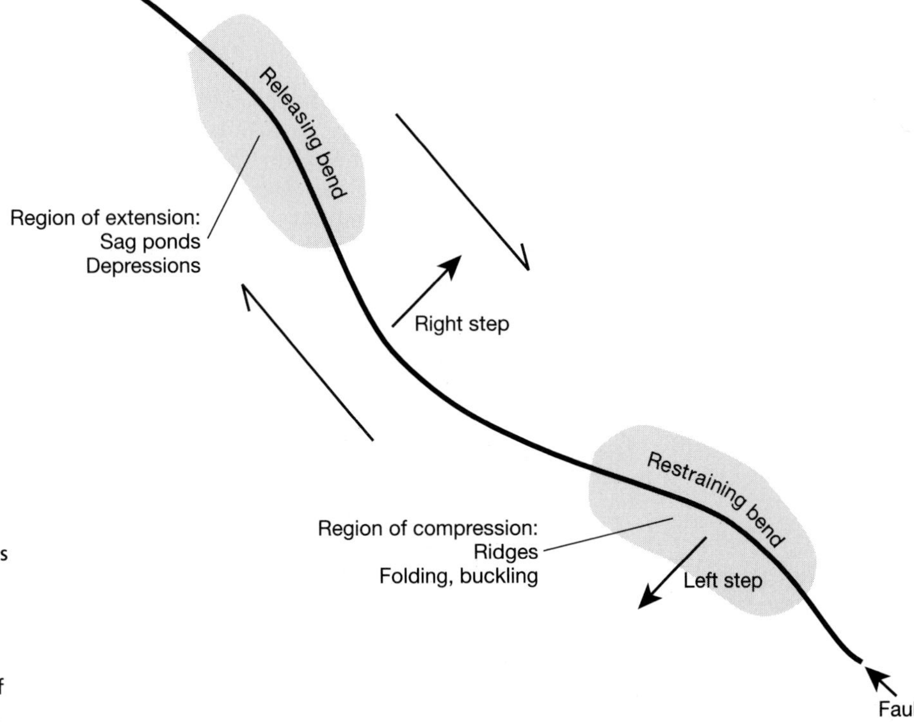

Fig. 14-3 Bends in the faults of the San Andreas system create areas of compression and extension. Bends and their effects are seen at all scales along the faults of the San Andreas system.

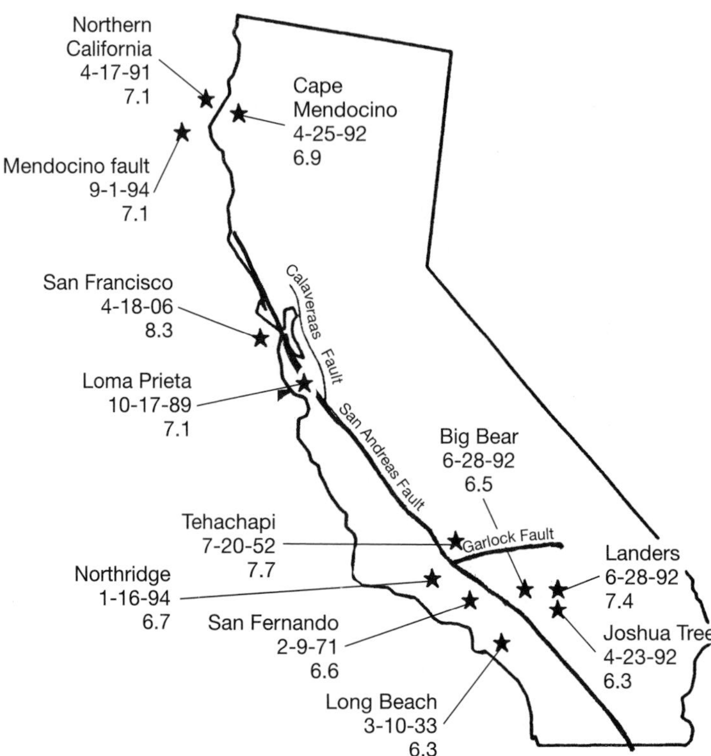

Fig. 14-4 Map showing the epicenters and moment magnitudes of major California earthquakes, 1906-1994. (Source: Updated from *Geotimes,* 1993.)

The contrasts in the appearance of the San Andreas system along its length correspond with contrasts in earthquake behavior. Since 1769, 117 earthquakes with magnitudes greater than or equal to 6 have occurred on the San Andreas system (Fig. 14-4). Some parts of the San Andreas system experienced no large earthquakes, whereas others ruptured in great earthquakes of magnitude 8.0 or greater. An understanding of each segment of the San Andreas fault and its subsidiaries is critically important for earthquake prediction.

GEOLOGIC HISTORY OF THE SAN ANDREAS SYSTEM

Fault Offset and Slip Rates

At the Skinner Ranch in Marin County, a dairy farmer's fence and his barn straddled the San Andreas fault, running almost perpendicular to the fault trace. On the morning of April 18, 1906, this fence was instantaneously separated as the 1906 earthquake ruptured the ground surface along the fault. When geologist Grove Karl Gilbert examined the area shortly after the earthquake, he measured 4.7 meters of displacement, or ***offset,*** of this fence (Fig. 14-5). When he estimated the 1906 offset of the fence, Gilbert assumed that no displacement had occurred from the time that the fence was built until the 1906 earthquake. Other objects sitting on the fault trace showed similar offsets: the corner of the barn's foundation, a line of raspberry bushes, and a path. All of the objects recorded the same direction of offset, with the western side of the fault moving northwest relative to the eastern side, the characteristic right-lateral motion of

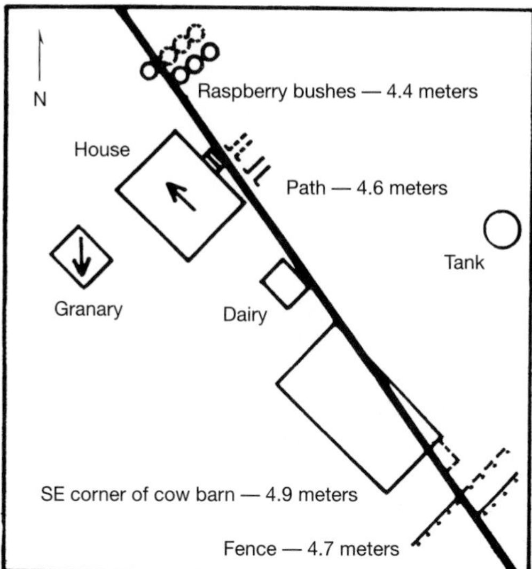

Fig. 14-5 The 1906 offset of a fence at Skinner Ranch, Marin County. Geologist G.K. Gilbert measured the displaced features shortly after the earthquake. Location is in the vicinity shown in Fig. 13-12. Broken lines show the positions of bushes, fences, and path before the earthquake. (Source: Modified from Lawson, A. 1908.)

all movements along the San Andreas. Today some of the offsets at the old Skinner Ranch, reconstructed from Gilbert's photographs and notes, can be seen along the earthquake trail at the Point Reyes National Seashore's Visitor Center.

Since 1906, no movement has occurred along the northern part of the 1906 rupture, and the fence is still offset by 4.7 meters. If the offset were averaged for the 90 years since the earthquake, an average rate of 5.2 centimeters per year would be obtained. Of course, the locked segments of the San Andreas fault system do not move at an average pace: at Skinner Ranch, the rate of offset is normally zero, punctuated by meters of movement within seconds. However, an understanding of the long-term rate of offset is important for two reasons. First, it provides geologists with a clue to the history of motion along the plate boundary. Second, if the long-term offset rate is known, segments of a fault which lag behind the long-term rate can be identified as "overdue" for significant movement—that is, a major earthquake.

Considering that great, ground-rupturing earthquakes occur only every century to every few centuries, the average rate of 5.2 centimeters per year at Skinner Ranch between 1906 and 1996 is far too high to be a representative long-term rate. A more reliable estimate of the fault's "average" behavior would be made over a longer time period and incorporate several earthquakes.

To determine the long-term rate of offset, the same principles can be applied to geologic features that have been separated by an active fault. A topographic feature such as a stream channel can be offset in the same way as Skinner's fence (Fig. 14-6). However, determining the rate of offset of this natural feature involves much more uncertainty, because the original course of the stream crossing the fault was probably not straight. Matching up the deposits of a particular stream on either side of the fault can be difficult if several streams are flowing in the area. Finally, establishing the precise ages of the deposits or landforms can also pose problems.

Despite all of these uncertainties, geologists working in the vicinity of Skinner Ranch have used offset stream deposits to conclude that offset or ***slip rate*** has been between 2.1 and 2.7 centimeters (21 to 27 millimeters) per year during the past 1800 years (Fig. 14-7). If all of this displacement occurs during 1906-type earthquakes,

A

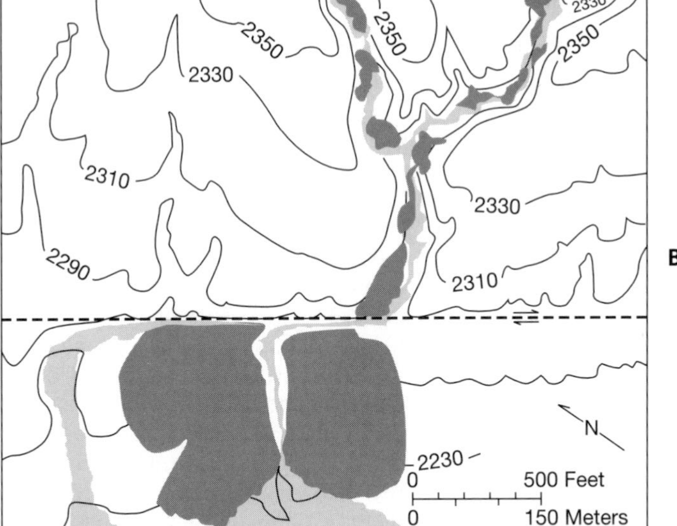

B

Fig. 14-6 **A,** Offset of Wallace Creek along the San Andreas fault, Carrizo Plain. **B,** Stream deposits used to calculate the offset are shaded. (Contour line elevations are given in feet.) (Source: Wallace, R. 1990. U.S. Geological Survey, Professional Paper 1515.)

then eight or nine earthquakes would be needed to create the total displacement of the stream deposits. This suggests that a great earthquake would be expected about every 190 to 260 years along the northern San Andreas fault. At a site near Cajon Canyon, 75 kilometers east-northeast of Los Angeles, alluvial sediments offset by the San Andreas fault have been matched across the fault. After dating those offset sediments, geologists calculated an average slip rate of about 2.5 centimeters per year for this part of the San Andreas fault.

As discussed in Chapter 1, the relative motion between the Pacific and North American Plates has been calculated at about 4.9 centimeters per year during the past 3 to 4 million years. This rate is obviously much greater than the 2.5 to 3 centimeters per year documented along the San Andreas fault. Even when the slip rates along subsidiary faults of the San Andreas system are added to the San Andreas slip rate, only about 35 millimeters (3.5 centimeters) per year can be accounted for in central California. Geologists and geophysicists believe that the "missing" right-lateral plate

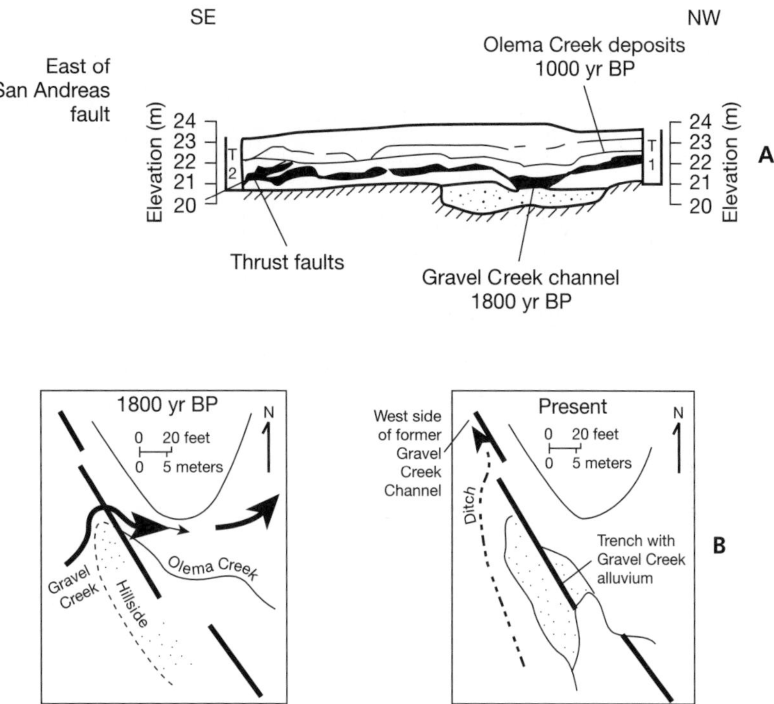

Fig. 14-7 Slip rates can be calculated from offset stream deposits if the offset channels can be matched across the fault and dated. **A,** The deposits of Gravel Creek as they appeared in a trench excavated on the eastern side the San Andreas fault near Skinner Ranch, Marin County. Similar deposits were identified and dated on the western side of the fault. **B,** The reconstructed positions of Gravel Creek, based on correlation and dating of the alluvial deposits. (*BP* = before present.) (Source: Niemi, T, and Hall, N.T. 1993. *Geology.* Geological Society of America.)

motion is occurring on faults east of the San Andreas system—along the eastern side of the Sierra Nevada and in the Basin and Range Province (see Chapter 7).

How Old Is the San Andreas Fault? Long-Term Slip Rates

As right-lateral motion continues on the San Andreas fault system, geologic features that originally straddled the fault traces become separated by greater distances. Older geologic units are offset more than recent deposits and landforms because they have been subjected to a longer period of displacement. Many geologists working in California are actively searching for rocks, faults, and other geologic features that can be matched up across the faults of the San Andreas system. These studies enable more refined estimates of slip rates to be made, even allowing geologists to see changes in the rate of slip along the system. It is also possible to determine which faults in the system are more recent breaks and thus to detect the birth of new fault strands and the deactivation of older ones. Studies of older offset rocks and features have also enabled geologists to determine the maximum amount of displacement on the San Andreas fault.

One geologic feature that has been matched across the San Andreas fault is the Pinnacles volcanic field, located on the west side of the fault in central California (Figs. 14-8 and 14-9). The volcanic rocks at Pinnacles National Monument are a series of high-silica volcanic flows that are overlain by pyroclastic rocks, including distinctive layers of perlite, pumice tuff, and breccias (see Chapters 5 and 12). On the eastern side of the San Andreas fault, 315 kilometers southeast of the Pinnacles near Gorman, is the Neenach Volcanic Formation, where a sequence of volcanic rocks remarkably similar to those at the Pinnacles is found (Fig. 14-9). Radiometric dating has shown that the rocks were erupted in early Miocene time, about 23.5 million years ago.

Fig. 14-8 Silicic volcanic rocks at Pinnacles National Monument. (Photo by author.)

Along the entire San Andreas fault proper, all units older than the Pinnacles-Neenach volcanic fields show about the same amount of offset, approximately 315 kilometers. The consistent amount of offset of older materials leads geologists to conclude that the modern San Andreas fault began its history of right-lateral faulting after 23.5 million years ago. Since that time, it has moved a total of 315 kilometers, or at a long-term average slip rate of 1.34 centimeters per year.

The Pelona and Orocopia schists are distinctive older rocks that have been used to calibrate the offset across the southern San Andreas fault system. These are sedimentary and igneous rocks that were metamorphosed during Mesozoic thrust faulting. Today the schist is exposed in the Sierra Pelona, Orocopia, and Chocolate Mountains (Fig. 14-10). The axis of a fold that cuts through the Pelona Schist on the western side of the San Andreas can be precisely matched to the same fold axis in the Orocopia schist on the eastern side of the fault. Presently this feature is offset 220 kilometers across the southern San Andreas fault, almost 100 kilometers less than the offset of much younger features like the Pinnacles-Neenach volcanics. Geologists infer that this strand of the fault must be a much more recently formed break than that to the north.

Similar studies of offset rates and amount on other members of the San Andreas system indicate that right-lateral motion began on the ancestral faults of the modern San Andreas about 28 million years ago. The birth of the transform plate boundary followed the completion of the subduction of the Farallon Plate, as discussed in Chapter 12 (see Fig. 12-1).

Topography, Fault Characteristics, and Earthquake History Along the San Andreas System

When viewed from an airplane, almost any part of the San Andreas fault appears on the ground as a sharp linear scar that cuts across all features of the landscape (Figs. 14-1 and 14-11). The fault displaces ridges, valleys, and streams faster than they can be smoothed by erosion and deposition. In some areas, elongated depressions or ***sag ponds*** line up along the fault zone. Along other stretches, ridges are pushed up parallel to the fault like the diggings of giant gophers. The stream channels crossing the fault are often completely disrupted, some to the extent that they have become disconnected (Fig. 14-12). In fact, the landforms formed in active transform fault zones are important characteristics used to identify such faults (Fig. 14-13).

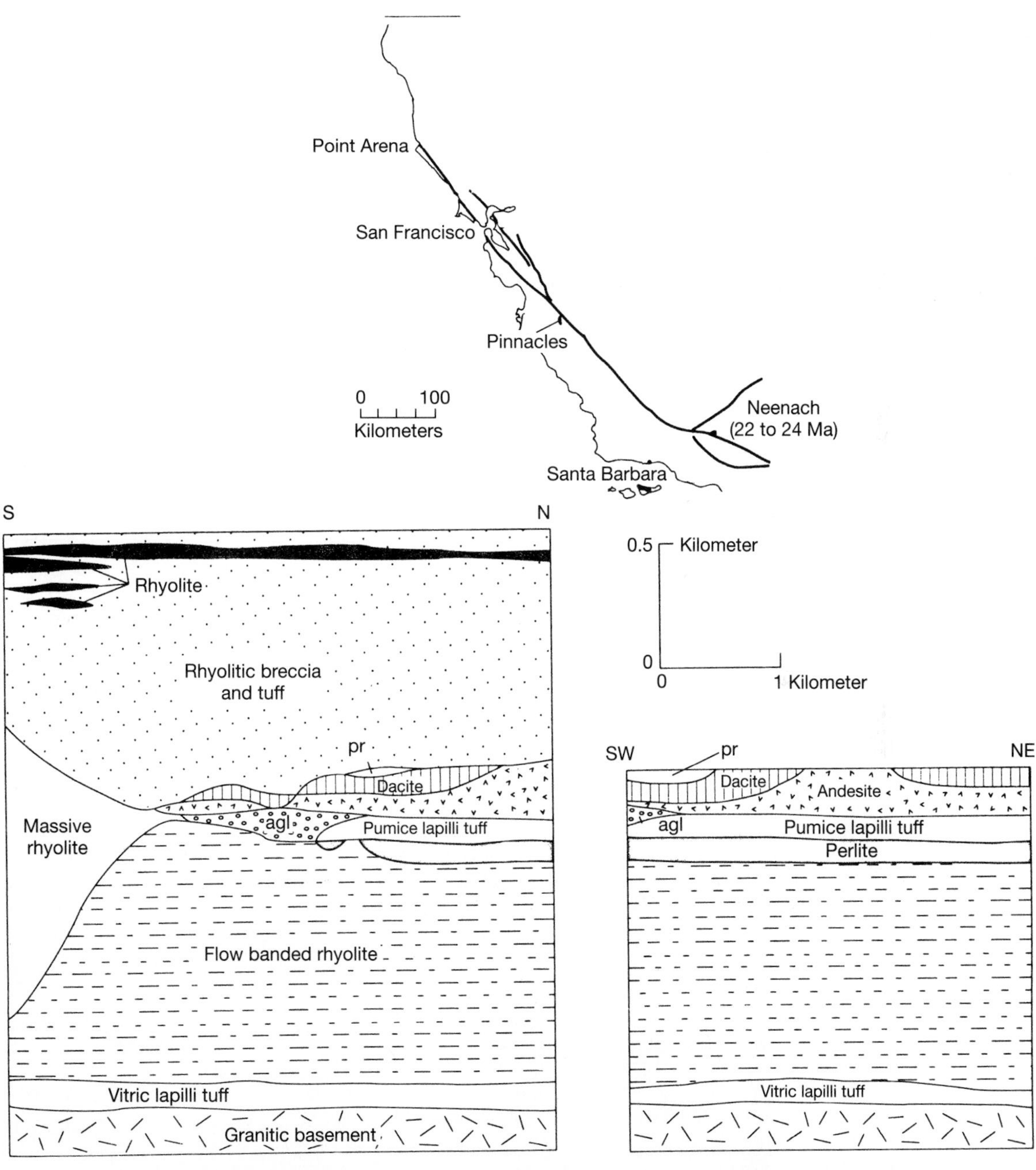

Fig. 14-9 Comparative cross sections of volcanic rocks seen at Pinnacles and Neenach volcanic fields, showing the remarkable similarity of units. (MA = million years ago.) (Source: Irwin, W.P. In Wallace, R. 1990. U.S. Geological Survey, Profession Paper 1515.)

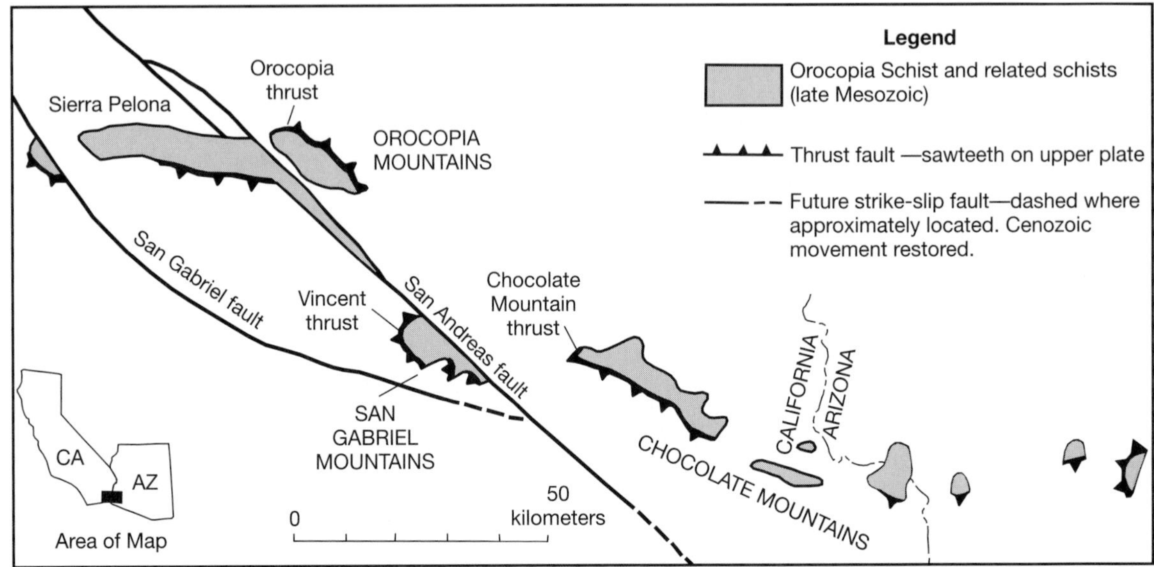

Fig. 14-10 Pelona and Orocopia schists and their offset along the San Andreas fault in southern California. (Source: Ehlig, P. In Geological Society of America. *The Cordilleran Orogen.* Decade of North American Geology, Vol. G-3, p. 236.)

Fig. 14-11 Sharp linear breaks in the topography mark the trace of the San Andreas fault, Carrizo Plains. A sag pond can be seen along the left fault trace. (Source: Wallace, R. U.S. Geological Survey.)

Fig. 14-12 Offset streams, Carrizo Plains. Wallace Creek (see Fig. 14-6) appears in the center of the photo. (Source: Wallace, R. U.S. Geological Survey.)

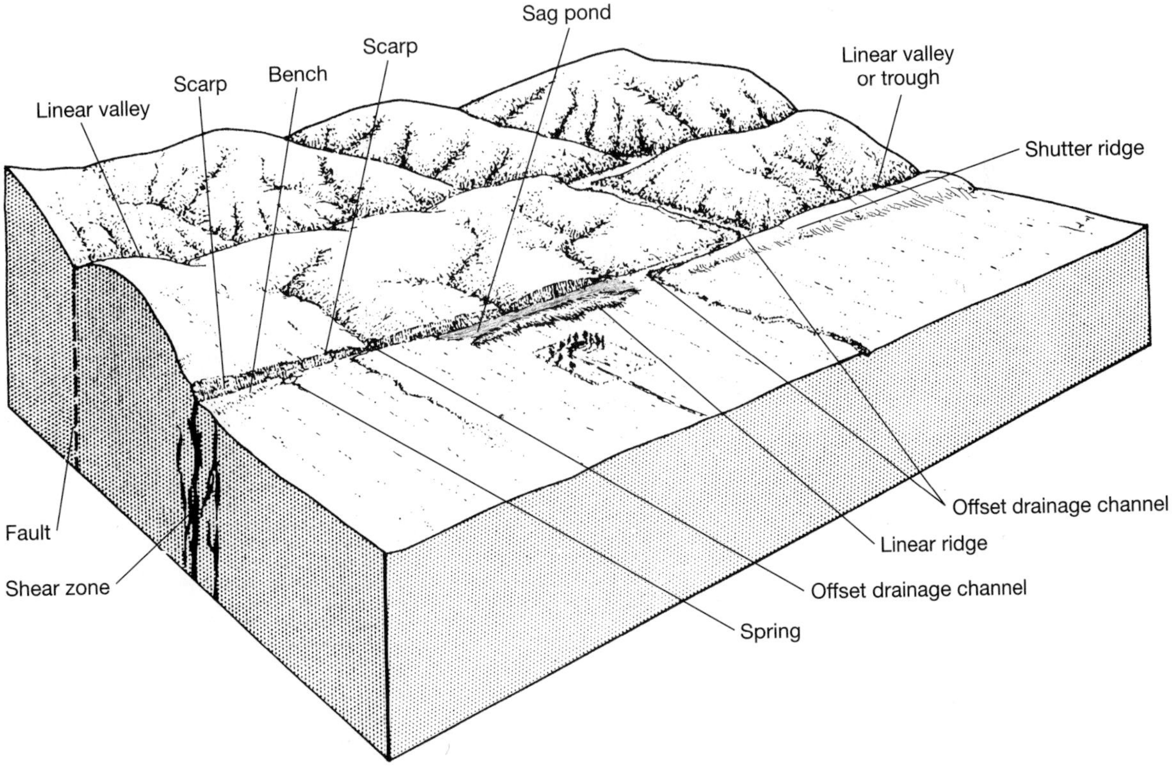

Fig. 14-13 Common landforms along the San Andreas fault system. (Source: Wallace, R. 1990. U.S. Geological Survey, Professional Paper 1515.)

The San Andreas fault system is a fundamental force shaping California's geology today. For that reason, it is important to understand the fault's history and the differences between different segments. A tour along the San Andreas fault system from Cape Mendocino to the Gulf of California highlights some characteristics of each section of the fault.

THE NORTH COAST AND THE MENDOCINO TRIPLE JUNCTION

Few Californians living outside northwestern California realize that the Cape Mendocino region is one of the most seismically active parts of the state. Because of the interactions among three crustal plates, a multitude of active faults cross northwestern California, each capable of generating earthquakes. The Pacific, North American, and Gorda Plates meet near Cape Mendocino at the Mendocino Triple Junction (see Chapter 1). The triple junction marks the northwestern end of the San Andreas fault system; north of this boundary, the Gorda Plate is being subducted beneath North America along the Cascadia subduction zone (Fig. 14-14). Because the exact configuration of the faults that define the plate boundaries is complex, the triple junction is an area rather than a single geographic point (Fig. 14-15). Most of the Cascadia subduction zone lies offshore, as do the associated faults and folds above it. However, southern Humboldt County is unique in that many of the active folds and faults can be followed from the offshore region onto land. This evidence suggests that the subduction zone itself extends on land, with the trench partly subducted beneath the Cape Mendocino region.

Earthquakes are generated along the northern San Andreas fault, the Mendocino fault, along several offshore faults within the Gorda Plate, and along thrust faults north of the triple junction within the North American Plate (Fig. 14-15). Moderate earthquakes from these sources are common, shaking the residents of Eureka and surrounding areas of Humboldt County on a regular basis. Because of the low population in Humboldt County, these earthquakes have resulted in less damage than if they had occurred in California's urban areas. However, a potentially deadly threat is posed by another earthquake source: recent geologic investigations indicate a

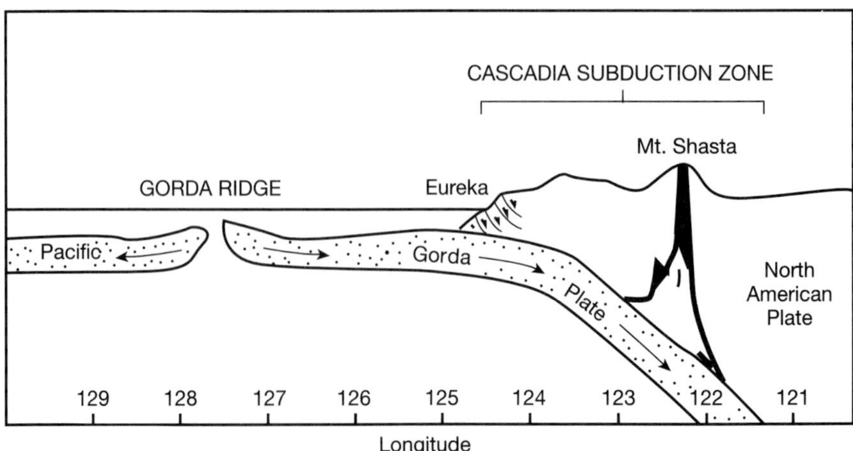

Fig. 14-14 Cross section of the Cascadia subduction zone. (Source: Dengler, L., and others. *California Geology,* 43[3], 1992.)

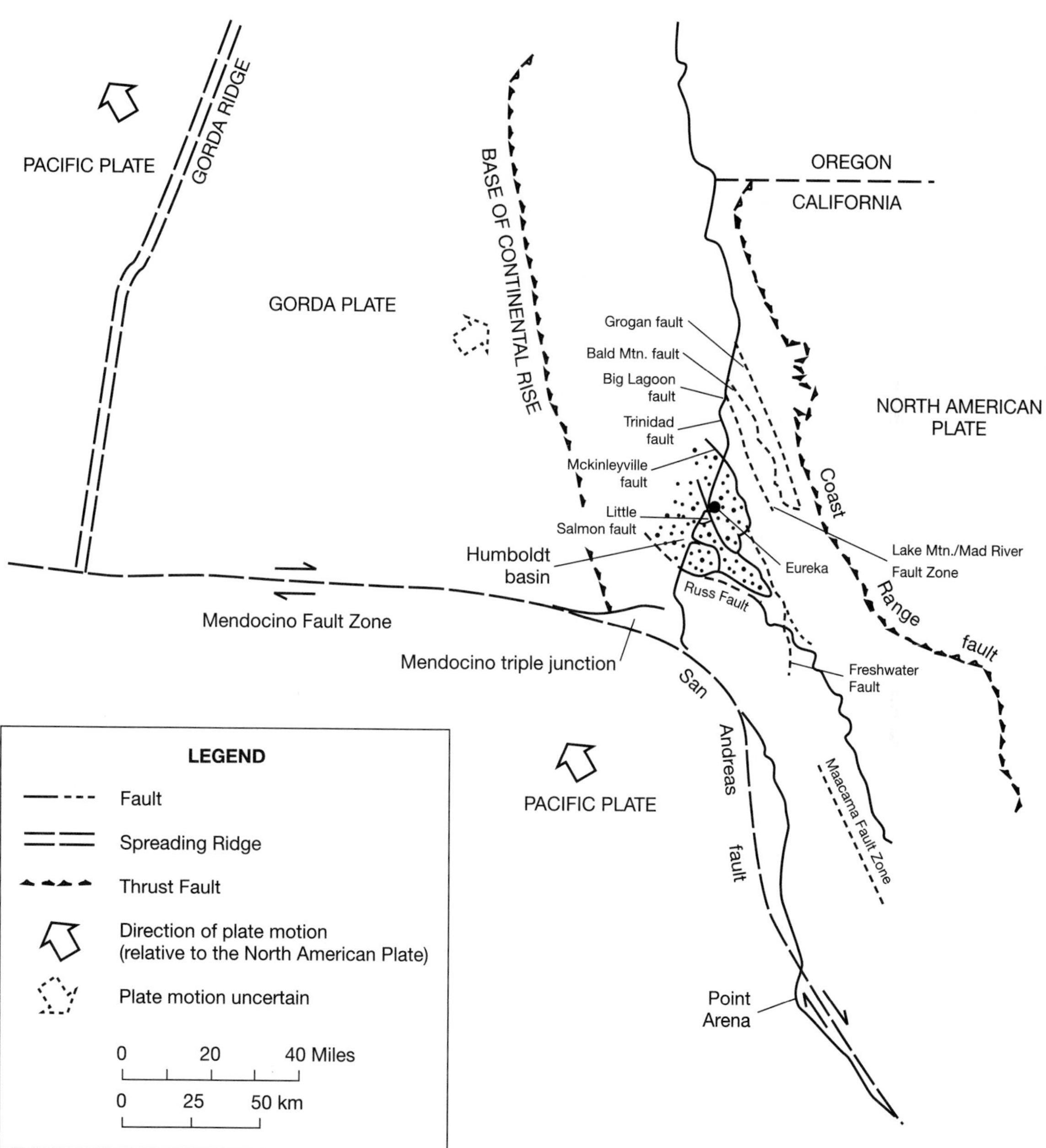

Fig. 14-15 Principal faults and other features of northwestern California. (Source: Modified from Sarna-Wojcicki, A., and others. 1982. *Friends of the Pleistocene Guidebook.*)

Fig. 14-16 Damage caused by the 1992 earthquakes at Ferndale, Humboldt County. (Source: National Geophysical Data Center.)

potential for great earthquakes generated by rupture along the subduction zone north of Cape Mendocino. The magnitude of a great earthquake would be between 8.5 and 9 and could rupture the entire Cascadia subduction zone, from Cape Mendocino as far north as British Columbia. The intervals between such earthquakes are thought to be long—from several hundred to as much as a thousand years. Nevertheless, the last of these earthquakes is believed to have occurred almost 300 years ago, so significant strain has accumulated in the rocks since that time.

The active faults and the earthquake history of northwestern California show the effects of both convergent and right-lateral plate motion. Active faults are numerous both on land and offshore, and Humboldt County experiences frequent earthquakes. North of Cape Mendocino, most features result from compression, whereas features produced by lateral motion are more common in the south. However, because the change in plate motions occurs over a broad area, active lateral and compressional features can be found throughout northwestern California.

Recent Earthquakes

On April 25 and 26, 1992, three earthquakes with magnitudes between 6.0 and 7.0 shook the region of Humboldt County near Cape Mendocino. The first and largest quake (M_W 7.0), centered just north of the small town of Petrolia, caused about $48 million in damages and injured 356 people. Geologists believe the event to be the first historic earthquake produced by thrust faulting along the Cascadia subduction zone. The earthquake triggered numerous landslides and shook houses off their foundations (Fig. 14-16); it also generated a 1-meter-high tsunami at Crescent City. The two earthquakes of April 26, with magnitudes of 6.0 and 6.5, resulted from right-lateral movement along a fault within the Gorda Plate. These events were felt as far away as San Francisco. The epicenters for the two later earthquakes were offshore, west of Cape Mendocino and north of the triple junction (Fig. 14-17). Only 8 months earlier, four large earthquakes with magnitudes between 6.3 and 7.1 had shaken the same area. Three of the 1991 events were located in the Gorda Plate north of the triple junction.

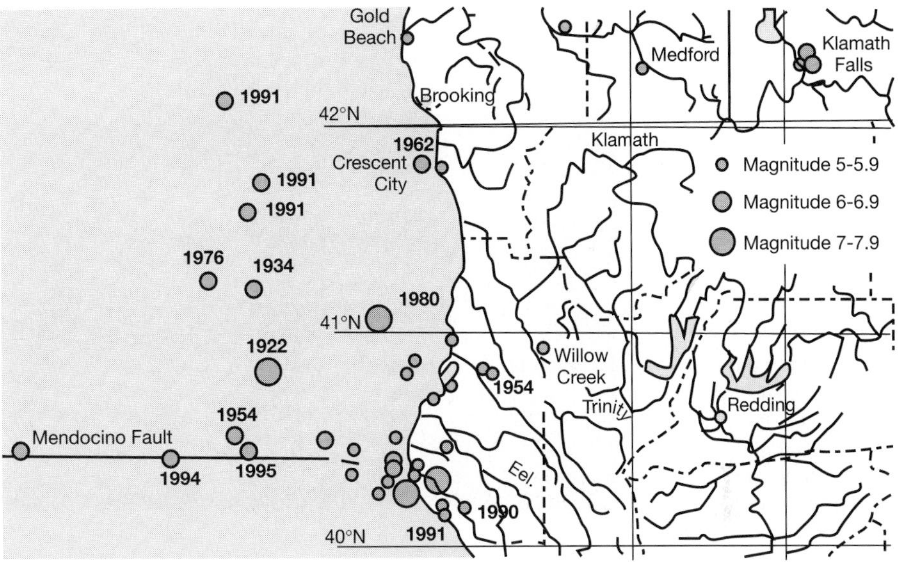

Fig. 14-17 Epicenters of recent earthquakes in northwestern California. (Source: Dengler, L., and others. 1993. *Living on Shaky Ground.* Arcata, Calif.: Humboldt State University Earthquake Education Center pamphlet.)

The fourth and largest earthquake, on August 17, was centered on land about 11 kilometers south of Petrolia near the town of Honeydew. The Honeydew earthquake damaged roads, bridges, and buildings and was felt in San Francisco.

Along the coast near Petrolia, local fishermen noted that the tides were extremely low in the days following the April 25 and 26 earthquakes. Along the beach, bivalves and other organisms normally below the tide level were suddenly lifted high and dry, and after about 2 weeks, these communities died (Fig. 14-18). Based on the elevation of these organisms above their normal habitat, researchers concluded that between 1 and 1.5 meters of uplift occurred during the April 26 earthquake. Careful documentation of this earthquake-triggered uplift, called ***coseismic uplift,*** has provided geologists with a model for similar events preserved in the geologic record as deposits and fossils along the California coast.

Convergence

At the surface above the Cascadia subduction zone is an area of intense compression where rocks, young sediments, soils, and landforms are being squeezed as the crust shortens. This area is known as a ***fold-and-thrust belt*** and is marked by numerous thrust and reverse faults and folds (see Fig. 14-15). As the three plates converge, compression is occurring in a northeast direction, so that the faults and the hinges of folds (*fold axes*) are oriented northwest. Sediments may be folded downward as well as upward when the crust shortens and buckles, as discussed further in Chapter 16.

One of the large folds in Humboldt County is the Freshwater syncline, a downward fold which runs northwest along the northern edge of Humboldt Bay and the mouth of the Mad River. The downstream end of the Mad River is a tidal channel or ***slough*** surrounded by salt marsh. North of the modern slough are layers of sediment deposited by the river during the past 1600 years (Fig. 14-19). Two types of deposits are present

Fig. 14-18 Tidal flats uplifted by the 1992 earthquake, Petrolia. (Source: Dengler, L., Humboldt State University.)

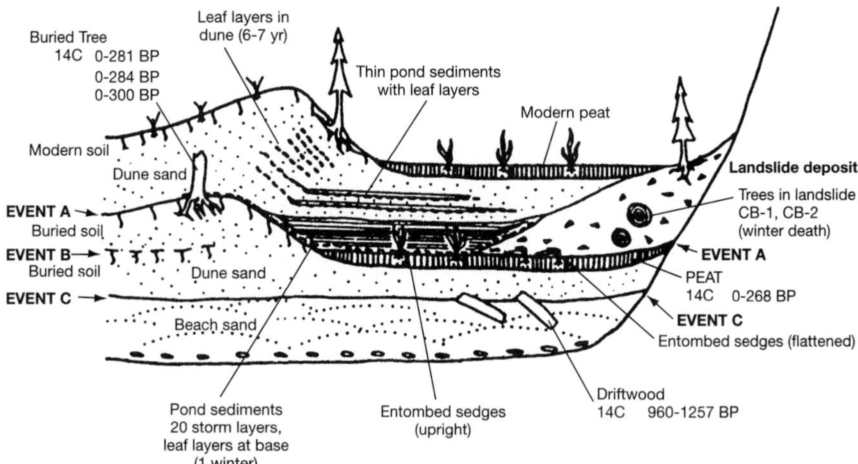

Fig. 14-19 Salt marshes drowned by an ancient earthquake, Clam Beach, north of the Mad River slough, Humboldt County. (Source: Carver, G., and Burke, R.M. 1982. *Friends of the Pleistocene Field Trip Guidebook.)*

within the layered section: estuarine mud layers formed in a tidal slough and peat layers formed by decomposed salt marsh plants. The salt marsh plants are buried upright by the overlying mud layers. At the edges of the marsh, redwood trees have been abruptly killed by burial in mud. After carefully describing the layers, geologists have concluded that the area experienced sudden subsidence (down-dropping) during earthquakes. The subsidence caused the marsh plants and nearby trees to be drowned. Mud then accumulated in the newly formed estuary until salt marshes were reestablished. Radiocarbon dating of the peat layers indicates that ***coseismic subsidence*** took place during earthquakes about 300, 1100, and 1600 years ago. Similar buried salt marshes found along the Oregon and Washington coasts are of about the same age. The similarity in ages along the entire length of the Cascadia subduction zone has led geologists to conclude that great earthquakes, with magnitudes as high as 9.0, have ruptured the entire zone in an earthquake similar to the 1964 Alaska earthquake. The most recent of these earthquakes may have occurred as recently as about 300 years ago.

THE NORTHERN SAN ANDREAS FAULT ZONE AND THE SAN FRANCISCO BAY AREA

From the Mendocino triple junction south to Point Arena, the San Andreas fault zone lies offshore. The Garberville and Maacama fault zones, which parallel the northernmost San Andreas, are also thought to be active, but have not slipped during historic time (see Endpaper 2). At Point Arena (see Figs. 14-2 and 14-15), the fault emerges from the Pacific and can be traced to Fort Ross, where it runs offshore. Further south, Tomales Bay is one of the San Andreas fault's most striking landforms, separating the Point Reyes Peninsula from the "mainland" of Marin County along a narrow slash that extends south of Tomales Bay to Bolinas Bay. From Bolinas Bay the fault lies offshore, west of San Francisco, until it intersects the San Francisco Peninsula and remains on land for the rest of its length.

From Point Arena south to San Juan Bautista and Hollister, the faults of the San Andreas system cross through the San Francisco Bay Area, home to 6 million people. In the same segment, movement on the San Andreas system is shared among four major right-lateral fault zones. All four cross through urbanized areas, and all show evidence of recent, damaging earthquakes (Fig. 14-20). The San Andreas fault northwest of San Francisco was the epicentral region of the great 1906 earthquake. Along much of this segment, the San Andreas is marked by linear valleys. The Rodgers Creek fault north of San Pablo Bay, the Hayward fault along the eastern margin of San Francisco Bay, and the Calaveras fault in the southeastern Bay Area (see Fig. 14-20) all display offset streams, sag ponds, and other characteristic features of active transform faults.

In addition to the "master" faults of the San Andreas system in the Bay Area, several reverse faults produced by compressional forces are thought to be active. These faults are oriented roughly parallel to the San Andreas system. In the Santa Clara Valley, these faults are found along the low mountains on the sides of the Almaden and Santa Clara valleys (Fig. 14-21). Geologists believe that the complexity of the San Andreas fault system in the Bay Area is at least partly caused by the slight bend in the orientation of the fault in the Santa Cruz Mountains.

Earthquake History

On the warm afternoon of October 17, 1989, 62,000 baseball fans filled Candlestick Park to watch the San Francisco Giants and the Oakland Athletics play the third game of the World Series. Thousands more had left work early to watch the contest at home, leaving the freeways between San Francisco and Oakland emptier than on a normal Tuesday at 5 o'clock. It would prove a fateful coincidence that saved lives. At 5:04 PM, the stands rocked and shook, and a cheer went up from the crowd. Two thousand feet overhead, the Goodyear blimp lost its TV link to the stadium. Within seconds, the fans fell silent, realizing that this was no swell of foot-stomping. As power to the stadium lights was lost, the sky darkened. Above, the blimp operator saw a flash as transformers exploded under a nearby highway. During the next hour, portable radios and the blimp transmitted the news: part of the Bay Bridge had fallen, fires were ablaze, power lost, and people dead. Remarkably, there was no panic in Candlestick Park that night as fans left the stadium, not knowing what damage they would find at home. It was the first time that a major league baseball

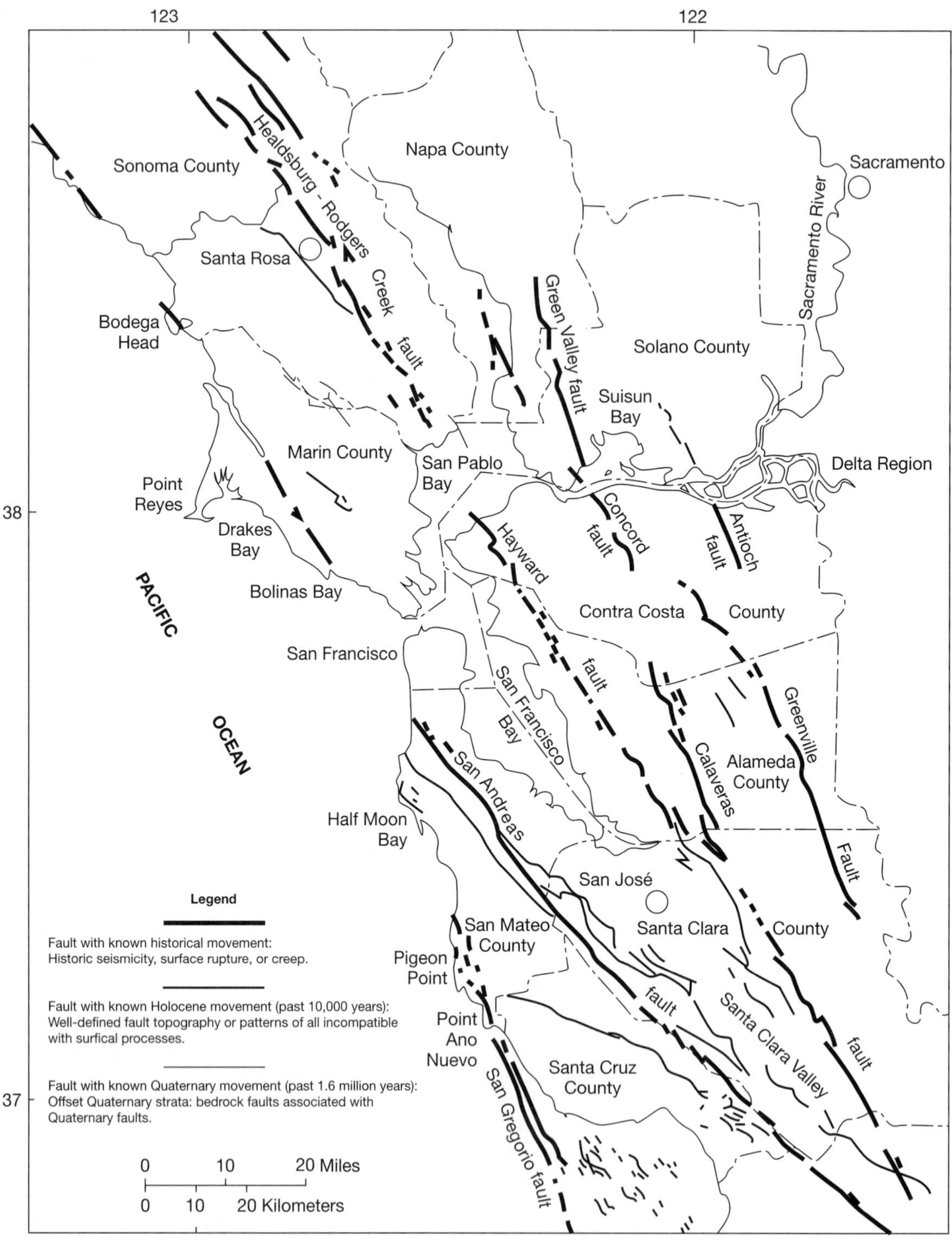

Fig. 14-20 Faults of the San Andreas system in the San Francisco Bay Area. Only the onland traces are shown here, but the faults continued offshore, as shown in Fig. 14-2 and Endpaper 2. (Source: †Wahrhaftig, C., and Sloan, D. 1989. *Geology of San Francisco and Vicinity.* Washington, D.C.: American Geophysical Union, Field Trip Guidebook T105.)

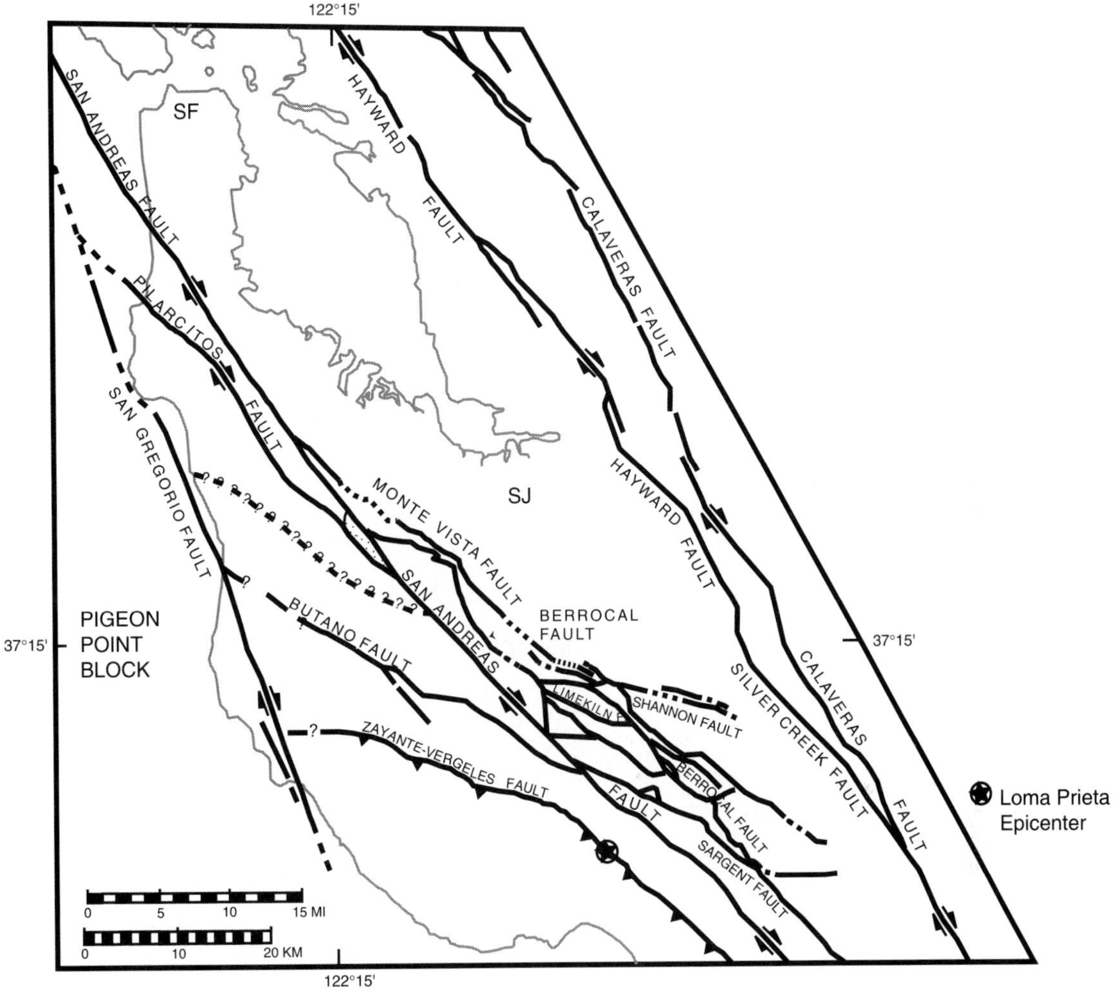

Fig. 14-21 Active reverse faults in Santa Clara County are associated with the San Andreas fault system. (Source: McLaughlin, R. U.S. Geological Survey; unpublished.)

damage they would find at home. It was the first time that a major league baseball game was canceled on account of an earthquake.

The Loma Prieta earthquake, named for the peak near its epicenter, was the most damaging earthquake on the San Andreas system since the great earthquake of 1906. (The 1994 Northridge earthquake, discussed in Chapter 13, did not occur on a right-lateral fault.) The Loma Prieta earthquake left 62 people dead, 3757 injured, and more than 12,000 homeless. Estimated losses from the earthquake were more than $6 billion. The lives of most of the six million people living in the Bay Area were affected for days, weeks, and months after the quake. Now, 8 years later, some structures stand unrepaired, reminders of the Loma Prieta earthquake. The earthquake had a local magnitude (M_L) of 7.1 (M_W of 7.0) and was centered in the southern Santa Cruz mountains along a remote stretch of the San Andreas fault. The ruptured segment was approximately 40 kilometers long, and both right lateral and vertical movement occurred (Fig. 14-22).

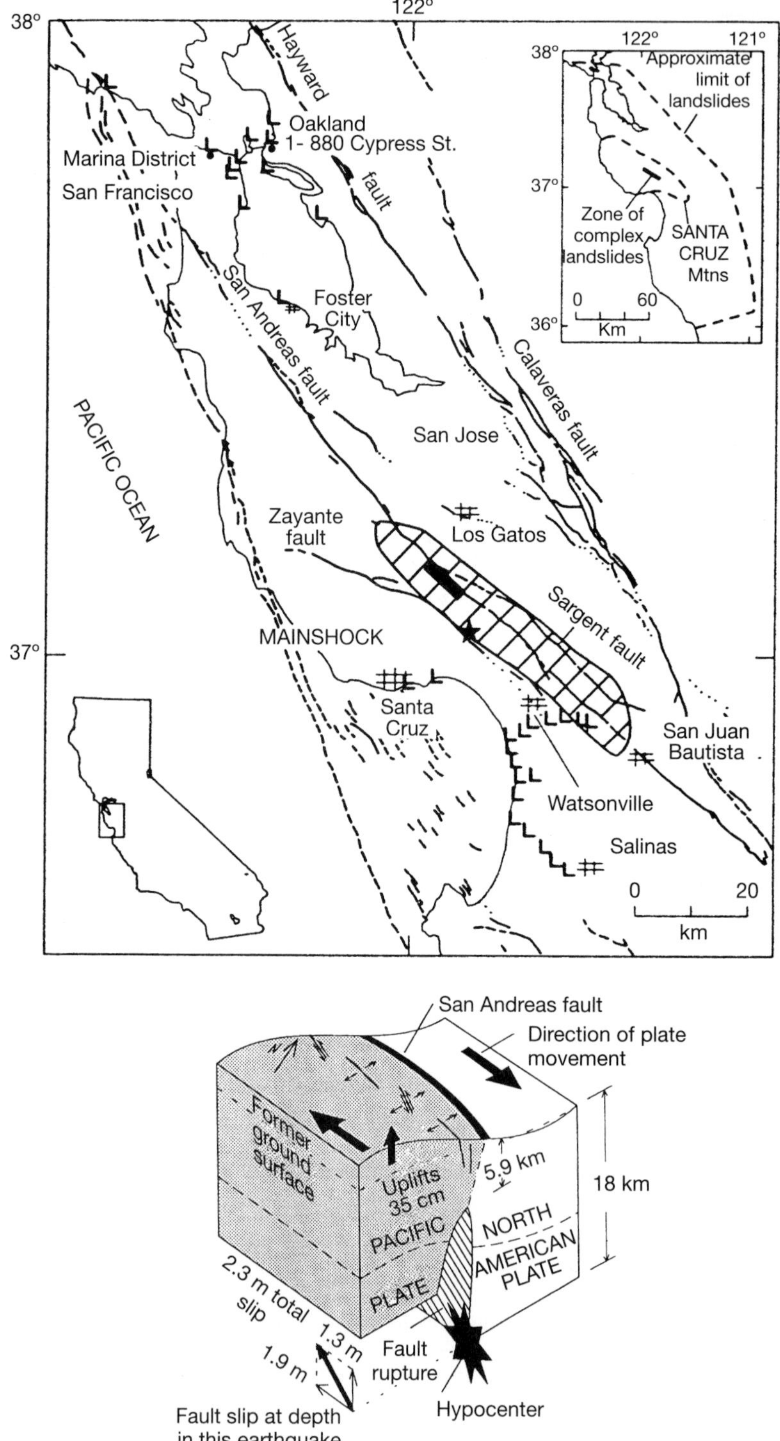

Fig. 14-22 Diagram and map showing the rupture zone of the 1989 Loma Prieta earthquake. (Source: Rymer, M. U.S. Geological Survey.)

Fig. 14-23 Ground cracking near the epicenter of the Loma Prieta earthquake; Summit Road. (Source: Wilshire, H. U.S. Geological Survey.)

Unlike the 1906 event, this earthquake did not rupture the ground surface along the fault, although cracks formed because of landsliding and severe shaking along ridges near the epicenter (Fig. 14-23).

As would be expected, communities near the epicenter were hard-hit by the 1989 earthquake. Unreinforced brick buildings crumbled, and wooden structures were knocked off their foundations (Fig. 14-24). In some towns, almost every house lost its chimney. Almost 100 kilometers north of the epicenter, unexpected devastation and loss of life also struck. In the Marina District of San Francisco, entire apartment buildings collapsed (Fig. 14-25), and several burned in a major fire. Ironically, the disastrous ground shaking in the Marina District can be traced back to the 1906 earthquake (Fig. 14-26). To celebrate the city's recovery from the great earthquake, San Francisco held an exhibition in 1913, having created the exhibition grounds by filling in a lagoon on the edge of San Francisco Bay (Fig. 14-27). Some of the artificial fill was actually rubble cleaned up after the 1906 earthquake. When the artificially filled land failed during the 1989 earthquake, liquefaction brought up old debris from 1906 along with the wet, sandy fill.

Across the Bay in Oakland, the top of a double-decked section of Interstate 880 collapsed in sections like dominoes, crushing 41 commuters beneath. If traffic had been as heavy as normal, hundreds more could have died here. The heavy damage in this area was the result of design defects. One end of the collapsed section was anchored in soft sediments that amplified the ground shaking.

Fig. 14-24 House thrown from its foundation near the epicenter of the 1989 earthquake. (Source: Nakata, J. U.S. Geological Survey.)

Fig. 14-25 Collapsed apartments, Marina District, San Francisco, 1989.(Source: Meyer, C. U.S. Geological Survey.)

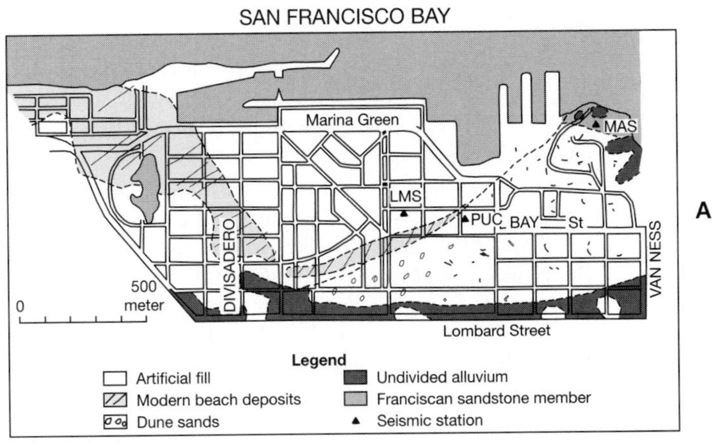

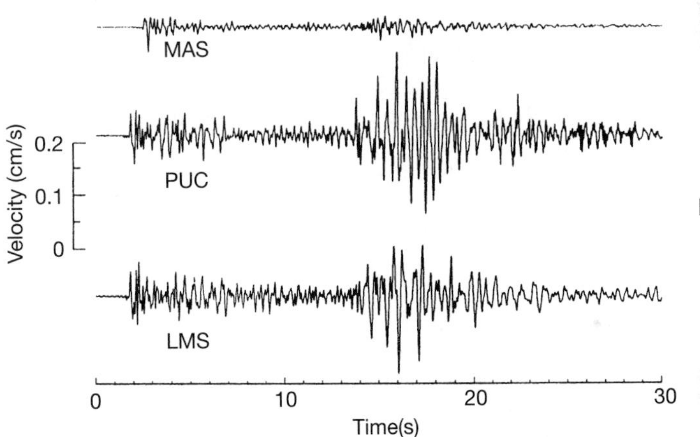

Fig. 14-26 **A,** Map showing the extent of filled land and the severity of ground shaking near the Marina District during the 1989 earthquake. **B,** Differences in the amplitude of aftershocks recorded at seismometers on bedrock *(MAS)*, dune sand *(PUC)*, and filled land *(LMS)*. (Source: U.S. Geological Survey.)

Fig. 14-27 A view of San Francisco looking toward the Golden Gate, ca. 1870. Arrow shows the lagoon that was later filled in to become San Francisco's Marina District.

1906 Earthquake

San Francisco in 1906 was a prosperous city of 400,000 inhabitants. It was a major center for world trade and banking, as well as a city of beautiful hills overlooking the ocean and San Francisco Bay. But at 5:12 AM on April 18, a great earthquake brought disaster to San Francisco and the surrounding areas. Centered on the San Andreas fault west of the city, the earthquake registered about $M_L 8.2$. It was felt as far away as Los Angeles and central Nevada. Many residents died in their beds as unreinforced masonry buildings collapsed on them; many others were killed by falling debris as they ran outside (Fig. 14-28). Enormous fires swept through the city unchecked because water mains had broken, leaving fire hoses useless. Firefighters eventually controlled the fires by dynamiting entire blocks to make fire breaks. As many as 2000 people are believed to have died in San Francisco and 189 people were killed in surrounding areas (Fig. 14-29). People lived in tent camps for months while soldiers kept order in the ruined city.

The 1906 earthquake is one of the world's most famous disasters, partly because of the lessons learned by geologists, seismologists, and engineers who studied its effects. Geologist Andrew Lawson compiled a comprehensive report published in 1908 detailing the earthquake phenomena observed throughout the area. One important finding was the correlation between geologic materials and the intensity of ground shaking in different parts of San Francisco (Fig. 14-30). The severe damage in areas that had been reclaimed by filling in San Francisco Bay was a pattern that was repeated in 1989 in the Marina District (see Fig. 14-25).

Geologists measured and photographed surface offset along the rupture zone, carefully documenting the fact that the entire northwestern 430 kilometers of the San Andreas fault had ruptured at once. The southern end of the 1906 rupture zone was near the mission at San Juan Bautista, where a wall collapsed. At the time, the evidence of lateral offset was puzzling—the plate tectonics theory would not be accepted until 60 or more years later. However, after studying the 1906 earthquake, seismologist H.F. Reid developed the model that is still the basis for our understanding of earthquakes (see Chapter 13).

Fig. 14-28 Golden Gate Avenue in San Francisco, 1906, after the earthquake but before the fire that destroyed many of the buildings. (Source: Branner Collection.)

Fig. 14-29 Santa Rosa City Hall after the 1906 earthquake. (Source: Branner Collection.)

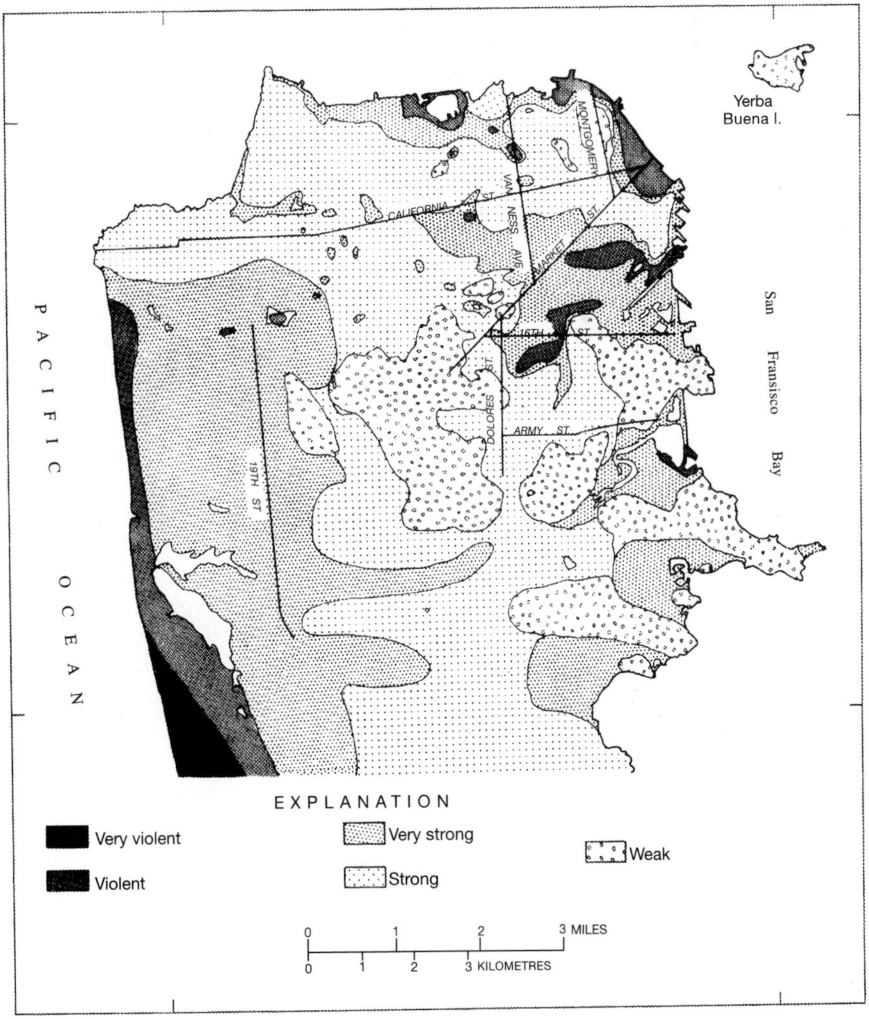

Fig. 14-30 A 1908 map showing the severity of ground shaking during the 1906 earthquake. (Source: U.S. Geological Survey, Professional Paper 941-A.)

CENTRAL CALIFORNIA: THE CREEPING SEGMENT AND THE PARKFIELD AREA

Between San Juan Bautista and Parkfield (Endpapers 2 and 3), the San Andreas system behaves very differently from its neighboring segments. This central segment experiences numerous small earthquakes, usually with magnitudes less that 4.0 (Fig. 14-31). The Calaveras fault and the Hayward fault east of San Francisco Bay show a similar pattern. Seismologists believe that strain energy is being released continuously in small amounts, so the likelihood of a great earthquake on these seismically active segments is small. However, the 1984 magnitude 6.2 earthquake on the Calaveras fault and the larger 1868 Hayward fault earthquake are proof that damaging moderate earthquakes are possible.

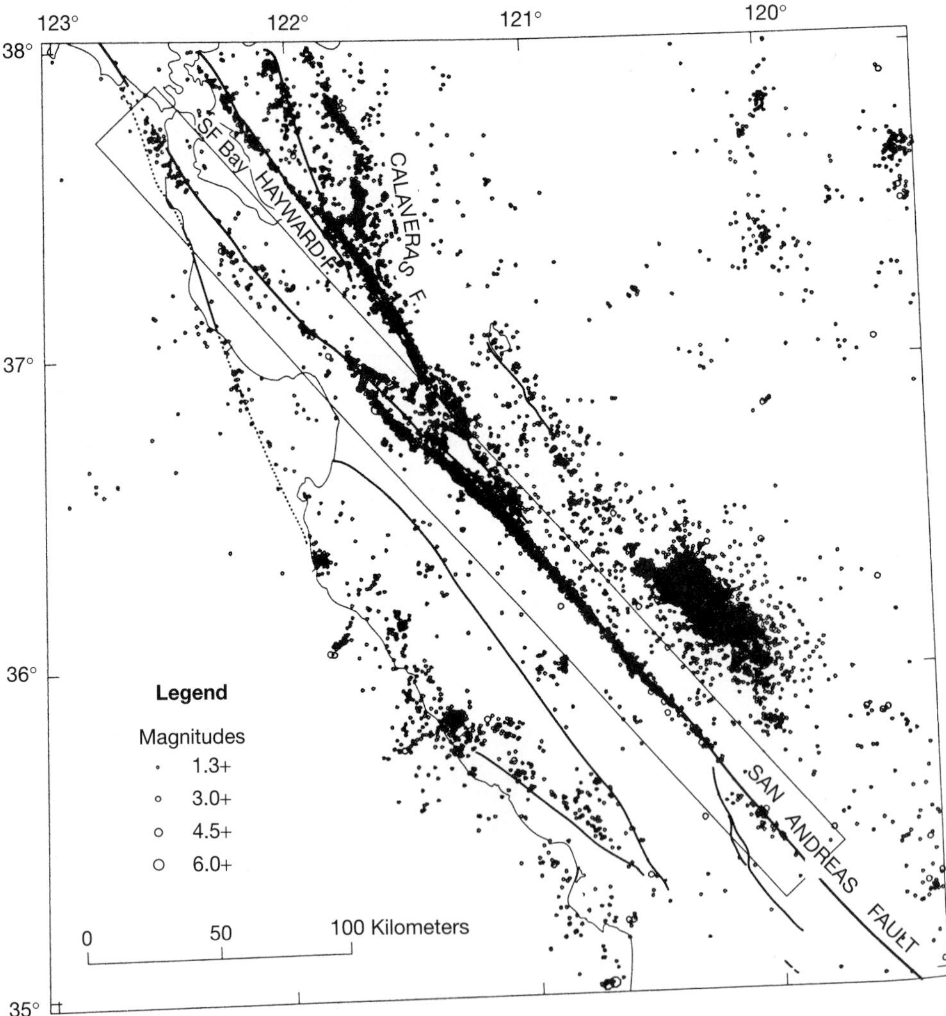

Fig. 14-31 Seismicity of the central San Andreas fault system showing creeping segments of the Hayward, Calaveras, and San Andreas faults. The creeping segments are those with solid clusters of earthquake epicenters along the faults. (Source: Wallace, R. 1990. U.S. Geological Survey, Professional Paper 1515.)

Relatively constant, slow displacement causes ***fault creep*** along the central segment of the San Andreas system. The continuous offset displaces sidewalks, curbs, and other cultural features along the faults (Color Plates 22 and 23). In the town of Hayward in the San Francisco Bay area, cracks, buckles, and small offsets produced by creep on the Hayward fault can be traced through the downtown area (Fig. 14-33). The average rate of creep on the San Andreas fault southeast of Hollister is about 1 centimeter per year. Following the 1989 Loma Prieta earthquake north of the creeping segment of the San Andreas fault, a "burst" of several millimeters of creep was recorded (Fig. 14-34).

Fig. 14-32 An aerial photograph showing the San Andreas fault in Cienega Valley. The main fault trace runs from left to right, just below the center of the photo, and through the Cienega Winery buildings (see Plates 22 and 23).

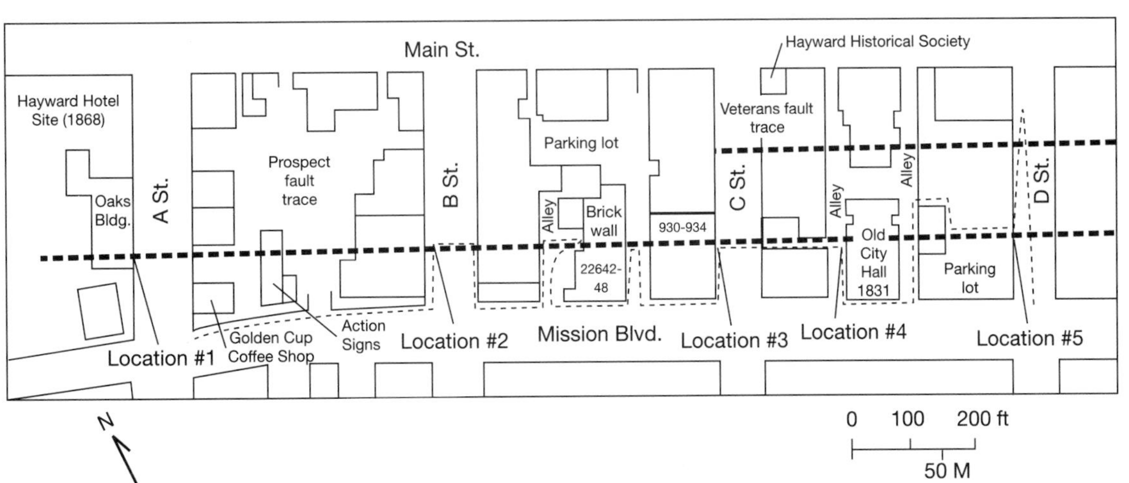

Fig. 14-33 Trace of the Hayward fault through downtown Hayward, Alameda County. (Source: Hirschfield, S. 1987. Field Guide to the Hayward Fault. Hayward, Calif.: California State University Hayward.)

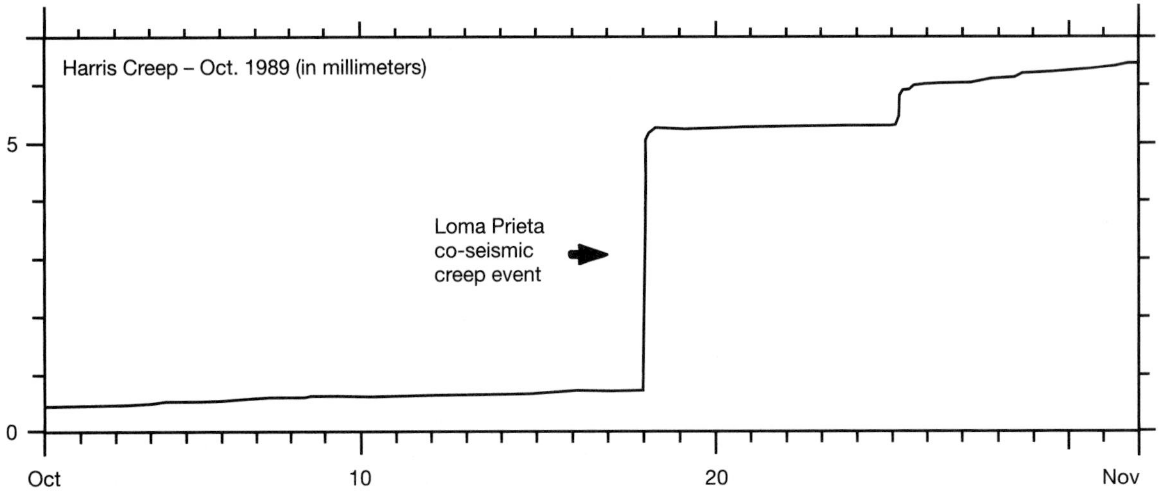

Fig. 14-34 Creep rates measured on the San Andreas fault about 3 kilometers southeast of Hollister, before and after the 1989 earthquake. (Source: U.S. Geological Survey.)

At the southern end of the creeping segment near the town of Parkfield, a 30-kilometer-long stretch of the San Andreas fault has experienced magnitude 6.0 earthquakes at 22-year average intervals since 1881. This regular pattern led seismologists to predict an earthquake for Parkfield in the late 1980s, but the event has not occurred as of 1996.

THE CARRIZO PLAIN AND THE FORT TEJON EARTHQUAKE

The unpopulated grasslands of the Carrizo Plain provide excellent opportunities to view the effects of the San Andreas fault on the landscape. Along this straight, relatively simple segment of the fault, one can view the offset streams, compressional ridges, linear valleys, and sag ponds that characterize a transform fault on land (see Figs. 14-1, 14-6, 14-11, and 14-12). On a clear day, travelers flying on commercial airliners between the San Francisco Bay area and the greater Los Angeles area are often treated to a birds-eye view of this section of the San Andreas fault. Knowledgeable pilots frequently point out the fault to passengers as they fly over it.

The 1857 Fort Tejon Earthquake

On January 9, 1857, an enormous earthquake ruptured the San Andreas fault from Parkfield through the Big Bend segment and southeast at least to Wrightwood, a total distance of at least 360 kilometers (Fig. 14-35). Fort Tejon, a military post at the southernmost end of the Carrizo Plain, was one of few population centers near the epicenter. There the ground shook for 1 to 3 minutes. Hundreds of antelope stampeded in terror, and water sloshed out of a sag pond along the fault and stranded fish to die on dry land. One eyewitness, a cowboy working in the southern Carrizo Plain, reported that a round sheep corral was instantly bent into an S-shape by right-lateral slip of several meters. The "earthquake crack" along the fault trace could be seen for years by travelers and early settlers.

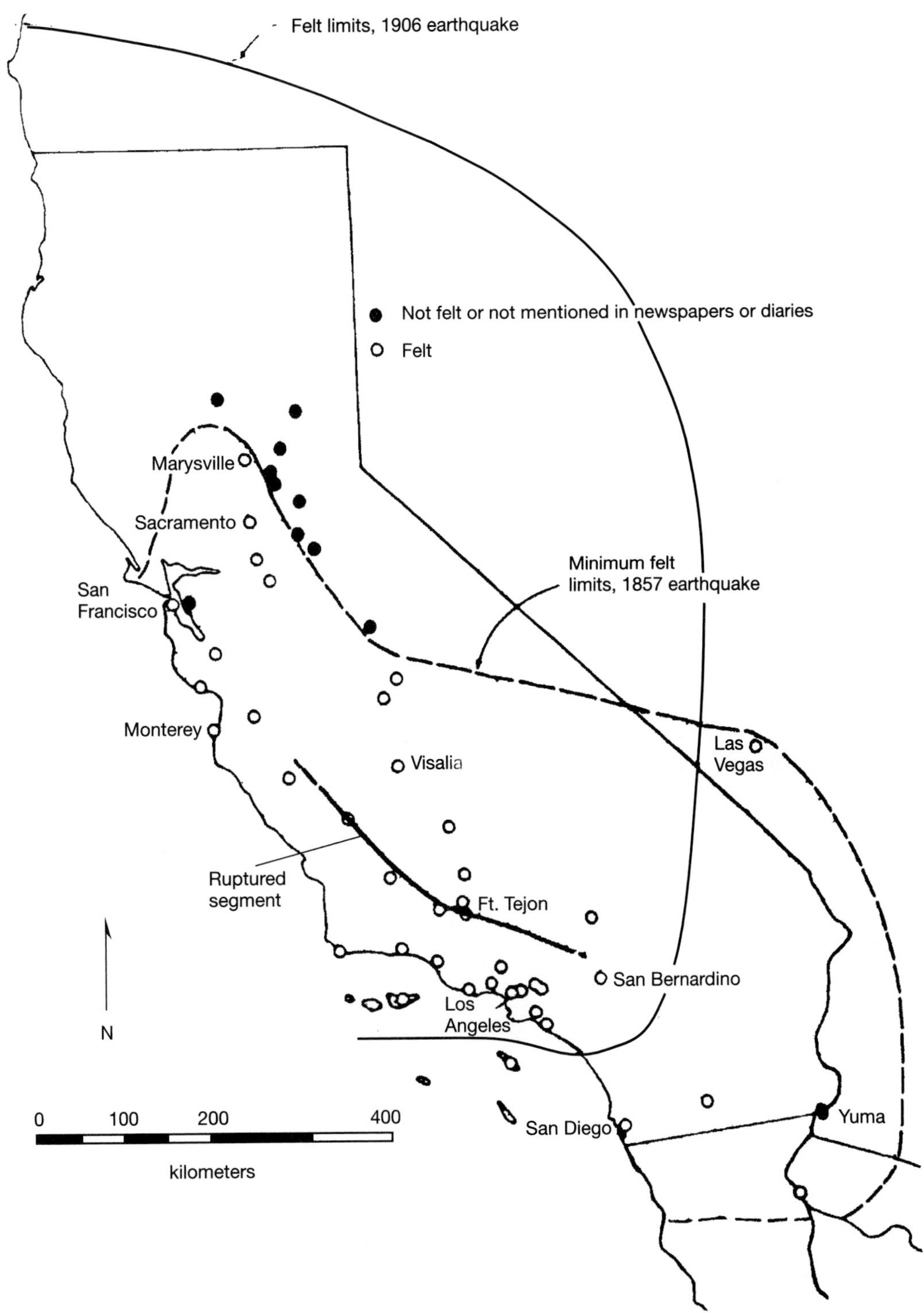

Fig. 14-35 Map showing the fault rupture produced by the 1857 Fort Tejon earthquake. (Source: Sieh, K. Seismological Society of America Bulletin, Vol. 5.)

The earthquake produced as much as 9 meters of offset in the Carrizo Plain and 3 to 4 meters in the Mojave Desert. The average displacement was between 4.5 and 4.8 meters. Seismologists estimate the magnitude of the Fort Tejon Earthquake as slightly larger than the 1906 earthquake, at an approximate magnitude of 8.2. Since 1857, this segment of the San Andreas has remained locked.

THE BIG BEND: THE SAN ANDREAS FAULT SYSTEM IN THE TRANSVERSE RANGES AND THE SAN GABRIEL AND SAN BERNARDINO MOUNTAINS

From its junction with the Garlock fault, the San Andreas fault makes a marked bend to the southeast for about 120 kilometers (Fig. 14-36). This segment of the fault is appropriately referred to as the Big Bend. The complexity found along all bending segments of the fault is maximized along the Big Bend. Because of the significant component of compression in this region, rocks are actively being squeezed and uplifted (Fig. 14-37). As a result of the compression, spectacular mountain ranges have been

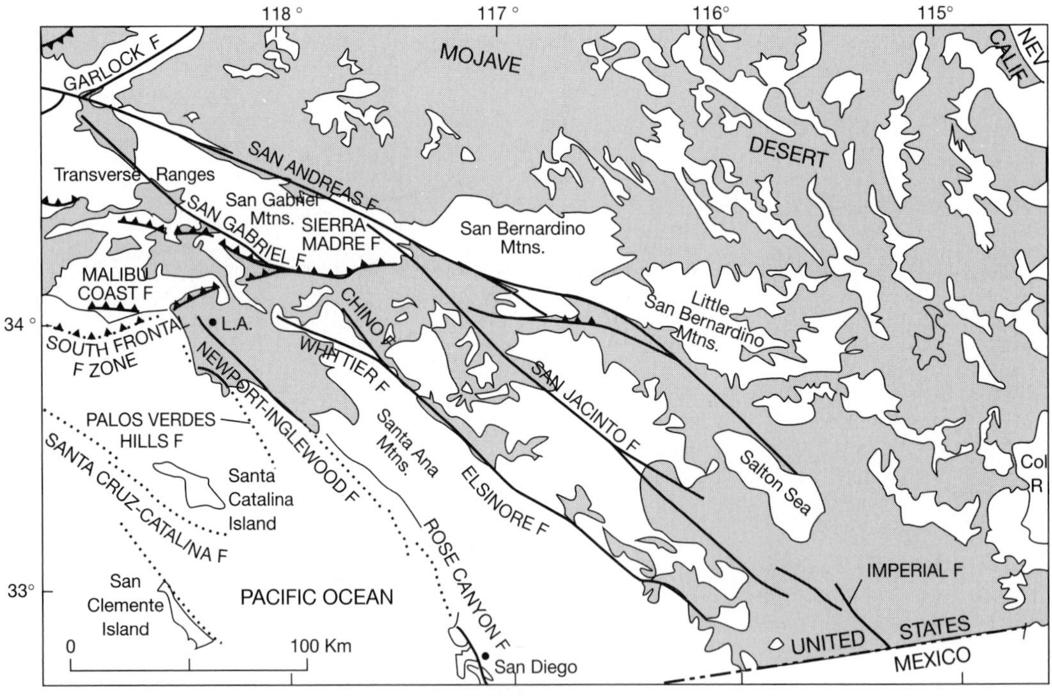

EXPLANATION

Alluvial and estuarine deposits (Quaternary)—chiefly basin fill; may also include some deposits of Pliocene age

Bedrock (Tertiary and older)—varied rock types

Fault, exhibiting evidence of Quaternary displacement—sawteeth on upthrown block of reverse or thrust fault; dotted when concealed by water

Fig. 14-36 Major faults of the San Andreas system in southern California.

thrown up along the margins of the fault—the San Gabriel, San Bernardino, and San Jacinto ranges have peaks that reach over 3000 meters in elevation. The steep mountain slopes have shed enormous quantities of debris that are spread across the range fronts in large alluvial fans (see Chapter 6 and Fig. 6-11) The fans, increasingly occupied by high-density housing and commercial developments, are in many places cut by active range-front faults of the San Andreas system. Many of the range-front faults are reverse faults caused by the compression in this area. Reverse faulting and folding in the Transverse Ranges and the Los Angeles basin are discussed in more detail in Chapter 16, and the Garlock fault is discussed later in this chapter.

The right-lateral motion along the San Andreas system is taken up by a number of fault strands in the Big Bend segment. The San Andreas fault itself defines the eastern end of the system, and it is presently the most active fault strand. As discussed at the beginning of this chapter, Mesozoic rocks are offset about 220 kilometers across the southern San Andreas fault. In the northern part of this segment, the San Gabriel fault can be traced through the eastern Transverse Ranges and through the western San Gabriel Mountains near Pasadena. In the Ridge Basin, the Mint Canyon Formation, a 12- to 13-million-year-old sedimentary unit, contains cobbles of unusual volcanic rocks (Fig. 14-38). Well to the southeast in the Chocolate Mountains, the source for these volcanic clasts can be found. These relatively young rocks are offset as much as the older rocks across the San Gabriel and San Andreas faults, about 220 kilometers. This indicates that the faults did not begin movement until after the Mint Canyon Formation was deposited 12 to 14 million years ago.

Geologists have estimated that the San Gabriel fault has offset the Pelona Schist and other older rocks about 50 kilometers. Today the San Gabriel fault is inactive.

Fig. 14-37 Folded and faulted rocks in the San Andreas fault zone, San Bernardino County. (Source: Tarbuck, E. Central Illinois University.)

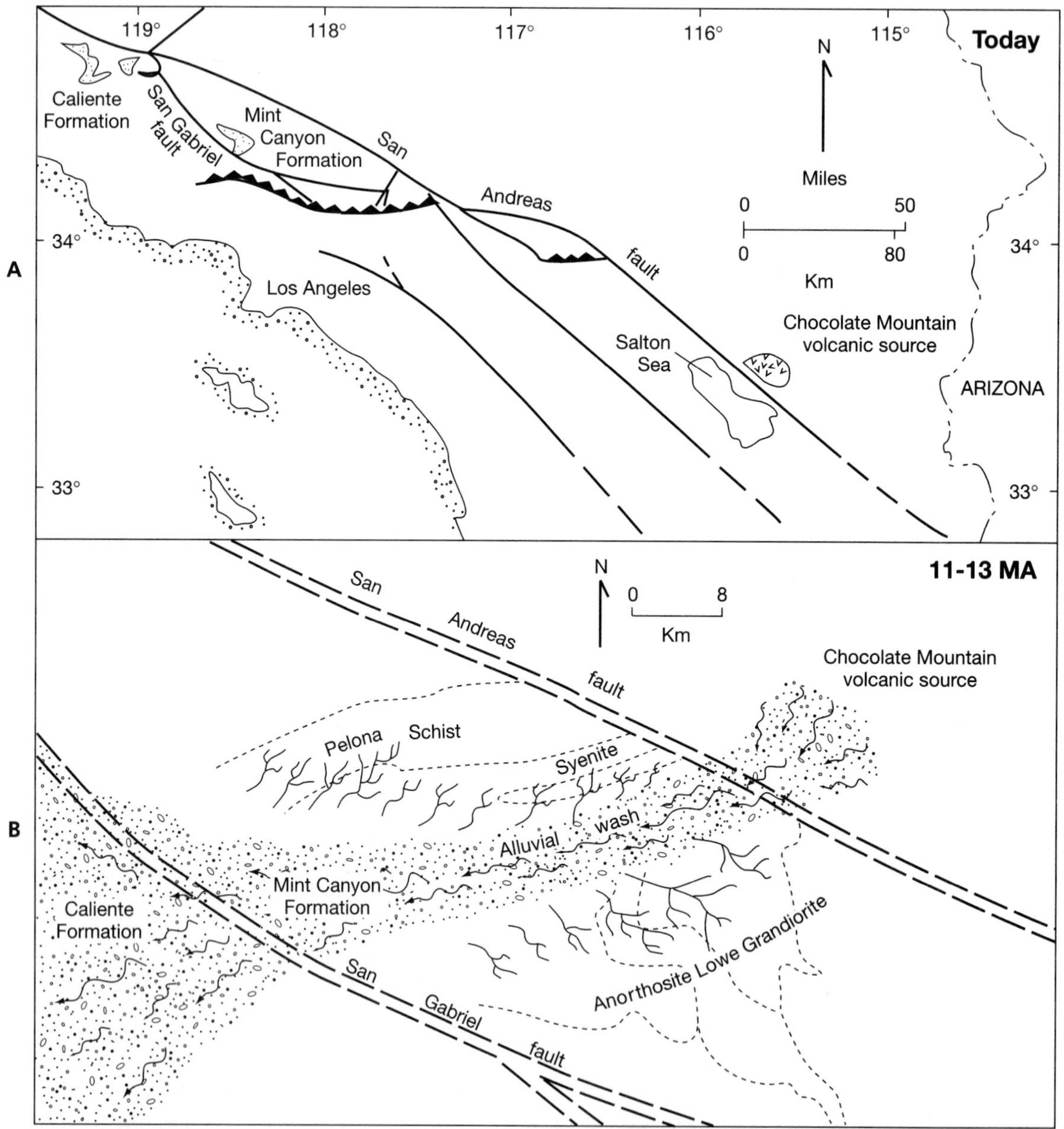

Fig. 14-38 **A,** Present location of the Mint Canyon Formation, a sedimentary unit containing volcanic clasts, and its volcanic source terrane. **B,** The reconstructed drainage pattern before the units were offset along the San Andreas system. (Source: California Department of Conservation, Division of Mines and Geology. From Crowell, J., ed. 1975. *The San Andreas Fault in Southern California.* Special Report 118.)

Sediments younger than about 3 million years old show no displacement across the fault, and it is seismically quiet. Geologists believe that the San Andreas fault to the east is presently accommodating the slip formerly taken up by the San Gabriel fault.

South of the San Gabriel Mountains, near Cajon Pass and San Bernardino, the San Andreas fault system splays into several complex strands. The San Andreas fault zone consists of three subparallel strands which bend eastward along the San Bernardino Mountains: the Mission Creek, San Gorgonio Pass, and Banning faults (see Fig. 14-2 and Endpaper 2). The San Jacinto fault zone trends southeast from Cajon Pass, and further west, the Elsinore fault parallels the San Jacinto's trend. The San Jacinto and Elsinore faults are discussed in the following section.

Pallett Creek and Earthquake Frequency

Along the bank of Pallett Creek, about 55 kilometers northeast of Los Angeles, is a marsh that has been cut by a strand of the San Andreas fault. Deposits here record amazing evidence of past earthquakes on the San Andreas fault in southern California. At least 12 earthquakes have broken the sediments at Pallett Creek during the past 1700 years at this site, giving geologists the best information about earthquake history along the San Andreas (Figs. 14-39 through 14-41). Clusters of earthquakes, two or three within 100 to 200 years, are followed by 200 to 300 years without large earthquakes (see Fig. 14-41). The average recurrence interval for this segment of the fault is about 132 years, and 139 years have passed since the great Fort Tejon earthquake of 1857. The established recurrence times between earthquakes are one major reason for assigning a high probability of a large earthquake on the southern San Andreas fault within the next 30 years.

Recent Earthquakes

Three major earthquakes associated with this part of the San Andreas system—Long Beach 1933, San Fernando 1971, and Northridge 1994—have been described in Chapter 13. Although none of these events was centered on the San Andreas fault itself, all are the indirect result of the same transform motion, and all caused major damage in the greater Los Angeles area. This segment of the San Andreas fault was part of the rupture zone during the 1857 Fort Tejon earthquake discussed above. At a site near Wrightwood, geologists have found evidence of earlier fault rupture in an 1812 earthquake (Fig. 14-40). There the rings of trees along the San Andreas fault show evidence of simultaneous damage, and geologists are able to date the event by radiocarbon dating of the damaged section of the wood cores.

Since 1857, the segment of the San Andreas fault along the Big Bend north of Cajon Pass has been seismically quiet. South of Cajon Pass, the 1948 Desert Hot Springs earthquake (M_L 6.5) ruptured the Mission Creek fault, and the 1986 North Palm Springs earthquake (M_L 5.9) ruptured the Banning fault. The area near San Gorgonio Pass, at the southern end of the Big Bend, shows southern California's highest level of background seismicity.

Whittier Narrows 1987

Two earthquakes within the first 4 days of October 1987 shook residents of the northern Los Angeles, particularly those living in Whittier. The larger of the two events had a magnitude of 5.9 (M_W and M_L) and was centered 16 kilometers east of

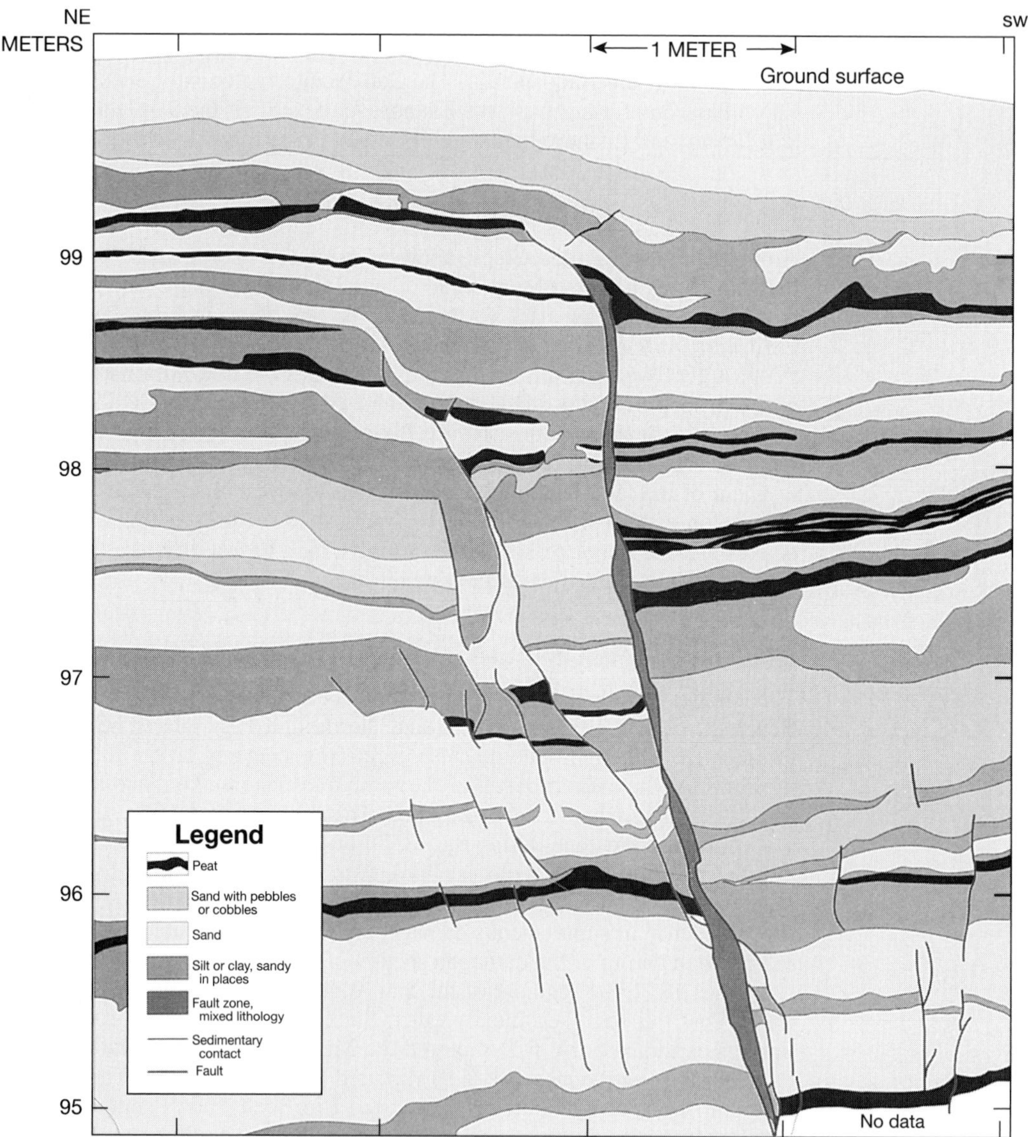

Fig. 14-39 Offset sedimentary layers, including peat beds, at Pallett Creek. (Source: Sieh, K. 1990. In Wallace, R., ed. *The San Andreas Fault System.* U.S. Geological Survey, Professional Paper 1515.)

Fig. 14-40 The San Andreas fault zone exposed in the banks of Swarthout Creek near Wrightwood; view to the northwest. The fault ruptures are marked by tapes with the date of the earthquake producing each strand shown at the top of the tape. To the left (southwest) of the fault, deposits consist of alternating dark peat beds and light debris-flow deposits, similar to the deposits at Pallett Creek. Massive gravelly sediment from nearby hillslopes appears to the right of the fault. (Source: Fumal, T. U.S. Geological Survey.)

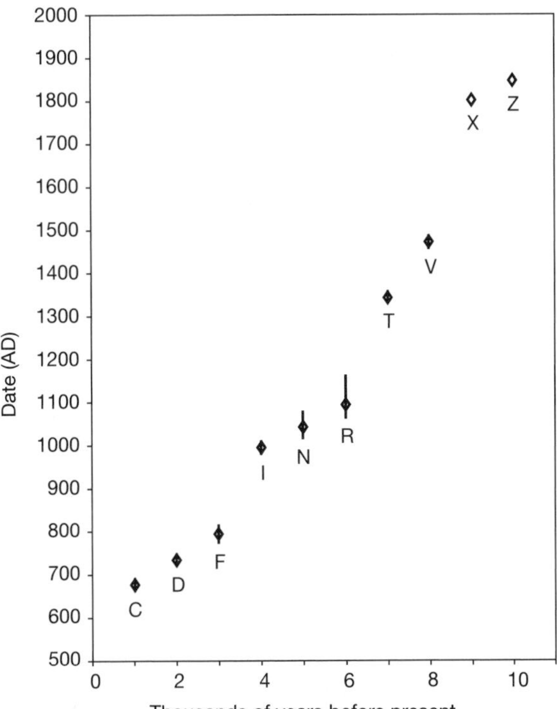

Fig. 14-41 Timing of earthquakes at Pallett Creek. (Source: Sieh, K., and others. 1989. *Journal of Geophysical Research*.)

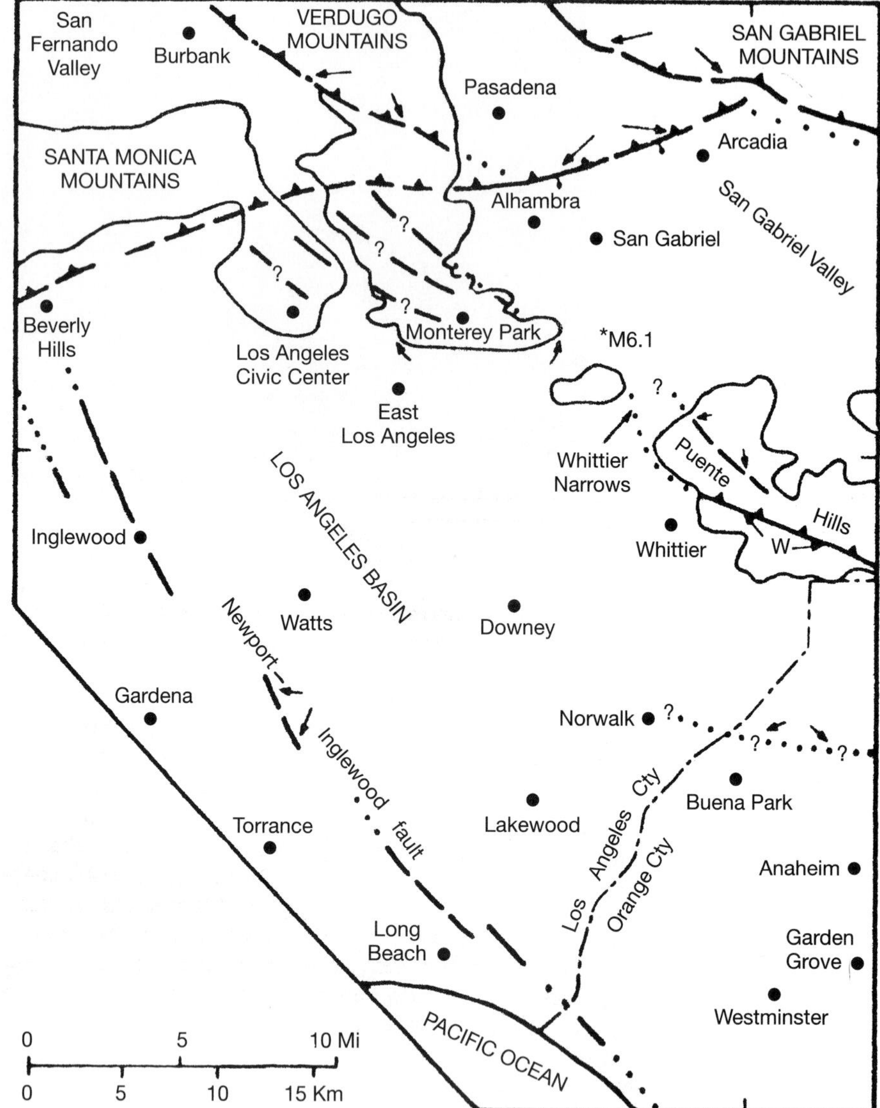

Fig. 14-42 Epicenter of the 1987 Whittier Narrows earthquake. *N* = Norwalk; *R* = Raymond; *SM* = Sierra Madre; *VE* = Verdugo; *W* = Whittier; *WH* = Workman Hill. (Source: California Department of Conservation, Division of Mines and Geology. Weber, F. *California Geology,* 40[12], 1987.)

the Los Angeles Civic Center (Fig. 14-42). Together, the two earthquakes damaged 10,500 structures and caused at least $215 million in damages. They were the first damaging earthquakes in the Los Angeles area since the San Fernando earthquake of 1971 (discussed in Chapter 13).

Like the 1971 San Fernando and 1994 Northridge earthquakes, the 1987 Whittier Narrows earthquake occurred on a fault that is only subtly expressed at the Earth's surface. The southeastern Whittier fault was known to offset 2000-year-old sediments, and further southeast, the Whittier fault joins the active Elsinore fault. However, in the vicinity of Whittier, no evidence was found of a recently active

Fig. 14-47 Surface rupture produced by the 1987 Superstition Hills earthquake. Note canteen at left for scale. (Source: Rymer, M. U.S. Geological Survey.)

The southern San Andreas fault, which forms the northeastern edge of the Salton Trough (see Fig 14-36 and Endpaper 2), has had no earthquakes in historic time, although its southern end showed minor sympathetic slip during recent earthquakes on the Imperial and San Jacinto faults. Geologists estimate that this segment has experienced four ground-rupturing earthquakes since 1000 AD, with the last event occurring about 300 years ago. These earthquakes have displaced sediments of ancient Lake Cahuilla. The southern San Andreas, Imperial, and San Jacinto faults all exhibit fault creep of a few millimeters per year.

West of the Salton Trough, the Elsinore fault is another northwest-trending member of the southern San Andreas system. The Elsinore fault cuts through the plutonic crystalline rocks of the Santa Ana mountains. To the northwest in the Los Angeles Basin, the Elsinore fault splits into two strands, the Whittier fault and the Chino fault (see Fig. 14-36). In the Peninsular Ranges, the Elsinore fault has a slip rate of about 5 millimeters per year. The northern part of the fault is more active, with only small earthquakes documented on the southern Elsinore fault. A magnitude 6 event on the central Elsinore fault was recorded in 1910.

Together the faults of the southernmost San Andreas system account for right-lateral displacement of about 45 to 50 millimeters per year between the North American and Pacific Plates. The San Jacinto fault accounts for as much as 12 millimeters per year, and the southern San Andreas fault south of Cajon Pass between 25 and 30 millimeters per year. Further west, slip on the southern Elsinore fault is about 5 millimeters per year, and the Newport-Inglewood and Palos Verdes faults 0.6 and 3 millimeters per year, respectively.

THE GARLOCK FAULT AND THE EASTERN CALIFORNIA SHEAR ZONE

The Garlock fault trends east-northeast from Tejon Pass for 250 kilometers, forming the northern border of the Mojave Desert Province (see Fig. 14-36). It is an unusual fault for California because its motion is left-lateral (Fig. 14-48). No significant

Fig. 14-48 High-altitude U-2 photo of the Garlock fault. (Source: USAF and Clark, M. U.S. Geological Survey.)

earthquakes have occurred on the Garlock fault during historic time, but the western part of the fault exhibits low-level seismicity and aseismic creep. Based on the displacement of young alluvial deposits, geologists estimate that about 8 meters of left-lateral offset have occurred on the western Garlock fault during the past 10,000 years. Along the eastern Garlock fault, offset lake shorelines and other geomorphic features suggest that earthquakes have produced slip of as much as 7 meters during the past 10,000 years.

A zone of active faults known as the Eastern California Shear Zone runs parallel to the San Andreas fault through the central Mojave Desert (Figs. 14-49 and 14-50). Many of the faults are young normal faults, but right-lateral motion also occurs on several, including the Johnson Valley, Landers, Homestead Valley, Emerson, and Camprock faults. The pattern of activity on these faults, which form a north-trending zone (see Fig. 14-50), indicates that they could correctly be considered part of the boundary between the Pacific and North American Plates. Geologists currently refer to these faults collectively as the Eastern California Shear Zone. Several moderate earthquakes have ruptured the faults of the Eastern California Shear Zone, including the 1975 Galway Lake (M_L 5.2), the 1979 Homestead Valley (M_L 5.6) and the April 1992 Joshua Tree (M_L 6.1) events (see Fig. 14-50). The Joshua Tree earthquake is now considered a preshock of the 1992 Landers earthquake.

Landers and Big Bear Earthquakes of 1992

Two earthquakes on separate faults shook the southern California desert on June 28, 1992. The Landers earthquake, which occurred at about 5 AM and had a moment magnitude of 7.5, ruptured parts of the Eastern California Shear Zone. Because it

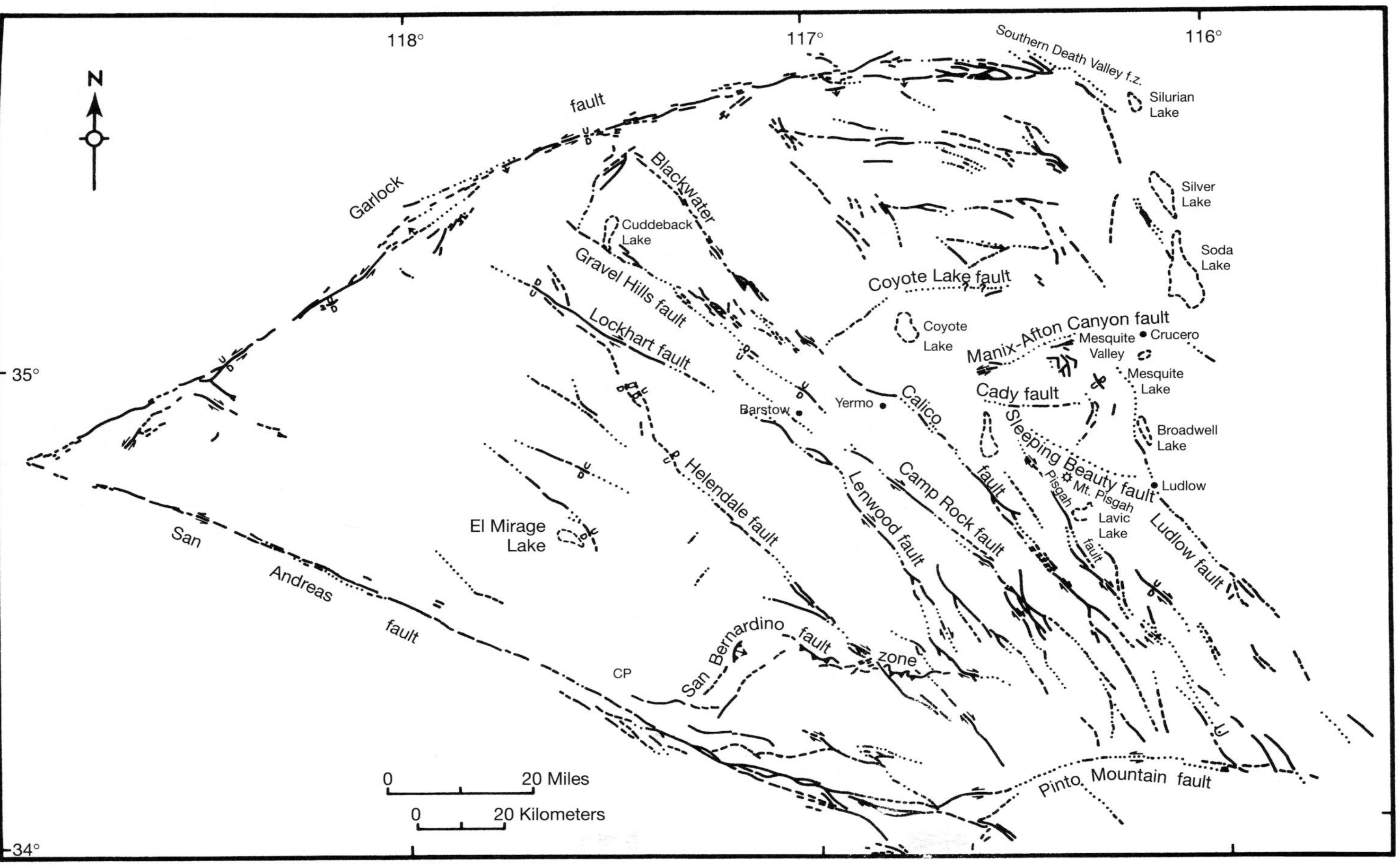

Fig. 14-49 Active faults in the Mojave Desert. (Source: Dokka, R, and others. 1979. Geological Society of America Field Trip Guidebook.)

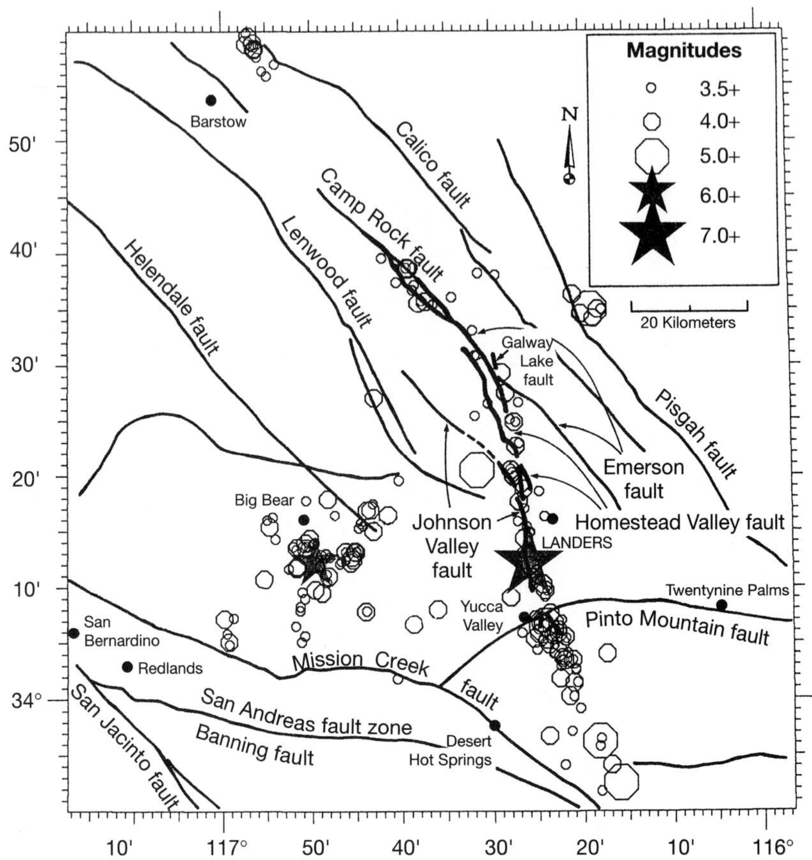

Fig. 14-50 Map showing earthquake epicenters and fault rupture shading produced by the 1992 Big Bear and Landers earthquakes along the Eastern California Shear Zone. (Source: Modified from California Department of Conservation, Division of Mines and Geology. The Landers–Big Bear earthquakes. *California Geology* 46(91), 1993.)

was centered in a sparsely populated area of the Mojave desert, the Landers earthquake caused only one fatality. It ruptured the ground surface along an 80-kilometer stretch across the desert floor. The rupture occurred along parts of four previously mapped faults (see Fig. 14-50). Both right-lateral displacement of up to 6 meters and vertical offset of as much as 1.2 meters were observed by geologists (Fig. 14-51). The Landers earthquake was a reminder that not all of the transform motion between the Pacific and North American Plates takes place on the San Andreas system and an important confirmation of the existence of the Eastern California Shear Zone.

Three hours later, a magnitude 6.5 earthquake took place on a buried, northeast-oriented fault near Big Bear Lake. This earthquake took place on a left-lateral fault, but no surface rupture was found. Seismologists believe that the Landers earthquake may have triggered the Big Bear event. Many seismologists are also concerned that the Landers and Big Bear earthquakes have brought the nearby segment of the San Andreas fault closer to failure.

Fig. 14-51 Vertical fault scarp along the Johnson Valley fault produced by the 1992 Landers earthquake. (Source: Rymer, M. U.S. Geological Survey.)

FURTHER READINGS

DENGLER, L., CARVER, G., AND MCPHERSON, R. 1992. Cape Mendocino. *California Geology* 45(2):40-51.

EISMAN, D.B. Dust from antelope. *California Geology* 25(8):171-173, 1972.

IACOPI, R. 1971. *Earthquake country.* Menlo Park: Lane Publishing Co., 160 pp.

JENNINGS. C.W. 1994. *Fault map of California.* California Division of Mines and Geology, scale 1:750,000 (Endpaper 2).

The Landers-Big Bear earthquakes. 1993. *California Geology* 46(1).

PETERSON, M.D., AND WESNOUSKY, S.G. 1994. Fault slip rates and earthquake histories for active faults in southern California. Bulletin of the Seismological Society of America, 84(5):1608-1649.

WALLACE, R.E., ed. 1990. *The San Andreas Fault System, California.* U.S. Geological Survey, Professional Paper 1515, 283 pp.

WEBER, H.F., Jr. 1987. Whittier Narrows earthquakes. *California Geology* 40(12):275-281.

CHAPTER

15

The California Coast

The western edge of the North American continent meets the Pacific Ocean along 1800 kilometers (1100 miles) of California coastline, a varied landscape where change is continuous. Tectonics, waves, and wind combine forces to create beautiful scenery and a wealth of geologic features along the coastal zone. California's beaches are among its most valuable resources, enjoyed by both residents and tourists for recreation and relaxation.

The California coastline can be broadly divided into two distinct sections—north and south of Point Conception (Fig. 15-1). Here, at the northern end of the Transverse Ranges, the coastline changes direction from its overall northwesterly trend. The southern coast heads eastward and then gradually returns to its north-westerly direction. As a result, the coast south of Point Conception is broadly curved. The shift in the orientation of the coast brings warmer-water currents to southern California's coastal areas (see Fig. 15-1). Because of its orientation and the presence of a broad offshore shelf, the southern coast is also somewhat protected from winter storm waves, which usually approach the shore from the northwest.

WAVES

With the exception of tsunamis (see Chapter 13), all ocean waves are generated by winds blowing over the water surface. Waves radiate outward from storm areas, traveling hundreds or thousands of kilometers on the open ocean without losing much energy. During winter, most waves that approach the California coast are generated by storms in the northern Pacific Ocean near the Gulf of Alaska. These waves approach the coast from the northwest. During late summer and fall, storms off Baja California and the Mexican mainland generate waves that approach the coast from the south.

In the open ocean, the surface water moves in circular orbitals to transmit the waves, leaving the ocean bottom unaffected. When waves reach the shallow water at the coastline, they begin to "feel" the bottom (Fig. 15-2). At this depth, the bottom of the wave slows down. The top of the wave overruns the base, and the wave breaks. The exact depth at which waves break depends on the size of the wave: large waves will break farther from the shore, in deeper water, than small waves will. The zone of breaking waves along a coastline, the ***surf zone,*** is constantly subjected to wave attack.

Fig. 15-1 The California coastline. Note the difference in the coast's orientation north and south of Point Conception.

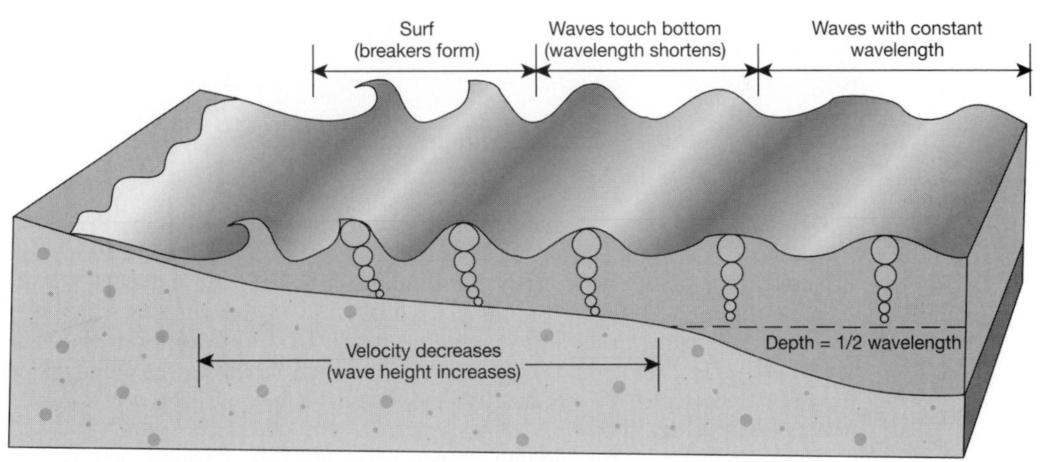

Fig. 15-2 The circular motion of water in ocean waves. (Source: Tarbuck, E., and Lutgens, F. 1993. *The Earth,* 4th ed. Upper Saddle River, N.J.: Prentice Hall, Inc.)

Fig. 15-3 Waves breaking along a long, straight beach, Point Reyes National Seashore. (Source: †Skapinksky, S. San José State University.)

Fig. 15-4 Wave refraction around rocky points, Sonoma County. (Photo by author.)

The continental shelf is relatively narrow along most of the coast north of Point Conception. As a result, the ocean bottom drops off quickly from the beach into deep water. Even large waves arrive close to the shoreline before breaking. Wave energy along unprotected coastlines is high because of the steep, narrow shelf. Wave energy is particularly high on straight beaches that face northwest, because they directly face incoming winter waves. The open beach on the Point Reyes Peninsula, unprotected by any nearby land masses, is one of California's highest energy beaches (Fig. 15-3).

Rocky points that protrude into the ocean are attacked by waves more fiercely than embayments in the shoreline. A wave front is slowed down as it approach the rocky headlands, but continues into the embayment for a greater distance before breaking. As a result, the wave fronts are bent or ***refracted*** (Fig. 15-4). As the pro-

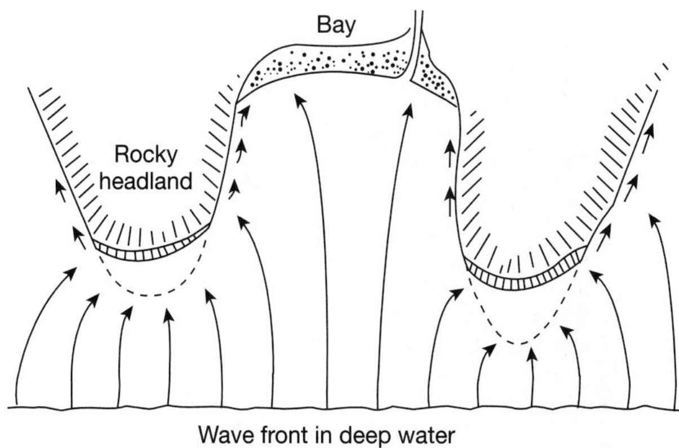

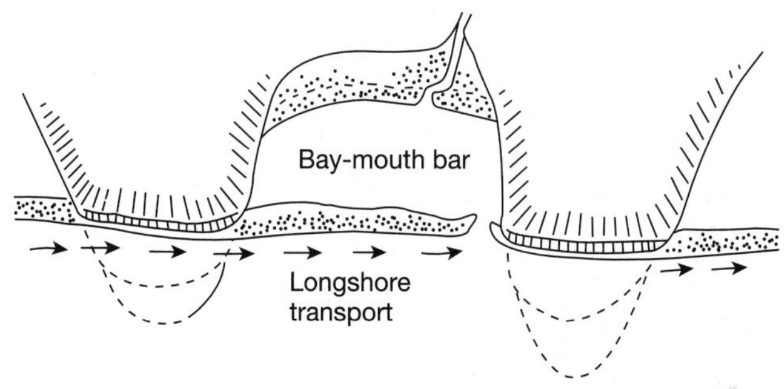

Fig. 15-5 Diagrams illustrating how coastlines straighten over time, as headlands are eroded and bays filled in by wave refraction and longshore transport. (Source: Bascom, W. 1980. *Waves and Beaches.* Garden City, N.Y.: Anchor Books, p. 15.)

truding points are vigorously eroded by waves, the protected embayments tend to fill in with beach sediment. Over time, the irregularities in the coastline are removed and the coast tends to straighten (Fig. 15-5).

BEACHES AND LONGSHORE TRANSPORT

The southern California coast is particularly famous for its long, smooth stretches of sand. Most of California's beaches are narrow strips of sand at the base of seacliffs (Fig. 15-6), but some, like those at the mouth of the Santa Clara River near Ventura, are backed by coastal lowlands (see below). Although most California beaches are composed of sand, gravel beaches can be found in many areas (Fig. 15-7). Most gravel beaches form in areas where the nearby seacliffs contain conglomerate, as is the case just north of Encinitas in southern California. Beach sediments, gravel as well as sand, are typically well rounded and sorted by the tumbling action of the waves (see Chapter 2).

A beach is a narrow strip of sand parallel to the coastline. Above the water level is the beach berm, where sand has been carried to shore by waves. Below the water, out to about 10 meters of water depth, are elongate piles of sand parallel to the beach, called ***offshore bars*** (Fig. 15-8). Both the beach berm and the offshore bars are part of the beach system, and the sand on a beach is in almost constant motion. Evidence for this can be seen by comparing the same beach in winter and in

Fig. 15-6 Narrow beach and steep cliffs at Point Conception. Note the wave-cut platform in the surf zone; here vertical beds of the Monterey Formation are exposed at low tide. (Source: Sarna-Wojcicki, A. U.S. Geological Survey.)

Fig. 15-7 A gravelly pocket beach in northern California. (Photo by author.)

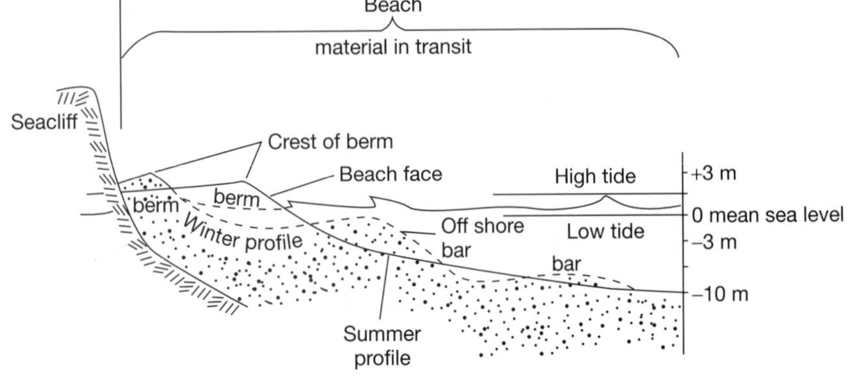

Fig. 15-8 Profiles of a typical beach showing the seasonal movement of sand. During winter, storm waves move much of the sand to offshore bars. (Source: Bascom, W. 1980. *Waves and Beaches.* Garden City, N.Y.: Anchor Books, p. 251.)

Fig. 15-9 **A,** Winter beach and **B,** summer beach, Santa Cruz. The photos were taken at low tide, when the wave-cut platform is exposed in the surf zone. The houses sit on a marine terrace. (Source: **A,** Bradley, W. **B,** author.)

summer. During the winter storm season, large waves attack the beach. The waves carry off much of the sand to the offshore bars. The beach berm becomes narrower. During summer, the beach widens as the summer waves carry sand from the bars up onto the beach (Fig. 15-9).

Most beach sand is supplied from rivers that empty into the ocean. Eroding seacliffs supply a lesser amount, although they may be important local sources. From the river mouths, sand is carried along the coast by ***longshore currents.*** Longshore currents result from the interaction between approaching waves and the coastline. Waves approach the coast from the direction of wave-generating winds. In California, the largest waves originate from winter storms in the northern Pacific,

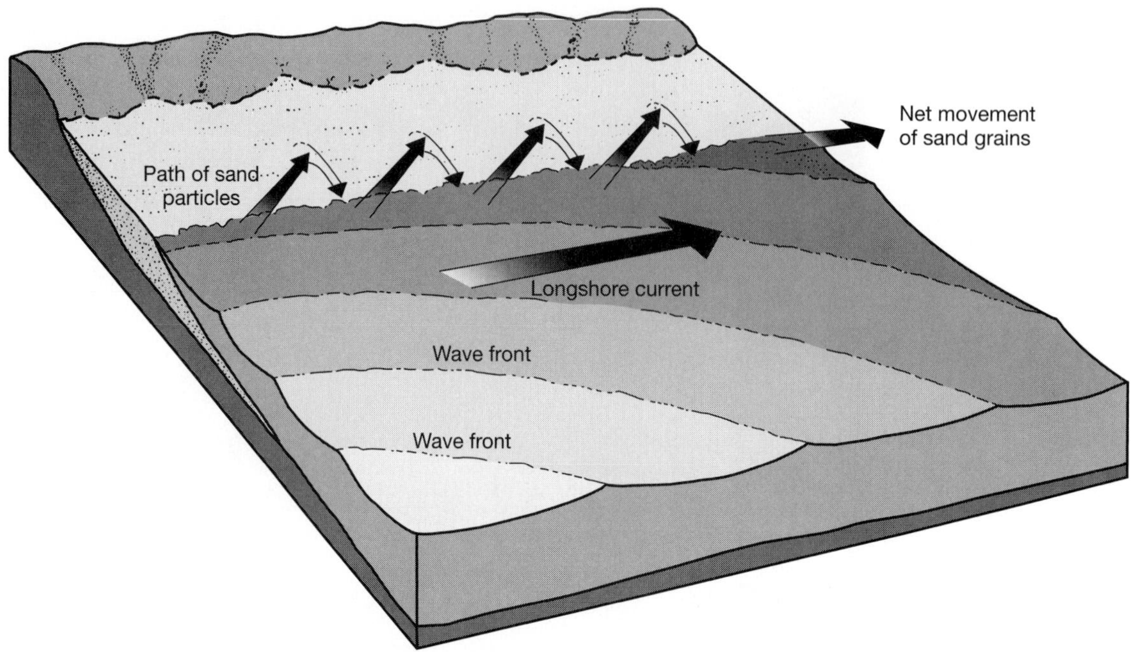

Fig. 15-10 Movement of sand along a beach by longshore currents. The longshore currents are created by obliquely breaking waves. These currents transport large quantities of sand along the beach and in the surf zone. (Source: Tarbuck, E., and Lutgens, F., 1993, *The Earth,* 4th ed. Upper Saddle River, N.J.: Prentice Hall, Inc.)

and they approach the coast from the northwest. As they approach the coastline, the waves are refracted, but they generally arrive at the surf zone at an angle to the beach. When the waves break, the water runs up the beach face at right angles to the line of approach (Fig. 15-10). The water returns down the beach face at right angles to the shoreline. Unless the wave fronts and the shoreline are exactly parallel, the water is displaced a small distance parallel to the shore (see Fig. 15-10). Sand moving with the waves is also carried parallel to the shore; this type of sand movement is known as ***longshore transport.***

Longshore currents operating in the shallow waters adjacent to California's beaches effectively distribute sand along the coast. Because the largest waves approach the California coast from the northwest, the overall direction of longshore transport is generally from north to south. North of Point Conception, the overall direction of sand transport is from north to south because the coastline faces southwest in most areas. South of Point Conception, sand is carried east, and then southeast because of the bend in the coastline (Fig. 15-11). Oceanographers have confirmed the direction of longshore transport by measuring the movement of sand in the surf zone. They have also identified the sources of many beach sands by determining their mineral content and matching the composition with rocks found in the drainage basins of coastal streams.

The local direction of longshore transport may be complex, because transport depends on the geometry of the coastline as well as on the direction of wave approach. For example, in Monterey Bay, longshore currents move both north and

Fig. 15-16 Marine terraces near Rincon Point, Ventura County. (Source: Sarna-Wojcicki, A. U.S. Geological Survey.)

MARINE TERRACES

Marine terraces are one of the most striking features of emergent coasts (Fig. 15-16). At some localities—San Clemente Island or Palos Verdes, for example—multiple terraces are cut against the hillslopes, their levels rising from the shoreline like flights of stairs. Terraces are conspicuous along most of California's coast; only where the coast is low-lying are they completely absent.

Marine terraces are ancient marine shorelines, or standlines. They consist of a relatively flat ***wave-cut platform,*** eroded by waves in the surf zone, and a cover of beach sand (Fig. 15-17; see also Figs. 15-6 and 15-9). Fossils, including microfossils and mollusk species like clams, may be preserved in the terrace deposits. Many geologists believe that the most extensive wave-cut platforms are formed during times when worldwide sea levels are high. The line where the platform meets the steep seacliff behind it is called the shoreline angle. As tectonic uplift causes the coastline to rise from the sea, new platforms are cut at lower levels, while the former platforms are lifted above sea level to become terraces (Fig. 15-18).

Because they represent former sea-level positions, marine terraces provide excellent indicators of uplift rates. Using radiometric techniques to date fossils in the marine terraces, it is possible to determine the rate of uplift since the terrace formed. Geologists have calculated the uplift rates along the California coast using the widespread 125,000-year-old terrace cut during the last global high stand of sea level. At most places, the rates are 0.1 to 0.4 meters per 1000 years. Along most of California, the 125,000-year-old terrace is found at 12.5 to 50 meters elevation. However, the rates are much greater in areas of rapid uplift. Both the north coast near Cape Mendocino (see Chapter 14) and the western Transverse Ranges near Ventura

Fig. 15-17 An uplifted wave-cut platform is overlain by sandy beach sediments. The plaform is cut on dark-colored sedimentary rocks in the lower cliffs. Fitzgerald Marine Preserve, San Mateo County. (Photo by author.)

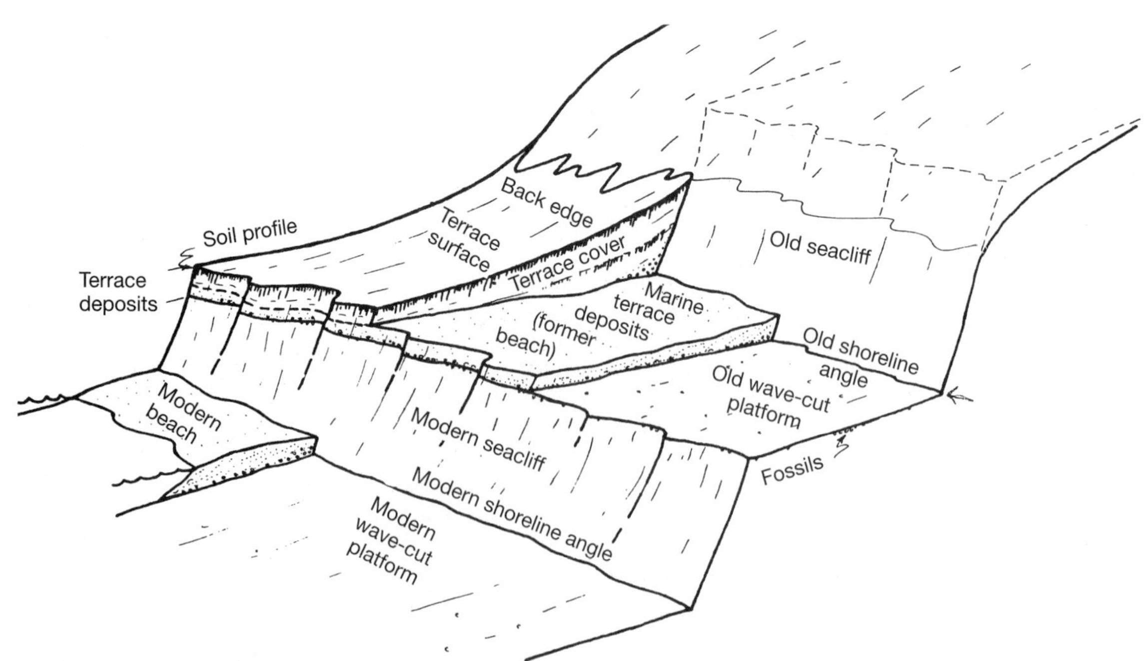

Fig. 15-18 Scematic diagram illustrating marine terraces and seavliffs along an emergent coast. (Source:LaJoie, K. 1972. U.S. Geological Survey, *Friends of the Pleistocene Field Trip Guidebook.*)

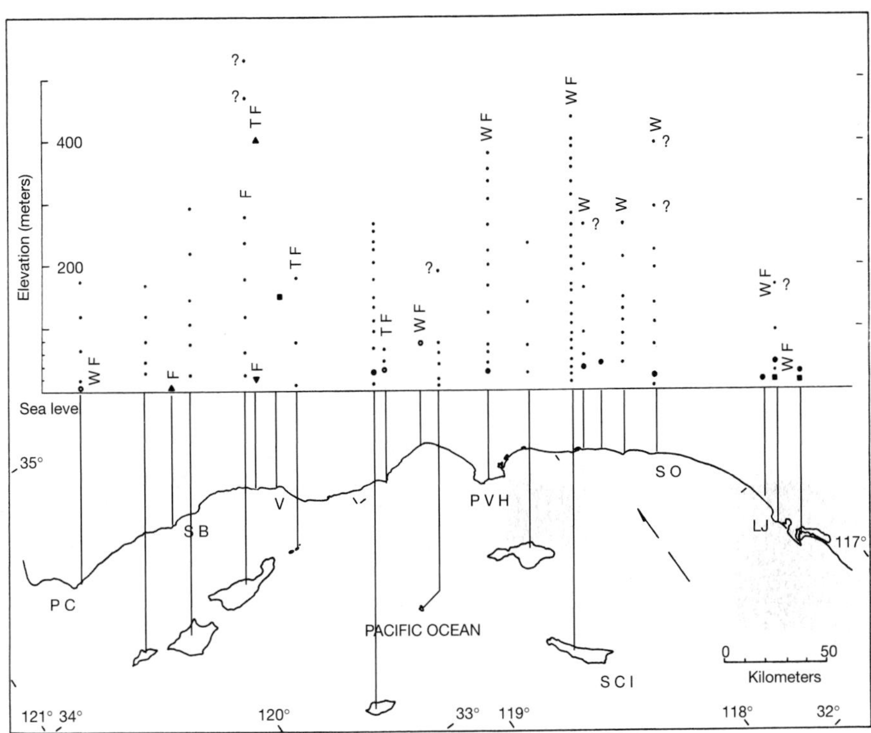

Fig. 15-19 Schematic profiles showing the elevations of terraces along the southern California coast. The map in the lower portion of the figure shows the locations of the terrace profiles. The symbols in the profiles portray both the elevations and the ages of terraces found at each site. Note that the lower terraces are youngest. Places where terraces are warped *(W)*, faulted *(F)*, or tilted *(T)* are denoted by letters on the profiles. (Source: LaJoie, K., and others. 1979. *Quaternary Terraces and Crustal Deformation in Southern California.* Geological Society of America Field Trip Guidebook.)

Legend

Map names	
P C	Point Conception
S B	Santa Barbara
V	Ventura
P V H	Palos Verdes Hills
S C I	San Clemente Island
S O	San Onofre
LJ	La Jolla

Terrace Ages	
●	Undated
○ ●	120,000
■	85,000
▲	45,000
▼	5000

(see Chapter 16) are rising rapidly. This is not surprising, considering that these are the regions most strongly affected by compression. At Ventura, a fold is lifting marine terraces at an impressive rate of 13 meters per 1000 years. There, a 40,000-year-old terrace is found 400 meters above sea level (see Fig. 15-16; see Chapter 16).

Terraces also record any tilting, folding, and faulting that have occurred since they formed. The shoreline angles can be thought of as geologic "level lines" (Fig. 15-19). Wherever the shoreline angles are warped, tectonic activity must have deformed them. Geologists have used California's marine terraces extensively when evaluating the long-term seismic safety of nuclear reactors, refinery facilities, and other critical projects.

In the San Diego area, ancient beach ridges—linear piles of sand up to 30 meters high—are preserved on most of the 14 marine terrace levels (Fig. 15-20). The ridges are probably the ancient equivalent of the sandbars and spits that enclose Mission Bay and San Diego harbor today.

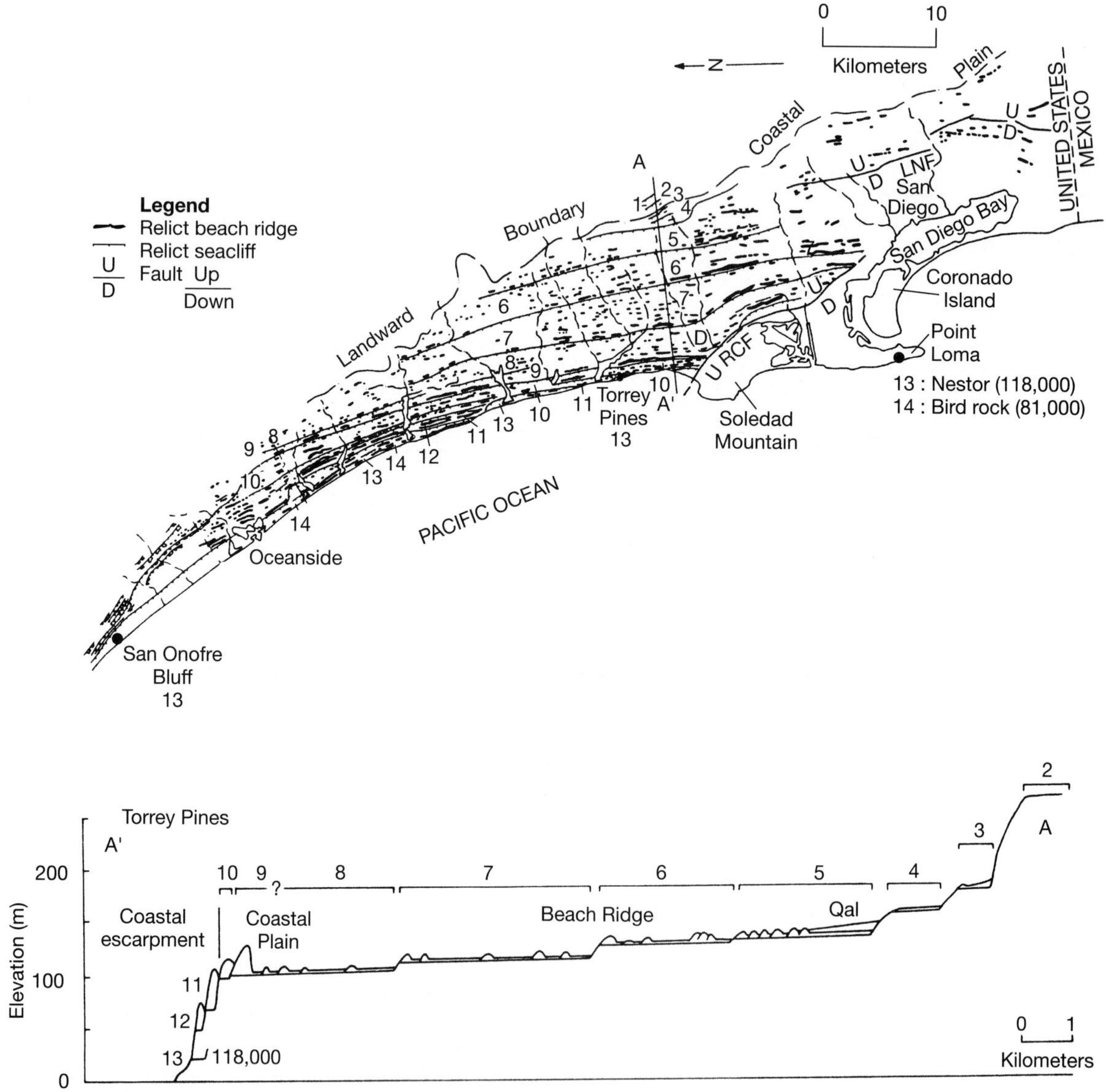

0 10
Kilometers
N
Legend
Relict beach ridge
Relict seacliff
U D Fault Up Down
Coastal Plain
Landward Boundary
A
A'
U
D
LNF
San Diego
San Diego Bay
Coronado Island
Point Loma
13 : Nestor (118,000)
14 : Bird rock (81,000)
U RCF
Soledad Mountain
Torrey Pines
13
PACIFIC OCEAN
Oceanside
San Onofre Bluff
13
UNITED STATES
MEXICO
Torrey Pines
A'
A
Elevation (m)
200
100
0
Coastal escarpment
Coastal Plain
Beach Ridge
Qal
118,000
0 1
Kilometers

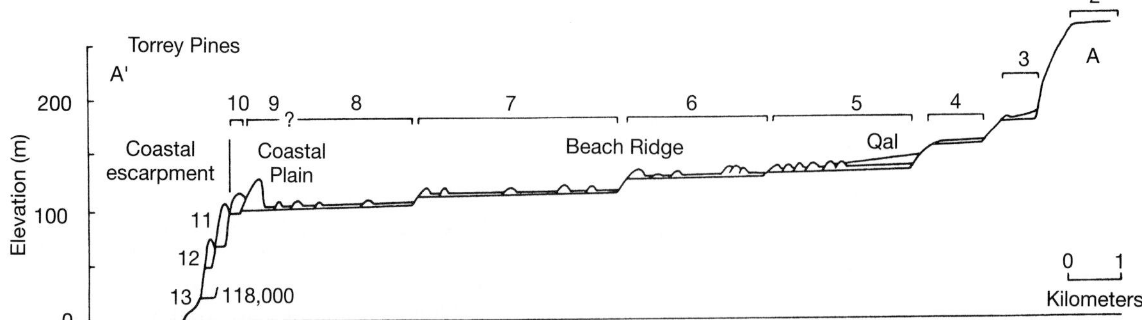

Fig. 15-20 Ancient beach ridges along the San Diego County coast. The cross section shows the elevations of beach ridges at Torrey Pines. San Onofre Bluff is shown as locality *SO* on Fig. 15-19. (Source: LaJoie, K. 1991. In *Nonglacial Quaternary of the U.S.* Geological Society of America, Decade of North American Geology, Vol. K-2.)

ESTUARIES, COASTAL WETLANDS, AND SUBMERGED COASTS

Although most of California's coastline has been tectonically uplifted during the past few million years, there are areas of the coast that have sunk relative to sea level. In these areas, known as ***submerged coastlines,*** sea level has risen sufficiently to inundate the lower parts of the landscape. In many of these areas, tectonic uplift has not been rapid enough to elevate the river valleys. As a result, tidal marshes, lagoons, or estuaries have formed at the mouths of rivers. San Francisco Bay is an excellent example of an estuary formed by the "drowning" of a river valley (see Chapter 12). Coastal wetlands have also developed in areas that have been tectonically down-dropped. As we will see in Chapter 16, down-warped areas are found in regions of compression. Two areas where coastal wetlands are found in tectonic depressions are along the Cascadia subduction zone (Fig. 15-21) and in the Transverse Ranges near Carpinteria (Fig. 15-22).

Estuaries are important breeding and feeding areas for a variety of animal species, including seabirds and a variety of fish. Estuaries are particularly important nurseries for juvenile salmon and steelhead, fish species that spend part of their life cycle in fresh water and part in the open ocean. After spawning in rivers, juvenile fish increase their chances of survival by feeding and growing in the estuary before entering the open ocean to face their adult habitat. Two of California's protected National Marine Sanctuaries have been established in estuaries and their surrounding wetlands. Elkhorn Slough is located in Moss Landing along the central coast of Monterey Bay (see Fig. 15-12), and the Tijuana Slough forms the Mexican border at the mouth of the Tijuana River. Despite the untreated sewage that flows down the river from Tijuana, the slough and the surrounding marshes are a rich habitat for many species. In contrast, the filled, paved, and diked wetlands of Mission Bay in San Diego or San Francisco Bay are human habitats created with much greater impact on other species.

Fig. 15-21 Coastal wetlands and estuary at the mouth of the Mad River, Humboldt County. (Source: Dengler, L. Humboldt State University.)

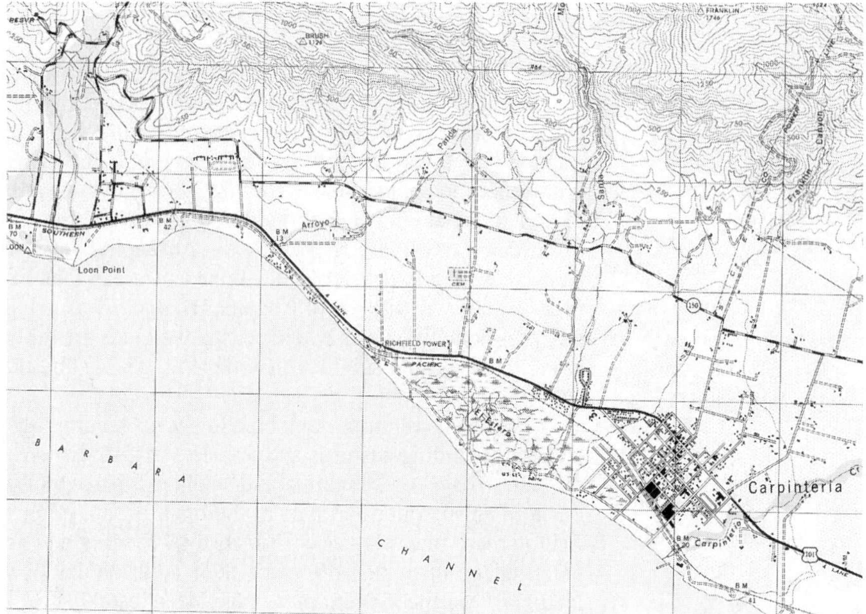

Fig. 15-22 A portion of a 1944 Carpinteria topographic map showing the coastal wetlands near Carpinteria. (Source: U.S. Geological Survey.)

COASTAL EROSION

Tremendous energy is focused on a narrow strip of rock and sediment, as waves pound ceaselessly against the shoreline for years. Waves generated by large storms are particularly effective agents of erosion. For example, at Redondo Beach in 1953, observers saw large storm waves throw 5-centimeter rocks more than 30 meters into the air. Coastal erosion is a natural process along all shorelines, particularly during storms. Along emergent coasts, waves undermine the lower parts of seacliffs, causing large blocks to fall into the ocean. Rocky points are eroded to form sea stacks, which are eventually removed by the waves. Over time, the entire shoreline retreats inland.

Rates of coastal erosion are determined by several factors, including the orientation of the coast, the height of storm waves, and the erodibility of the rocks along the shore. Natural erosion rates may be increased by human activities that interrupt the longshore transport of sand. An example of this is the Half Moon Bay Harbor in San Mateo County. Half Moon Bay's distinctive curved shape results from the refraction of waves around Pillar Point and the distribution of sand southward along the coast (Fig. 15-23). From 1959 to 1961, a breakwater was constructed to protect the harbor from summer waves approaching from the southwest. The breakwater disrupted the southward flow of sand, with the predictable result that the waves south of the breakwater are actively eating away at the cliffs to replenish the system with sediment. Between 1962 and 1985, the coast retreated by as much as 2.5 meters per year, a fourfold increase over the prebreakwater rates. Portions of roads, including State Highway 1, disappeared with the coastal area.

The town of Encinitas, north of San Diego, has lost several buildings, streets, and railroad lines to coastal erosion. During the winter of 1941, large storm waves along the base of the cliffs and heavy rains on the hills above combined forces to toss a four-

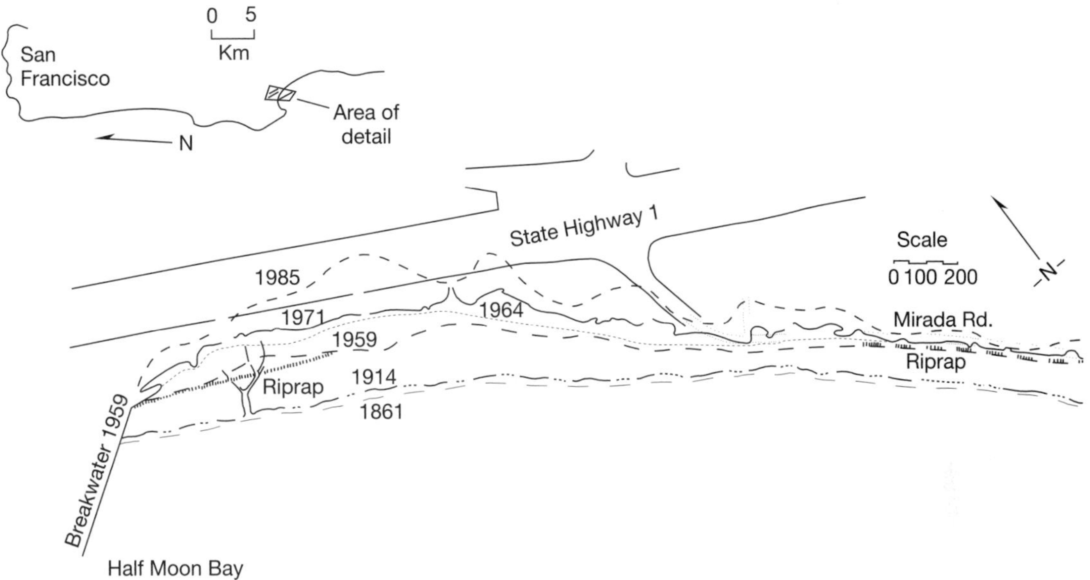

Fig. 15-23 Progressive coastal erosion and shoreline retreat in Half Moon Bay near El Grenada, San Mateo County. (Source: La Joie, K. 1972. *Friends of the Pleistocene Field Trip Guidebook.*)

Fig. 15-24 Drainage lines and erosion control structures, Santa Cruz county coast. (Photo by author.)

story-high, relatively new temple into the ocean as the cliffs collapsed. In 1977, similar storms toppled miles of cliffs and destroyed the train tracks along the terrace.

Coastal erosion poses problems because people build supposedly permanent structures along mobile coastlines. The U.S. Army Corps of Engineers has estimated that 86 percent of California's coast is experiencing significant erosion. Along 124 kilometers of coastline, coastal erosion is threatening houses, roads, and other structures (Fig. 15-24). Erosion is typically episodic, with major cliff retreat, landsliding, and sand removal taking place during large storms. Many years may pass between cliff-eating storms, making the threat of coastal erosion fade in the memories of communities and developers. In some of California's coastal counties, the California Coastal Zone Commissions and other civic agencies are attempting to restrict coastal development to reduce the risk posed to property by future coastal erosion. However, because much of the coastline is already developed, coastal erosion will continue to cause property damage and loss.

FURTHER READINGS

BASCOM, W. 1980. *Waves and Beaches.* Garden City, N.Y.: Anchor Books, 366 pp.

CALIFORNIA COASTAL COMMISSION. 1994. *California Coastal Resource Guide,* 2nd ed. Berkeley: University of California Press, 384 pp.

DESANTIS, M. 1985. *California Currents: An Exploration of the Ocean's Pleasures, Mysteries, and Dilemmas.* Novato, Calif.: Presidio Press, 238 pp.

EMERY, K.O. 1960. *The Sea Off Southern California.* John Wiley & Sons, 364 pp.

GRIGGS, G. *Living With the California Coast.*

KAUFMAN, W., AND PILKEY, O. 1979. *The Beaches Are Moving.* Garden City, N.Y.: Anchor Press, 326 pp.

CHAPTER

16

The Transverse Ranges, the Los Angeles Basin, and the Offshore Islands

Compression and Rapid Change

Even a quick look at a California relief map (Endpaper 4) reveals how the Transverse Ranges got their name: they run transverse (crosswise) to almost all of California's other mountains and valleys. Several rugged mountain ranges make up the Transverse Ranges: the most prominent are the Santa Ynez Mountains, the Santa Monica Mountains, the San Gabriel Mountains, and the San Bernardino Mountains. The highest peaks in the ranges are San Gorgonio Peak in the San Bernardino Mountains (elevation 3500 meters) and San Antonio Peak in the San Gabriel Mountains (elevation 3072 meters). On a clear day, the Transverse Ranges form an imposing fringe around the flat valleys that are sandwiched against them (Fig. 16-1). Two of the largest of these valleys, the San Fernando Valley and the Los Angeles Basin, are home to many of California's residents. Like the mountains, this part of the California coastline also runs east-west (see Chapter 15). A person standing at the ocean's edge in Santa Barbara or Ventura to watch the waves is actually facing south (see Fig. 16-1).

The geologic process creating the distinctive features of the Transverse Ranges Province is compression. Because of the Big Bend in the San Andreas fault, the Pacific and North American Plates converge in this region (Fig. 16-2; see also Endpaper 2 and Chapter 14). As a result, the entire region is being squeezed together, with the maximum compression oriented in a north-south direction. The mountains, the basins, and many of the faults crossing the province are all expressions of this squeezing. Most trend east-west, perpendicular to the direction of maximum compression. As the crust is compressed, it is being squeezed and thickened. This process is expressed at the surface, where some areas are being uplifted to form mountains, while others are being pushed down to form basins.

Fig. 16-1 Landsat satellite image of the Los Angeles basin. The Los Angeles basin and the San Fernando Valley to the north, are seen as smooth plains. Fault-bounded mountain ranges ring the valleys. (Source: Image processing by Rymer, M. U.S. Geological Survey.)

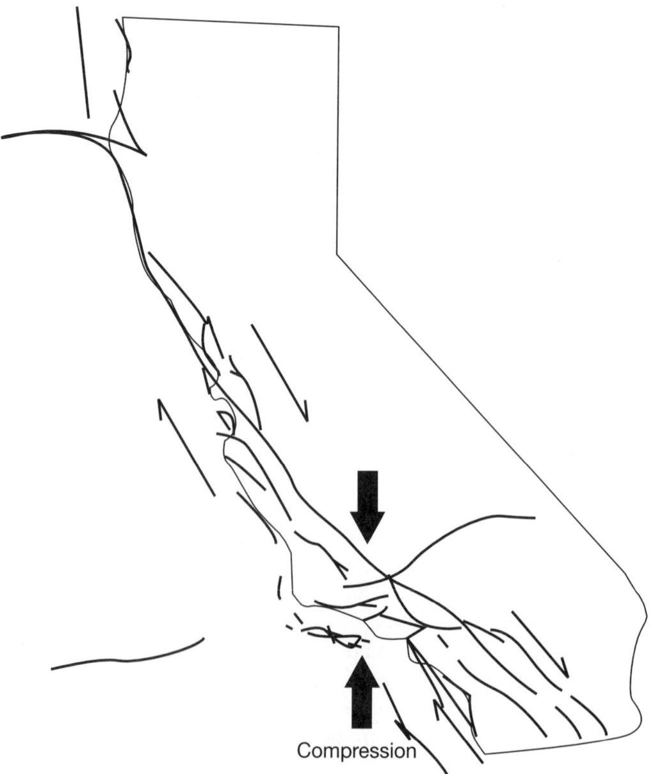

Fig. 16-2 Map showing the region of compression along the Big Bend in the San Andreas fault. (Source: LaJoie, K., and others. 1979. Geological Society of America Field Trip Guidebook.)

The Transverse Ranges are one of the Earth's most rapidly uplifting areas. The steep, rugged mountains along the Big Bend are some of the evidence of rapid uplift. The stepped marine terraces near Ventura and Palos Verdes (Chapter 15) provide further proof of rapid uplift: the terraces there are rising by as much as 5 to 10 millimeters per year. Sudden, dramatic uplift takes place during some earthquakes. During the 1994 Northridge earthquake, the Santa Susana Mountains rose 70 centimeters, and during the 1971 San Fernando earthquake, the same mountains rose about 2 meters.

REVERSE FAULTS AND FOLDS

Not surprisingly, many of the faults in the Transverse Ranges Province are the type produced by compressional forces. When blocks of rocks are compressed, they buckle together, with one block sliding over the other (Fig. 16-3). Both vertical movement and shortening of the crust occur across the resultant fault, known as a ***reverse fault.*** If a reverse fault is only slightly inclined, it is called a ***thrust fault.*** Earthquakes on reverse faults may generate strong vertical motion as well as horizontal motion.

Many of southern California's reverse and thrust faults have not ruptured the sediments or rocks at the surface. In many cases, the ground surface is buckled where the fault projects to the surface (Fig. 16-4), but some buried faults have no surface expression (see Color Plate 24). Geologists have recently recognized that these buried, blind thrust or reverse faults pose significant hazards. The 1994 Northridge, 1987 Whittier Narrows, and 1971 San Fernando earthquakes were all

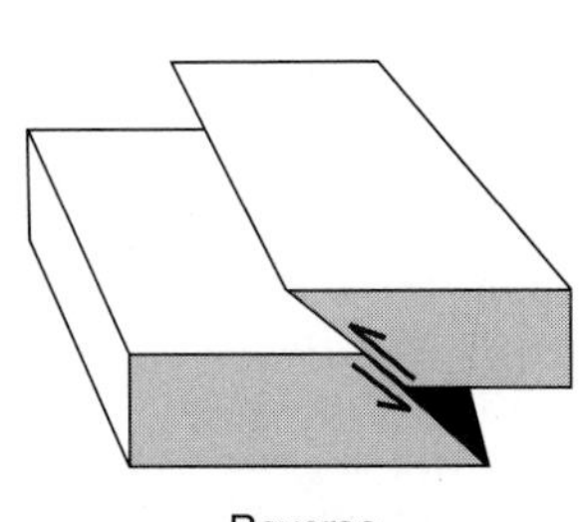

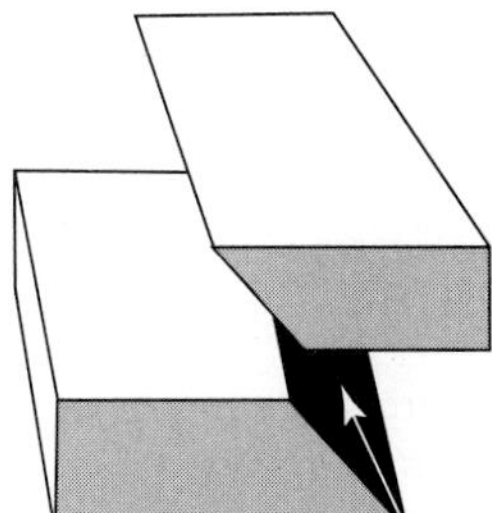

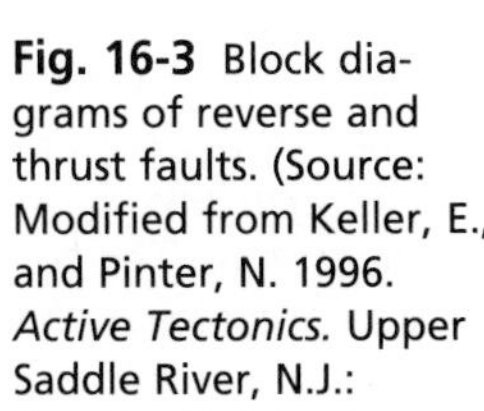
Fig. 16-3 Block diagrams of reverse and thrust faults. (Source: Modified from Keller, E., and Pinter, N. 1996. *Active Tectonics.* Upper Saddle River, N.J.: Prentice Hall, Inc.)

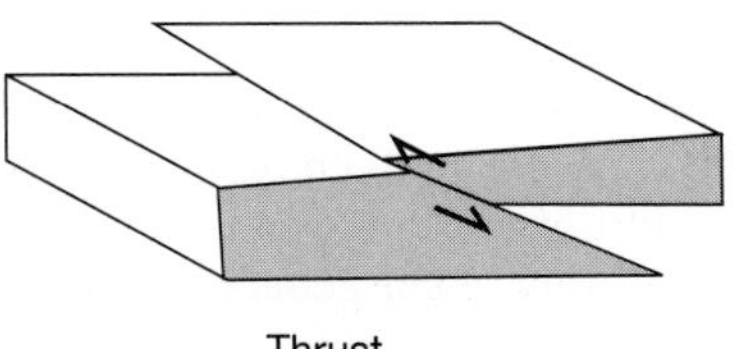

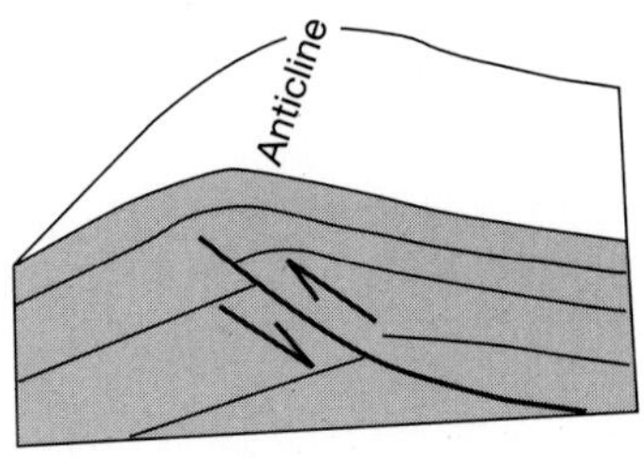

Fig. 16-4 View to the north across the Ventura fault in Ventura. The steep rise in the road shows the surface expression of the fault. A trench excavated by geologists across the fault trace is marked by the linear mound of sediment on the right side of the photo. (Source: Sarna-Wojcicki, A. U.S. Geological Survey.)

centered on reverse faults buried beneath the Los Angeles Basin (Fig. 16-5; see also Chapters 13 and 14). The 1983 Coalinga earthquake (magnitude 6.5) was also centered on a blind fault, one of several bordering the western edge of the San Joaquin Valley.

When rocks and sediments are compressed, they may also be crumpled together without rupturing. When this happens, folds are created (Fig. 16-6). It is important to realize that both folding and reverse faulting can be produced by compression and that both phenomena result in crustal shortening. A fold created by an upward buckling of rock or sediment is an ***anticline*** (Fig. 16-7 and Color Plate 27), whereas a downward-buckled fold is a ***syncline***. In many regions where rocks are folded, multiple anticlines and synclines are aligned with their hinge lines, or fold axes, perpendicular to the direction of maximum compression. The east-west alignment of reverse faults and fold axes in the Transverse Ranges is good evidence that both were created by north-south compression.

Southern California's ***basins*** are regions where the crust has been tectonically depressed in the center and pushed up on all sides, forming a low-lying bowl or basin. Southern California's basins are the result of down-buckling of the crust, just as many of the hills and mountains have resulted from up-buckling. The basins are typically surrounded by mountain ranges or lower hills, regions where the crust is tectonically uplifted. For example, the Santa Ynez, Topatopa, and Santa Monica Mountains (see Figs. 16-1 and 16-5) are large, complex anticlines in the western Transverse Ranges. The borders between basins and mountains are typically active faults—either reverse faults or transform faults of the San Andreas system. The correlation between the topography and geological structures is strong evidence that tectonic forces are active.

Some folds in the Transverse Ranges appear to be growing rapidly. For example, the Ventura Avenue anticline, just north of the town of Ventura, has tilted very young

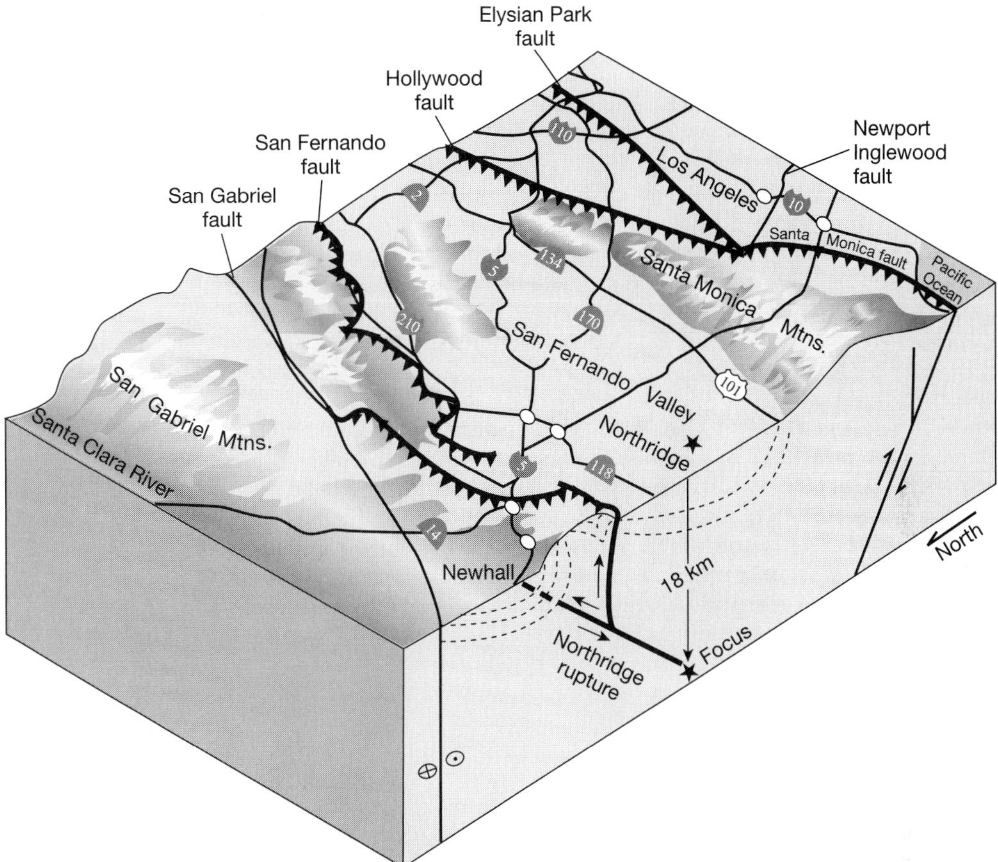

Fig. 16-5 A cross sectional view of the Northridge area showing reverse faults beneath the San Fernando Valley and the northern Los Angeles basin. (Source: Williams, P. Lawrence Berkeley Laboratory. In Learning from Los Angeles. *Earth Magazine*, September 1994.)

sedimentary layers, only a few hundred thousand years old, by as much as 30 degrees (see Figs. 16-7 and 16-8). Repeated surveys indicate that the valley of the Ventura River, near the axis of the anticline, rose about 0.5 centimeters per year between 1920 and 1968. Even the surface of California Route 33 is buckled slightly where it crosses the fold. Further south, the 200,000-year-old San Pedro sand is uplifted in the Palos Verdes and San Joaquin Hills and buckled beneath the area between them (Fig. 16-9). The total difference in elevation of the once-flat formation is now tens of meters.

The Transverse Ranges Province—including the Channel Islands as well as the entire Los Angeles Basin—is crisscrossed by reverse faults and folds. Trending east-west, these faults and folds intersect the northwest-trending right-lateral faults of the San Andreas system, like the Newport-Inglewood fault zone (see Endpaper 2 and Fig. 14-36). Not all of the Transverse Range's reverse faults are buried, and along some, rugged mountain ranges have been pushed up. For example, at the northern edge of the province, the Santa Ynez fault runs along the northern slope of the Santa Ynez

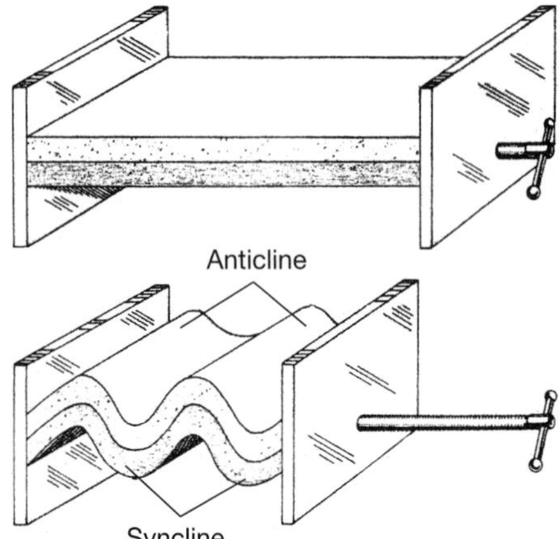

Fig. 16-6 An anticline and a syncline *(lower figure).* The vise shows the direction of maximum compression. (Source: Coch, N., and Ludman, A. 1991. *Physical Geology.* New York: Macmillan.)

Fig. 16-7 The Ventura Avenue anticline, near Ventura (see this photo in color in Plate 27). The view is toward the west-northwest along the axis of the fold. (Source: Sarna-Wojcicki, A. U.S. Geological Survey.)

Mountains (see Fig. 14-36). The San Cayetano and Oak Ridge faults (see Fig. 16-13), two steep reverse faults dipping in opposite directions, have sandwiched and down-dropped the Santa Clara Valley between the Topatopa Mountains on the north and the Santa Susana Mountains on the south. The Sierra Madre fault zone (see Fig. 14-36) forms the southern boundary of the San Gabriel Mountains. At the southern edge of the Transverse Ranges, the Malibu Coast, Santa Monica, and Raymond Hills reverse faults all extend to the surface, where they cut across young sediments.

Other fault and fold zones are marked by low hills crossing southern California's basins—for example, the Puente Hills, the San Pedro Hills, and the Repetto Hills. In the urbanized basins, subtle ramps along streets mark the surfaces of folds and reverse faults. On older maps, these changes in slope sometimes appear as aligned marshy areas or small hills (Fig. 16-10). At some localities in southern California, folding at or near the surface may be the expression of reverse faulting in the deeper layers beneath. The less consolidated sediments above a buried reverse fault may be buckled into an anticline. Geologists believe that many of southern California's faults are still growing upward and have not yet broken the younger sed-

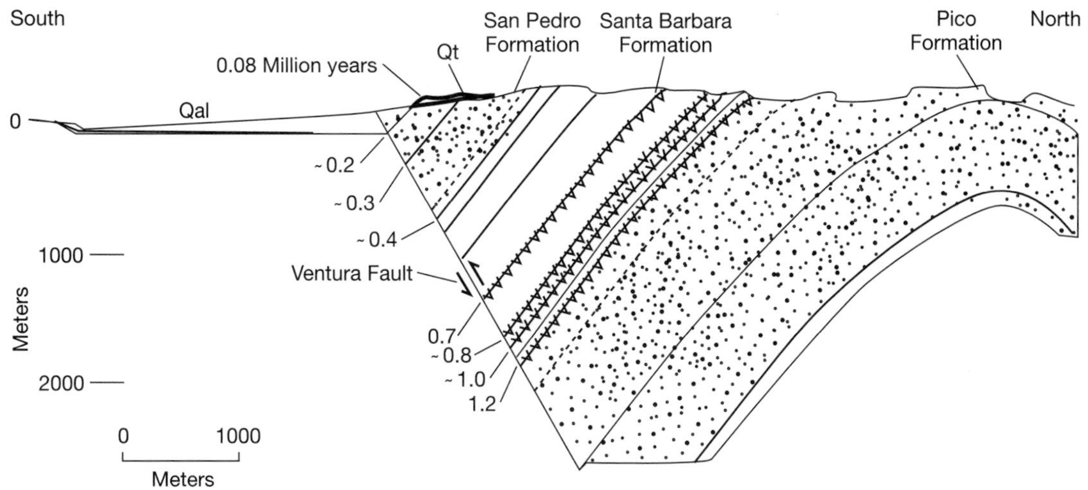

Fig. 16-8 A cross section showing the Ventura fault (see Fig. 16- 4) and the southern limb of the Ventura anticline. Volcanic ashes used to date the sedimentary layers are shown by V's in the diagram. The ages of units are shown along the left of the cross section in millions of years. (Source: Sarna-Wojcicki, A., and Yerkes, R. U.S. Geological Survey.)

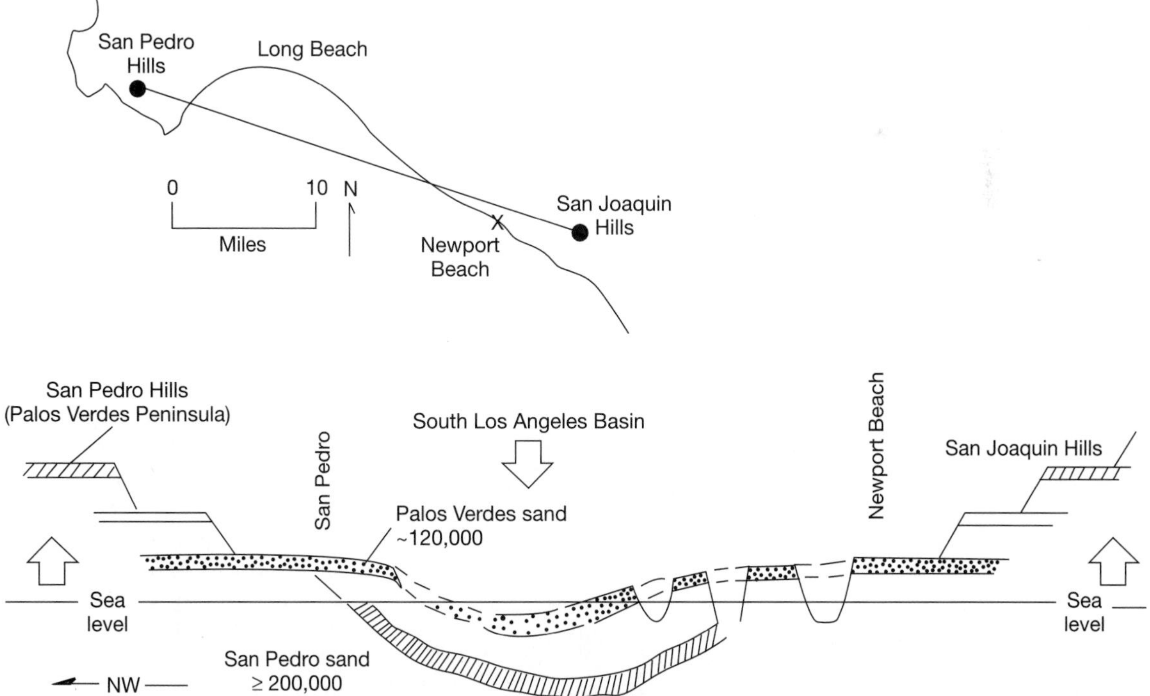

Fig. 16-9 Diagram of a syncline in the south Los Angeles basin. Very young sedimentary units are uplifted in the San Joaquin Hills and the San Pedro Hills, at the ends of the cross section. Beneath the southern Los Angeles Basin, the same units are down-warped below sea level. (Source: LaJoie, K., and others. 1979.)

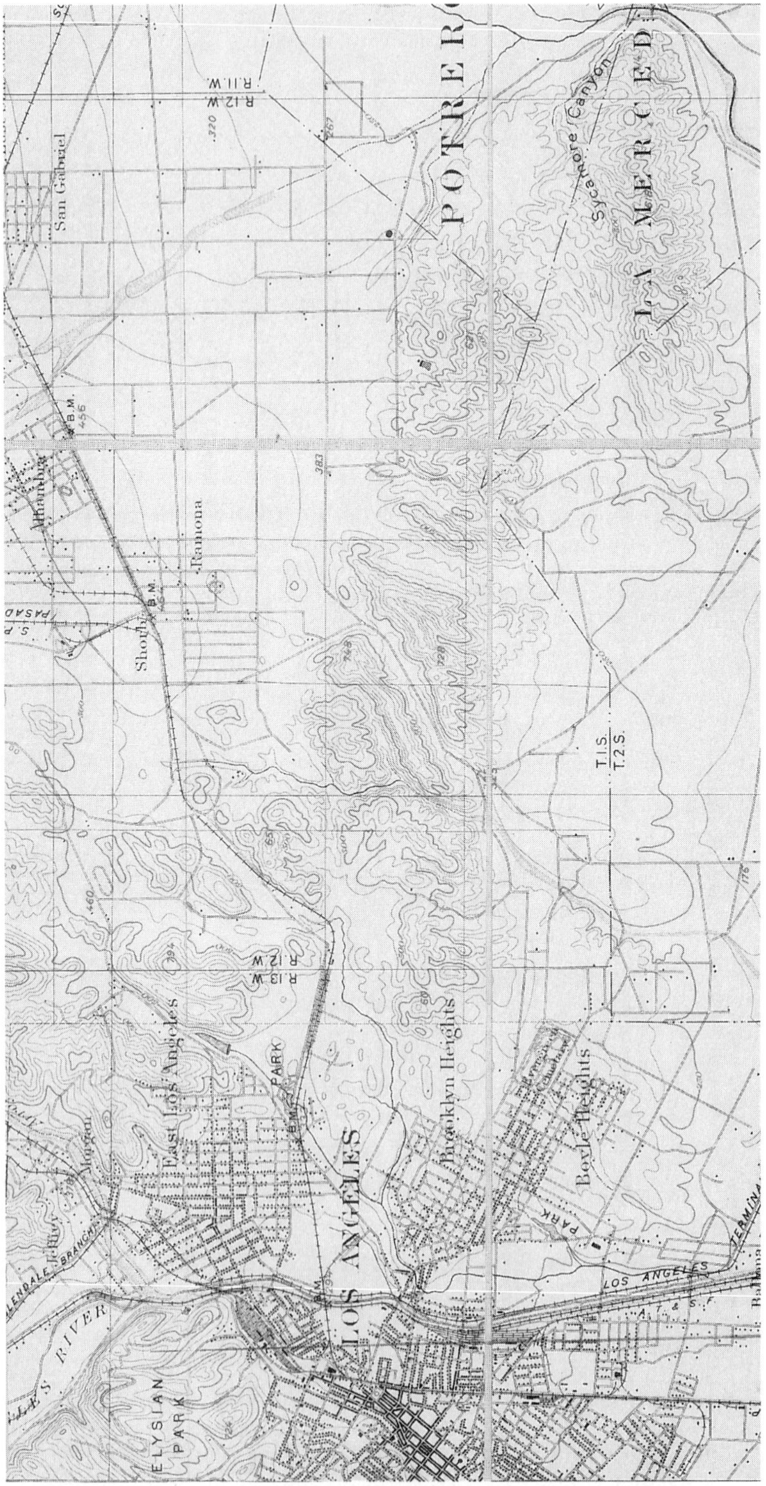

Fig. 16-10 A portion of the 1900 Pasadena topographic map near the present community of Monterey Park. Because the map was made before extensive urbanization of the Los Angeles Basin, one can clearly see the low hills marking the surfaces of recent anticlines. Today, Interstate 10, the San Bernardino Freeway, runs east-west along the western part of the railroad line shown in the northern part of the figure running through Shorb.

imentary layers near the surface. Recent earthquakes on blind reverse faults, most notably the 1994 Northridge and 1987 Whittier Narrows earthquakes discussed in Chapter 13, have prompted geologists to reexamine southern California's folds in light of their connection to buried faults.

BASINS OF THE TRANSVERSE RANGES PROVINCE

Another dramatic effect of compression in the Transverse Ranges is the formation and rapid filling of deep basins during the past 4 million years. As the San Andreas fault system evolved and the Big Bend formed, the basins subsided and began to fill with sediment (Fig. 16-11). About 4 million years ago, the rate of subsidence, or depression, of the basins increased, as did the rate of uplift in the surrounding mountains.

The basins of the Transverse Ranges are in various stages of being filled with sediment. The Santa Cruz and San Nicolas offshore basins are entirely below sea level. By contrast, the Los Angeles Basin is filled to the brim, so that its surface lies above sea level. Beneath its relatively flat surface lies a complex depression containing up to 5500 meters of sediments deposited during the past 4 million years (Fig. 16-12) and up to 3600 meters of older Miocene sediment beneath that. The Los

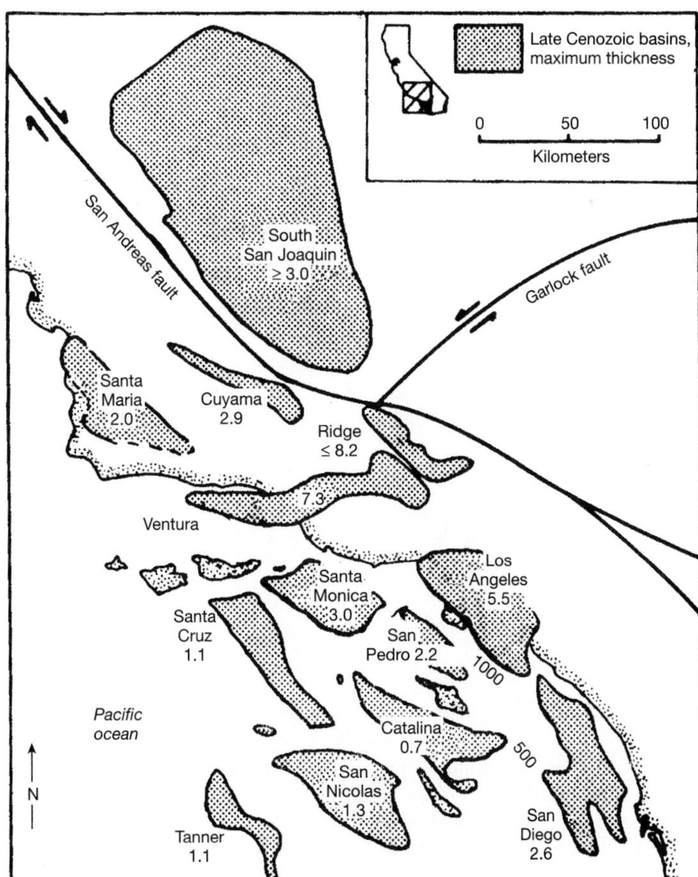

Fig. 16-11 Deep basins of southern California. The numbers show the maximum thickness (in kilometers) of sediments filling the basins during the past 4 million years. (Source: Yeats, R. 1991. In *The Cordilleran Orogen.* Geological Society of America, Decade of North American Geology, Vol. G-3, p. 336.)

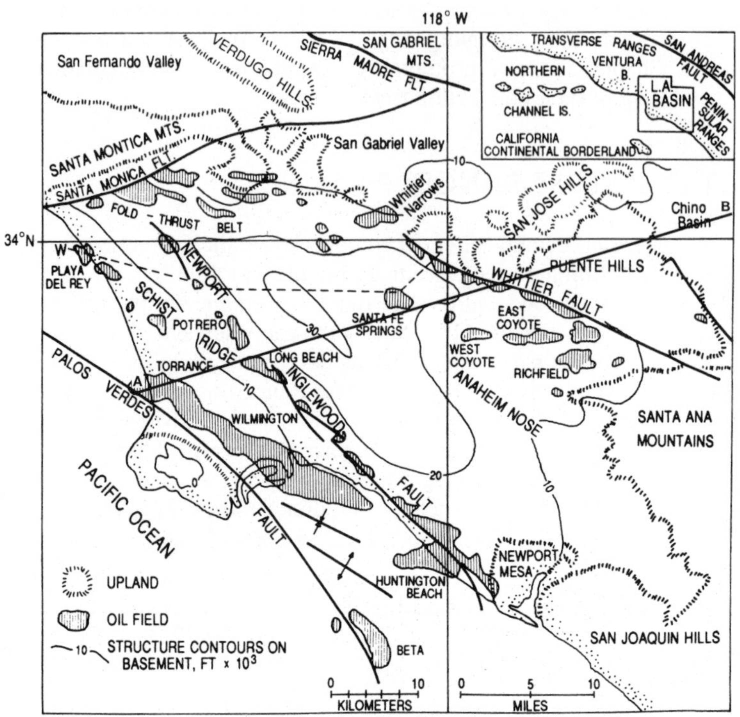

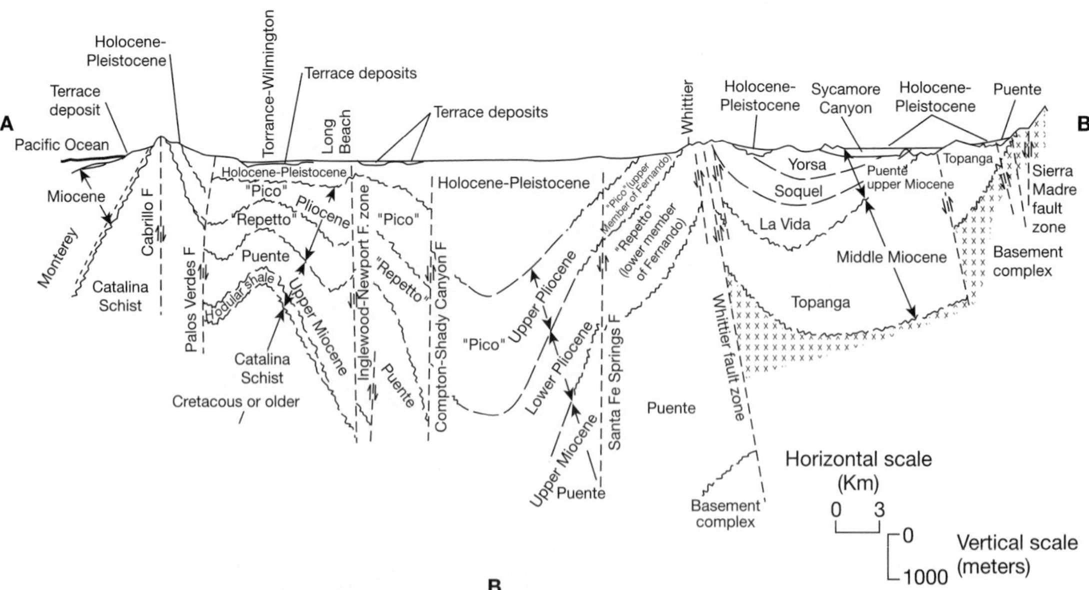

Fig. 16-12 Above, map, and below, cross section of the Los Angeles Basin. The shaded areas on the map show oil fields. (Source: Map, Yeats, R. 1991. In *The Cordilleran Orogen.* Geological Society of America, Decade of North American Geology, Vol. G-3, p. 339; cross section, California Oil and Gas Report. California Division of Mines and Geology, Publication TR12, Vol. 2.)

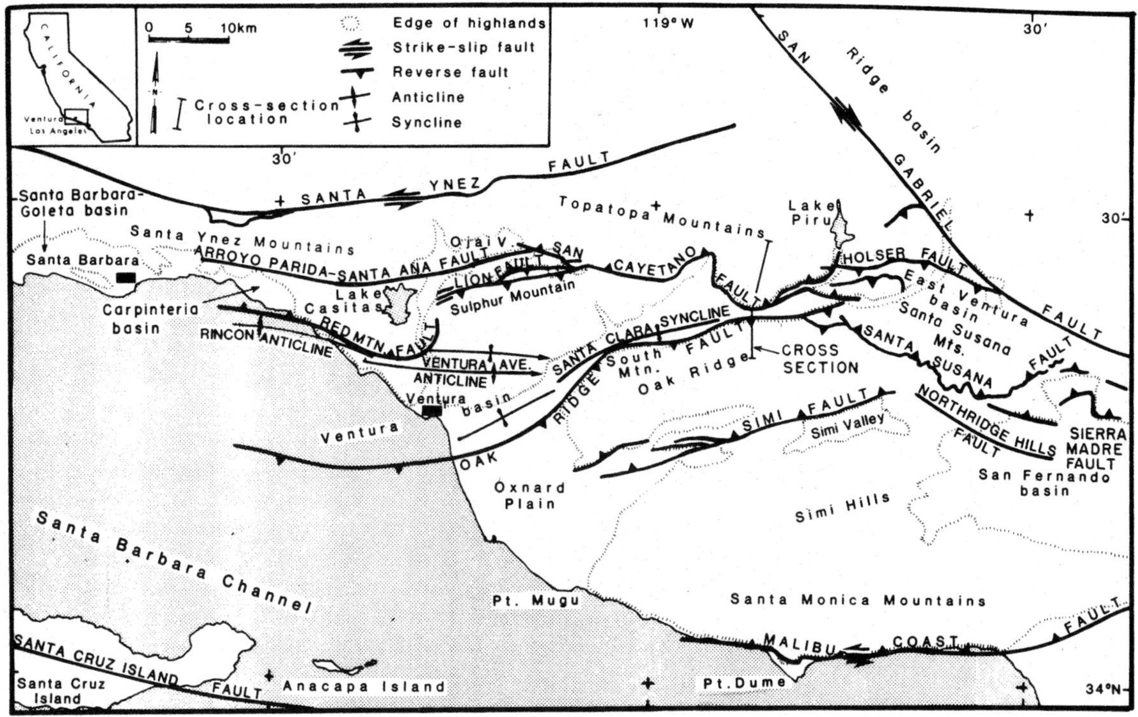

Fig. 16-13 Map of the Ventura Basin area. (Source: Yeats, R. 1991. In *The Cordilleran Orogen.* Geological Society of America, Decade of North American Geology, Vol. G-3, p. 335.)

Angeles Basin is so deep that geologists are uncertain about what rocks lie at its base. Most believe that the basin sediments overlie the Mesozoic Catalina Schist, which is exposed on Santa Catalina Island (see below) and in the Palos Verdes Hills.

The Los Angeles Basin is also broken by faults and folds. Large fault zones slice the basin into four well-defined blocks (see Fig. 16-12). On the west, the Palos Verdes fault zone separates the uplifted Palos Verdes Hills from the lower Long Beach–Wilmington area. The Newport-Inglewood fault zone, a member of the San Andreas system, runs northwest through the basin (see Chapter 14). Further east, the Whittier-Elsinore fault cuts across the basin, separating the Puente Hills and the San Gabriel Valley from Los Angeles. Anticlines and buried faults cross through the basin within these major blocks.

Ventura Basin

The Ventura Basin is sandwiched between the Santa Ynez and the Santa Monica Mountains, as if caught between the jaws of a giant vise. As the two ranges move together, sediments caught in the trough are being deformed and pushed up into mountains (Fig. 16-13). The western Ventura Basin, beneath the Santa Barbara Channel, lies below sea level west of the city of Ventura. Onshore, the surface of the Ventura Basin forms the Santa Clara Valley and the Oxnard Plain. Beneath

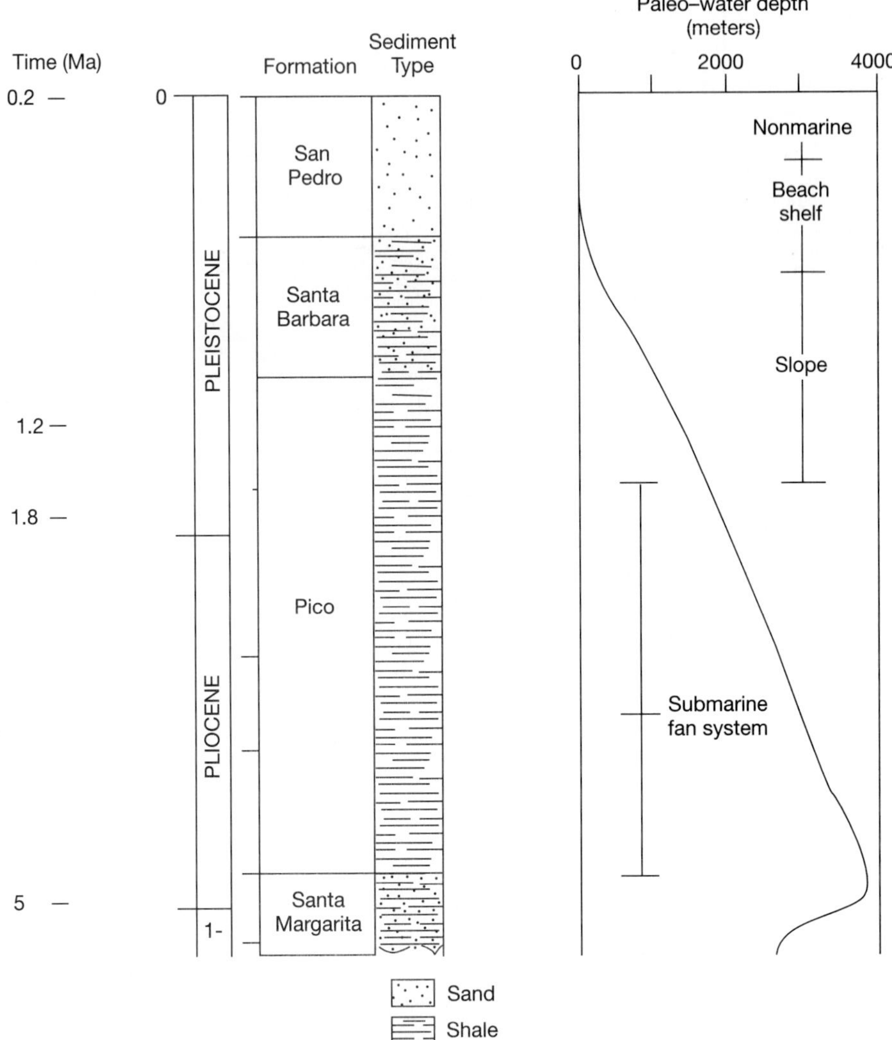

Fig. 16-14 The graph on the right shows the ancient water depths during the filling of the Ventura Basin, based on the species of foraminifera found in the sedimentary formations. (Source: Ingle, J. 1981. IGC Field Trip Guidebook.)

those smooth plains lie up to 10,000 meters of sediment, 7300 meters of it deposited during the past 4 million years. The Ventura is the deepest of all the basins of the Transverse Ranges, and it contains the world's thickest section of Pleistocene marine deposits.

The Ventura Basin began to form in early Pliocene time, about 4 to 5 million years ago. Using both the types of sediment being deposited in the basin and the microfossils, mostly foraminifera, geologists can reconstruct the changes in water depth during the basin's history. At first, the Ventura Basin was subsiding faster than it was filling with sediment. As a result, the water depth increased (Fig. 16-14, beginning at the water depth curve). The sediments and fossils found in the older Ventura

Fig. 16-15 Turbidites in the Pico Formation near Pitas Point. (Source: Sarna-Wojcicki, A. U.S. Geological Survey.)

Basin formations are typical of deep marine conditions. During the time that the Pico Formation was deposited, the basin was receiving sediments from deep-water turbidity currents (see Figs. 16-14 and 16-15). Then, between about 3.1 and 1.8 million years ago, 2600 meters of sediment were deposited in the basin. During this period, the rate of sedimentation was greater than the rate of tectonic subsidence, so the depth of water in the basin decreased. The evidence of shallower marine waters is clear: the fossils represent organisms that inhabit shallower water, and the sediments were deposited in a more nearshore environment. By the time the Santa Barbara and San Pedro Formations were deposited, between about 1.2 and 0.5 million years ago, shallow marine conditions prevailed in the central Ventura Basin (Fig. 16-16).

Between 200,000 and 400,000 years ago, the central Ventura Basin emerged above sea level and rivers began to deposit nonmarine sediment. The river sediment contains distinctive granitic pebbles and cobbles derived from the San Gabriel Mountains to the east. Reverse faulting and folding continued during and after sedimentation in the Ventura Basin. The older basin sediments are disrupted by numerous faults and folds, like the Ventura Avenue anticline discussed above. Continued compression is squeezing the deposits, uplifting them so that they are exposed in the hills surrounding the Santa Clara Valley (Fig. 16-17). Deposition in the eastern Ventura Basin ceased earlier than in areas to the west. Folded, faulted, and uplifted Ventura Basin sediments can now be seen in the Santa Susana Mountains (see Fig. 16-13). These sediments are now being eroded and redeposited on the Oxnard Plain and in the Santa Barbara Channel to the west.

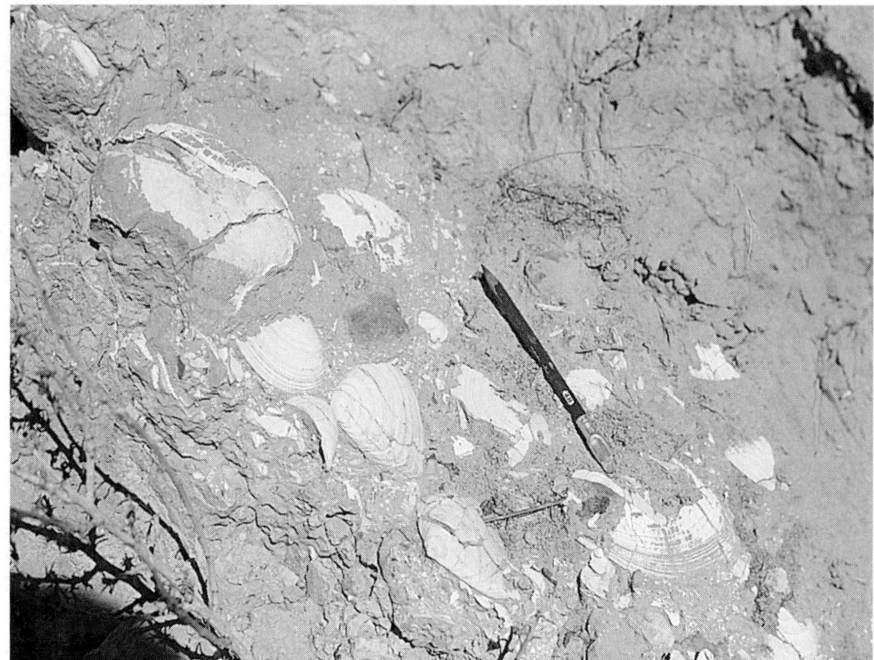

Fig. 16-16 Shallow-water fossils in the Santa Barbara Formation near Ventura. (Source: Sarna-Wojcicki, A. U.S. Geological Survey.)

Fig. 16-17 Uplifted sedimentary rocks of the Ventura Basin, near Rincon Point. (Photo by author.)

THE MONTEREY FORMATION

One of California's most widespread and famous sedimentary formations is the Miocene Monterey Formation (also called the Modelo Formation in southern California by some authors). Today it is exposed throughout coastal California, from Point Arena south to Dana Point, as well as on the offshore islands (Fig. 16-18). Its surface area occupies as much as 100,000 square kilometers of offshore and onland California. Because it was the source rock for much of California's petroleum (see below), geologists have taken a particular interest in the conditions that created the Monterey formation.

The Monterey Formation accumulated in marine basins along the California coast between about 23 and 12 million years ago. The environment of deposition was similar to today's southern California borderland or the Gulf of California. Like the sediments in the Ventura Basin, the Monterey Formation's sediments reflect deposition in a variety of marine conditions. During the time of deep-water deposition, when water depths ranged from 1000 to 2000 meters, fine-grained shale and siliceous (silica-rich) sediment accumulated by the settling of fine sediment from the overlying waters. During this time, the basins were not receiving large amounts of sediment from the coastal areas.

Much of the silica-rich sediment in the Monterey Formation is ***diatomite,*** a fine-grained sediment composed of the skeletal remains of marine diatoms. Diatoms are single-celled marine plankton. Today diatoms flourish in offshore waters where upwelling of deep water provides large amounts of nutrients. When the diatoms die, their siliceous cells settle to the bottoms of the marine basins. Today, diatomaceous sediments are accumulating in deep ocean basins where little or no land-derived sediment is deposited. These conditions apparently characterized the Miocene basins during the time that the Monterey diatomite formed, beginning about 12 to 13 million years ago. Throughout coastal California, one can find distinctive, white outcrops of ***diatomite*** (Figs. 16-19 and 16-20). Some of the diatomite has been altered to harder siliceous rocks—chert and porcelainite. Diatomite can also form in lakes, and beds of diatomite can be seen in the sediments that accumulated in many of California's Pleistocene lakes (see Chapter 6).

Pure diatomite has many commercial uses, including many types of filters, absorbents, and insulation. Diatomite filters are used in brewing, sugar refining, and water treatment. The diatom skeletons are ideal for these purposes because they are hollow and of variable shapes; diatomite filters contain about 90 percent open space and 10 percent solids. California supplies over 60 percent of the world's diatomite, and much of that is mined from the Monterey and Sisquoc Formations near Lompoc in Santa Barbara County.

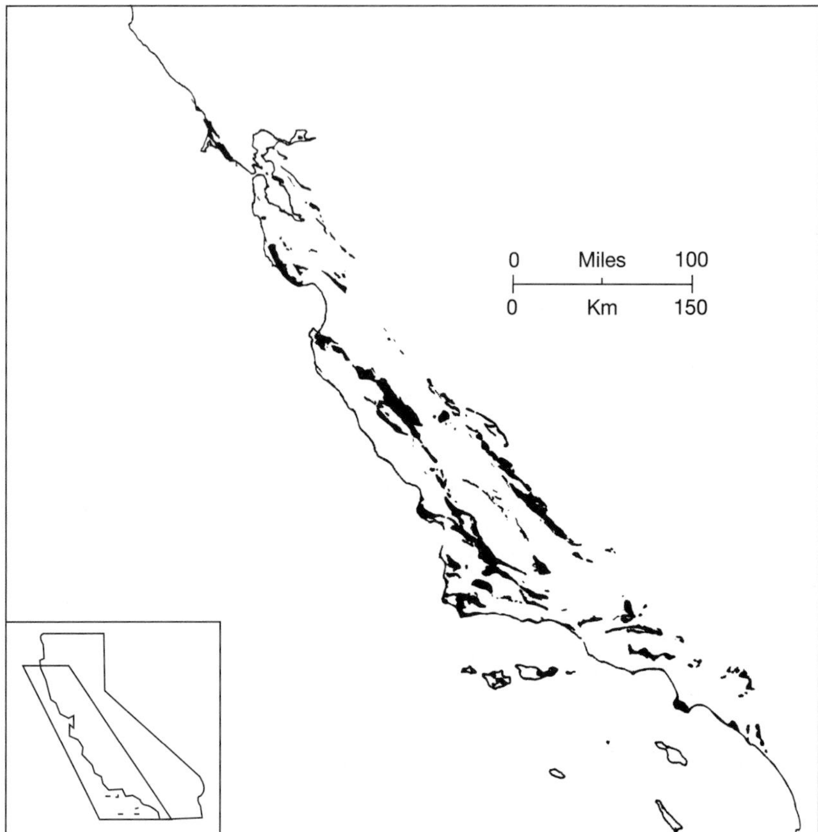

Fig. 16-18 The distribution of the Monterey Formation and similar siliceous rocks in California. (Source: Pisciotto, K., and Garrison, R. 1981. *The Monterey Formation and Related Siliceous Rocks of California.* Society of Economic Paleontologists and Mineralogists, p. 97.)

Fig. 16-19 Diatomite beds in the Monterey Formation, Lompoc. (Source: Sarna-Wojcicki, A. U.S. Geological Survey.)

Fig. 16-20 Microphotograph of diatom in the Monterey Formation. (Source: Grefco Corporation.)

OLDER SEDIMENTARY ROCKS

Marine sedimentary rocks of late Mesozoic and early Cenozoic age can be seen in the Santa Ynez, Topatopa, Santa Monica Mountains and the Santa Ana Mountains (see Figs. 16-12 and 16-13). Like the younger sedimentary rocks in the region, they have been uplifted, folded, and faulted during the more recent periods of compression. Careful mapping and correlation have enabled geologists to reconstruct the original depositional sequences of sediments from these deformed rocks. Based on the similarities between sequences in different ranges, geologists concluded that the marine sediments accumulated along the continental margin before it was sliced into separate basins. The Cretaceous and early Cenozoic marine sedimentary rocks are also quite similar to those seen in the Great Valley Sequence to the north, suggesting that all of these rocks formed in the forearc basin along the western margin of North America before the San Andreas transform developed (see Chapter 11).

A thick sequence of Mesozoic and early Cenozoic sedimentary rocks can be viewed in the Santa Ynez Mountains north of Santa Barbara. The light-colored and resistant Matilija Sandstone forms the most rugged slopes in the area. Another younger sandstone, the Coldwater Sandstone, also forms rugged slopes along the southern Santa Ynez Mountains. Less resistant, more thinly bedded shale units such as the Eocene Cozy Dell Formation are found on less-prominent slopes (Fig. 16-21).

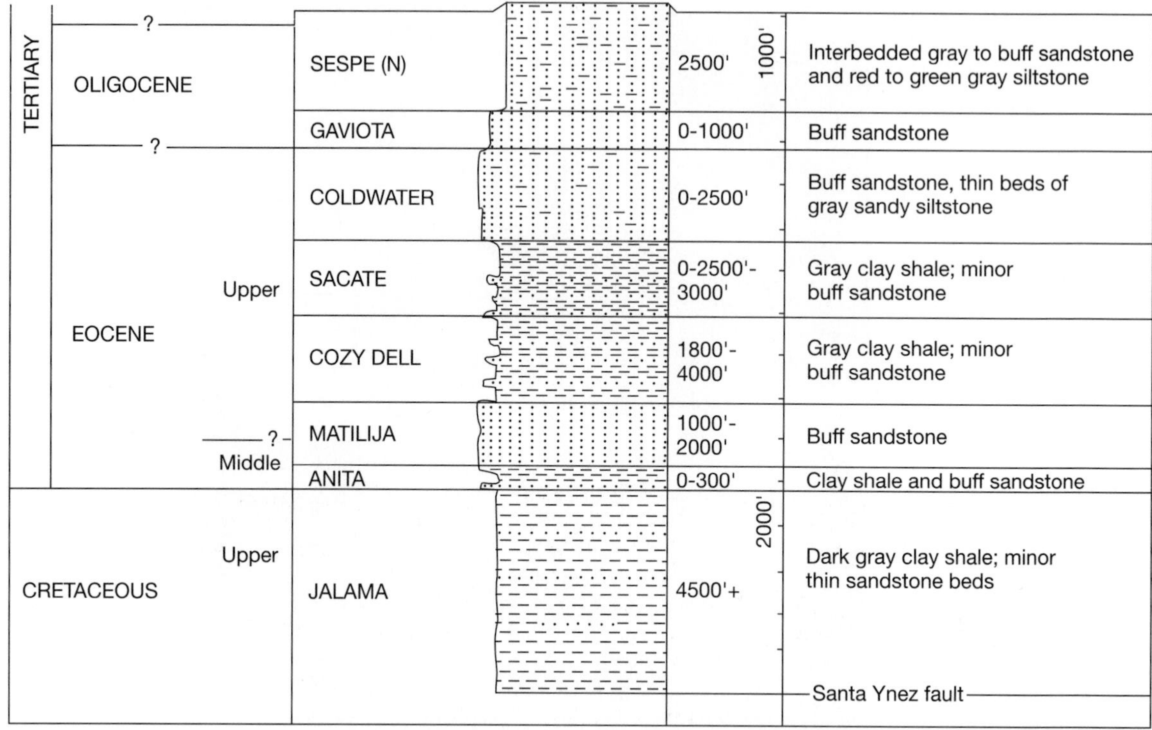

Fig. 16-21 A simplified stratigraphic section of early Cenozoic formations in the Santa Barbara area. Note that thicknesses are given in feet. (Source: California Division of Mines and Geology, Bulletin 186.)

During the Oligocene Epoch, the Transverse Range region apparently emerged from the sea, as indicated by the presence of nonmarine sedimentary rocks of that age. These distinctive rocks are the maroon, red and green Sespe Formation, which contains a mixture of sandstone, siltstone, and conglomerate. The sediments of the Sespe Formation and the Vasquez Formation to the east, were deposited in ancient alluvial fans along the flanks of mountains that have long since disappeared from the landscape. Marine conditions returned to the region during Miocene time (see Fig. 16-24) and persisted until the most recent emergence began during Pliocene time. Some areas, like the Ventura Basin, emerged from the sea only a few hundred thousand years ago. During Miocene time, today's Santa Monica Mountains were also the site of numerous active volcanoes, as indicated by thick sequences of submarine Miocene volcanic rocks (Figs. 16-22 and 16-23, *A* and *B*). The active volcanoes extended to the northern Channel Islands, as discussed below.

One of the more dramatic conclusions geologists have made in recent years is that the entire Transverse Range Province has been rotated in a clockwise direction during the past 15 million years. This conclusion is based on two different lines of geologic evidence. First, major formations and boundaries of Miocene and older rocks can be matched up with those in the southern San Joaquin Valley by "unrotating" the province, as shown in Fig. 16-25 and 16-26. Of course, before making

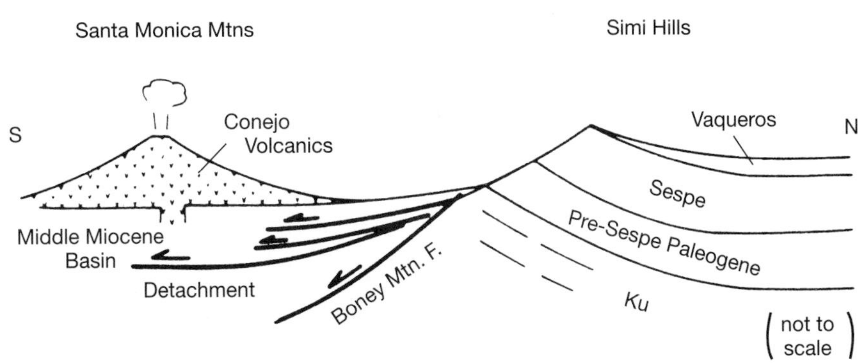

Fig. 16-22 A schematic cross section showing the inferred location of the volcanoes in the Santa Monica Mountains during Miocene time. (Source: Yeats, R. 1991. In *The Cordilleran Orogen,* Geological Society of America, Decade of North American Geology, Vol. G-3, p. 333.)

Fig. 16-23 **A,** Outcrop and **B,** closeup of Miocene pillow basalts in the Santa Monica Mountains. (Source: Wentworth, C. U.S. Geological Survey.)

Fig. 16-24 Rugged cliffs underlain by the San Emigdio Formation, near San Emigdio Creek. (Source: Brabb, E. U.S. Geological Survey.)

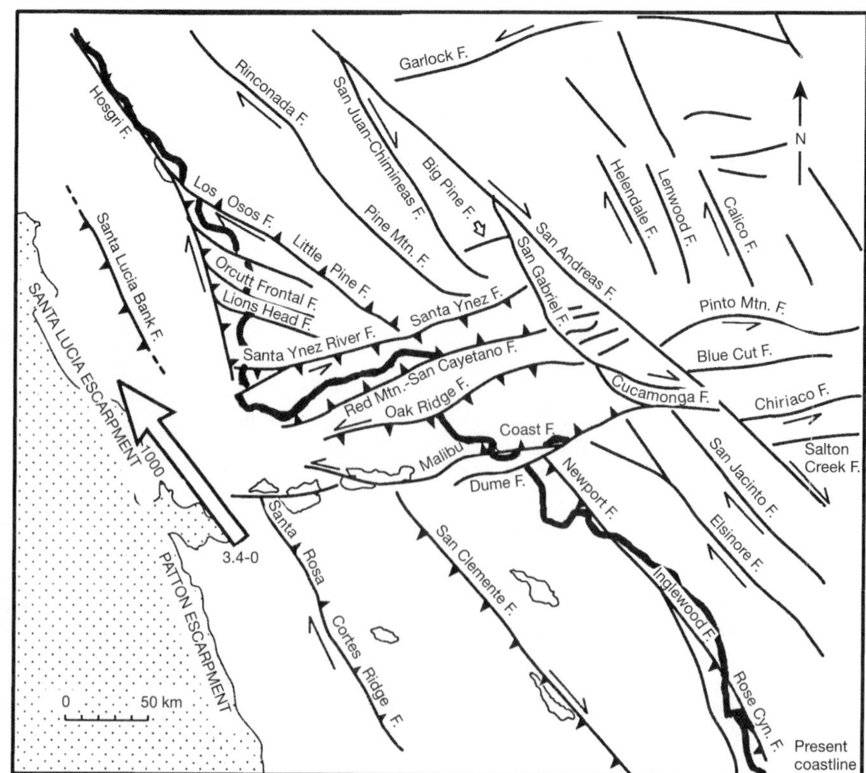

Fig. 16-25 Map showing major faults of the western Transverse Ranges about 3 million years ago. (F = fault.) (Source: Luyendy, B. 1991. Geological Society of America Bulletin, Vol. 103[11].)

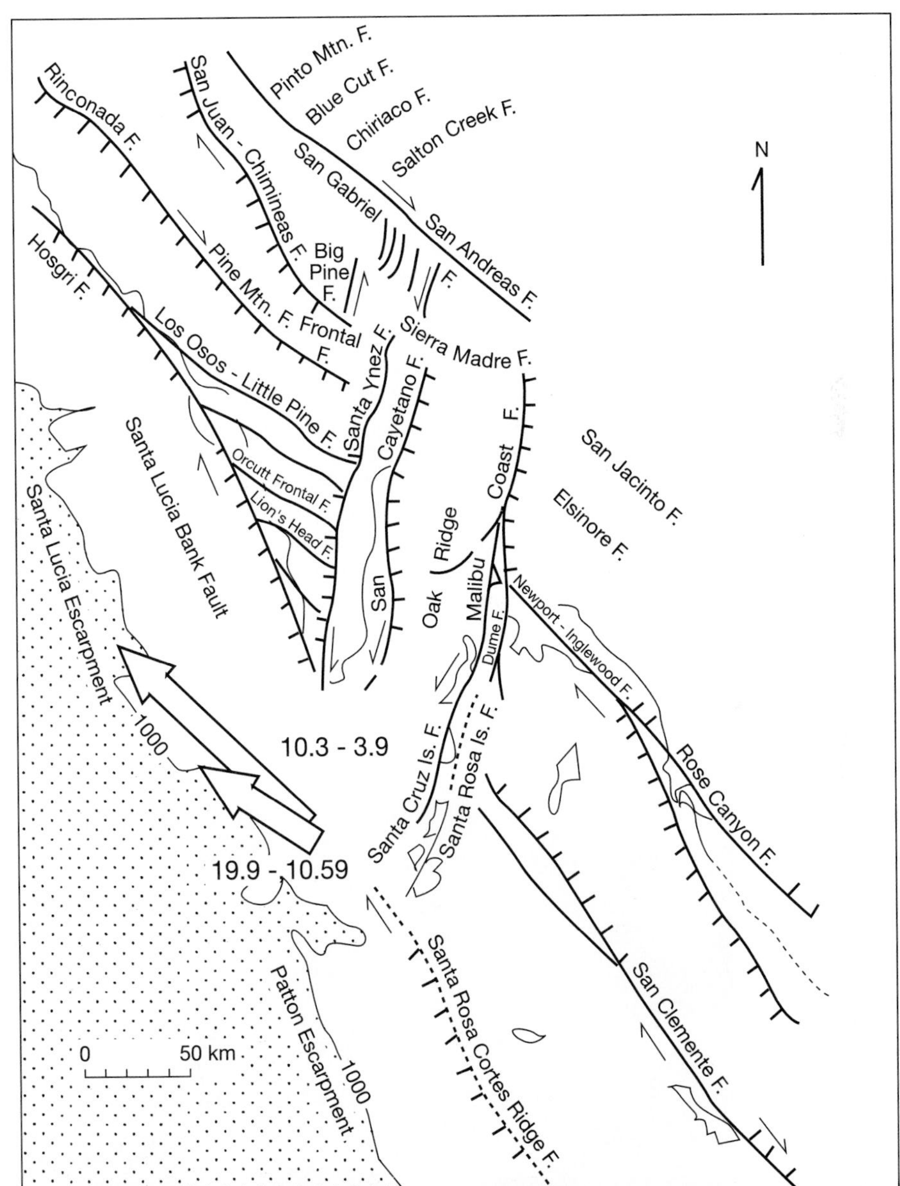

Fig. 16-26 Map showing the major faults of the western Transverse Ranges in Miocene time, about 13 million years ago, before clockwise rotation. (F = fault.) (Source: Luyendy, B. 1991. Geological Society of America Bulletin, Vol. 103[11].)

this match, it is also necessary to "undo" some of the very recent transform faulting and folding. The second line of evidence that supports the rotation model comes from studies of the Earth's magnetic field as recorded in these Transverse Range rocks. The direction of magnetic north appears in the proper position only if these older rocks are rotated back to their restored positions.

OFFSHORE ISLANDS

Eight islands and numerous smaller rocks are found off the southern California coast (Figs. 16-27 and 16-28). The four northern islands form a chain at the southern boundary of the Santa Barbara Channel, while the four southern islands are a more scattered group lying southwest of Los Angeles and Long Beach. Five of the islands—San Miguel, Santa Rosa, Santa Cruz, Anacapa, and Santa Barbara—make up Channel Islands National Park, established in 1980. Santa Catalina Island is more developed and a very popular summer resort, and San Clemente Island a military reservation. Numerous sea stacks are scattered throughout the islands.

Juan Rodriquez Cabrillo, a Portuguese explorer who traveled up the California Coast in 1542 and 1543, is thought to be buried on San Miguel Island. The islands were also visited by Spanish explorer Sebastian Vizcaino in 1602. He named Santa Barbara, San Clemente, and the Santa Catalina Islands. Both parties found well-established groups of Indians living on the Islands and making frequent trading trips to the mainland. About 17 kilometers north of Avalon, on Santa Catalina Island, Native Californians quarried steatite, an impure variety of talc. They used the steatite to make distinctive bowls, which they traded on the mainland.

The Channel Islands were isolated from each other about 12,000 to 10,000 years ago when sea level rose as the world's climate warmed and ice sheets melted (see

Fig. 16-27 Aerial photograph of San Nicolas Island (see Fig. 16-28 for location). The small island is oriented northwest-southeast and is about 15 kilometers long. Note the prominent sand dunes along the northwestern end at the left of this photo. (Source: Muhs, D. U.S. Geological Survey.)

Chapter 12). The rising sea level also isolated the plant and animal communities of the islands. Many species of plant and animals are found only on the offshore islands (see the box on p. 45), and some are found only on a single island. Before the sea-level rise, at least some of the northern islands were connected, forming a single large island named Santarosae by geologists.

Marine terraces are prominent on the offshore islands, particularly on San Clemente, which may have the most complete sequence of marine terraces in California (Fig. 16-29). Twenty-five terrace levels are preserved on San Clemente Island. The highest terrace, about 580 meters above sea level, is about 2.8 million years old. As discussed in Chapter 15, the terraces attest to the tectonic uplift of the islands. On Santa Catalina Island, terraces are not nearly as prominent as on the other islands, but narrow terrace benches do mark former shorelines.

The four northern Channel Islands are the western extension of the Santa Monica Mountains. Many of the rocks, folds, and faults of the Santa Monica Mountains can be matched up with those seen on the islands. For example, the Santa Monica fault runs along the south side of the Santa Monica Mountains, continuing west along the south side of the northern Channel Islands (see Endpaper 2).

The northern Channel Islands are composed mainly of volcanic rocks. Pillow basalt, indicating submarine lava flows, are found on Santa Barbara Island. On Anacapa Island, andesitic breccia and tuff are present, in addition to basalt. Volcanic rocks are also present on San Clemente, Santa Catalina, and Santa Cruz Islands. The volcanic rocks on the islands are of Miocene age, similar to the thick sequences of volcanic rocks found in the Santa Monica Mountains (see Figs. 16-22 and 16-23). Most are characteristic of rocks produced by submarine eruptions.

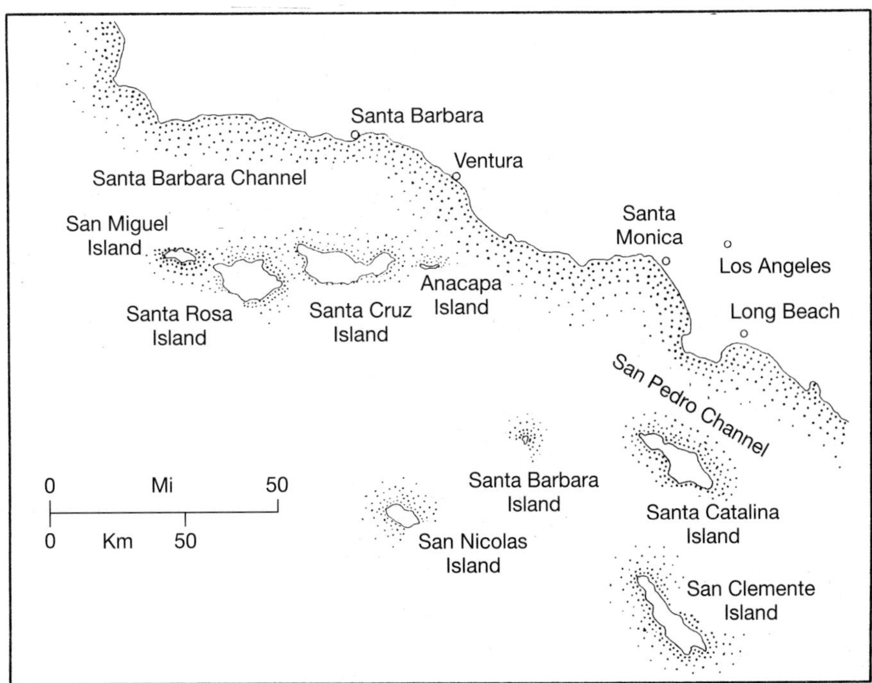

Fig. 16-28 Map of southern California's offshore islands and the Santa Barbara channel. (Source: Moon Publications, Chico, California.)

Fig. 16-29 At least eight marine terraces can be seen on this aerial photograph of San Clemente Island near Eel Point. The terrace edges make scalloped lines parallel to the shoreline. The long runway is approximately 1.6 kilometers long. (Source: NOAA, 1971.)

The Catalina Schist is a distinctive rock found on Catalina Island and in the Palos Verdes Hills, but not on the other Offshore Islands. The Catalina Schist contains blueschist minerals formed at great pressures in a subduction zone, probably at depths of 20 to 45 kilometers. The Catalina Schist is similar to blueschist found in the Franciscan Complex in the Coast Ranges, and geologists believe that the two units may be related.

BASEMENT ROCKS OF THE TRANSVERSE RANGES

A look at the California geologic map (Endpaper 1) reveals that many of the ranges contain large areas of granitic and metamorphic rocks. The basement rocks of the Santa Ynez Mountains are oceanic, part of the Franciscan Complex. Geologists believe that the Catalina Schist, another subduction-related unit, may be the basement rock beneath the western Los Angeles Basin and the adjacent offshore area. In the other ranges in the region, as well as on the islands, granitic plutonic rocks and associated metamorphic rocks appear to be the typical basement rocks. Together with the Mesozoic and early Cenozoic marine sedimentary rocks discussed above, they are comparable to the Great Valley and its underlying granitic basement. Some of California's oldest known rocks can be found in the San Gabriel Mountains. These include gneiss and anorthosite, a type of plutonic rock. The ancient gneiss formations are dated between 1.0 and 1.7 billion years old, and the anorthosite is about 1.2 billion years old. Similar ancient metamorphic rocks are present in the southern San Bernardino Mountains. Other crystalline rocks in the San Gabriel and San Bernardino Mountains include younger metamorphic rocks, such as the Mesozoic Pelona Schist (Fig. 16-30) discussed in Chapter 14, and Mesozoic granitic rocks.

Fig. 16-30 A polished sample of Pelona Schist from the San Bernardino Mountains. (Source: Morton, D. U.S. Geological Survey.)

CALIFORNIA OIL

On August 17, 1769, Gaspar de Portola's expedition encountered a group of Chumash Indians who were caulking their sea-canoes with tar, or brea, collected from a nearby pool. Portola's men named the site Carpinteria (carpenter's shop). Carpinteria, located along the Ventura County coast, is one of several places in the western Transverse Ranges where tar oozes from the ground in natural seeps (Fig. 16-31). Much later, in 1892, prospector Edward Doheny would look out the window of his Los Angeles boardinghouse and notice a wagon loaded with chunks of brea. Southern Californians used it to waterproof their roofs and fenceposts, and later to lubricate machinery. On a brilliant hunch, Doheny figured that the ***brea*** oozing from the ground might in fact be a signal of oil riches beneath. He and partner Charles Canfield leased a lot east of Los Angeles, dug a 155-foot-deep shaft by hand, deepened it to 460 feet with a sharpened eucalyptus drill, and struck oil. Thus began the big oil boom in the heart of Los Angeles.

The world's first oil well had been drilled in Pennsylvania in 1859, starting the systematic search for petroleum. In the early 1860s, prospectors drove tunnels into the side of Sulphur Mountain near Ojai to extract oil from sand there. In 1866, the first commercial oil well in southern California was completed in Ventura County. In the 1870s, wells were drilled near Newhall in Los Angeles County, one of them yielding 30 barrels per day from a depth of 300 feet. This well, designated Pico Number 4, was later deepened, and its production increased to 150 barrels per day. As a result, California's first oil pipeline was built. The pipeline carried oil from Pico Canyon 5 miles to a refinery near Newhall. Offshore wells were drilled from piers at Summerland east of Santa Barbara as early as 1896.

Fig. 16-31 Natural tar seeps along Highway 150 near Ojai. Note can for scale. (Photo by author.)

In *The Great Los Angeles Swindle,* Jules Tygiel writes:

> The Doheny discovery inaugurated an oil boom in the heart of downtown Los Angeles. During the next 5 years derricks "as thick as the holes in a pepper box" sprouted along a narrow 2-mile residential strip. Gardens, palm trees, and even homes disappeared as steam-powered rigs and primitive hand-driven drills probed the earth. Most found oil at shallow depths and produced a paltry 10 barrels daily. But by the turn of the century, thousands of wells in the Los Angeles field yielded almost 2 million barrels a year.

By 1900, more than 1000 oil companies were drilling in California, producing about 12,000 barrels per day. The first major oil boom in southern California began when geologists realized the connection between anticlines and oil fields, and by 1930, most of California's major oil fields associated with surface faults or folds had been discovered. During the 1920s, California's oil production reached a peak of 850,000 barrels per day, much of it from the Long Beach and Huntington Beach fields. Wells pumped oil from beneath the streets of Los Angeles, Wilmington, Huntington Beach, and other cities (Fig. 16-32). In 1914, a publishing company offered a 25- by 100-foot plot of land in Huntington Beach to buyers of its encyclopedia. When the Huntington Beach oil field was discovered under the "encyclopedia lots," those who had kept up their property taxes got sizable monthly royalty checks.

The early oil boom brought a wave of fortune-seekers to southern California, and with them came scandal, swindle, and speculation. The oil boom overlapped with the land and water wars (see Chapter 10) and the early movie-making boom, making Los Angeles a colorful place during the early twentieth century.

Tygiel, J. 1994. *The Great Los Angeles Swindle.* Oxford, England: Oxford University Press, p. 21.

Fig. 16-32 Early oil field in the Signal Hills. (Source: Shell Oil Company.)

Fig. 16-33 Pumping oil wells in Taft, Kern County. (Photo by author.)

TODAY'S OIL PRODUCTION

In 1994, California ranked fourth among U.S. oil-producing states, after Alaska, Texas, and Louisiana. A total of 344.5 million barrels of oil were produced from 214 oil fields, half from fields in Kern County (Fig. 16-33). These numbers seem large compared with the amount of oil produced during California's early oil years, but even this amount is very small compared with the United States' annual use of about 6 billion barrels. During the past decade, new wells drilled offshore along the southern California coast have accounted for an increasing amount of California's production (Fig. 16-34 and Table 16-1).

Fig. 16-34 Offshore oil platform in the Santa Barbara channel. (Source: U.S. Geological Survey.)

TABLE 16-1 SELECTED CALIFORNIA OIL FIELDS

Field	Year Discovered	Cumulative Production (millions of barrels)	Number of Producing Wells in 1994
Wilmington	1932	2,436,452	1456
Midway-Sunset	1894	2,238,723	10,330
Kern River	1899	1,474,280	7381
Huntington Beach	1920	1,089,370	443
Elk Hills	1911	1,054,609	1037
South Belridge	1911	983,345	4519
Ventura	1919	930,252	419
Long Beach	1921	921,009	297
Coalinga	1890	820,223	1922
Hondo[1]	1969	139,368	28
Point Arguello[1]	1981	76,320	34

[1]Offshore field.
Source: California Department of Conservation, Division of Mines and Geology.

FORMATION OF PETROLEUM POOLS

California's geologic history during Cenozoic time has supplied the ideal conditions for the formation of oil and natural gas resources. The basins of the Transverse Ranges, including offshore basins and the Los Angeles Basin, have the exact combination of rock types and geologic structures necessary for the development of petroleum fields, as outlined below.

Source Rocks

Petroleum, which includes oil and natural gas, is a combination of ***hydrocarbons,*** chemical compounds of hydrogen and carbon. Studies of the hydrocarbon compounds found in petroleum have shown that they are derived from organic, or living, matter. Many people share the misconception that this living matter was the skeletons of dinosaurs, but the prime source of petroleum is marine organisms, particularly plankton like the diatoms found in the Monterey Formation. The Monterey Formation has been considered a prime source for California's hydrocarbons. The siliceous sedimentary rocks in the Monterey Formation contain between 1 percent and 5 percent organic carbon, and the phosphate-rich rocks in the Monterey contain 5 percent to 20 percent organic carbon. Other Cenozoic marine sedimentary formations containing organic material are also suitable sources for hydrocarbons.

Heating

The next necessary step in forming a petroleum pool is to heat the source rocks sufficiently to form hydrocarbon compounds. Heating occurs as sediments are buried by younger deposits. At depths of about 500 meters, heat and pressure are great enough to transform organic matter to hydrocarbon compounds. The ideal temperatures for hydrocarbon formation are between 50° and 80° C. If the rocks are heated to higher temperatures, which would result from deep burial, igneous activity, or subduction, the hydrocarbons are converted to other organic compounds or burned off. Rocks that have been heated sufficiently to convert organic matter to hydrocarbon compounds are known as oil shales.

Migration to Reservoir Rocks

For petroleum to concentrate in a pool, the hydrocarbon compounds must next migrate from the source rock into rocks that can become saturated with petroleum. The rocks or sediments capable of carrying and holding large amounts of petroleum are the same materials that make good aquifers. That is, reservoir rocks must be permeable. Porous rocks like sandstone and highly fractured rocks make suitable reservoir rocks. A petroleum-saturated reservoir rock is referred to as a "pool," but it is not an underground lake of oil. Like an aquifer, a petroleum reservoir is a rock or sediment whose pore space is saturated with oil, natural gas, or both. Many of southern California's reservoir rocks are late Cenozoic sandstone formations, but fractured Monterey Shale can also be a reservoir for significant petroleum.

Traps

To accumulate as a pool or reservoir of petroleum, the hydrocarbon compounds must be trapped in the reservoir. Oil geologists estimate that more than 90 percent of all the petroleum formed escapes to the Earth's surface, where it oxidizes and escapes into the atmosphere or water. Geologic traps may be formed in a variety of ways. Faults and anticlines are two common types of ***structural traps*** found in California. If a reservoir rock, such as a permeable sandstone, is arched into an anticline *and* sandwiched between less permeable rocks, such as shale, petroleum is trapped at the crest of the fold. Because oil is lighter than water and natural gas is lighter than oil, a layered pool often forms (Fig. 16-35). Faults also form effective structural traps. The traps in many of California's oil fields are formed by a combination of reverse faults and folds (Fig. 16-36). Petroleum traps may also form where permeable reservoir rocks form lenses within less permeable rocks; known as ***stratigraphic traps,*** these are also common in southern California.

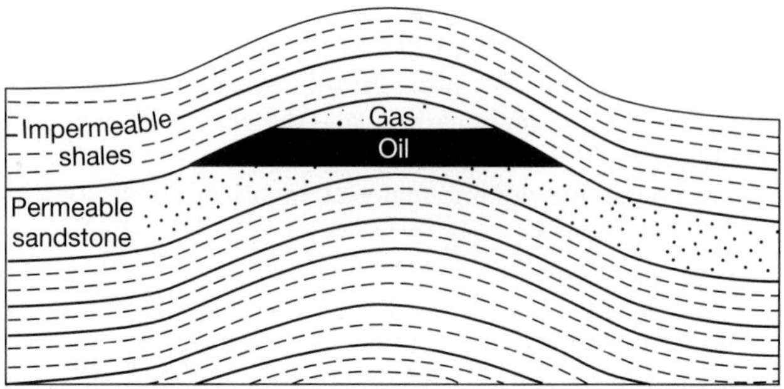

A simple fold trap

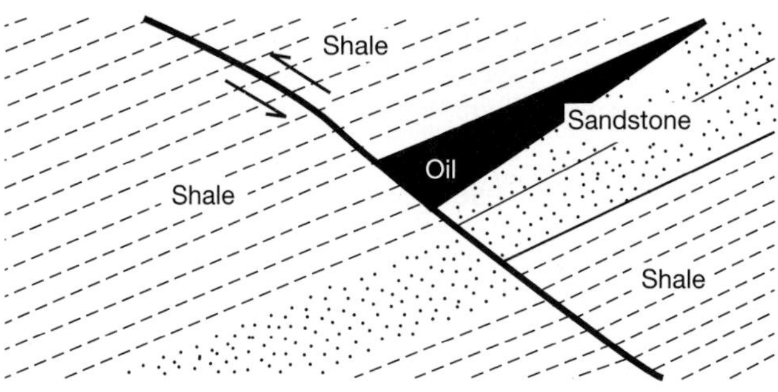

Reverse fault trap

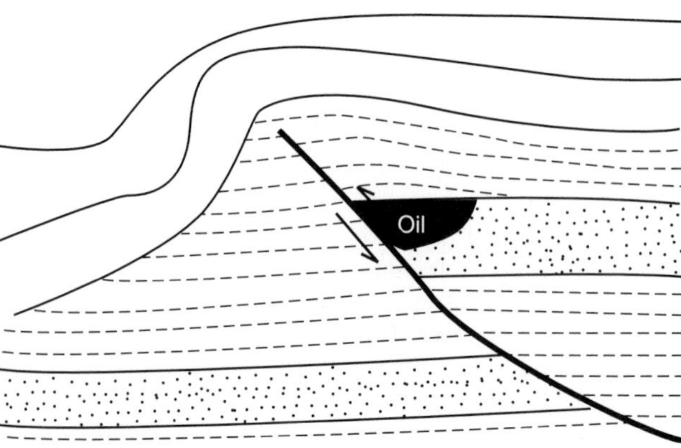

Anticline over blind reverse fault

Fig. 16-35 Diagrams of petroleum traps. (Source: Craig, J. and others. *Resources of the Earth.* Upper Saddle River, N.J.: Prentice Hall, Inc.)

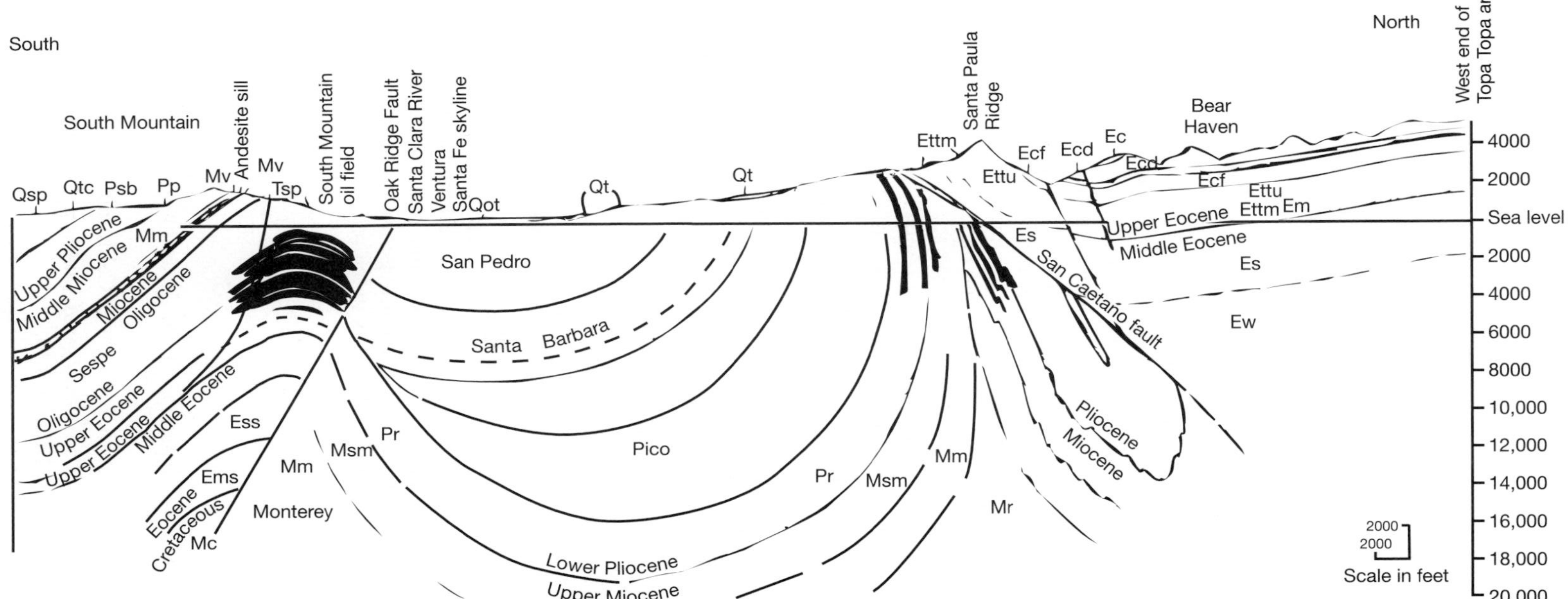

Fig. 16-36 Cross section of the central Ventura Basin, showing the oil fields associated with the Ventura anticline on the south and the San Caetano fault on the north side. See Fig. 16-13 for location. Note that the scale is in feet. (Source: Bailey, E., and Jahns, R. 1954. *Geology of Southern California.* California Division of Mines and Geology, Bulletin 170.)

Many of southern California's blind reverse and thrust faults are expressed as anticlines at the surface (see Chapter 13). Because petroleum is trapped in many of these structures, oil wells commonly tap into the anticlines above the blind faults. Although petroleum geologists long recognized that the folds are the surface expression of buried faults, they did not recognize the earthquake potential of those faults until the 1987 Whittier Narrows earthquake.

ENVIRONMENTAL CONCERNS

Because tar naturally seeps from the ground at several sites beneath the Santa Barbara Channel and along the southern California coast, oil on southern California's beaches can be a natural phenomenon. The 1969 Santa Barbara oil spill, however, was the result of a blowout at an offshore oil well in the Santa Barbara Channel. The well had not been properly capped, and the pressurized oil and gas literally exploded into the sea. The accident blackened Santa Barbara's beaches and wildlife with crude oil, provoking public outcries for greater environmental protection. Since then, the oil industry has had no major accidents along the coast or on the offshore platforms.

Ground subsidence following removal of oil is a major concern, particularly in coastal areas. In the period between 1933 and 1953, the ground above the Wilmington oil field sank as much as 6 meters. Parts of central Long Beach subsided 25 centimeters. Groundwater contamination associated with oil-field operations is also a concern in some parts of the region.

During the 1994 Northridge earthquake, 19 oil fields in northern Los Angeles and eastern Ventura county sustained some damage. At the Aliso Canyon oil field, closest to the epicenter, a string of well casing 1400 feet below the surface was damaged. Farther south, a well in a residential area of Hermosa Beach, which had already been leaking small amounts of oil at the surface, began to leak significant amounts of oil. Damage at the other sites was confined to pipelines, tanks, roads and power lines. For a short time after the earthquake, greater flow rates were noted at a number of sites in southern California.

It is geology and not coincidence that has brought beaches and oil together in southern California. Both onshore and along the continental borderland, the region is dotted with producing oil fields, pipelines, refineries, storage facilities, and tanker ports. However, the same areas are southern California's most valuable recreation areas and wildlife habitat. Protection of those resources makes environmentally sound management of petroleum operations an important priority. In addition, the certainty of future earthquakes in the region poses additional risks from petroleum-related accidents.

FURTHER READINGS

DAVIDSON, K. September 1994. Learning from Los Angeles. *Earth Magazine,* pp. 40-48.

GORE, R. April 1995. Living with California's faults. *National Geographic,* pp. 2-35.

JAHNS, R., ed. 1954. *Geology of Southern California.* California Division of Mines and Geology, Bulletin 170.

MORTON, D.M., AND YERKES, R.F., eds. 1987. *Recent Reverse Faulting in the Transverse Ranges.* U.S. Geological Survey, Professional Paper 1339, 203 pp.

SHARP, R.P., AND GLAZNER, A.F. 1993. *Geology Underfoot in Southern California.* Missoula, Montana, Mountain Press, 224 p.

STEIN, R.S., AND YEATS, R.S. Hidden earthquakes. *Scientific American,* June 1989, pp. 48-57.

TYGIEL, J. 1994. *The Great Los Angeles Swindle: Oil Stocks and Scandal During the Roaring Twenties.* Oxford, England: Oxford University Press, 398 pp.

CHAPTER

17

The Peninsular Ranges

SETTING

The Peninsular Ranges Province is dominated by a series of northwest-oriented mountain ranges extending from the Baja California peninsula north to the Transverse Ranges. Only the northern 240 kilometers of the Peninsular Ranges lie in California; they extend south of the border for about 1200 kilometers, forming the long crest of the Baja peninsula (Fig. 17-1). The highest peak of the Peninsular Ranges is San Jacinto Peak in the San Jacinto Mountains (3296 meters). Like the Sierra Nevada, the Peninsular Ranges are steeper on the eastern side than on the west.

Because it comes at the close of this tour of California's geologic highlights, the discussion of the Peninsular Ranges may seem brief to the reader who turns here first. Many aspects of Peninsular Ranges geology have been discussed in earlier chapters. Technically speaking, the Los Angeles Basin and the southern offshore islands are part of the Peninsular Ranges Province, although they are discussed in Chapters 15 and 16. The Colorado Desert and the Salton Trough are discussed in Chapters 6, 7, and 14. Finally, the major active faults of the Peninsular Ranges, including the Elsinore and San Jacinto fault zones, are parts of the San Andreas system discussed in Chapter 14.

THE PENINSULAR RANGES BATHOLITH

The similarity between the Peninsular Ranges and the Sierra Nevada extends beneath the landscape. A look at the geologic map (Endpaper 1) reveals that Mesozoic plutonic rocks make up much of the Peninsular Ranges Province. As discussed in Chapter 8, the Sierra Nevada batholith is part of a great Mesozoic chain that extends along much of western North America. The Peninsular Ranges batholith is part of this chain, and like the Sierra Nevada, it represents the roots of a magmatic arc formed as a result of active subduction along western North America. The age of the Peninsular Ranges batholith is also similar to Sierra Nevada. Plutonic rocks in the batholith crystallized between 140 and 80 million years ago, with the largest volume of magma intruded between 120 and 90 million years ago.

Geologists have found systematic variations in the types of plutonic rocks within the Peninsular Ranges batholith. On the western side of the batholith in

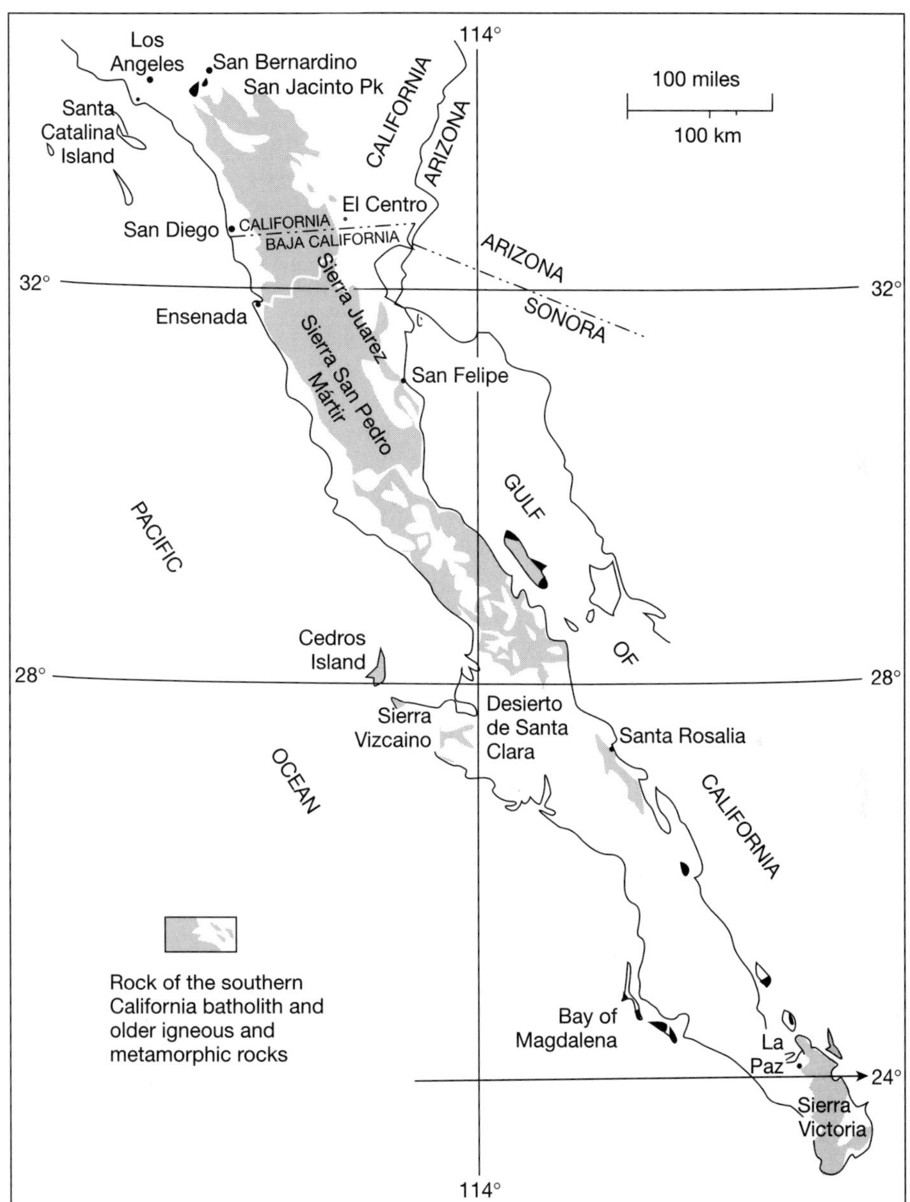

Fig. 17-1 Map showing the extent of the Peninsular Range batholith in southern California and Baja California, Mexico. (Source: Jahns, R. 1954. *Geology of Southern California,* Bulletin 170, p. 30. Used with permission of the California Department of Conservation, Division of Mines and Geology.)

southern California, more mafic plutonic rocks, including gabbro, are found along with granitic rocks. Because of its composition, the dark-colored gabbro is sometimes (incorrectly) called "black granite," which confuses those who have learned that granite is light-colored because mafic minerals are rare (see Chapter 8). In San Diego County, gabbro is quarried for use as building stone.

On the eastern side of the batholith, plutonic rocks are more silica-rich granodiorite and ***tonalite*** (see Fig. 8-11). The relative abundance of rare earth elements sharply differs in the two areas, indicating that the magmas formed from different source rocks. The boundary between the eastern and western belts of plutonic rocks

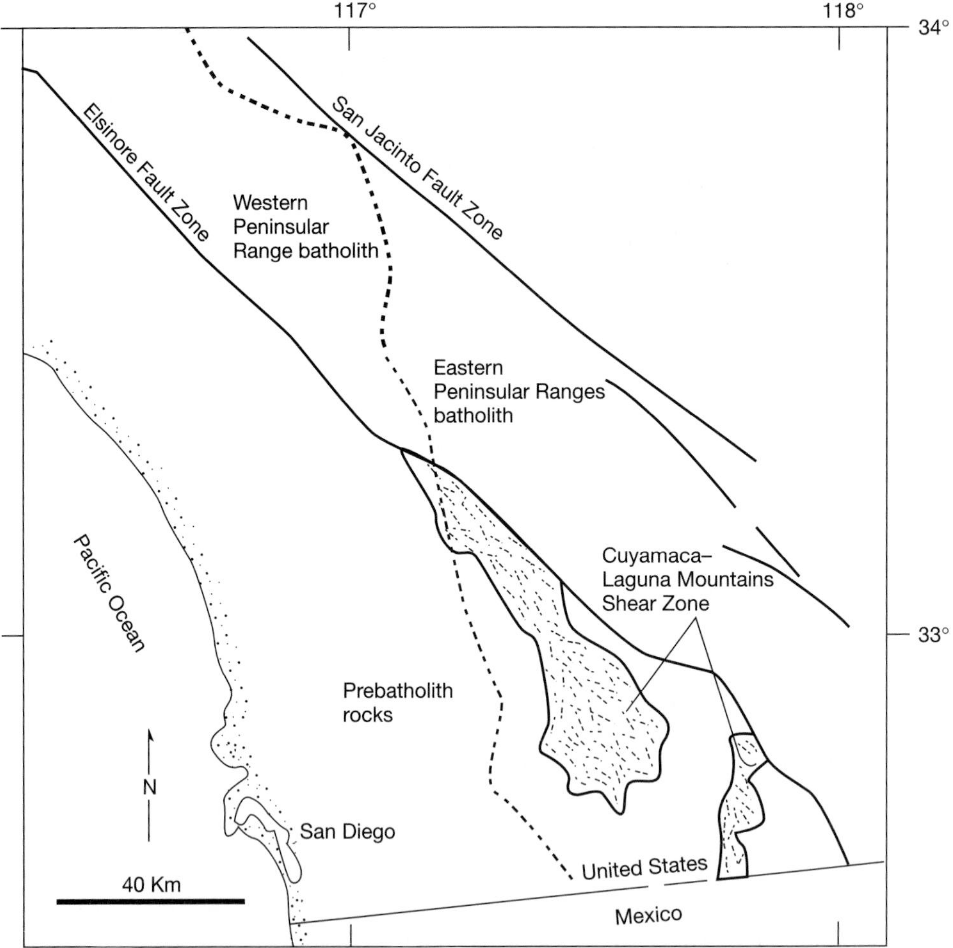

Fig. 17-2 The eastern and western parts of the Peninsular Ranges batholith and the Cuyamaca–Laguna Mountains shear zone, which lies along the boundary. The approximate boundary between the eastern and western batholiths is shown by the dashed line. (Source: Walawender, M., and others. 1991. A synthesis of recent work in the Peninsular Ranges batholith. In *Geological Excursions in Southern California and Mexico.* Geological Society of America, Field Trip Guidebook.)

is a north to northwest-trending zone near the center of the batholith (Fig. 17-2). In general, the eastern plutons are also younger than those to the west; they are dated between about 100 and 95 million years old.

All of this evidence has led geologists to believe that the formation of the Peninsular Ranges batholith took place in two phases. In Early Cretaceous time, collision and accretion of an oceanic volcanic arc generated the western belt of plutonic rocks. Later in Cretaceous time, the magmatic activity shifted eastward. The magmas changed composition because continental rocks were being melted and incorporated into the magma (see Chapter 8). This eastward shift was probably caused by a tectonic change along the continental margin.

GEMS OF SAN DIEGO COUNTY

One interesting feature of the Peninsular Ranges batholith is a belt of gem-bearing ***pegmatite*** dikes found in the mountains northeast of San Diego. Pegmatite dikes are coarse-grained, silica-rich plutonic rocks whose crystals grow to unusually large sizes because of the large amount of water vapor and boron present in the magmas. Most pegmatite dikes in San Diego County are a few meters thick, although some are as thick as 30 meters (Fig. 17-3). Near the center of the dikes are pocket zones containing crystals, some of which are unusually large and beautiful (Fig. 17-4). The pegmatite

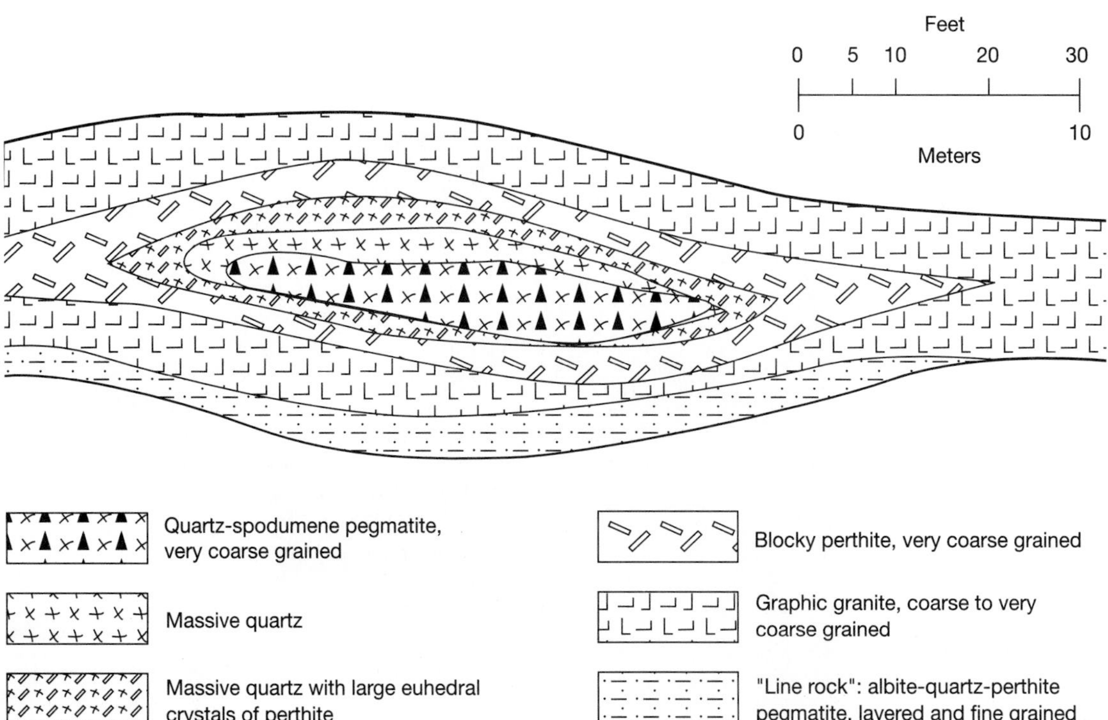

Fig. 17-3 The mineral zones in a typical vein of pegmatite. (Source: Jahns, R. Gem pegmatites in San Diego County. *California Geology* 30(2):44, 1977. Used with permission of the California Department of Conservation, Division of Mines and Geology.)

dikes of San Diego County occur in a 30-kilometer-long zone, mostly within the gabbro and tonalite plutons of the Peninsular Ranges batholith. The dikes crystallized in fractures in the older plutonic rocks, late in the formation of the batholith, about 90 to 100 million years ago. Many of the pegmatite dikes are well exposed on the hillsides. They stand out as light-colored ledges because of their resistance to erosion.

The most common minerals in pegmatite dikes are quartz, feldspar, and large, platy muscovite mica crystals. However, the uncommon chemistry of pegmatites gives rise to 100 different minerals, some rare and very beautiful. The most famous gems of the Pala area are the pink, green, and "watermelon"—pink and green—***tourmaline*** crystals (Fig. 17-4). Beautiful specimens of many other gems have also been mined from the area, including black tourmaline, kunzite, beryl, topaz, and garnet.

Native Americans in the Pala area were the first to collect tourmaline crystals, which have been found in a number of burial sites. In the late 1800s, prospectors began to excavate crystals, and between 1900 and 1912, mining flourished. Many of the gems were shipped to China and other overseas countries. Crystals have been mined from three areas in San Diego County—the Pala, Rincon, and Mesa Grande

Fig. 17-4 Watermelon tourmaline crystals from the Pala District, San Diego County. (Source: Van Pelt, H., and Van Pelt, E., for Pala International.)

districts. At the Stewart mine, lepidolite was mined for its lithium content until the late 1920s, when the lepidolite zones were mined out. More recently, gem tourmaline, kunzite, and beryl, including its blue variety, aquamarine, have been mined from underground operations at the Stewart and Tourmaline mines. The gem mines are not open to the public, but examples of many of the crystals can be viewed in museums, including the mineral museum of the California Division of Mines and Geology in Mariposa.

PREBATHOLITHIC ROCKS OF THE PENINSULAR RANGES

A complex assortment of rocks older than the batholith can be found in the Peninsular Ranges. The intrusion of the batholith, as well as later tectonic events, resulted in the rocks being sheared and metamorphosed (Fig. 17-6). However, careful mapping and detailed studies of rock composition have led geologists to group these prebatholithic rocks into three general zones. Like the metamorphic belts in the Sierra Nevada province, the zones are elongate and trend generally northwest (Fig. 17-5).

Metavolcanic rocks can be found in a discontinuous belt that runs along the coast. These rocks, named the Santiago Peak Volcanics, consist mostly of andesitic

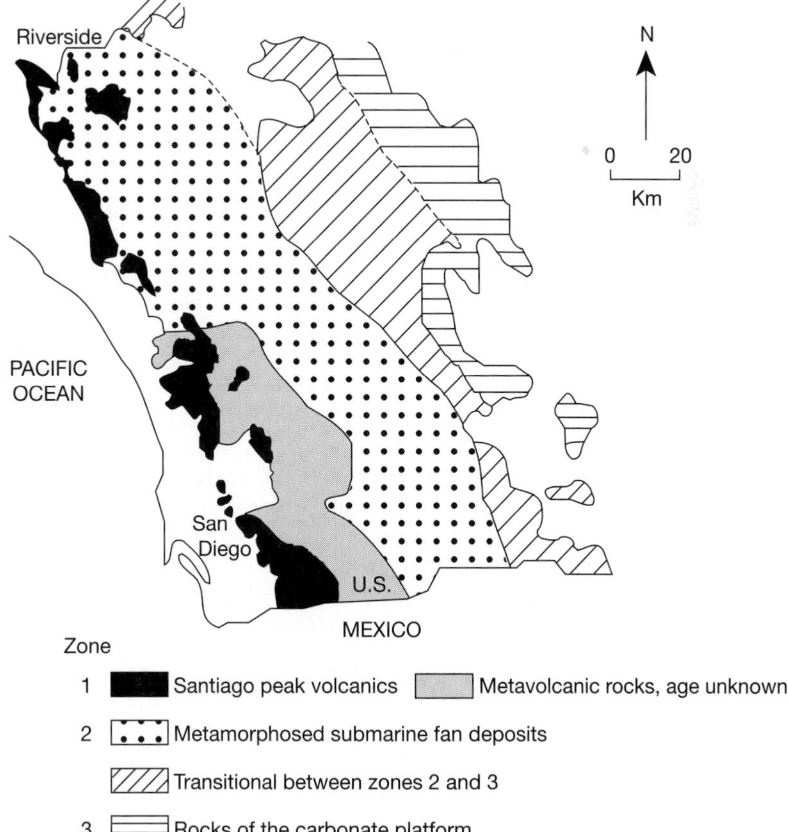

Fig. 17-5 Belts of pre-batholith metamorphic rocks in the Peninsular Ranges. The belts are generalized from outcrops of metamorphic rock. On this map, plutonic rocks and younger rocks and sediment are not shown. (Source: Metamorphic and tectonic evolution of the northern Peninsular Ranges batholith. In Ernst, W., ed. *Metamorphism and Crustal Evolution of the Western United States.* After Jahns, R. Upper Saddle River, N.J.: Prentice Hall, Inc.)

Fig. 17-6 Outcrop of high-grade gneiss in the Sierra San Pedro Martir, northern Baja California (see Fig. 17-1). Metamorphism occurred during intrusion of the Peninsular Ranges batholith. (Source: Gastil, G. San Diego State University.)

and silicic flows, tuff, and pyroclastic material, as well as sedimentary rocks rich in volcanic fragments. They can be seen east of Del Mar, in the Santa Ana mountains. Both marine and plant fossils of Cretaceous age have been found in sedimentary rocks of the sequence. The likely source of the Santiago Peak Volcanics was an andesitic volcanic arc at or near the continental margin.

East of the Santiago Peak Volcanics is a zone of metasedimentary rock preserved within the central part of the batholith. Geologists recognize the metamorphosed sandstone and shale in this zone as turbidite sequences originally deposited on submarine fans (see Chapter 12). The turbidites must have accumulated in one or more ocean basins, but their exact origins are unknown. The metasedimentary rocks include the Bedford Canyon Formation of the northern Santa Ana Mountains, the Julian Schist, named for the town of Julian, and French Valley Formation found in the eastern part of the zone.

The third zone of prebatholithic rocks in the Peninsular Ranges can be matched with similar rocks seen throughout southeastern California. Rocks in this zone are the metamorphosed remnants of the stable carbonate platform that existed along western North America before Mesozoic time (see Chapter 7). In the Peninsular Ranges, these rocks are well exposed in the San Jacinto and Santa Rosa Mountains, west of Borrego Valley, and in the Coyote Mountains. Like the roof pendants in the Sierra Nevada, the original sedimentary rocks have been metamorphosed to marble and schist.

CRETACEOUS SEDIMENTARY ROCKS OF THE SAN DIEGO AREA

Sedimentary rocks of Cretaceous age are exposed along the uplifted cliffs of the San Diego County coastline and in road cuts in the greater San Diego area. These rocks were deposited on the plutonic rocks of the Peninsular Ranges batholith and on the

PART

California Geology: An Integrated View

The final two chapters of this book allow the reader to bring together the material presented in the previous chapters into a unified picture of California's geology. Chapter 18 summarizes important events in California's geologic history, presented in chronologic order, with emphasis on the role of plate tectonics and the important aspects of individuals provinces. Finally, Chapter 19 discusses the role that geology plays in the lives of all Californians.

CHAPTER

18

The Evolution of California Through Geologic Time

After looking at the highlights of California's provinces, we can see a unifying theme in California's geology: *complexity caused by tectonics.* The complexity has long been recognized by geologists attempting to decipher California's geologic history. Beginning with the Gold Rush, the search for mineral and petroleum resources spurred the preparation of accurate and detailed geologic maps in most of California. By the 1960s, the nature and approximate age of many of the rock formations and the geometry of most of the faults and folds were fairly well established. Interpretations of California's geologic history, however, still contained a great number of uncertainties.

During the past 25 years, the concept of plate tectonics has given geologists a framework for interpreting the complicated assortment of California rocks and faults. Advances in technology have enabled us to better identify and date geologic materials at the surface and to collect information about the Earth's interior. At the same time, societal concerns about seismic risks and other geologic hazards have led the public to provide the necessary resources to support geologic and geophysical investigations. Today researchers are recognizing pieces of the tectonic puzzle with each new study. In turn, assembly of the pieces is shedding new light on California's tectonic setting during different periods in geologic time.

Because of the great complexity of California geology, our understanding of the evolution of California through geologic time is nowhere near complete. Nevertheless, it is useful to review California's history and synthesize the tectonic setting of all of the provinces at different points in geologic time.

CALIFORNIA'S EARLIEST HISTORY: BEFORE ABOUT 600 MILLION YEARS AGO

Rocks older than about 600 million years old are relatively rare in California. They are found only in southeastern California—in the southern Panamint and Nopah Ranges near Death Valley, in some ranges of the Mojave Desert, and in parts of the San Gabriel and San Bernardino Mountains (Endpaper 1). Because these rocks are found in only a few places and because they have been subjected to many geologic events since their formation, little is known about the earliest geologic environment of southeastern California. Based on their composition, these rocks are thought to be the remnants of very old terranes that were accreted early in the history of the North American Plate.

The oldest rocks in southeastern California, called the ***crystalline basement,*** are granitic plutonic rocks and metasedimentary rocks, mainly gneiss, schist, and marble. Radiometric dates (see Chapter 3) indicate that they are about 1.4 to 2.4 billion years old. From the known distribution of these basement rocks at the surface as well as their presumed extent at depth, geologists have reconstructed the western edge of North America as it was about 1 billion years ago (Fig. 18-1). However, the exact configuration of the margin is not well known.

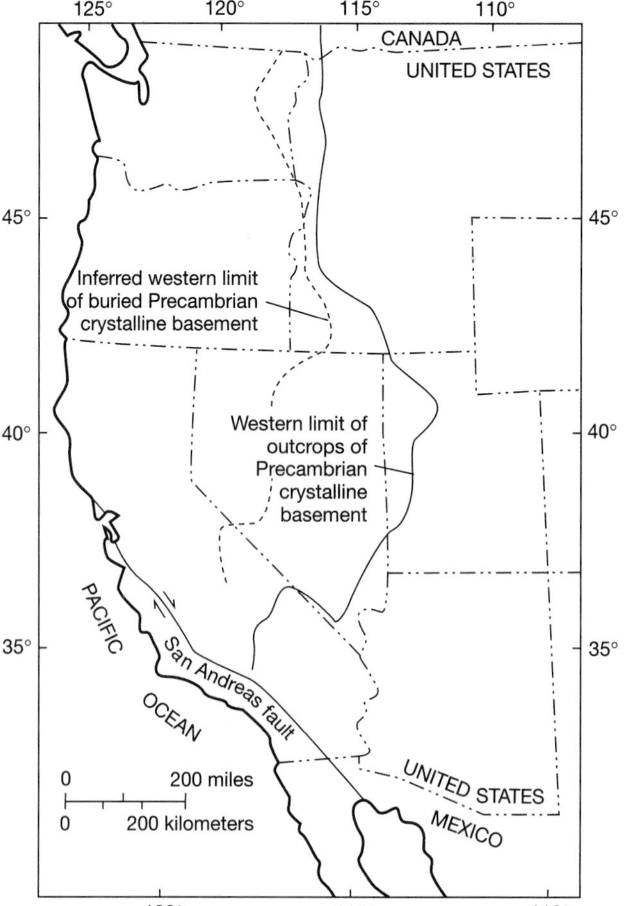

Fig. 18-1 Presumed extent of the crystalline basement in the western United States, based on known distribution of rocks older than 600 million years old. (Source: Stewart, J.H., and Suczek, C.A. 1977. Cambrian and Late Pre-Cambrian Paleogeography and Tectonics in the Western U.S. Los Angeles: Pacific Section of the Society of Economic Paleontologists and Mineralogists.)

About 900 million years ago, sediments were deposited on the crystalline basement rocks. Today these sediments are represented by the Kingston Peak Formation in the Death Valley region (Chapter 7) and by more widespread formations in northeastern Washington, southern Idaho, and northwestern Utah. The Kingston Peak Formation contains gravels, and the large size of the clasts indicates that they must have from a landscape with considerable relief. Some of the cobbles in the formation appear to have been striated by glaciers (see Chapter 8). Elsewhere in the western United States, sedimentary rocks of this age are associated with volcanic rocks. The exact positions of the continents, including the ancient plate that included North America during this early period, is not well known.

LATE PROTEROZOIC AND PALEOZOIC CALIFORNIA: THE PASSIVE MARGIN

About 600 million years ago, sediments began to accumulate along the margin of ancient North America in response to the sinking of the continental edge. Geologists believe that the sinking was a response to rifting within the ancient North American Plate (Fig. 18-2). This ancient plate was part of a megacontinent called *Rodenia* by

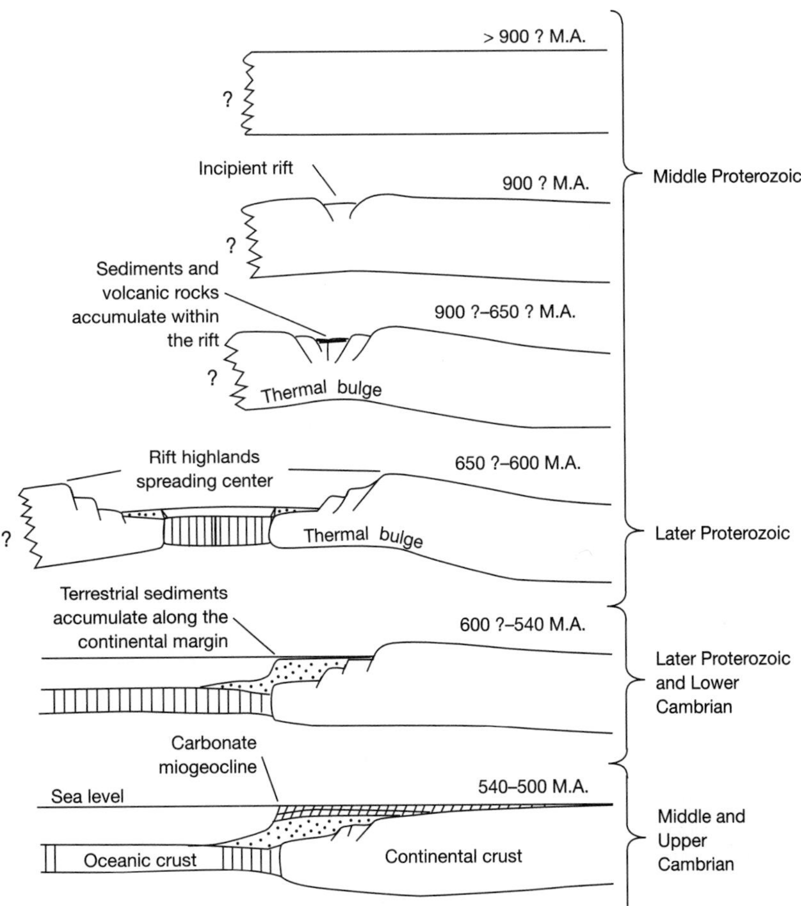

Fig. 18-2 Thermal expansion, rifting, subsidence, and creation of a passive continental margin. (Source: Stewart, J.H., and Suczek, C.A. 1977. Cambrian and Latest Pre-Cambrian Paleogeography and Tectonics in the Western U.S. Los Angeles: Pacific Section of the Society of Economic Paleontologists and Mineralogists.)

some geologists. Sedimentary rocks in several areas of southeastern California, including the San Bernardino Mountains and Death Valley regions, and in other areas in Washington, Oregon, Utah, and Nevada appear to have formed in shallow marine waters during the initial subsidence. Quartz sandstone and associated shale, later metamorphosed to quartzite and phyllite, are the most common sedimentary rocks deposited about 600 million years ago in late Proterozoic time.

By early Paleozoic time, a stable continental shelf had developed along the continental margin of western North America (see Chapter 7). During the next 350 million years, thousands of meters of Paleozoic sedimentary rocks, mainly carbonates, accumulated on the carbonate platform. Geologists refer to this type of sedimentary environment—found along a tectonically quiet (passive) margin—as a ***miogeocline.*** Today the rocks formed in the miogeocline can be seen in the ranges of the Basin and Range Province and the Mojave Desert (see Fig. 7-25). They are also preserved as metamorphic roof pendants in part of the eastern High Sierra (see Chapter 8).

In general, the section of Paleozoic sedimentary rocks in southeastern California and adjacent parts of Nevada and Arizona increase from southeast to northwest. Assuming that the sediment accumulations decrease toward the ancient shoreline, geologists can reconstruct the continental margin. Based on the trends observed in Paleozoic sedimentary rocks, geologists believe that the continental margin faced toward the northwest in most areas (Fig. 18-3).

Southeastern California's relatively quiet Paleozoic tectonic setting was interrupted during the Devonian and Mississippian periods, by an episode of plate collision and mountain building. This event is preserved in Paleozoic rocks in western

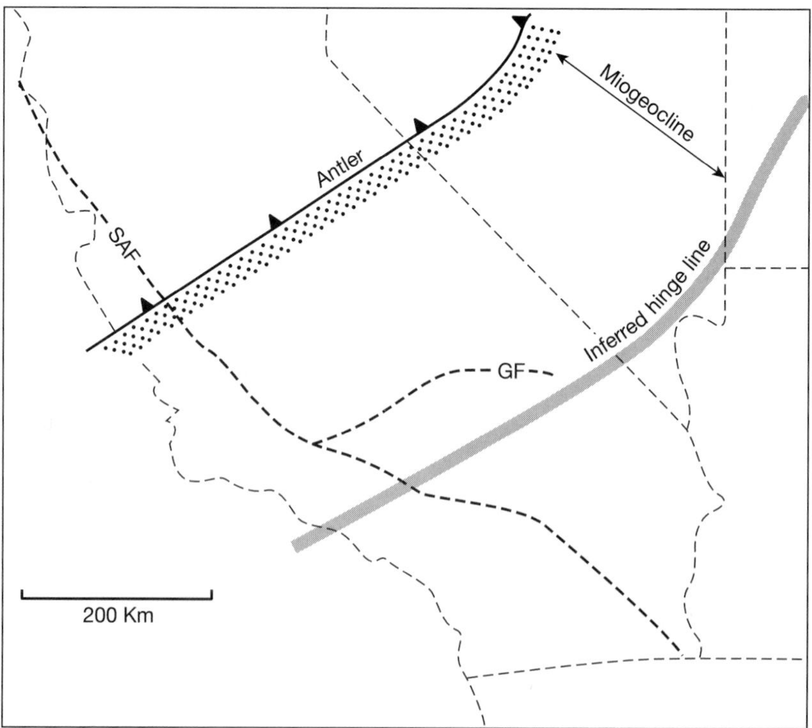

Fig. 18-3 Reconstruction of the Paleozoic continental margin showing the position of the miogeocline and the Antler uplift.

Nevada and some adjacent parts of eastern California. The disturbance resulted in the accumulation of sediments shed from newly uplifted mountains (see Chapter 2) and caused preexisting sedimentary sequences to be folded and faulted. Rocks seen today in the northern Inyo Mountains and the Death Valley area were affected by this episode of disturbance, and the deformation is well preserved in Nevada, where geologists have named the ancient uplifted area the Antler Mountains (see Fig. 18-3).

Beyond the continental margin, other oceanic rocks were forming west of Paleozoic North America. Remnants of these rocks would be accreted to North America during Mesozoic time, 100 to 200 million years later (see Figs. 8-24 and 9-1). Some of the rocks represent subduction-zone complexes, while others are remnants of oceanic volcanic terranes. During Mississippian and Permian time, a belt of island arcs lay west of the North American continent. Volcanoes from this chain erupted large volumes of andesite, and coral reefs ringed the volcanic islands. Today the ancient island arc system is preserved in the Paleozoic rocks of the northern Sierra Nevada and the eastern Klamath Mountains of California (see Chapter 9).

MESOZOIC CALIFORNIA: A TIME OF BUILDING

Early in Mesozoic time, about 200 million years ago, the ancient continental plate known as ***Pangaea*** began to break up. The breakup of Pangaea created the North American Plate as we recognize it today. Between about 100 and 50 million years ago, North America moved westward, and a rift developed along the eastern edge of the continent. This rift would eventually evolve into the northern Atlantic Ocean, splitting the North American Plate from the western edge of the European continent. Along western North America a north-south subduction zone developed, roughly in the same position as today's Sierra Nevada. This event gave rise to California's most important geologic building period.

During Mesozoic time, belts of oceanic rock were added to California by a series of collisions and accretions along the active western margin of the continent. The earliest rocks added to North America were Paleozoic sedimentary and volcanic rocks. These were accreted to North America during early Mesozoic time. Following each episode of accretion the subduction zone shifted westward, and accretion of a younger terrane began. The westernmost, youngest belt of accreted rocks in the Sierra Nevada is of middle to late Jurassic age, about 160 million years old (see Chapters 8 and 9).

Inland from the Mesozoic subduction zones, magma rose above the down-going plates, forming chains of andesitic volcanoes and granitic plutons beneath them. At its maximum development, the Mesozoic magmatic arc extended along western North America from Baja California in Mexico to British Columbia. In California, the arc was located approximately in the position of today's Sierra Nevada (see Fig. 12-12). Plutonic rocks representing the roots of the Mesozoic volcanic chain are found today throughout the Klamath Mountains, Sierra Nevada, Basin and Range, Mojave Desert, and Peninsular Ranges provinces (see Endpaper 1). Mesozoic volcanic rocks, originally erupted along the arc, are also preserved in placed in the same provinces. The greatest volume of magma was generated during the subduction of the Farallon Plate.

By about 100 million years ago, the subduction zone had shifted westward to the approximate position of today's Coast Ranges (Fig. 18-4). Pieces of the Farallon Plate were being accreted along western California as it was being subducted

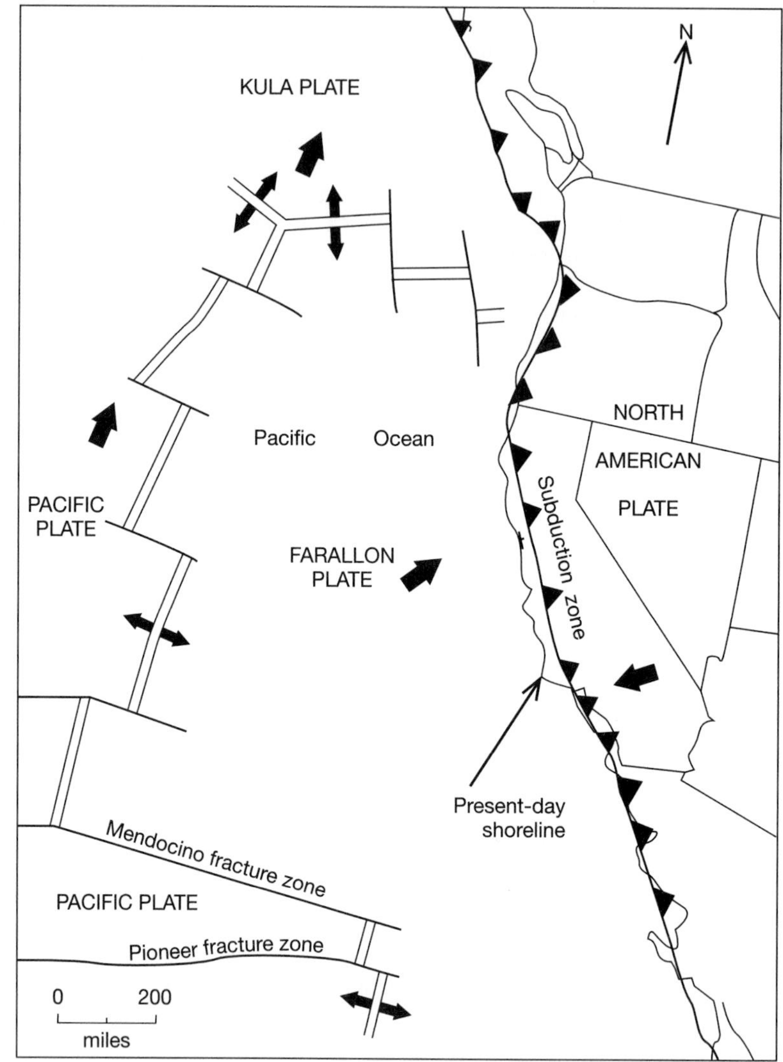

Fig. 18-4 Position of the subduction zone about 100 million years ago (Source: *National Geographic.*)

beneath North America (Fig. 18-5). Millions of years later, these pieces would be recognized by geologists as the Franciscan complex of the Coast Ranges (see Chapter 12).

Between the late Mesozoic trench and the Sierran volcanic arc, a marine forearc basin extended along most of California's length (see Chapter 11). Sediment that would become the Cretaceous Great Valley Sequence, including marine sedimentary rocks found today in the Transverse Ranges (see Chapter 16), accumulated in this linear basin. The sedimentary rocks of the Great Valley Sequence accumulated on the Coast Range ophiolite, which we now recognize as representing the crust and upper mantle beneath the ocean floor (see Chapter 9). Sedimentary rocks of the Great Valley Sequence contain fragments of the immense volcanoes that once sat at

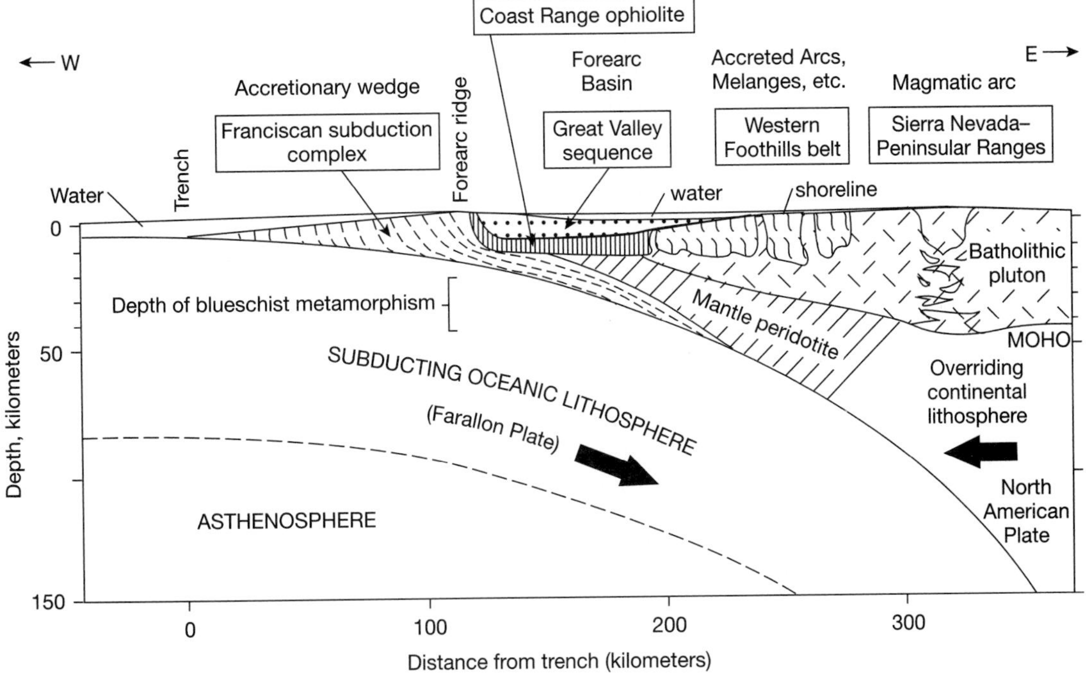

Fig. 18-5 A general cross section across California about 100 million years ago. (Source: Crouch, J., and Soppe, J. 1993.Geological Society of America Bulletin, Vol. 105(11). After Hamilton, W. In Geological Society of America Memoir 172, 1989.)

the top of the Sierra Nevada batholith, providing further support for the model of Mesozoic subduction and accretion (see Fig. 12-12).

The major tectonic events of the Mesozoic era in California had drastic effects on the rocks that were already there. Folding and faulting accompanied the collision events of the Mesozoic, with the result that older rocks were folded and faulted during each episode. In many areas, Paleozoic and Mesozoic rocks were also metamorphosed during the intrusions of magma and episodic compressional events. Geologists attempting to understand California's pre-Mesozoic history have to "see through" these later changes to determine the original nature of Paleozoic and older rocks.

CENOZOIC CALIFORNIA: TRANSFORM MOTION BEGINS

About 28 million years ago, the Pacific and North American Plates made direct contact for the first time when the closest part of Farallon Plate was completely consumed beneath North America (Fig. 18-6 or see Fig. 12-1). This event created the San Andreas transform boundary. North and south of the transform, subduction of the Farallon Plate continued, and as increments of the Farallon Plate were consumed in the subduction zones, the transform margin lengthened (see Chapter 14). The transform margin became a dominant factor in California's history. Throughout California, even in provinces not directly at the plate boundary, geologic events of the past 28 million years have been strongly influenced by the San Andreas system.

The development of the San Andreas transform had major impacts on the western edge of California. Large blocks have shifted by as much as 315 kilometers along the right-lateral faults of the San Andreas system (see Chapter 14). For example, rocks of the Salinian block, which probably formed near the southern end of the Coast Ranges, moved northwestward along the San Andreas fault to their present position along the central California coast (see Fig. 12-4). Right-lateral movement along the transform boundary also brought crystalline basement rocks from southeastern California to their present positions in the San Gabriel and San Bernardino Mountains (see Endpaper 1).

During the early history of the San Andreas system, large volumes of sediment accumulated in marine basins along the continental margin. Geologists believe that these basins developed because extension accompanied lateral motion along the transform boundary (Fig. 18-7). They interpret the thick sequences of sedimentary rocks as evidence for transtensional tectonics. For example, the San Joaquin Basin was a very deep marine basin between 35 and 15 million years ago, and a shallow marine embayment persisted there until 2.5 to 3 million years ago (see Chapters 11 and 12).

One major consequence of the development of the transform margin was the shutting off of magmatic activity beneath North America. In areas east of the growing transform margin, no oceanic plate was being subducted; as a result, no magma was generated. After about 28 million years ago, magmatic activity began to cease inland of the transform margin (see Chapter 12). Where remnants of the

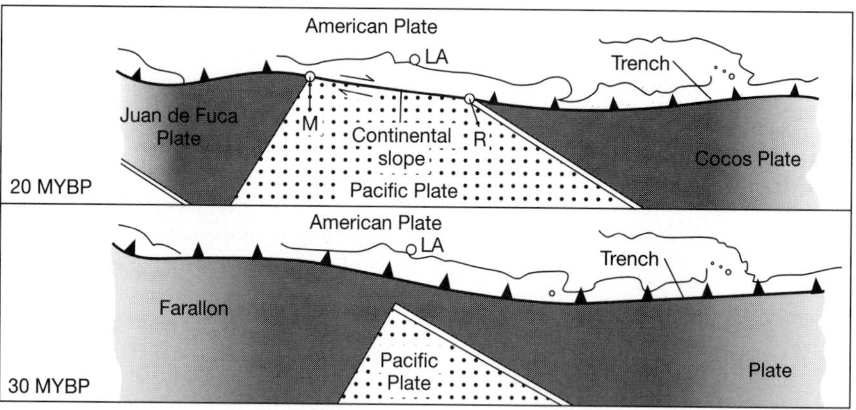

Fig. 18-6 The initial contact between the Pacific and North American Plates about 28 million years ago. (Source: Dickinson, W. 1981. In Ernst, W.G. *The Geotectonic Development of California.* Upper Saddle River, N.J.: Prentice Hall, Inc.)

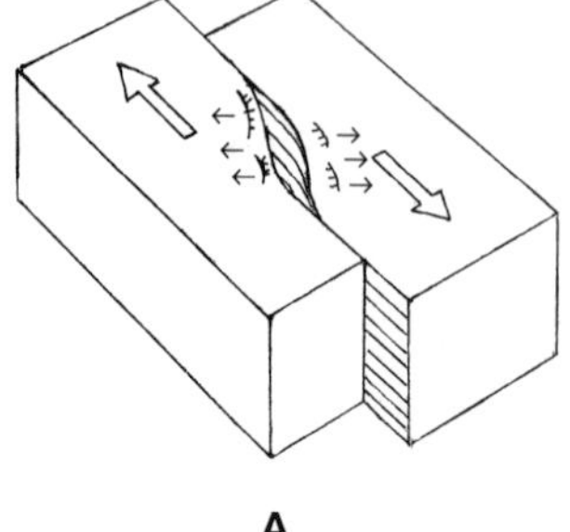

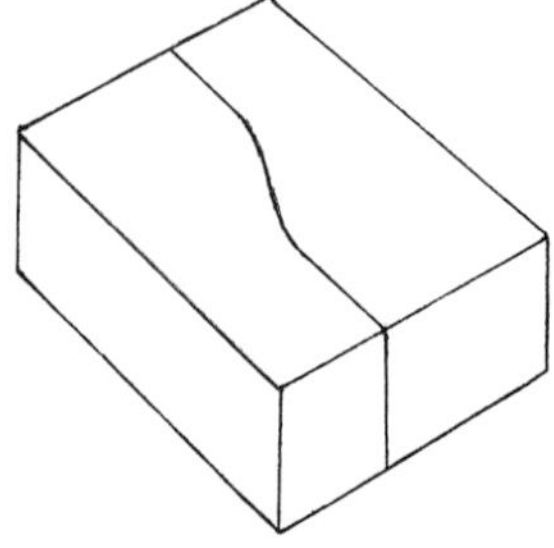

Fig. 18-7 Transtension may be created if extension accompanies transform motion. As a result, a pull-apart basin forms in the region where the crust is extended (**A**) (*shown by arrows*). (Source: Creely, R.S. 1997.)

Farallon Plate and East Pacific Rise still exist today, north of the Mendocino triple junction and south of Baja California, Mexico, subduction-related volcanic activity persists on the North American Plate (see Chapter 5). East of the plate boundary in the Basin and Range Province, crustal extension is responsible for recent volcanic eruptions in Death Valley, along the eastern side of the Sierra Nevada, and in the Modoc Plateau (see Chapters 5 and 7).

The boundary between the Farallon and Pacific Plates—the East Pacific Rise—was an oceanic spreading ridge. As the Farallon Plate was being consumed by subduction, the East Pacific Rise approached closer to western North America. When the Farallon Plate was completely subducted, the East Pacific Rise itself would have encountered the subduction zone about 28 million years ago. Some geologists believe that this encounter may have triggered a period of volcanic activity close to the plate boundary. They hypothesize that the mid-Cenozoic volcanic rocks seen today in the Coast Ranges (see Chapter 12), near Point Conception, and in the Santa Monica Mountains (see Chapter 16) resulted from arrival of the East Pacific Rise at the western margin of the North American Plate.

In early Miocene time, about 22 to 17 million years ago, extension of the crust began along a roughly north-south zone that included parts of what is now the Mojave Desert and Basin and Range. Large crustal blocks were pulled apart along detachment faults (see Chapter 7). In some highly extended terranes, detachment faults completely removed the overlying rocks to expose the metamorphic rocks of the lower plate. Today these ***metamorphic core complexes*** can be seen in a belt that stretches from Canada to Mexico, including many that form the dome-shaped ranges of southeastern California (Fig. 18-8).

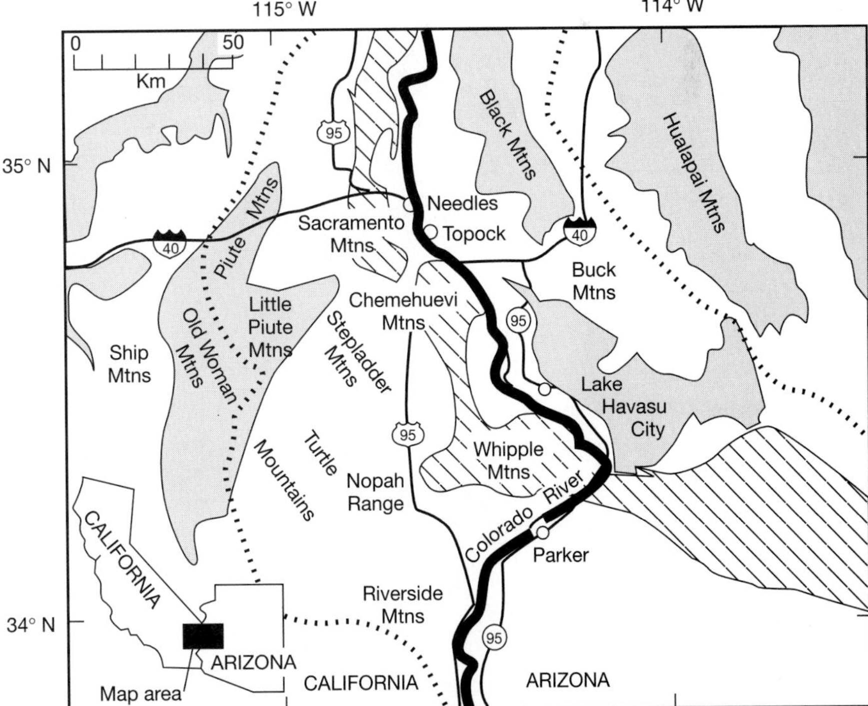

Fig. 18-8 Mountain ranges (*shaded areas*) and metamorphic core complexes (diagonally striped areas) in southeastern California. (Source: Nielson, J., and Beratan, K. 1995. Geological Society of America Bulletin, Vol. 107(2).)

Geologists have uncovered evidence that rocks in both the Transverse Ranges and the Mojave Desert Province have been rotated in a clockwise direction during Miocene time. Rotation in these areas was probably a side effect of right-lateral shearing along the plate boundary.

CALIFORNIA DURING THE PAST 3 TO 4 MILLION YEARS

During the past few million years, the transform boundary between the Pacific and North American Plates has continued to lengthen. Today it reaches along most of California in the form of the San Andreas fault system (see Endpaper 2). Right-lateral motion along the active faults of the San Andreas system is carrying the Pacific Plate northwest and the North American Plate to the southeast. North of Cape Mendocino, remnants of the Farallon late—known as the Gorda and Juan de Fuca Plates—are still being subducted beneath North America. As a result, a magmatic arc is still being generated north of Cape Mendocino, and the Cascades volcanic chain marks the surface above it (see Chapter 5).

About 3 to 4 million years ago, the Pacific Plate shifted its direction of motion slightly, taking a more northerly path (Fig. 18-9). The change in motion created greater convergence between the Pacific and North American Plates, and this change

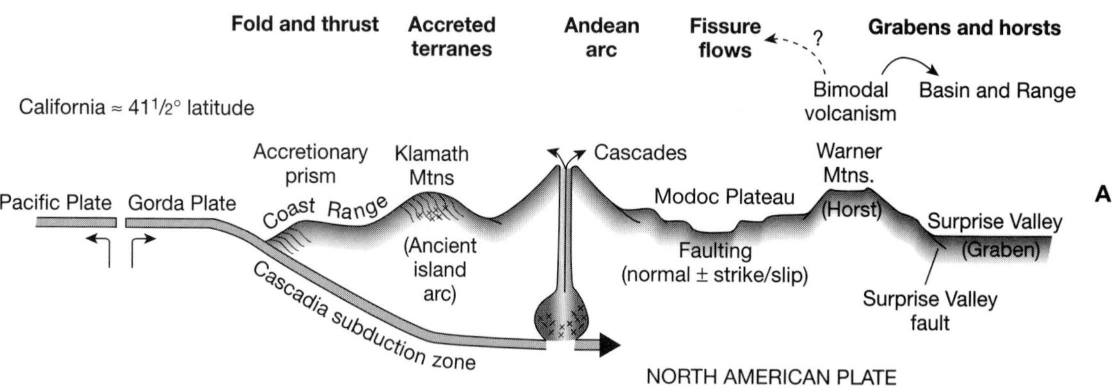

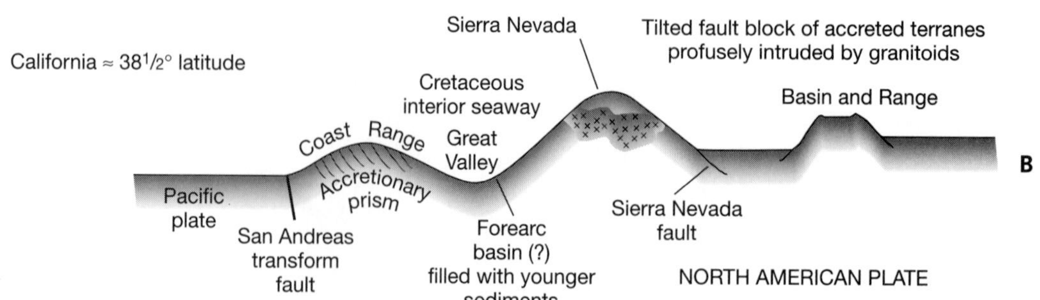

Fig. 18-9 Schematic cross sections across **A,** northern and **B,** central California, showing the tectonic framework of the provinces. (Source: Danielson, J., and May, A. 1992. Weed, Calif.: Shasta Valley College.)

had many effects in California, including an increase in the rate of right-lateral motion along the San Andreas system. During roughly the same period, Baja California began to rift apart from the Mexican mainland, forming the Gulf of California.

Most of California's major mountain ranges were elevated about 3 to 5 million years ago in response to increased compression along the plate boundary. The present Sierra Nevada range rose in about the same position as the Mesozoic chain of volcanoes, which had been eroded to form a lower landscape during early and middle Cenozoic time. In coastal southern California, reverse faults and folds developed, forming deep basins as well as newly uplifted mountains. Beginning about 4 million years ago, basin subsidence in southern California increased, also in response to increased compression. Along the coast, flights of marine terraces add to the evidence of ongoing uplift, particularly where compressional forces are strengthened near the Big Bend and the Mendocino triple junction (see Chapter 15). Finally, earthquakes on both reverse and transform faults serve as constant reminders that both compression and lateral plate motion are active in California today. Further east, in the Mojave Desert and Basin and Range Provinces, crustal extension continues, as evidenced by active normal faults. Many faults in southeastern California also have right-lateral displacement, indicating that they too are part of the transform boundary.

FURTHER READINGS

ERNST, W.G. 1981. *The Geotectonic Development of California.* Upper Saddle River, N.J.: Prentice Hall, Inc., 704 pp.

MCPHEE, J. 1994. *Assembling California.* New York: McGraw-Hill Book Co.

CHAPTER

19

Geology and California Citizens

GEOLOGIC HAZARDS

One of the most important benefits of understanding California's geologic history is an increased knowledge of geologic hazards. After decades of careful research, much of which was not specifically directed at these practical goals, geologists have come to better understand the processes that create these hazards. A basic understanding of earthquakes, volcanic eruptions, floods, and landslides can benefit all Californians who seek to avoid becoming a victim of one of these hazards. Landslides are discussed below, and the hazards of and processes that cause volcanic eruptions, floods, and earthquakes are discussed in Chapters 5, 10, and 13, respectively.

Earthquakes

As we have seen in Chapters 13 and 14, earthquakes are an unavoidable aspect of life in California. During the past 25 years, scientists have made considerable progress in understanding the sources and mechanisms of earthquakes, but exact predictions are yet to come. The Alquist-Priolo Act prohibits new construction on active faults, and the Uniform Building Code is designed to make buildings earthquake-resistant. However, protection is far from complete. For example, the patterns of ground shaking are not well identified in most areas, and seismologists are just beginning to assess the risks posed by buried faults. It is therefore important that all citizens be informed about the extent of active faulting in their area. Because earthquakes are possible at any time, particularly along the Cascadia subduction zone, the San Andreas fault system, and in the Transverse Ranges, it is vital for all Californians to be prepared for earthquakes. Preparations should consider all the places where people spend their days and nights, including their homes, the workplace, and schools.

The box on pp. 439-440 depicts an earthquake scenario and offers some guidelines for preparing yourself and your family for such an event. This list is not all-inclusive, but is intended to prompt you to consider measures that would be appropriate for your family in such an event. You should, of course, be guided by directives from local authorities during any emergency.

THE NEXT BIG ONE

A typical work day begins. You're sitting in your office at the computer, checking your e-mail. Your commute was only about 30 minutes this morning, not too bad. Your husband, Wendell, is in New York on a business trip; your daughter, Austin, is at her elementary school; and your son, Bruce, is on his way to the community college on his bike. Before you left for work, you gave Mom her medication and made her comfortable on the TV room sofa. Buff the cocker spaniel is in the back yard.

Suddenly, your computer starts shaking—and so does everything else. A long minute later, you find yourself under your desk and realize that you are OK. You crawl out from under the desk and step over fallen file cabinets to go in search of office mates. You learn that a 7.5 magnitude earthquake has struck by listening to a co-worker's car radio—your car is buried somewhere in the rubble of the parking garage. No telephones are working, not even cellular phones. Three million people are trying to use them simultaneously. All power is out.

You run through a mental checklist of your family's whereabouts:

Wendell—safe, but frantic with worry
Austin—you *think* the school sent home a handout about emergency procedures
Bruce—was he in traffic? Did he avoid the underpass? Power lines?
Mom—will anyone know to check on her? Can she get out of the house? What about her medicine?
What about Buff?

This isn't the earthquake you had planned for—the one when everyone was at home together, able to grab flashlights and the emergency kit. The freeways are completely clogged, and many overpasses are down. Your only way home seems to be on foot, but you figure it will take you 2 days to walk 25 miles.

The reference list at the end of Chapter 13 includes a number of resources to aid in preparing for future earthquakes. In addition to these resources, the front sections of most telephone directories contain earthquake information on first aid and survival. The guides present planning suggestions, as well as emergency procedures to follow during and after earthquakes. Because most Californians have at least one telephone book on hand at all times, this is probably the most accessible planning resource. The following list has been compiled from various sources.

Before the Next Earthquake

- Discuss earthquake safety with children, caretakers, and family members.
- Calmly go over possible earthquake scenarios for different times and places.
- Evaluate your home and workplace for earthquake hazards; identify safe places within each.
- Correct hazards or reduce the potential for hazards where possible.
- Make sure that *all* adults and teens are able to turn off gas, electricity, and water in your residence.
- Plan and practice evacuation of your residence and workplace.
- Discuss reunion and communication plans with loved ones.
- Plan for care of elderly or disabled family members and pets in the event of your absence.
- Take a basic first aid training course.
- Store emergency supplies at home, at work, and in all your cars.
- Know the emergency plans of day care centers, schools, and nursing homes where your loved ones may be during an earthquake.
- Organize the people in your workplace and your neighborhood for emergency response preparedness.
- Keep lists of important personal, financial, medical, and household information in safe and readily accessible places.

During an Earthquake

- Duck, cover, and hold if inside.
- Get into the open if outside.
- If driving, stop ***if*** you are in a safe place. ***Do not stop*** under bridges, power lines, or overpasses.
- If in a wheelchair, stay in it, move to a safe place, lock the wheels, and cover your head.
- Do not rush the exits in crowded public places.

Continued

THE NEXT BIG ONE—cont'd

Following an Earthquake

- Put on sturdy shoes if possible; get a flashlight if it is dark.
- Check for injuries and treat appropriately.
- Make a safety check of immediate surroundings and prepare for likely aftershocks.
- Evacuate to a safe place; implement reunion plans if appropriate.
- Check food, water, and medical supplies.
- Do not use electrical appliances if you suspect there is leaking gas.
- Do not use the telephone except for extreme emergencies.

Volcanic Eruptions

The 1980 eruption of Mt. St. Helens in Washington (see Chapter 5) was an effective reminder of the possibility of eruptions from other Cascades volcanoes, including Mt. Shasta and Lassen Peak. Recent seismicity, ground bulging, and carbon dioxide emissions in the Mammoth area have also alerted scientists to the possibility of future eruptions there. As discussed in Chapter 5, geologists are in the process of evaluating volcanic hazards in the Pacific northwest, including those posed by California's active centers. Continued monitoring of these volcanoes should continue to be an important and effective part of mitigating the hazard posed by future eruptions.

Because of the location of California's active volcanoes, the most populated areas of the state are fortunately not at risk for experiencing catastrophic pyroclastic flows, debris flows, or blasts (Fig. 19-1). However, ash from major eruptions from the Lassen and Long Valley areas has blanketed much of California in the past (see Chapter 5). A future eruption from one of these areas could have destructive effects, particularly if east-blowing Santa Ana winds happened to be active during the eruption. Near the active volcanoes, hazards can be reduced by enactment of zoning ordinances based on the volcanic hazard maps (see Chapter 5), as well as by careful development of evacuation plans.

Floods

Although they may be unaware of the fact until a flood occurs, many Californians live on floodplains (see Chapters 10 and 11). Millions of dollars have been spent to reduce flood hazards by constructing dams, levees, and flood-control channels, but it is not economically possible to eliminate all flood risk in most areas. Furthermore, flood elimination may not be desirable in areas where stream channels and ***riparian*** forests provide recreation for humans and habitat for other species. Many communities have recently adopted floodplain management plans that allow these areas to be preserved (Fig. 19-2).

The most effective way of reducing the risk of flooding is to avoid building on flood-prone areas. In addition to low-lying floodplains and coastal wetlands, other high-risk areas in California include desert washes, alluvial fans, and debris-flow fans at the mouth of steep canyons (see Chapter 6).

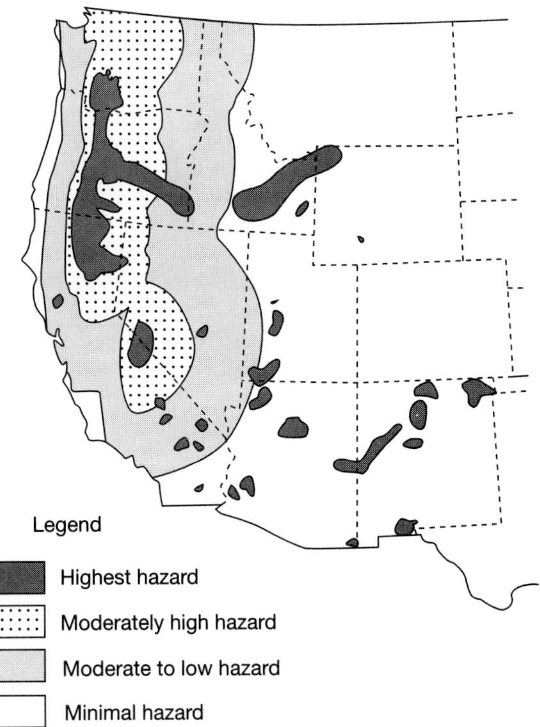

Fig. 19-1 Map showing relative volcanic hazards in the western United States. (Source: U.S. Geological Survey. 1981. Professional Paper 1240-B.)

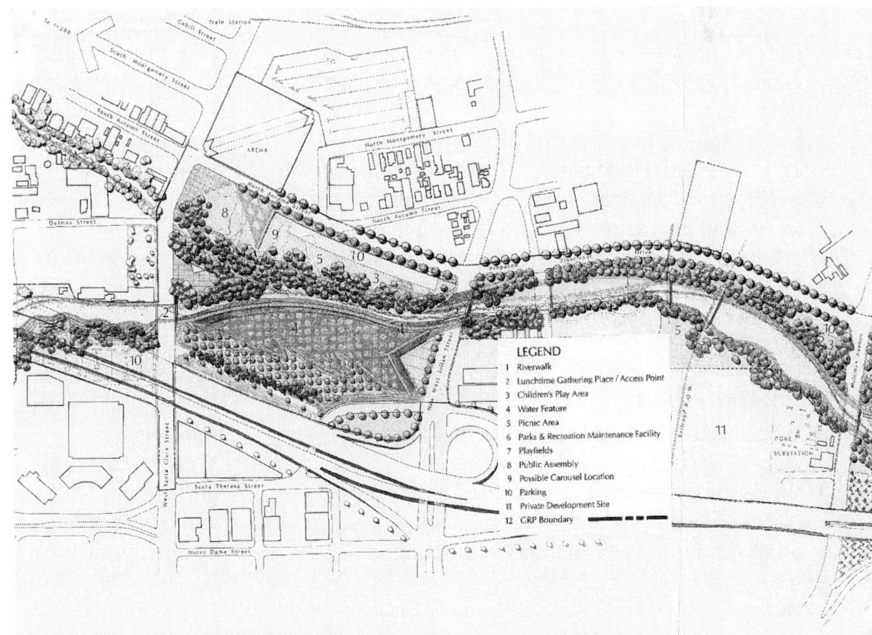

Fig. 19-2 Map showing the planned riparian corridor and park for the Guadalupe River in downtown San José. (Source: San José Department of Parks.)

The box below offers suggestions for preparing for and getting through a flood safely.

Landslides

Landslides are an all-too-familiar event to many of California's hillside residents. In many cases landsliding damages roads, generating constant repair work and annoying traffic diversions. However, landslides in California have also dammed rivers, destroyed hundreds of homes, and killed several people during the past 30 years.

Landsliding is the downhill ***mass movement,*** driven by gravity, of soil, rock, or a combination of the two, called ***debris.*** Slow mass movement, called ***creep,*** occurs on virtually all hillslopes, but in some areas, movement is rapid enough to send discrete masses of material down the hillsides. When this happens, a visible landslide scar is

---

PLANNING AHEAD FOR A FLOOD

Northern California's many beautiful rivers are enjoyed by water enthusiasts and anglers across the state. But when heavy rains come, those same rivers have been known to escape their banks, endangering lives and property. In fact, northern California experienced heavy flooding in 1986 and as recently as 1995, 1996, and 1997. When heavy rains fall, the rule for being safe is simple: ***plan ahead.***

Before the Storm

- Determine whether you are in a flood area by contacting your local insurance agency or planning office.
- Store supplies in a convenient location: first aid kit and essential medicines, nonperishable foods, nonelectric can opener, radio, flashlights, batteries, and instructions on how and when to turn off your utilities.
- Decide where your family will go in case you must evacuate. Inform family and friends of the location.

When a Storm Warning Is Issued

- When flash flooding is ***possible*** in a designated area, a Flash Flood Watch is issued.
- When flash flooding is ***reported*** or is ***imminent,*** a Flash Flood Warning is issued.
- Store drinking water in closed containers.
- If advised by local authorities to leave your home, follow your planned route.
- Before leaving, turn off all electrical circuits at the fuse panel, gas service at the meter, and water at the main valve.

During the Storm

- Avoid areas prone to sudden flooding: dry creek beds, culverts, and river banks.
- A car can be swept away in as little as 12 inches of running water. If your car stalls, abandon it immediately and seek higher ground.
- Flood water is often contaminated with sewage. Keep children away from flood water and storm drains.

After the Storm

- Rely on utility company crews to turn on all services.
- Do not use fresh food or electrical and gas appliances that have come in contact with flood waters.
- Follow local instructions or purify water before drinking.
- Wash homes down with soap and water and disinfect them with a chlorine bleach solution of 1 tablespoon to 1 gallon of water.
- Contact your insurance agency immediately.

Source: Modified from KOVR 13 Northern California Online; © 1997 KOVR 13, *http://www.kovr.com/almanac/flood.htm.*

left in the area where debris was detached (Fig. 19-3). Landslide scars are typically bowl-shaped, with a steep slope at the head of the slide area. Downhill of the scar the debris may come to rest on the lower slope, leaving a characteristic lumpy and disrupted mass (see Fig. 19-3). In most areas, geologists can recognize landslide-prone areas by mapping landslide scars and deposits with the use of aerial photographs.

Geologists and engineers classify different types of landslide features according to the depth of the material that fails, amount of water involved, rate of movement, and type of movement that occurs (Fig. 19-4). Landslide classification is important, because the risks posed by various types of landslides are different (see below). In addition, classification is important in deciding whether stabilization measures will be effective.

Of the many types of landslides, several are particularly common in California. ***Rockfalls,*** or the free-fall of loosened rock from cliffs, occur in the steep cliffs of the Yosemite area (Fig. 19-5) and in other very steep terrain. Rockfalls do not require water for their initiation, and they can occur without warning, posing a threat to unsuspecting hikers.

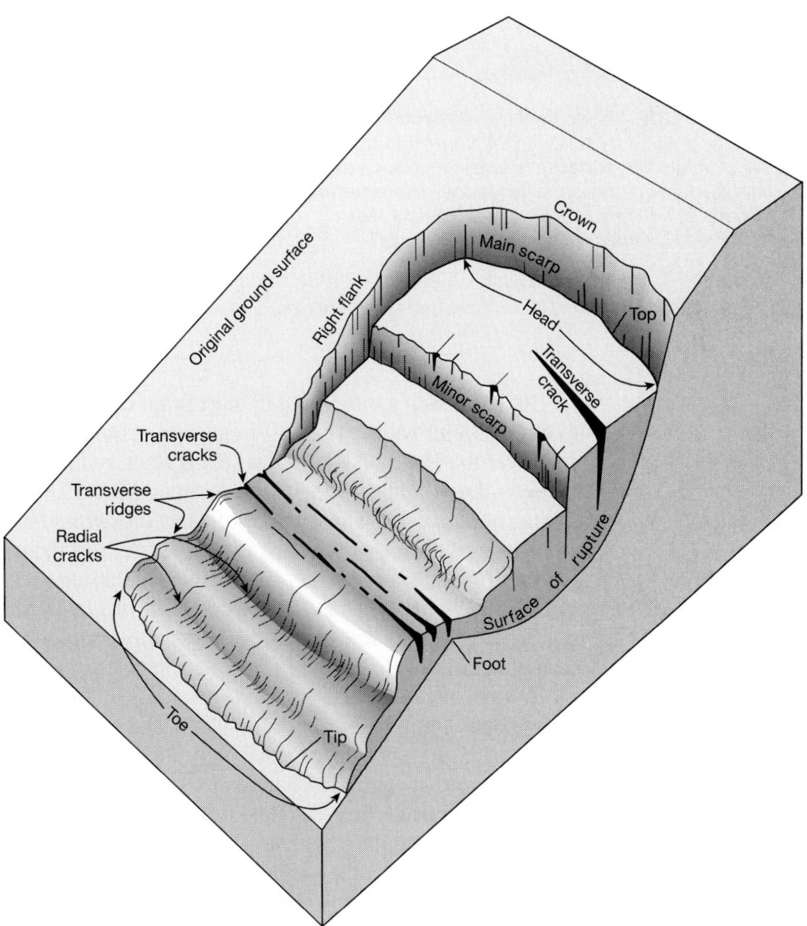

Fig. 19-3 Diagram showing features of a typical landslide. (Source: Highway Research Board, Special Report 29, 1958.)

FALL (ROCKFALL)

SLIDE (ROCK SLUMP)

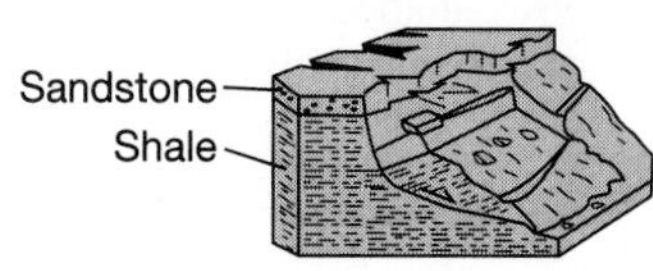

FALL—Mass travels most of the distance in free fall, by leaps and bounds, and rolling of bedrock or soil fragments.

ROTATIONAL SLIDE—Movement involves turning about a point (surface of rupture is concave upward).

SLIDE—movement of material by shear displacement along one or more surfaces or within a relatively narrow zone.

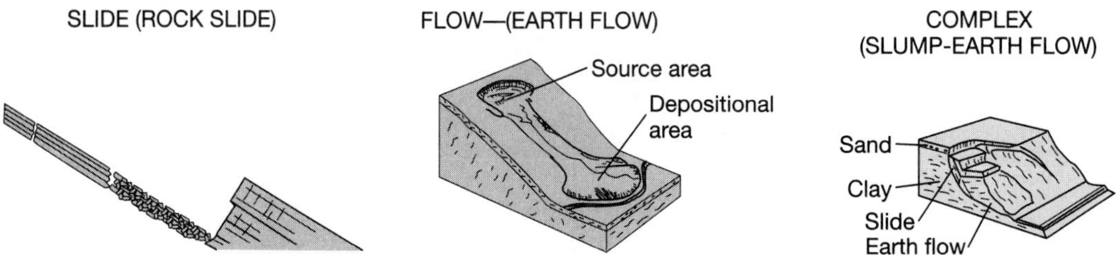

TRANSLATIONAL SLIDE—Movement is predominantly along planar or gently undulatory surfaces. Movement frequently is structurally controlled by surfaces of weakness, such as faults, joints, bedding planes, and variations in shear strength between layers of bedded deposits, or by the contact between firm bedrock and underlying detritus.

Fig. 19-4 Diagram showing different landslide types. (Source: U.S. Geological Survey. 1981. Professional Paper 1240-B; modified from Varnes.)

In July 1996, a massive slab of granitic rock suddenly broke loose from the cliffs above Yosemite Valley. The 120-meter-high chunk fell almost 800 meters to the valley floor near the Happy Isles area, where it shattered into debris that covered parts of the valley with rock dust up to 15 centimeters thick. One Yosemite hiker was killed, and several were injured. The number of casualties would have been higher if the rockfall had occurred during daylight hours rather than in the early evening. Rockfalls are a natural process of erosion on steep cliffs, including areas that are shaped by exfoliation (see Chapter 8). The sudden crashing of giant rock slabs is the result of slow, gradual weathering and loosening, and the exact moment of failure is unpredictable. Earthquakes may also trigger multiple rockfalls in mountainous areas.

Debris avalanches are relatively small, shallow slides that are also common on steep slopes held up be moderately resistant rocks such as sandstone. They commonly occur at the heads of steep canyons (Fig. 19-6). On slopes underlain by weak rocks such as serpentinite or mudstone, landslides tend to be deeper and to cover a larger area. These types of landslides, called ***slumps,*** are especially common in the melange units of the Franciscan Complex in the Coast Ranges (Fig. 19-7), although they occur in many other areas as well. At the downhill end of many slumps, the landslide debris loses its coherence, becoming an ***earthflow.***

Debris flows are a particularly dangerous form of mass movement (see the box on p. 216). Debris flows are created when rock and soil become completely satu-

Fig. 19-5 Rockfall in Yosemite National Park, 1996. (Source: National Park Service.)

Fig. 19-6 Debris avalanches near Ventura. (Source: Sarna-Wojcicki, A. U.S. Geological Survey.)

Fig. 19-7 Complex slump-earthflow, Humboldt County. (Photo by author.)

rated, sometimes becoming so mobile that they can attain speeds of hundreds of kilometers per hour. Debris flows form in steep stream channels, sometimes downhill of rockfalls or debris avalanches. In many mountainous areas of California, debris flows pose serious hazards to residents living in mountain canyons or on fans at the canyon mouths. Careful examination of these fans often reveals that they contain abundant materials carried by debris flows, sometimes even house-sized boulders (see Chapter 6).

The triggering mechanisms for slope failures are present in many of California's hilly areas. Because of active tectonics, steep slopes are abundant throughout the state. Tectonic activity has uplifted naturally weak materials, such as the unconsolidated sediments or marine sandstone and shale—materials that in most areas would be found beneath lowlands. In addition, shearing and faulting have weakened rocks along fault zones. As we have seen in Chapter 10, California regularly experiences severe winter storms, even though average rainfall is low during most years. Many landslides are generated during these wet winters. If streams are swollen by floodwaters, rivers may undermine the lower areas of unstable slopes, causing them to slide (Fig. 19-8). Along the coast, storm waves may undercut unstable sea cliffs, especially if high tides occur at the same time. In addition to weak rocks, steep slopes, and severe storms, California's hillslopes are also subjected to earthquakes. Ground shaking during earthquakes may trigger hundreds of rockfalls or shallow landslides, as was the case during the 1994 Northridge earthquake (see Fig. 13-1).

Landsliding is a natural process on California's hillslopes, and destruction by landsliding has predictably increased as hillsides have been developed. One of many notorious California landslides is the Portuguese Bend slide in the Palos Verdes Hills. In this area, portions of the marine terrace are sliding into the Pacific Ocean (Fig. 19-9). Between 1956 and 1959 alone, the Portuguese Bend slide caused more than $10 million in damages to homes, roads, and other structures.

In many counties, development has been restricted on landslide-prone hillslopes. In these areas geologists have mapped vulnerable areas by recognition of

Fig. 19-8 Landslides triggered by stream undercutting, Humboldt County. (Photo by author.)

Fig. 19-9 Landslide along the southern California coast near Palos Verdes Hills. (Source: Bradley, W.)

Fig. 19-10 Landslide stabilization along Highway 1, Marin County coast. (Source: Caltrans and Israel, K. U.S. Geological Survey.)

landslide deposits and scars, and local agencies have enacted ordinances against developing landslide-prone areas. However, many unstable slopes have already been developed. In some of these areas, engineers have installed drains, built retaining walls, and planted vegetation to stabilize the hillslopes, with varying success (Fig. 19-10). To avoid purchasing or building a home on an unstable hillslope, a prudent Californian seeking residence in any hilly area would be well advised to obtain a slope-stability evaluation from a geologist.

EARTH RESOURCES

California's geologic history has created significant Earth resources in addition to the geologic hazards discussed above. As a state with more than one automobile per citizen, California is highly dependent on petroleum, and most citizens are aware of this dependency. However, few people realize how important other geologic materials are in meeting our everyday material needs. Minerals are necessary ingredients for everything from cosmetics to spark plugs, and rock materials are components of most modern building materials (Table 19-1). In addition to providing these commodities, California's geologic history has also endowed the state with a rich and varied landscape, a resource to be enjoyed by all citizens.

The mineral industry has played an important role in California's history, beginning with the Gold Rush (see Chapter 8). Because of plate tectonics, California has a wide variety of rocks containing a wealth of mineral resources. In some areas, minerals bearing metallic elements such as copper, gold, and tin have been profitably extracted by underground and surface mining. Nonmetallic mineral resources such as borate or gypsum have been mined by surface mining other regions.

We have investigated only a few of these resources, including evaporite minerals (Chapter 6), gold (Chapter 8), chromite (Chapter 9), mercury (Chapter 12), diato-

Table 19-1 CALIFORNIA'S MINERAL PRODUCTION

Material	Type	Examples of mines	Uses	Production (in millions of dollars)[3]
Asbestos[1]	Silicate mineral	San Benito County	Refractory material	5.0
Borates[1]	Evaporite minerals	Boron, Inyo	See Chapter 6	462.0
Cement	Limestone	Permanente, Santa Clara	Construction, masonry	522.0
Clays	Silicate minerals	Ione, Sacramento	Ceramics, cement, cosmetics	32.0
Crushed rock	Commonly igneous	Numerous	Concrete, other construction	307.0
Diatomite[1]	Silica from diatoms	Lompoc, Santa Barbara	Filters (see Chapter 16)	90.0 (1989 value)
Dimension stone	Hard and durable rock	Fresno County	Building facades, counters	6.5
Gemstones	Tourmaline, beryl	Pala, San Diego	Jewelry	0.6
Gold[2]	Metal	McLaughlin, Lake; Mesquite, Imperial	Electronics and other (see Chapter 8)	354.0
Gypsum	evaporite	Plaster City, Imperial	Wallboard, plaster	11.0 (1989 value)
Magnesium compounds	Sedimentary minerals	Bissel, Kern	Chemical industry	NA
Mercury[2]	Metallic mineral	McLaughlin, Napa, Lake, Yolo	Instrumentation, gold mining (see Chapter 12)	NA
Perlite,[2] pumice,[2] cinders	Volcanic rocks	Inyo, Modoc, Shasta counties	Absorbents, abrasives (see Chapter 5)	6.5
Rare earths[1]	Igneous rocks	Mountain Pass, San Bernardino	Televisions, superconductor (see Chapter 7)	24.0 (1989 value)
Sand and gravel	Alluvium	Numerous	Construction	548.0
Silver	Metal	Produced with gold	Photographic and other industries	2.0
Sulfur	Native element or sulfate		Fertilizer	51.0 (1989 value)
Tungsten[1]	Metal, associated with granitic rocks	Pine Creek, Inyo	Alloy	0 in 1995; mine reopened in 1996
Zeolites	Silicates in altered volcanic rock	Barstow, San Bernardino	Ion exchange, absorbers and filters	NA

NA = Data not available; [1]California largest producer in U.S; [2]California second to fifth largest producer in U.S. [3]1995 figures, unless otherwise noted.

Data sources: Tanner, A., and Kohler-Antablin, S. 1995 California mineral industry survey. *California Geology,* p. 143; Tooker, E.W., and Beeby, D.J. 1989. *Industrial Minerals in California: Economic Importance, Present Availability, and Future Development.* U.S. Geological Survey, Bulletin 1958, 127 pp.; Langer, W.H., and Glanzmen, V.M. 1993. *Natural Aggregate: Building America's Future.* U.S. Geological Survey, Circular 1110, 39 pp.

We have investigated only a few of these resources, including evaporite minerals (Chapter 6), gold (Chapter 8), chromite (Chapter 9), mercury (Chapter 12), diatomite (Chapter 16), and lithium and gemstones (Chapter 17).

Mineral discoveries in other countries, combined with strict environmental regulations and the high cost of labor in the United States, have reduced the U.S. production of minerals during the past 30 years. Nevertheless, California continues to be a supplier of many mineral resources, despite the availability of many commodities from foreign suppliers. In 1995 the industry produced \$2.7 billion of nonfuel mineral commodities.

Aggregates, Limestone, and Other Rock Materials

Although they are less glamorous than gold, copper, or gemstones, ordinary rock materials are very important resources in California. Sand, gravel, and crushed rock, known as ***aggregates,*** are essential as foundation materials for roads, commercial buildings, and residences (see Table 19-1). They are also essential for making concrete, which is a mixture of cement and aggregate. Cement is also made of geologic materials. Portland cement contains about 60 percent lime, which comes from carbonate sedimentary rocks (see Chapters 2 and 7) or from crushed shells. The lime is mixed with silica and alumina, which come from clay, sand, or shale, and a small amount of iron oxide and gypsum. Because of California's geologic history, carbonate rocks are rare except in the sparsely populated desert area. As a result, relatively small bodies of carbonate rock in coastal California—for example, the Franciscan limestone block near Cupertino in Santa Clara County—are valuable resources.

Aggregates are extracted from the Earth by surface mining. Sand and gravel are mined from river beds (Fig. 19-11), and crushed rock is quarried from rock outcrops. Ironically, meeting the demand for aggregates in California has threatened their easy availability. Supplies of these materials are limited in developed areas. In addition, communities are unwilling to have active quarries next to residential and commercial developments. Developers must obtain these

Fig. 19-11 Gravel mining operations. (Source: Langer, W. U.S. Geological Survey.)

materials from more rural areas. Although the materials can be quarried at relatively low cost, transporting them to construction sites is very expensive. As California's population continues to expand, aggregates will become an increasingly valuable commodity.

Petroleum

As we have seen in Chapter 16, petroleum is a major geologic resource in the San Joaquin Valley, the Transverse Ranges, and in the offshore basins bordering the continent (see Table 16-1). At the present time, many of California's known oil fields are producing at low levels because of the lower cost of imported petroleum. However, as global supplies of oil decrease during the next several decades, production from California fields may increase, particularly from offshore platforms. Natural gas, which is plentiful in California's Cenozoic marine sedimentary rocks, may also be a significant source of revenue in the future.

Soils

As we have seen in Chapter 11, fertile soils are one of California's most important resources. However, as California's population increases, residential areas are constantly encroaching onto the rich alluvial valley floors. At the same time, agricultural productivity is also threatened by limited water supplies (see Chapter 10), groundwater contamination, salinization of soils (see Chapter 11), and air pollution from nearby urban areas. If agriculture is to continue to be important in the state's economy, then preservation of agricultural resources must be a priority, particularly in the Central Valley, Salinas Valley, and Imperial Valley areas.

CALIFORNIA'S LANDSCAPE

Because of its complex geology and varied climates, much of California's landscape is dramatically beautiful. The scenery varies from rocky deserts to alpine meadows and from forested rivers to coastal sea cliffs (see the Color Plates). Despite being home to 31 million people, California has abundant space that remains open for its residents to enjoy. One important side effect of recent uplift along the San Andreas fault system was the creation of rugged mountains, natural open spaces for California's residents. With the exception of Sacramento, California's major urban areas are naturally confined, because rugged mountains enclose the urbanized valleys. The mountains have prevented limitless urban sprawl from overtaking the state, and the creation of national, state, and local preserves guarantees that some of these areas will be left undeveloped for the enjoyment of future Californians.

The California landscape provides opportunities for a wide range of outdoor recreational activities. Equipped with even a small amount of knowledge about California geology, one can learn much from the rocks and landforms in different parts of the state while enjoying these activities. Investigations into California's geologic history provide learning for students of all ages and levels of expertise, particularly if they have an opportunity to visit geologically important localities in the field. To this end, it is my hope that those who read all or parts of this book are only at the beginning of their study of California geology.

FURTHER READINGS

BAILEY, E., AND HARDEN, D. 1974. *Map Showing Mineral Resources in the San Francisco Bay Area: Present Availability and Future Demand.* U.S. Geological Survey, Miscellaneous Investigations Map I-909.

BLAIR, M., AND OTHERS. 1979. *Seismic Safety and Land-Use Planning—Selected Examples From California.* U.S. Geological Survey, Professional Paper 941-B, 82 pp.

HAYS, W.W. 1981. *Facing Geologic and Hydrologic Hazards: Earth-Science Considerations.* U.S. Geological Survey, Professional Paper 1240-B, 109 pp.

LANGER, W.H., AND GLANZMEN, V.M. 1993. Natural Aggregate: Building America's Future. U.S. Geological Survey, Circular 1110, 39 pp.

TANNER, A., AND KOHLER-ANTABLIN, S. California mineral industry survey. *California Geology,* p. 143, 1996.

TOOKER, E.W., AND BEEBY, D.J. *Industrial Minerals in California: Economic Importance, Present Availability, and Future Development.* U.S. Geological Survey, Bulletin 1958, 127 pp.

TYLER, M. 1995. *Look Before You Build: Geologic Studies for Safer Land Development in the San Francisco Bay Area.* U.S. Geological Survey, Circular 1130, 54 pp.

Glossary

A horizon The top layers of a soil where organic matter accumulates.

ablation The loss of mass from a glacier, mainly by melting.

accreted terrane A portion of a plate added to a larger block of crust along a convergent plate boundary.

accumulation area The part of a glacier where ice builds up.

acre-foot The amount of water needed to cover an acre of land to a depth of 1 foot; about 325,851 gallons.

active fault A fault capable of generating earthquakes today; *syn. capable fault.*

aftershock Smaller earthquake that follows the largest earthquake of a series in the same area.

aggregate Crushed rock, gravel, and sand used in construction.

alluvial fan A fan-shaped landform along a mountain front formed by the buildup of stream sediment and debris flow deposits.

alluvium Sediment deposited by a river.

alpine Pertaining to mountains.

Alquist-Priolo Earthquake Fault Zoning Act 1972 California law that prohibits construction within 50 feet of an active fault. (Before 1994, *Special Studies Zone Act.*)

amplitude Height of a wave.

andesite Volcanic rock with intermediate silica content.

anticline An upward-buckling fold.

appropriation rights The system of water law whereby water is allocated in the order of first historic use.

aqueduct An artificial channel constructed to carry water.

aquiclude Water-excluding layer.

aquifer Water-carrying layer.

aquitard Water-retarding layer.

arroyo Canyon.

artesian Condition in which groundwater is held under high pressure.

ash Fine-grained volcanic material erupted into the atmosphere.

asthenosphere The region of the mantle about 100 to 250 kilometers beneath Earth's surface where rocks are plastic and easily deformed.

attenuation The decrease in the amplitude and energy of seismic waves with distance from the epicenter.

axis The central hinge line of a fold.

B horizon The layers of a soil where weathering products accumulate.

bajada A broad area where alluvial fans overlap along the edge of a mountain range.

basalt Mafic volcanic rock.

basaltic *See mafic.*

base level The lowest point in a stream system.

basement The complex of older rocks, usually igneous and/or metamorphic, which are overlain by younger rocks and sediments in an area; *syn. crystalline basement.*

basin A bowl-shaped depression.

batholith Very large body of plutonic igneous rock.

bedding The layering in sedimentary rocks.

bimodal volcanism Eruptive pattern producing both high-silica and mafic rocks.

blind fault A fault that does not offset materials at the surface.

blueschist Foliated metamorphic rock containing diagnostic minerals that form under conditions of high pressure and low temperature in a subduction zone.

body wave Earthquake wave that travels through the earth's interior.

breccia Coarse-grained rock composed of broken angular rock fragments.

caldera A bowl-shaped depression formed by the collapse of a volcanic cone following a major summit eruption.

carbonate Mineral or rock containing CO_3.

carbonate platform A shallow marine shelf along the edge of a continent where most sediment is contributed by carbonate-secreting organisms like coral.

carbonatite Unusual plutonic rock composed mostly of carbonate minerals.

Cenozoic The geologic era between 66 million years ago and the present.

channel Pathway carved by a stream to carry runoff.

chert Fine-grained sedimentary rock composed of silica, usually the skeletons of radiolaria.

chromite A mineral containing chromium, iron, and oxygen.

cinder cone A small, steep-sided volcano built by the accumulation of mafic pyroclastic material ejected from a single vent.

cinnabar Mercury sulfide, the most common mercury ore.

cirque Bowl-shaped depression at the head of a mountain valley, created by glacial erosion.

clastic Rock texture in which fragments are held together by a cementing mineral.

clay 1. Type of sheet silicate minerals containing water. 2. Sediment composed of particles with diameters less than 2 microns.

composite volcano *See stratovolcano.*

composition The type and proportion of constituent elements or minerals in a rock.

compression Stress resulting from forces directed toward each other.

confined aquifer A water-bearing layer sandwiched between less permeable layers.

contact Boundary between two rock types.

contact metamorphism Changes in rocks adjacent to intruding magma.

convection The movement of liquid and gas in which hot, less dense materials rise and colder, denser materials sink.

converge Come together.

convergent boundary The line where two plates meet as they move toward each other.

core The inner part of the Earth, divided into the outer liquid and solid inner core, both of which are composed mostly of iron and nickel.

coseismic Taking place during an earthquake.

crest The imaginary line running along the top of a mountain range.

cross beds Distinctive sedimentary layers, running criss-cross to each other, resulting from shifting currents or winds.

cross section A diagram showing a slice beneath the surface along a line and depicting the arrangement of rocks beneath the line.

cross-cutting relations Fundamental principle that states that a feature cutting through a rock must be younger than the rock it cuts.

crust The thin, outer layer of the Earth, composed mostly of silicate minerals; upper part of Earth's lithosphere.

crystalline basement *See basement.*

crystalline The state of matter in which the atoms are arranged in a regular, repeating pattern.

dacite Volcanic rock type with silica content intermediate between andesite and rhyolite.

debris avalanche A very rapid, tumbling movement of rock and soil from a steep slope or a cliff.

debris Broken rock and soil material.

debris flow A water-laden movement of unconsolidated soil and rock material, including large blocks, down a channel.

deformation A general term for folding, faulting, and other changes in rocks, sediments, and the land surface that take place in response to stress.

deposition The settling of sediment.

desert pavement A veneer or armor of stones on the desert floor.

desert varnish Thin, dark coating on the surfaces of rocks formed by the gradual buildup of manganese and iron oxide.

detachment fault A large, gently dipping fault, along which a large crustal block moves over deeper rocks.

diatom Single celled plankton whose cell walls are composed of silica.

diatomite Siliceous rock composed of the remains of diatoms.

differential weathering Weathering that wears away rocks in an area at variable rates because of differences in the resistance of materials to erosion; *syn. differential erosion.*

dike A relatively narrow igneous body that cuts through the surrounding rocks.

dip The inclination of layered rocks, with 0 degrees being horizontal.

discharge The volume of water flowing in a stream per unit of time.

displacement The separation of formerly continuous features across a fault.

divergent boundary The line where two plates meet as they move apart from each other.

divide Imaginary line separating the drainage basins of two streams.

dome A hemisphere-shaped mountain.

drainage basin The area of land from which a given area collects runoff.

dredge field A pile of river sediment left after gold dredging operations.

dredge To mechanically dig up the bed of a stream, lake, or ocean to extract sediment, deepen a channel or harbor, or mine placer deposits.

dunite Ultramafic plutonic rock composed of the mineral olivine.

dynamic equilibrium A state of balance in which a system constantly adjust to disturbances to maintain the balance.

eclogite Metamorphic rock formed in the mantle, contains garnet.

emergent coast A coastline that has risen relative to sea level.

epicenter The point on the earth's surface directly above the focus of an earthquake.

epoch Subdivision of a geologic period.

era Major division of the geologic time scale: Paleozoic, Mesozoic and Cenozoic.

erosion The removal of solid particles and ions in solution from an area at the surface.

erratic An out-of-place rock transported by a glacier and left behind when the ice melted.

estuary A semienclosed coastal body of water where fresh and salt water mix.

evaporation Change of state from liquid to gas (vapor).

evaporite Mineral formed from solution as a result of evaporation.

exfoliation The peeling off of the outside surfaces of massive rocks such as granite.

extension Stretching that occurs in reponse to stresses directed away from each other.

extinction The disappearance of a species from the Earth.

fault An abrupt break in rocks or sediments along which movement has occurred.

fault creep Gradual movement along a fault that takes place without earthquakes; *syn. tectonic creep.*

fault scarp A break in the ground surface created by vertical movement on a fault.

fault segment A section of an active fault that has the same earthquake characteristics and rupture history.

fault strand A single break within a fault zone.

fauna The assemblage of animals in a given habitat.

felsic *See silicic.*

fissure A linear crack.

flash flood A sudden flood of water through a stream channel.

floodplain The part of a stream valley adjacent to the channel that is covered during floods.

focus The point where an earthquake originates.

fold A bending of rocks or sediments without faulting.

fold-and-thrust belt An area of subparallel faults and folds indicating active compression.

foliation Sheet-like layering in metamorphic rocks that results from the growth of minerals in a preferred orientation.

foraminifera Single-celled protozoans whose shells are constructed from calcium carbonate.

fore-arc basin A marine depression between a volcanic island arc and an oceanic trench.

formation A recognizable geologic unit that can be identified and mapped.

fossil The remains or traces of living organisms preserved in geologic materials.

fumarole A volcanic vent that emits only gas (fumes).

gabbro Mafic plutonic rock.

geobarometer A mineral whose presence indicates the pressure at which a rock was metamorphosed.

geomorphic province A region identified by its characteristic landscape, influenced by a combination of geology, climate, and topography.

geothermal Related to the heat of Earth's interior.

geothermometer A mineral whose presence indicates the temperature at which a rock was metamorphosed.

glacial outwash *See outwash.*

glacial polish Shiny rock surfaces created by the grinding of rock fragments contained within a moving glacier.

glacier A body of ice that moves over the land surface under its own weight.

glaucophane Blue silicate mineral of the amphibole family, found in blueschist.

gneiss High-grade metamorphic rock with characteristic light and dark banding .

graben A depressed segment of the Earth's crust bounded on at least two sides by faults; *see horst.*

graded bedding Sedimentary layer in which the particles decrease in size from bottom to top.

gradient The steepness of a hillslope or stream channel.

granite Silicic plutonic rock.

granitic rocks Igneous rocks with relatively high silica content, including granite, granodiorite, tonalite; *syn. granitoid.*

gravel Rounded rock particles of larger size than sand.

greenstone A type of metamorphosed volcanic rock, usually basalt, with a characteristic green color.

greywacke Type of sandstone containing abundant sand-sized rock fragments, typically gray or brown.

ground rupture Breaking of the surface along a fault during an earthquake.

groundwater Water stored underground in the spaces within rocks or sediments.

half-life The time required for one half of a given amount of a radioactive isotope to decay.

hanging valley A glacial valley whose mouth is much higher than the elevation of the valley it joins.

hardpan A soil layer formed by the buildup of carbonate, silica, or iron.

harmonic earthquakes Multiple, small to moderate earthquakes at regular time intervals, signalling the movement of magma rising toward the surface in an active volcanic area; *syn. harmonic tremors.*

hazard map A map portraying the relative risks from one or more geologic processes.

headwaters The highest-elevation areas of a stream system.

heap leaching A chemical process using cyanide to extract gold from quartz veins.

Holocene The most recent geologic epoch; the past 10,000 years.

horst A block of the Earth's crust separated by faults from adjacent blocks; *see graben.*

hot spot A stationary, localized zone of melting beneath the lithosphere.

humus Partially decayed plant and animal material; the organic part of soil.

hydraulic mining Washing of sediments from hillsides using high-pressure hoses to extract *placer* gold.

hydrocarbon compound A chemical compound composed of hydrogen and carbon.

hydrology The study of water.

hydrothermal Produced by hot fluids.

igneous Fire-formed; rocks crystallized from cooling magma.

incision Downcutting by a stream into its channel.

inclusion a fragment of older rock incorporated into a younger igneous rock; *syn. xenolith.*

index fossil A fossil that is very useful for geologic correlations.

infiltration The movement of water from the surface into soil or rock.

intensity A measure of ground shaking during an earthquake, based on types and extent of damage and human experience; *syn. Mercalli intensity.*

intermittent stream or **lake** One that has water only periodically.

invertebrate Lacking a spinal column.

inverted topography A landscape in which areas formerly in low-lying areas are now at the highest elevations, and vice versa.

isotope A form of a chemical element having the same number of protons, but a different number of neutrons in its nucleus, and hence a different atomic mass.

landslide A general term for gravity-driven movement of rock and soil down a slope

lava molten silicate material (magma) at the Earth's surface.

lava tube Opening in volcanic rock formed when fluid lava flows through a recently cooled channel of lava, evacuating the center of the channel.

levee A naturally formed low ridge built along the side of a stream channel, or an artificial barrier constructed to increase the capacity of a channel.

limestone Sedimentary rock composed of calcium carbonate.

liquefaction The process by which ground shaking transforms wet,sandy sediment into an unstable, dense fluid during an earthquake.

lithification The process by which sediments become sedimentary rocks.

lithology Type of rock or sediment in a geologic unit.

lithosphere The Earth's rigid outer 100 kilometers, including the crust and the outer mantle.
load The dissolved and solid material transported by a stream.
local magnitude One commonly used measurement of earthquake magnitude; *see magnitude.*
lode Type of mineral deposit in which a valuable commodity is concentrated in a vein.
long tom A large version of the rocker used to separate placer gold from river sediments.
longshore current The flow of ocean water parallel to the coast within the surf zone.
longshore transport The movement of sand by the longshore current.
Love wave Type of surface earthquake wave.
mafic Term describing igneous rocks or minerals with high proportions of iron and magnesium and relatively low silica; *syn. basaltic.*
magma differentiation Compositional changes in a magma caused by the early crystallization and isolation of early-formed minerals.
magma Molten silicate material.
magnitude A measure of earthquake size determined from seismographs.
mantle The thickest of Earth's layers, between the crust and the core; composed of more dense silicate minerals than the crust.
marble Nonfoliated metamorphic rock formed from carbonate sedimentary rocks.
marine terrace A wave-cut platform that has been uplifted above the modern shoreline; *see wave-cut platform, shoreline angle.*
mariposite Unusual apple-green silicate mineral in the mica family; contains chromium.
marker bed An easily identified and unique geologic layer that can be used to correlate rock or sediment sequences from one area to another.
mass movement The downslope movement of rock, soil, and debris under the influence of gravity.
matrix The fine-grained material in a rock that contains larger particles.
maximum contaminant level (MCL) The highest level of a substance allowed by law in drinking water.
Mediterranean climate A climate type characterized by warm, dry summers and mild, wet winters.
melange A mixture of rock types jumbled together by tectonic activity.
Mercalli scale *See intensity.*
Mesozoic The geologic era from about 235 million to about 66 million years ago.
metamorphic core complex Metamorphic rocks exposed in the lower plate of a detachment fault in a highly extended terrane.
metamorphic Rocks changed by the action of heat, pressure, and hot fluids.
metasedimentary Sedimentary rocks transformed by heat and pressure.
metavolcanic Volcanic rocks transformed by heat and pressure.
mineral A naturally occurring, inorganic, crystalline solid composed of one or more chemical elements.
miogeocline A wedge of shallow marine sediment that accumulated on the continental shelf and slope of a passive margin; *see passive margin.*
moment magnitude Measure of earthquake size related to the force couple across the fault rupture surface; *see magnitude.*
moraine A ridge of sediment left by a glacier.
Mother Lode The major gold mining district in California's Sierran foothills; *see lode.*
mud Fine-grained sediment composed of particles smaller than 1/16 mm.
mylonite Rock formed by intense grinding along the deeply buried parts of fault zones.
normal fault A fault that drops one side down relative to the other as a result of extension.
obsidan dome A steep-sided hill built by the eruption of high-silica lava.
obsidian Volcanic glass with high silica content.
offshore bar A submerged ridge of sand.
ooze Fine-grained deep-sea sediment composed mostly of the remains of marine animals or plants.
ophiolite suite Assemblage of rocks thought to represent ocean-floor sediments and the crust and upper mantle beneath them; *syn. ophiolite.*
ore A concentration of a mineral or rock that can be extracted at a profit.
original horizontality Fundamental principle of sedimentary layers that states that sediments are deposited in flat-lying layers.
orogeny The process of mountain foundation, especially by folding of the Earth's crust.
orographic effect The effect of mountain ranges on precipitation.
outwash Sediment carried by a stream flowing from a glacier.
overdraft The pumping of groundwater in excess of recharge rates.
P wave The primary, fastest wave generated by an earthquake, a compressional body wave.

paleogeography The ancient environments and landscapes reconstructed from the types and locations of rocks found in an area.

paleontology The study of fossils and the history of life.

paleoseismology The study of ancient earthquakes.

Paleozoic The geologic era from about 570 million to about 235 million years ago.

Pangaea The supercontinent that formed by the collision of all of the continents during late Paleozoic time.

panning Technique used to extract gold from river sediments by swirling the sediment in water using a round, shallow dish.

parent material The rock or sediment from which a soil formed.

pass A low point along the crest of a mountain range.

passive margin A continental edge far from active plate boundaries.

peak The highest point on a mountain; *syn. summit.*

peat A dark brown or black organic-rich material composed of partially decomposed wetland plants.

pegmatite Very coarse-grained igneous plutonic rock.

peridotite Ultramafic igneous rock composed of olivine together with plagioclase feldspar and pyroxene.

period Subdivision of a geologic era.

permeability The ability of a material to transmit fluid.

petroglyph A carving on a rock surface made by etching into the desert varnish.

petroleum Naturally occurring mixture of hydrocarbon compounds that includes natural gas, gasoline, and oil.

pillow lava Distinctive pillow-shaped basalt that forms when lava solidifies under water.

placer A type of mineral deposit in which a valuable commodity is concentrated in river or beach sediments.

plate tectonics The concept that the Earth's lithosphere consists of a number of rigid, mobile pieces (plates) riding over the more plastic asthenosphere.

playa Dry lake bed in a desert basin.

Pleistocene The geologic time period from about 1.8 million until about 10,000 years ago: the Ice age.

pluton An identifiable body of igneous rock crystallized beneath the surface from a single body of magma.

plutonic Rocks crystallized from magma beneath the earth's surface; *syn. intrusive.*

polymerize To form a complex molecule by the chaining together of simple molecules.

porosity The percentage of the total volume of a rock or sediment that is empty spaces.

porphyritic Texture describing igneous rocks with some large, early-formed crystals in a mass of finer grained crystals or glass.

precursor A change in geologic conditions that precedes a geologic event such as an earthquake or volcanic eruption.

Proterozoic The geologic time period from about 2.5 billion to about 570 million years ago.

protolith The parent rock type of a metamorphic rock.

pull-apart basin A depression created by crustal extension.

pumice Lightweight volcanic glass containing abundant vesicles.

pyroclastic Fire-broken; fragmented volcanic material ejected during explosive eruptions.

pyroclastic flow Hot volcanic flow containing fragmental materials.

quartz Common silicate mineral composed of silicon and oxygen.

quartzite Metamorphosed quartz sandstone.

quicksilver Mercury.

radioactivity Spontaneous decay of an unstable isotope; *see isotope.*

radiolarian A single-celled marine organism whose tests are composed of silica; *pl. radiolaria.*

radiometric dating Age dating of Earth materials using the known half lives and decay rates of isotopes; *see half-life, isotope.*

rain shadow A dry area on the downwind side of a mountain range, created because the mountain range forms a barrier to moisture-carrying clouds.

range-front fault A young fault along the edge of an uplifting mountain range.

rare-earth elements Fifteen elements of the periodic table valued for their uses in technology industries.

Rayleigh wave Type of surface earthquake wave.

recharge The addition of water to groundwater from the surface.

recurrence interval The characteristic time period between repeated geologic events such as floods, earthquakes, or volcanic eruptions.

refraction The bending of waves as they travel from one medium to another and change velocity.

regional metamorphism Changes in rocks over a large scale in response to major magmatic and/or tectonic events.

relative age dating Determination of the ages of geologic formations, materials, or events relative to each other.

resonance Amplifying effect produced when the natural vibration frequency of a material or structure is matched by the frequency of seismic waves.

resurgent dome A volcanic mound built in the center of a caldera by eruptions that follow the caldera formation.

reverse fault A fault along which one side is moved up over the other side as a result of compression.

rhyolite Silicic volcanic rock.

riparian Adjacent to a stream.

riparian rights A system of water laws whereby water is allocated to those owning land that borders a body of water.

roche moutonée An asymmetric bedrock dome created by flowing ice.

rock cycle A conceptual model used to illustrate the origins of and the relationships between earth materials.

rock fall Gravity-driven fall of loosened rock fragments from a steep cliff.

rocker A long box mounted on rockers and used to sort placer golds from river sediments; *syn. cradle.*

roof pendant A block of metamorphic rock, originally at the top of a magma chamber, at the edge of a plutonic body.

runoff Water that flows over the land surface.

rupture surface The portion of a fault that slips during and earthquake.

S wave The secondary body wave generated in an earthquake, a shear wave.

safe yield The amount of groundwater that can be pumped without exceeding the recharge to the system.

sag pond A small depression in the land surface along an active fault zone.

salinization The buildup of soluble minerals in the lower part of a soil.

salt-water intrusion The inflow of salt water into freshwater aquifers.

schist Strongly foliated metamorphic rock with visible platy minerals.

scoria Glassy volcanic rock low in silica that contains abundant vesicles.

sea stack A block of rock isolated from the shoreline by wave erosion.

sea-floor spreading The divergent motion and creation of oceanic crust caused by rising magma along divergent plate boundaries.

sediment Rock and mineral fragments derived from preexisting rocks.

sedimentary basin A low-lying area where sediments accumulate.

sedimentary Rocks formed at the earth's surface from sediment, organic remains, and/or minerals precipitated from solution.

seismograph Instrument for recording earthquake waves.

seismology The study of earthquakes.

seismometer Sensor used to measure earthquakes.

selenium A chemical element essential for living animals but toxic in large amounts.

serpentine (serpentinite) 1. A green, platy, hydrated silicate mineral containing magnesium. 2. Internal name for metamorphic rock composed of serpentine *synserpentinite.*

shale Fine-grained sedimentary rock composed of mud-sized particles; *syn. mudstone.*

shear stress Forces causing two blocks of rock or other material to move past each other.

sheeted dikes Vertical basalt dikes carrying magma to the ocean floor; part of an ophiolite suite; *see dike.*

shelf The gently sloping part of the ocean floor below the low-tide line adjacent to the shoreline.

shield volcano A large, low, round volcano built mostly of basalt flows.

shoreline angle The line where the a wavecut platform meets a seacliff.

silicate A mineral belonging to the group that contains the elements silicon and oxygen bonded together in a tetrahedron.

silicic A term describing igneous rocks or minerals with high proportions of silica and relatively low iron and magnesium; *syn. felsic.*

sinuous Snakelike, winding.

slate Fine-grained, foliated metamorphic rock that breaks into thin, regular layers.

slip rate The average rate of movement on an active fault.

slough A channel in a wetland area.

sluice box A long open box with a chain of baffles at the enddesigned to sort and concentrate *placer* gold from river sediment.

slump A type of deep landslide in which the upper part rotates as it moves downslope.

soil horizon A distinct soil layer whose properties differ from the layers above and below it.

soil profile The sequence of layers, extending from the surface to fresh rock or sediment, that result from the accumulation of organic material and the weathering of rock or sediment.

sorting A term used to describe how similar in size the particles in sediment or sedimentary rock are.

spatter cone Small volcanic cone formed when blobs of lava are thrown from a vent and pile up in a heap.

Special Studies Zone *See Alquist-Priolo Earthquake Fault Zoning Act.*

stage The height of water in a river at a given time.

stamp mill Device used to crush gold-bearing quartz.

strain energy The energy stored in rocks as they are subjected to stress.

stratigraphy The relationships and sequencing of layered rocks.

stratovolcano A large, steep-sided volcano built of both pyroclastic material and lava flows; *syn. composite volcano.*

stream terrace A remnant of a floodplain formed when a stream was flowing at a higher level.

stress The force per unit area acting on a rock; *syn. directed pressure.*

striations Small, parallel grooves created when a glacier flows over a bedrock surface.

strike The geographic orientation of an imaginary horizontal line drawn in the plane of an inclined rock layer.

strike-slip fault Fault with lateral motion; *syn. lateral fault.*

stromatolite Carbonate mounds built by organisms such as algae.

structurally controlled topography A landscape in which the mountains and valleys follow the trends of contacts and/or faults.

subduction complex The characteristic assemblage of rocks brought together along a subduction zone.

subduction The process by which an oceanic plate is driven beneath another plate into the mantle along a convergent boundary.

submarine canyon An underwater extension of a valley on land carved on the continental shelf.

submarine fan A fan-shaped feature on the ocean floor formed by deposits of turbidity currents.

submerged coast A coastal area that has been lowered relative to sea level.

subsidence Sinking.

superposition Fundamental principle of sedimentary layers that states that, in a sequence of undisturbed layers, the layers at the bottom are oldest.

surf zone The area along a coast where waves break.

surface waves Earthquake waves that travel on the earth's surface.

suspended sediment Solid particles carried within a body of flowing air or water.

syncline A down-buckling fold.

tephra Volcanic material thrown into the air during eruptions.

terrane A large block of crust with a distinct geologic character, originally part of the same crustal plate.

texture The shape, size, and orientation of particles in a rock, and the way they fit together.

thrust fault A type of reverse fault in which the fault surfaces is only slightly inclined; *see reverse fault.*

till Unsorted sediment transported by a glacier.

topographic map A map portraying the elevation and shape of the land surface.

topography The form of the land surface, including the steepness.

transform boundary The line where two plates meet as they move parallel to each other.

transpiration The process by which plants release water vapor into the atmosphere from their leaves.

transpression A combination of compression and transform motion.

transtension A combination of extension and transform motion.

trap A geologic barrier to the migration of petroleum.

trench (deep ocean) A very deep, long depression in the sea floor marking the surface of subduction zone.

tributary A stream that flows into a larger stream.

trilobite Arthropods extinct since the end of Paleozoic time.

triple junction The point at which three tectonic plates meet.

tsunami An ocean wave generated when an earthquake displaces the sea floor.

tufa Calcium carbonate formed along the edges of desert lakes where alkaline lake water mixes with springwater.

tuff Cemented (lithified) volcanic ash.

turbidite Rhythmically layered beds deposited by turbidity currents.

turbidity current A dense mixture of water and sediment that flows from the continental edge onto the deep ocean floor.

U.S. Geological Survey Federal agency in the Department of Interior responsible for the nation's geologic programs, including mapping, hazard studies, resource inventories, and water safety oversight.

ultramafic Describes igneous rocks with very high iron and magnesium and low (about 40%) silica.

ultrametamorphic Rocks subjected to such a high degree of metamorphism that they have partly melted and are transitional to igneous rocks.

Uniform Building Code (UBC) International standards for building construction adopted by California cities and counties.

Unreinforced Masonry (URM) Building Law 1986 California law mandating local governments to develop earthquake hazard mitigation programs for older brick buildings.

vein An igneous intrusion filling a crack; may contain ore minerals.

vent The opening in a volcano where lava erupts.

vesicle A hole in a volcanic rock made by escaping gas.

viscosity The resistance to flow in a fluid; stickiness.

volcanic Rocks crystallized from magma at the earth's surface.

water table The depth below which rock and sediment are saturated with groundwater.

watershed *See drainage basin.*

wave-cut platform A nearly horizontal surface eroded on bedrock by waves along an uplifting coast.

weathering The physical disintegration and chemical alteration of rocks at the earth's surface.

xenolith *See inclusion.*

zone of saturation The area beneath the surface where all available space in a sediment or rock is filled with water.

Place Index

A

B

C

D

N

O

P

Q

R

S

T

V

W

Y

Subject Index

A

B

C

D

N

O

P

Q

R

S

T

U

V

W

Y

Z

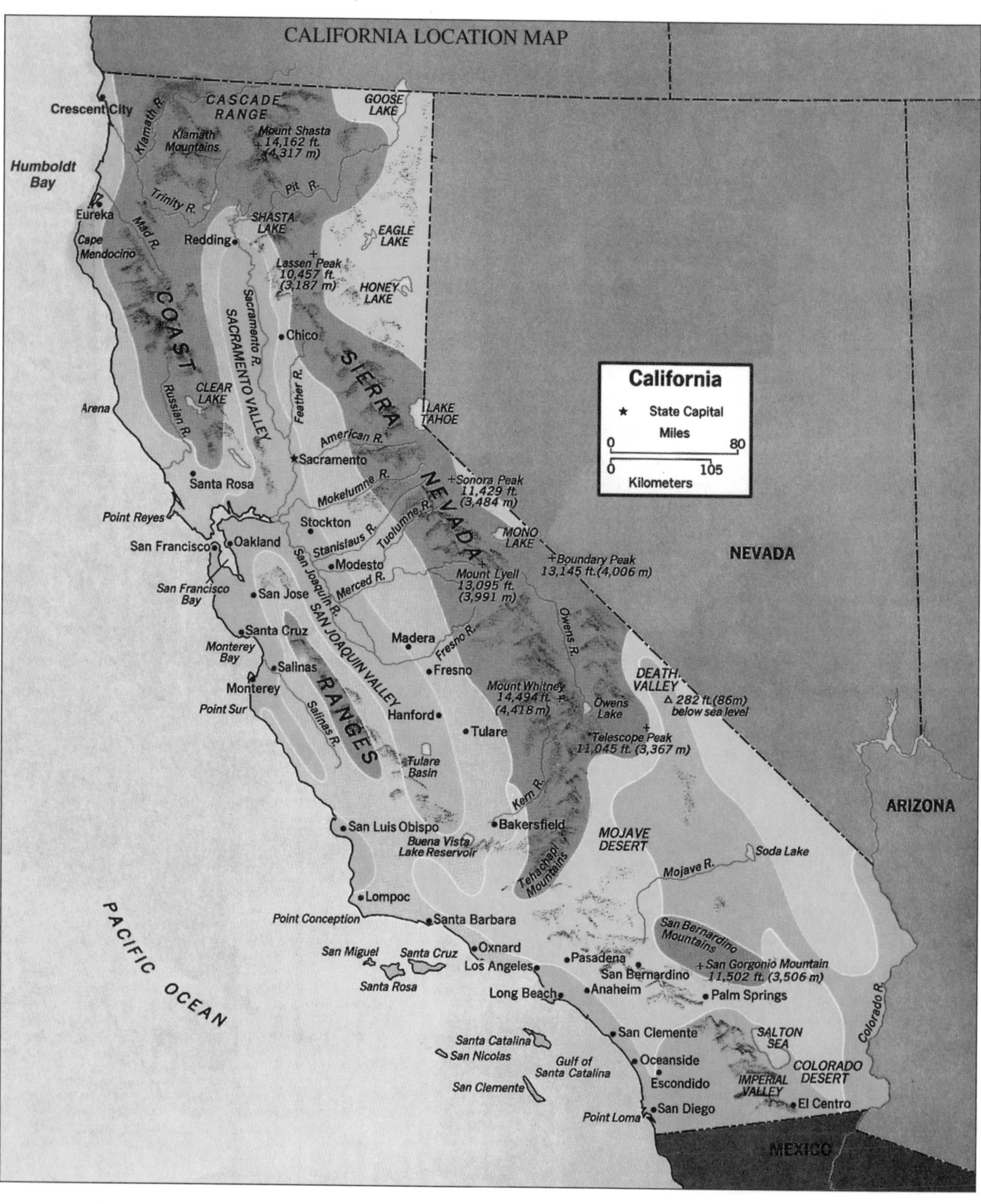

CALIFORNIA LOCATION MAP
California
State Capital
Miles
0
80
0
105
Kilometers
NEVADA
ARIZONA
MEXICO
PACIFIC OCEAN
Crescent City
CASCADE RANGE
GOOSE LAKE
Klamath R.
Klamath Mountains
Mount Shasta 14,162 ft. (4,317 m)
Humboldt Bay
Pit R.
Trinity R.
Eureka
Mad R.
SHASTA LAKE
EAGLE LAKE
Cape Mendocino
Redding
Lassen Peak 10,457 ft. (3,187 m)
HONEY LAKE
COAST
Sacramento R.
SACRAMENTO VALLEY
Chico
SIERRA
Feather R.
Russian R.
CLEAR LAKE
Arena
LAKE TAHOE
American R.
Sacramento
Santa Rosa
Sonora Peak 11,429 ft. (3,484 m)
Mokelumne R.
NEVADA
Point Reyes
Stockton
Tuolumne R.
MONO LAKE
Stanislaus R.
San Francisco
Oakland
Boundary Peak 13,145 ft. (4,006 m)
Modesto
San Joaquin R.
Mount Lyell 13,095 ft. (3,991 m)
San Francisco Bay
San Jose
Merced R.
SAN JOAQUIN VALLEY
Owens R.
Santa Cruz
Madera
Monterey Bay
Fresno R.
Salinas
Fresno
DEATH VALLEY
Monterey
RANGES
Mount Whitney 14,494 ft. (4,418 m)
Owens Lake
282 ft. (86m) below sea level
Point Sur
Salinas R.
Hanford
Tulare
Telescope Peak 11,045 ft. (3,367 m)
Tulare Basin
Kern R.
San Luis Obispo
Bakersfield
MOJAVE DESERT
Buena Vista Lake Reservoir
Soda Lake
Tehachapi Mountains
Mojave R.
Lompoc
Point Conception
Santa Barbara
San Bernardino Mountains
San Miguel
Santa Cruz
Oxnard
Los Angeles
Pasadena
San Gorgonio Mountain 11,502 ft. (3,506 m)
San Bernardino
Santa Rosa
Long Beach
Anaheim
Palm Springs
Colorado R.
San Clemente
SALTON SEA
Santa Catalina
San Nicolas
Gulf of Santa Catalina
Oceanside
COLORADO DESERT
Escondido
IMPERIAL VALLEY
San Clemente
San Diego
El Centro
Point Loma